The Interplay of Pesticides and Climate Change

Babafemi Raphael Babaniyi
Ebunoluwa Elizabeth Babaniyi
Editors

The Interplay of Pesticides and Climate Change

Environmental Dynamics and Challenges

Springer

Editors
Babafemi Raphael Babaniyi
Bioresources Development Centre
National Biotechnology Research And Development Agency (NBRDA)
Ogbomoso, Oyo State, Nigeria

Ebunoluwa Elizabeth Babaniyi
Department of Biology
Obafemi Awolowo University, Adeyemi College of Education
Ondo, Nigeria

ISBN 978-3-031-81668-0 ISBN 978-3-031-81669-7 (eBook)
https://doi.org/10.1007/978-3-031-81669-7

This Springer imprint is published by the registered company Springer Nature Switzerland AG
The registered company address is: Gewerbestrasse 11, 6330 Cham, Switzerland

Foreword

The global environmental landscape is undergoing profound transformations, with climate change at the forefront of challenges confronting ecosystems and human societies. At the same time, the widespread use of pesticides has been both a vital tool for agricultural productivity and a source of environmental and health concerns. *The Interplay of Pesticides and Climate Change: Environmental Dynamics and Challenges* provides a timely and crucial examination of how these two forces—pesticides and climate change—interact to shape our environment, food systems, and future sustainability.

In the first section, the book offers an in-depth introduction to pesticides, explaining their various categories, such as insecticides, herbicides, fungicides, and more specialized compounds. It goes beyond a simple classification to explore the specific environmental impacts of each type, including their potential to disrupt ecosystems, non-target species, and food chains. As readers navigate this section, they will not only gain a clearer understanding of how different pesticides are designed to function but also how they can contribute to unintended consequences when persistent in the environment.

A critical component of the book is its exploration of environmental persistence—the longevity of pesticides and their metabolites in soil, water, and air. With climate change altering precipitation patterns and hydrological cycles, the dynamics of pesticide runoff and leaching become increasingly unpredictable. The book thoughtfully examines how shifting environmental conditions enhance the potential for pesticide residues to persist and accumulate, affecting water quality, aquatic ecosystems, and even human health.

An important and often overlooked aspect is the role of adjuvants, substances added to pesticide formulations to enhance their effectiveness. The book investigates the types of adjuvants used in modern agriculture and their ecological consequences, particularly in the context of climate-induced environmental stress. This nuanced discussion highlights the need to consider not only the active ingredients of pesticides but also the auxiliary chemicals that influence their behavior in natural systems. One of the most significant challenges posed by the dual forces of pesticide use and climate change is how ecosystems respond to these combined stressors. The book presents a thorough analysis of how climate change exacerbates the effects of pesticide exposure, leading to potential declines in biodiversity, altered ecosystem functions, and compromised ecosystem services. However, it also presents strategies to build ecosystem resilience, offering hope for mitigating these negative impacts through adaptive management practices. Another major theme is the impact of pesticide runoff on water quality, a topic that is becoming increasingly relevant as climate change alters weather patterns, increasing the frequency of heavy rainfall events and affecting water availability. The book delves into the cascading effects of pesticide runoff, drift, and leaching into water bodies, underscoring the far-reaching consequences for aquatic ecosystems, drinking water sources, and overall ecosystem health.

Global trade further complicates the picture, as international supply chains and pesticide use patterns are influenced by both economic and environmental factors. This section uncovers how global commerce, intertwined with shifting climatic conditions, can drive pesticide demand in some regions and create cross-border environmental impacts in others. Through this lens, the book calls for collaborative international efforts to tackle the intertwined challenges of pesticide use and climate change. The intricate dynamics between pesticides and climate change are also explored through feedback loops. As climate change alters pest populations, the need for and effectiveness of pesticides evolve, creating cycles that can intensify the very conditions they are meant to control. Additionally, the book highlights how pesticide production, transport, and application contribute to greenhouse gas emissions, linking their use directly to the broader climate crisis.

The book also examines the ecotoxicological effects of pesticides, providing a comprehensive analysis of the direct and indirect impacts on non-target species, including pollinators, beneficial insects, and wildlife. As climate change modifies habitats and species behavior, the combined stress of pesticide exposure may amplify the threats to biodiversity, which in turn affects ecosystem resilience and food security. Through detailed case studies and examples, the book explores the mechanisms by which pesticides disrupt biological systems and offers mitigation strategies that emphasize the protection of biodiversity and ecosystem services.

In response to these challenges, the book investigates technological innovations aimed at reducing the environmental impact of pesticides and improving agricultural resilience to climate change. Emerging solutions such as precision agriculture, remote sensing, and ecofriendly pest control methods—including integrated pest management (IPM) and the development of biopesticides—are discussed as promising alternatives to conventional chemical-based agriculture. These innovations present a pathway toward more sustainable farming practices, with the potential to reduce the environmental footprint of pesticides. Another key focus is the impact of pesticides on soil health and its role in carbon sequestration. The book carefully explores how pesticides influence soil microbial communities, nutrient cycling, and the ability of soils to sequester carbon—a critical function in mitigating climate change. It presents strategies for enhancing soil resilience, including the promotion of sustainable farming techniques that protect soil health and improve its capacity to store carbon. However, the complexity of pesticide and climate change interactions goes beyond the field, extending to human health and community vulnerability. The book takes a close look at the disproportionate impacts on marginalized populations, particularly agricultural workers, indigenous groups, and low-income communities who are often most exposed to both pesticide residues and the effects of climate change. In addressing these inequities, the book offers strategies for enhancing community resilience and advancing environmental justice.

Finally, the book emphasizes the importance of public perception, education, and policy integration in shaping the future of pesticide use and climate action. It calls for greater awareness and engagement among farmers, policymakers, consumers, and researchers in adopting sustainable practices and promoting environmental stewardship. Through collaborative efforts between governments, industry, and civil society, the book envisions a future where the use of pesticides is better managed, environmental impacts are minimized, and agricultural systems are made more resilient to the changing climate.

The editors have done a great job by inviting 71 experienced experts from 9 different countries to contribute to this volume. Therefore, as you journey through the pages of this book, you will gain not only a deep understanding of the scientific and environmental issues at play but also an appreciation for the complex, interconnected solutions needed to address these global challenges. *The Interplay of Pesticides and Climate Change: Environmental Dynamics and Challenges* is a clarion call for action, offering both the knowledge and the tools necessary to chart a more sustainable path forward for our planet.

Professor of Environmental Chemistry
The Federal University of Technology
Akure, Nigeria

Ademola Festus Aiyesanmi

Preface

The world currently confronts unprecedented environmental challenges, with pesticide use and climate change emerging as critical issues that have profound implications. Although these phenomena originate from different sources, they are intricately linked in their effects on ecosystems, agriculture, and human health. The escalating demand for food security has led to the extensive application of pesticides, while climate change disrupts weather patterns and ecosystems, consequently affecting the efficacy of these chemicals. This intricate relationship necessitates thorough analysis and informed action.

The Interplay of Pesticides and Climate Change: Environmental Dynamics and Challenges aims to investigate the synergies and conflicts between pesticide use and climate change. It delves into how changing environmental conditions—such as rising temperatures, altered precipitation patterns, and shifting ecosystems—impact the persistence, behavior, and consequences of pesticides. Additionally, the book explores how these interactions affect agricultural productivity and environmental sustainability, presenting significant challenges to food security, biodiversity, and public health. Through a comprehensive examination of scientific research, case studies, and policy frameworks, this work provides a nuanced understanding of the complexities involved.

Moreover, the book emphasizes the urgent need for adaptive management strategies, regulatory reforms, and innovative agricultural practices that mitigate the adverse effects of pesticides while bolstering resilience against climate impacts.

This publication is designed for researchers, environmental scientists, policymakers, and all individuals concerned about the future of agriculture and environmental stewardship. By bridging two of the most pressing environmental issues of our time, it aspires to promote a more integrated approach to sustainable development and environmental protection.

We hope this book will inspire deeper reflection, foster interdisciplinary collaboration, and motivate action toward addressing the environmental dynamics shaped by the intersection of pesticides and climate change.

Ogbomoso, Nigeria — Babafemi Raphael Babaniyi
Ondo, Nigeria — Ebunoluwa Elizabeth Babaniyi

Acknowledgments

The Editorial team expresses profound gratitude to all contributors whose exceptional collaboration and timely responses were pivotal to the successful completion of this book. Their deep expertise in the subject matter not only enhanced the quality of each chapter but also brought valuable insights that enriched the overall work. We commend their unwavering dedication to meeting deadlines, which played a crucial role in the seamless progress of the project.

We are equally grateful to the reviewers for their thoughtful and invaluable feedback, which significantly contributed to improving the clarity and rigor of the content. Special thanks go to the Springer Editorial and Production team for their unwavering support, professionalism, and seamless collaboration throughout this endeavor. Their commitment to excellence ensured the project reached its full potential.

Ogbomoso, Nigeria — Babafemi Raphael Babaniyi
Ondo, Nigeria — Ebunoluwa Elizabeth Babaniyi

Contents

Contributors

Abdulhakeem Idris Abdulhakeem Department of Fisheries and Aquaculture, Federal University, Dutsin-Ma, Katsina State, Nigeria

Mustapha Abdulsalam Department of Microbiology, Skyline University Nigeria, Kano, Kano State, Nigeria

Taiwo Habib Adejumo Department of Chemistry, Federal University of Technology Akure, Akure, Ondo State, Nigeria

Ishola Jonathan Adekunle Department of Public Health, Obafemi Awolowo University, Ile-Ife, Osun State, Nigeria

Omole Emmanuel Adesuyi Department of Marine Biology, University of Padova, Padua, Italy

Stanley Ozoemena Agbo Environmental Toxicology Unit, Department of Animal and Environmental Biology, Federal University Oye-Ekiti, Oye-Ekiti, Ekiti State, Nigeria

Ewuola Akinola Ayoola Department of Chemistry, Osun State College of Technology, Esa Oke, Osun State, Nigeria

Olumide Akinrinola Department of Science Laboratory Technology, Osun State College of Technology, Esa-Oke, Osun State, Nigeria

Ulelu Jessica Akor Department of Animal Production, University of Ilorin, Ilorin, Kwara State, Nigeria

R. Arutselvan Regional Station, ICAR-Central Tuber Crops Research Institute (CTCRI), Dumuduma, Bhubaneswar, Odisha, India

Oni Ayodeji Christopher Bioresource Development Centre, National Biotechnology Research and Development Agency, Abuja, Nigeria

Babafemi Raphael Babaniyi Bioresources Development Centre, National Biotechnology Research And Development Agency (NBRDA), Ogbomoso, Oyo State, Nigeria

Gabriel Gbenga Babaniyi Department of Agricultural Development and Management (ADEM), Agricultural and Rural Management Training Institute (ARMTI), Ilorin, Kwara State, Nigeria

Natalia Bailon-Moscoso Universidad Técnica Particular de Loja, Loja, Ecuador

Piotr Barbaś Department Agronomy of Potato, Plant Breeding and Acclimatization Institute—National Research Institute, Serock, Poland

Mukul Machhindra Barwant Department of Botany, Sanjivani Rural Education Society (SRES), Sanjivani Arts, Commerce and Science College, Kopargaon, Ahilyanagar, Maharashtra, India

Department of Botany, Sanjivani Rural Education Society's, Sanjivani Arts Commerce and Science College Kopargaon, Ahmednagar, Maharashtra, India

B. H. Boje Gowd Center for Biotechnology, University College of Engineering Science and Technology, Jawaharlal Nehru Technological University Hyderabad, Hyderabad, Telangana, India

Fatimoh Abdulsalam Danjuma Ministry of National Guard Hospital, Riyadh, Saudi Arabia

Prity Das School of Pharmaceutical Technology, School of Health and Medical Sciences, Adamas University, Kolkata, West Bengal, India

Rakesh Das School of Pharmaceutical Technology, School of Health and Medical Sciences, Adamas University, Kolkata, West Bengal, India

Rodrigo Duarte-Casar Pontificia Universidad Católica del Ecuador Sede Manabí, Portoviejo, Ecuador

Ghadir A. El-Chaghaby Regional Center for Food and Feed, Agricultural Research Center, Giza, Egypt

Aliu Folorunso Patrick Bioresource Development Centre, National Biotechnology Research and Development Agency, Abuja, Nigeria

Manish Kumar Gautam Faculty of Pharmaceutical Science, Assam Down Town University, Guwahati, Assam, India

Katherine Georgina Menon Department of Microbiology, School of Allied and Healthcare Sciences Malla Reddy University, Hyderabad, Telangana, India

Alia Syafiqah Abdul Hamed Faculty of Ocean Engineering Technology, Universiti Malaysia Terengganu, Kuala Nerus, Terengganu, Malaysia

Nur Hidayah Centre for Earth Sciences and Environment, Faculty of Science and Technology, Universiti Kebangsaan Malaysia, Bangi, Selangor, Malaysia

Musa Ojeba Innocent Department of Microbiology, Skyline University Nigeria, Kano, Kano State, Nigeria

Routhu Jhansi Department of Microbiology, School of Allied and Healthcare Sciences Malla Reddy University, Hyderabad, Telangana, India

Abdulsalam Khadijah Department of Agricultural Science, University of Ilorin, Ilorin, Kwara State, Nigeria

Barbara Krochmal-Marczak Department of Plant Production and Food Safety, Carpathian State College in Krosno, Krosno, Poland

Alege Abdulraheem Lateefat Department of Microbiology, Kwara State University, Malete, Kwara State, Nigeria

Ayoigbala Monioluwa Lawal Department of Microbiology, Federal University Oye-Ekiti, Oye-Ekiti, Ekiti State, Nigeria

O. T. Lawal Department of Medical Biochemistry, Federal University of Technology Akure, Akure, Ondo State, Nigeria

Marlene Rojas Le-Fort Pontificia Universidad Católica del Ecuador Sede Manabí, Portoviejo, Ecuador

Miracle Uwa Livinus Department of Biochemistry, Skyline University Nigeria, Kano, Kano State, Nigeria

Debasree Lodh Department of Botany, Holy Cross College, Agartala, Tripura, India

V. Mageshwaran Microbial Technology Lab, ICAR-National Bureau of Agriculturally Important Microorganisms, Kushmaur, Mau, Uttar Pradesh, India

Sandip Mondal Amity Institute of Pharmacy, Amity University Kolkata, Kolkata, West Bengal, India

Muhammad Abdul Moneem El-Fostat Laboratory, Cairo Water Company, Cairo, Egypt

Rupali Munje Department of Zoology, Sanjivani Rural Education Society's, Sanjivani Arts Commerce and Science College Kopargaon, Ahmednagar, Maharashtra, India

Marell Navarro-Rojas Universidad Politécnica Salesiana, Cuenca, Ecuador

Uchechukwu Chinwe Nebo Department of Microbiology, Federal University Oye-Ekiti, Oye-Ekiti, Ekiti State, Nigeria

Abdulhameed Adewale Odeseye Department of Rural Development and Gender Issues (RUDEG), Agricultural and Rural Management Training Institute (ARMTI), Ilorin, Kwara State, Nigeria

Odangowei Inetiminebi Ogidi Department of Biochemistry, Faculty of Basic Medical Sciences, Bayelsa Medical University, Yenagoa, Bayelsa State, Nigeria

Olusola David Ogundele Chemical Science Department, Achievers University, Owo, Ondo State, Nigeria

Ige Emmanuel Ojo Department of Chemistry, School of Physical Sciences, Federal University of Technology Akure, Akure, Ondo State, Nigeria

Taiwo Hamidat Olaide Department of Chemistry, Federal University of Technology Akure, Akure, Ondo State, Nigeria

A. F. Ologundudu School of Computer, Mathematics and Natural Sciences, Morgan State University, Baltimore, MD, USA

Kehinde Oluwasiji Olorunfemi Department of Chemistry, Federal University of Technology Akure, Akure, Ondo State, Nigeria

Alabi Omotayo Oluwaseun Department of Chemistry, School of Physical Sciences, Federal University of Technology Akure, Akure, Ondo State, Nigeria

Daramola Oluwatosin Olaoluwa Department of Chemistry, Federal University of Technology Akure, Akure, Ondo State, Nigeria

O. O. Osemwegie Department of Food Science and Microbiology, Landmark University, Omu Aran, Kwara State, Nigeria

Varsha Sadashiv Patil Department of Botany, Shirdi Sai Rural Institute's, Arts, Science and Commerce College, Rahata, Maharashtra, India

P. Paul Vijay Center for Biotechnology, UCESTH, Jawaharlal Nehru Technological University Hyderabad, Hyderabad, Telangana, India

Piotr Pszczółkowski Research Centre for Cultivar Testing in Słupia Wielka, Słupia Wielka, Poland

Sayed Rashad Regional Center for Food and Feed, Agricultural Research Center, Giza, Egypt

Juan Carlos Romero-Benavides Universidad Técnica Particular de Loja, Loja, Ecuador

Barbara Sawicka Department of Plant Production Technology and Commodities Science, University of Life Sciences in Lublin, Lublin, Poland

Shalagha A. B. Sharma School of Biological Engineering, SIET, Meerut, Uttar Pradesh, India

Dipanwita Chaudhuri Sil Department of Botany, Holy Cross College, Agartala, Tripura, India

Balwant Singh Department of Botany, K. S. Saket PG College, Dr. Ram Manohar Lohia Avadh University, Ayodhya, Uttar Pradesh, India

Shivendra Singh Department of Agriculture, Udai Pratap College (An Autonomous Institution) Bhojubir, Varanasi, Uttar Pradesh, India

Shreya Singh Department of Agriculture, Ramlalit Singh Mahavidyalaya, Kailhat, Chunar, Mirzapur, Uttar Pradesh, India

B. Sri Sai Siddartha Naik Department of Agronomy, Acharya N.G. Ranga Agricultural University, Guntur, Andhra Pradesh, India

Tijani Abiola Tajudeen Department of Soil Science, Federal University of Technology, Minna, Niger State, Nigeria

Idowu Afeez Temitope Department of Medicinal Research, Kaohsiung Medical University Hospital, Kaohsiung City, Taiwan

Amarachi Nchekwube Ugwuja Department of Biology, Adeyemi Federal University of Education, Ondo, Ondo State, Nigeria

Kondakindi Venkateshwar Reddy Center for Biotechnology, University College of Engineering Science and Technology, Jawaharlal Nehru Technological University Hyderabad, Hyderabad, Telangana, India

Channe Yogita Department of Zoology, Sanjivani Rural Education Society's, Sanjivani Arts Commerce and Science College Kopargaon, Ahmednagar, Maharashtra, India

About the Editors

Babafemi Raphael Babaniyi is an accomplished academic with a strong background in chemistry. He earned his Bachelor of Science degree in Industrial Chemistry from Caritas University, Amorji-Nike, Emene, Enugu State, Nigeria, in 2008. Subsequently, he obtained a Master of Technology degree in Environmental Chemistry from the Federal University of Technology, Akure, Ondo State, Nigeria, in 2021. Currently, he is pursuing a PhD in Environmental Chemistry at the same institution. Babafemi has an impressive portfolio with over 40 publications, including book chapters and articles in internationally recognized journals. He serves as an Assistant Chief Research Officer at the Bioresources Development Centre within the National Biotechnology Research and Development Agency of Nigeria. His research interests encompass various aspects of environmental chemistry, focusing on green chemistry, phytoremediation, chemical biology, environmental science, bioremediation of contaminated soils, plant protection and animal health, environmental impact assessment, and biodegradable plastics.

Ebunoluwa Elizabeth Babaniyi is an educator with a solid academic foundation in the sciences. She earned her Bachelor of Science in Education (BSc Ed) in Biology from Adeyemi College of Education, Ondo, an affiliate of Obafemi Awolowo University, Ile Ife, Nigeria, in 2014. Prior to this, she obtained a National Certificate in Education (NCE) in Biology and Geography from the College of Education, Ikere-Ekiti, in 2009. Professionally registered with the Teachers

Registration Council of Nigeria (TRCN) since 2019, Ebunoluwa has accumulated significant experience in the field of education. Her contributions are further evidenced by her published research manuscripts, which reflect her commitment to advancing knowledge in her area of expertise.

Part I
Pesticides Overview

This part addresses the multifaceted realm of pesticides, wherein it scrutinizes a plethora of classifications such as insecticides, herbicides, and fungicides.

An Insight to Pesticides

Taiwo Hamidat Olaide, Babafemi Raphael Babaniyi, Taiwo Habib Adejumo, Kehinde Oluwasiji Olorunfemi, Olusola David Ogundele, and Olumide Akinrinola

1 Introduction

Environmental pollution has become a global issue, raising serious concerns about its harmful effects on human health and the stability of ecosystems (Khan et al., 2010; Nasrollahi et al., 2020). Although pollution largely stems from human activities, it is essential to weigh the associated benefits and risks rather than place blame outright (Ahmad et al., 2024; Mahmood et al., 2015). Pesticides, commonly used in agriculture to control pests, insects, and weeds, play a crucial role in both modern farming and health practices (Carvalho, 2017; Mahmood et al., 2015; Umetsu & Shirai, 2020). Each year, around 3 billion kilograms of pesticides is utilized worldwide, with an estimated cost of approximately 40 billion USD (Zhang et al., 2011). Various types of pesticides—including herbicides, insecticides, fungicides, and nematicides—are used extensively to boost crop yields, minimize harvest losses, and increase food supplies. However, the widespread and often unregulated use of these chemicals has led to significant environmental harm and negative health effects. These substances can accumulate in ecosystems, disrupting natural balances through bioaccumulation (Mahmood et al., 2015; Sharma et al., 2019). Due to their non-biodegradable nature, they persist in the environment for prolonged periods,

T. H. Olaide · T. H. Adejumo · K. O. Olorunfemi (✉)
Department of Chemistry, Federal University of Technology Akure, Akure, Nigeria

B. R. Babaniyi
Bioresources Development Centre, National Biotechnology Development Agency, Ogbomoso, Oyo State, Nigeria

O. D. Ogundele
Chemical Science Department, Achievers University Owo, Owo, Nigeria

O. Akinrinola
Department of Science Laboratory Technology, Osun State College of Technology, Esa-Oke, Osun State, Nigeria

B. R. Babaniyi, E. E. Babaniyi (eds.), *The Interplay of Pesticides and Climate Change*, https://doi.org/10.1007/978-3-031-81669-7_1

presenting considerable biohazards (van Dijk, 2010; Hashimi et al., 2020). The modern surge in chemical use has had a profound impact on the biosphere, affecting life quality (Pimentel, 2005). A wide array of chemical agents is used to control unwanted plant and insect species, with examples like dichloro-diphenyl-trichloroethane (DDT) and its metabolite dichlorodiphenyldichloroethylene (DDE) serving to manage vector-borne diseases such as malaria, dengue (WHO, 2012), leishmaniasis (Claborn, 2010), Japanese encephalitis (JE) (Simon et al., 2024), and schistosomiasis (Schmolke et al., 2010). Epidemiological research has shown that pesticide exposure can harm several organs, including the liver, brain, lungs, and colon. Studies have also linked pesticides to severe diseases like cancer, as they generate reactive oxygen species (ROS) that decrease antioxidant levels and weaken the body's defenses against cellular oxidative damage. Lipids, proteins, and nucleic acids are especially vulnerable to this oxidative imbalance, which interferes with cellular signaling and may cause chronic health issues, including reproductive disorders in both humans and animals (Ahmad et al., 2024; Moon et al., 2009; Oluwole & Cheke, 2009; Pesticide Action Network, 2010). The oxidative imbalance caused by pesticides contributes to various diseases and disrupts natural equilibrium.

While pesticide use has improved agricultural productivity, it has exacted a heavy toll on human health and the environment. Unrestricted application of these chemicals has been linked to numerous health risks, including organ damage, cancer, reproductive issues, and ecological imbalance (Abong'o et al., 2014; Buczyńska & Szadkowska-Stańczyk, 2005; Carvalho, 2017; Schreinemachers & Tipraqsa, 2012). The non-biodegradability of pesticides only intensifies these issues, as they linger in ecosystems and present long-term hazards. As agriculture advances, it is crucial to prioritize environmental conservation and human health protection.

2 Pesticides

Pesticides are a broad category of chemical substances, biological agents, or their mixtures introduced into the environment to prevent, manage, or eliminate populations of harmful insects, weeds, rodents, fungi, and other pests. These substances are essential in modern agriculture and public health, playing a central role in protecting crops and controlling disease vectors. Pesticides achieve pest control through various mechanisms, including attraction and toxic effects, to suppress or eradicate target organisms. Pests, which can include any species threatening food security, health, or human welfare, are countered using a wide range of pesticides, from natural compounds derived from organisms or environmental sources to synthetic chemicals (Mahmood et al., 2015; Sharma et al., 2019).

Pesticides are categorized by type, with common classes including organochlorines, carbamates, organophosphates, pyrethrins, and neonicotinoids, which collectively make up a large portion of currently applied pesticides (Foundation, 2018). These formulations consist of active ingredients along with inert substances, contaminants, and sometimes impurities. Once released, pesticides degrade into metabolites, which can, in some cases, be more toxic than the original compounds

(Carvalho, 2017). Notable for their mobility, persistence, and bioaccumulation potential, pesticides pose significant environmental risks (Fenik et al., 2011). Despite their effectiveness in pest control, the use of herbicides, fungicides, insecticides, acaricides, nematicides, antimicrobials, and rodenticides often comes with drawbacks. Non-selective pesticides may harm non-target species, and the rise of pesticide-resistant pests over time reduces the efficacy of these chemicals (Mahmood et al., 2015).

3 History of Pesticides

Crop cultivation has been a foundational element of human civilization for thousands of years, originating in the Fertile Crescent in Mesopotamia around 10,000 years ago. Throughout history, agriculture has faced continuous challenges from pests and diseases threatening food production. The first known use of pesticides dates back roughly 4500 years to the Sumerians, who used sulfur compounds to control insects and mites (Umetsu & Shirai, 2020). Additionally, the ancient Hindu scripture "Rig Veda," composed around 4000 years ago, mentions poisonous plants employed for pest control (Foundation, 2018).

The evolution of pesticide development can be divided into five phases: pre-1000 CE (early pest management practices), 1000–1850 (use of plant, animal, or mineral derivatives), 1850–1940 (introduction of inorganic products and industrial byproducts), 1940–1970 (adoption of synthetic organic compounds), and post-1970 (development of lower-risk synthetic organics) (Foundation, 2018; Mahmood et al., 2015; Sharma et al., 2019). Following World War II, the advent of organic chemistry significantly advanced pesticide science, marking the start of industrialized pesticide production. Over the last century, numerous pesticide companies have formed in Europe and the United States, with many consolidating in the twenty-first century. This growth in production has fueled global pesticide use, with Japan emerging as a leading producer.

The development of synthetic organic pesticides has focused on three main goals: creating highly effective pesticides that require minimal dosage, designing pesticides that readily degrade to reduce environmental persistence, and developing selective agrochemicals that target specific pests without affecting non-target species (Umetsu & Shirai, 2020). The first goal has resulted in reduced active ingredient quantities necessary for pest control, lessening environmental impact. The second goal has contributed to lower pesticide residue levels in both crops and the environment (Kraus, 1995). The third goal aims to produce compounds that specifically target pests while sparing beneficial organisms, leading to safer, eco-friendly pesticides (Hatfield, 2004).

These strategies have become increasingly important in recent years. For instance, pesticides in the 1930s and 1950s required much higher application rates—ranging from 1 to 10 kg/ha for active ingredients like dinitro-ortho-cresol (DNOC), thiuram, and DDT—resulting in significant environmental impacts. However, advancements in pesticide efficacy now allow for effective pest control at

Table 1 Evolution of pesticides usage

Year	Event
1867	Paris Green (form of copper arsenite) was used to control Colorado potato beetle outbreak
1885	Introduction of a copper mixture by Professor Millardet to control mildew
1892	Potassium dinitro-2-cresylate was produced in Germany
1939	DDT discovered by Swiss chemist Paul Muller; organophosphate insecticides and phenoxyacetic herbicides were discovered
1950s	Fungicides captan and glyodin and insecticide malathion was discovered
1961–1971	Agent Orange was introduced
1972	DDT officially banned
2001	Stockholm Convention

Source: Mahmood et al. (2015)

application rates as low as 10 g/ha (Umetsu & Shirai, 2020; Zadoks & Waibel, 2000). In the past decade, a variety of new pesticides have emerged, including fungicides, insecticides, nematicides, acaricides, herbicides, and biopesticides. Among the 105 chemical pesticides developed or in progress, many are considered safe for human health and environmentally friendly, comprising 43 fungicides, 34 insecticides/acaricides, 6 nematicides, 21 herbicides, and 1 herbicide safener. Although advancements in genomics, structure–activity relationship studies, and chemical biology have propelled pesticide research, only a few breakthroughs have reached widespread application (Joseph Shaba et al., 2019) (Table 1).

4 Production of Pesticides: A Multifaceted Process

The manufacturing of pesticides involves a sequence of carefully orchestrated steps, each essential to producing effective chemical formulations. This process begins with the precise selection of raw materials specific to the type of pesticide being manufactured (Brown, 2006; Llewellyn et al., 2016). These materials undergo synthesis to create the active ingredient, which is crucial for the pesticide's effectiveness. This synthesis stage typically involves intricate chemical reactions that require skilled expertise to achieve the desired compound (Alosaimy et al., 2023; Wang et al., 2015).

After the active ingredient is synthesized, it is formulated into a usable form, commonly as a liquid suspension or emulsifiable concentrate. This formulation step includes combining various inert ingredients such as solvents, emulsifiers, additives, and stabilizers—to enhance the pesticide's stability, effectiveness, and user-friendliness (Abubakar et al., 2020; Carvalho, 2017). Quality control measures are rigorously applied throughout production to ensure that the final product upholds high standards for purity, potency, and safety. These controls involve thorough testing of all materials and products to identify and address potential contaminants or inconsistencies. In the final stage of production, the concentrated pesticide is diluted to an appropriate concentration for its intended use, balancing effective pest control with minimized risks to human health and the environment (Abong'o et al., 2014).

5 Mode of Action of Pesticides

A pesticide's mode of action, or mechanism of action, refers to the specific biochemical interactions through which it disrupts essential biological processes in target organisms. This classification uniquely focuses on the biological pathways that pesticides affect, rather than simply categorizing them by pest type, physical properties, or chemical makeup (Joseph Shaba et al., 2019; Zoccali et al., 2009).

Pesticides can affect target organisms via two main routes: systemic and non-systemic. Systemic pesticides are absorbed by plants or animals and distributed internally, allowing them to impact areas that were not directly treated with the pesticide (Aktar et al., 2009). Conversely, non-systemic or contact pesticides act only when they come into direct contact with the pest, delivering effects at the point of application.

6 Classification of Pesticides

Pesticides can be categorized based on the specific types of pests they are designed to control. Examples include insecticides, fungicides, herbicides, rodenticides, biopesticides, and nematicides (Foundation, 2018) (Fig. 1).

(i) Fungicides are chemical agents designed to prevent and treat fungal infections in crops by targeting and eliminating harmful fungal pathogens. They are primarily classified into two types: inorganic and organic fungicides. Inorganic fungicides include substances like Bordeaux mixture and mercuric chloride, while organic fungicides encompass a wider variety of compounds, such as dithaneS-21, dithaneM-22, and dithaneZ-78 (all carbamates), as well as azoxystrobin, fenpropimorph, iprodione, tebuconazole, thiram, and ziram. Though ziram is effective against fungal pathogens, its toxicity poses considerable risks to aquatic zooplankton, impacting ecosystem health.

(ii) Herbicides and weedicides are chemical agents utilized to control unwanted plants or weeds in agriculture. These substances are classified based on their mode of action and application technique. Selective herbicides are designed to target specific plant species without affecting others, whereas non-selective herbicides eliminate all types of vegetation. Contact herbicides act on plant tissues immediately upon contact, while translocated herbicides are absorbed and spread throughout the plant. Herbicides can be applied to foliage or incorporated into the soil for effective weed control. Notable herbicide classes

Fig. 1 Different types of pesticides

include butachlor, chlorsulfuron, diuron, glyphosate, linuron, metsulfuron methyl, metamitron, triazines (like atrazine and simazine), carbamates (e.g., thiocarbamates, phenyl carbamates), and auxin derivatives (2, 4-D and 2, 4, 5-T). Agent Orange, a notorious herbicide mix of 2, 4-D and 2, 4, 5-T, was used during the Vietnam War. Certain herbicides, such as paraquat, have been associated with an increased risk of Parkinson's disease, and atrazine has demonstrated teratogenic effects, disrupting sex differentiation in frogs during metamorphosis.

(iii) Nematicides are chemical or biological agents used to control or repel nematodes-microscopic, worm-like parasites that damage crops and reduce yields. Chemical nematicides like Aldirab, an acetylcholine esterase inhibitor, are often applied to manage nematode infestations in crops such as tobacco. For example, "Meloidogyne incognita," a common nematode species, attacks tobacco roots, leading to severe yield losses. Biological control agents (BCAs) are also effective; "*Purpureocillium lilacinum*" targets "*M. incognita*" by releasing proteases and chitinase, enzymes that weaken the nematode eggshell to facilitate infection. Another potential BCA, fungi from the genus "*Paecilomyces*," are also promising in nematode control. Other nematicides, such as methyl bromide (MB), ethylene dibromide (EDB), and chloropicrin, are widely used, though their application is sometimes restricted due to environmental concerns. Soil steam sterilization (SSS), which uses heat to inactivate pathogens and nematodes in soil, provides a non-chemical control method by disrupting enzyme function in harmful organisms.

(iv) Rodenticides, commonly known as rat poisons, are substances formulated to control rodent populations. Examples include sodium fluoroacetate, warfarin, red squill, and zinc phosphide. Many of these agents function by disrupting the

Vitamin K cycle in rodents and other mammals, which prevents blood clotting and leads to fatal internal bleeding. Additionally, rodenticides like Vitamin D3, D2, and D1 can cause hypercalcemia (excessively high blood calcium levels) in rodents, leading to severe tissue damage and death. Strychnine, derived from the "Strychnos nux-vomica" tree, is a neurotoxin that causes asphyxiation by inducing muscle spasms. Chloralose, a chlorinated derivative of glucose, is used for controlling rodents and some bird species. Arsenic trioxide combined with copper acetate forms Paris Green, a multipurpose substance used as both a rodenticide and a blue colorant in fireworks.

(v) Insecticides are agents used to control or eliminate insect pests and can be grouped according to their mode of action. Stomach or alimentary canal poisons are ingested by insects, causing toxicity within their digestive systems; examples include azadirachtin from the neem tree (*Azadirachta indica*) and rotenone from the derris plant (*Derris elliptica*). Contact poisons, on the other hand, are lethal upon direct contact with the insect's body, disrupting either the nervous system or the development of the exoskeleton. Pyrethrum, derived from chrysanthemum flowers, and synthetic pyrethroids, which mimic pyrethrum's effects, are well-known contact insecticides.

Fumigants, another type of insecticide, function as gases or vapors that kill insects through inhalation. Commonly used in the treatment of stored grains and enclosed spaces, examples of fumigants include ethylene dibromide (EDB) and methyl bromide.

Insecticides also differ by origin. Natural insecticides, which are extracted from plants or other natural sources, are often seen as environmentally friendlier but may be less potent than synthetic alternatives. Examples of natural insecticides include azadirachtin, rotenone, and pyrethrum. Synthetic insecticides, such as organochlorines, organophosphates, carbamates, pyrethroids, triazines, cypermethrin, insect growth regulators (IGRs), azadirachtin, and hydroprene, are generally more powerful and effective, though they can pose higher risks to human health and the environment due to their persistence and toxicity (Fig. 2).

7 Advantages of Pesticide

Pesticides offer both immediate and long-lasting benefits. Primary benefits, such as the elimination of crop-damaging insects, are directly observable following pesticide application. Secondary benefits, which emerge from these primary effects, persist for extended periods (Mahmood et al., 2015). Pesticide use has garnered significant attention from various stakeholders, including farmers, scientists, governments, and agro-industries (Abubakar et al., 2020).

(i) Pesticides have been hailed as an effective tool in managing agricultural pests, leading to increased crop production. The application of pesticides has

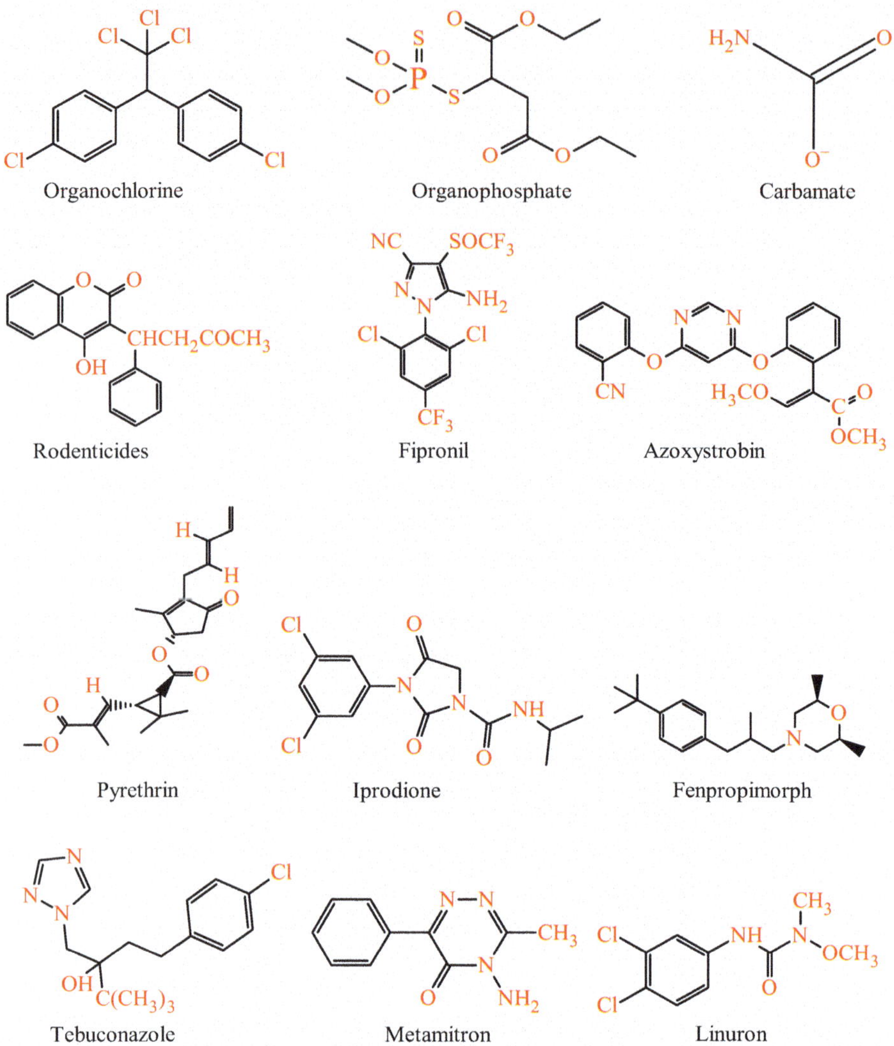

Fig. 2 2D Structure of some common pesticides such as insecticides, fungicides, and herbicide

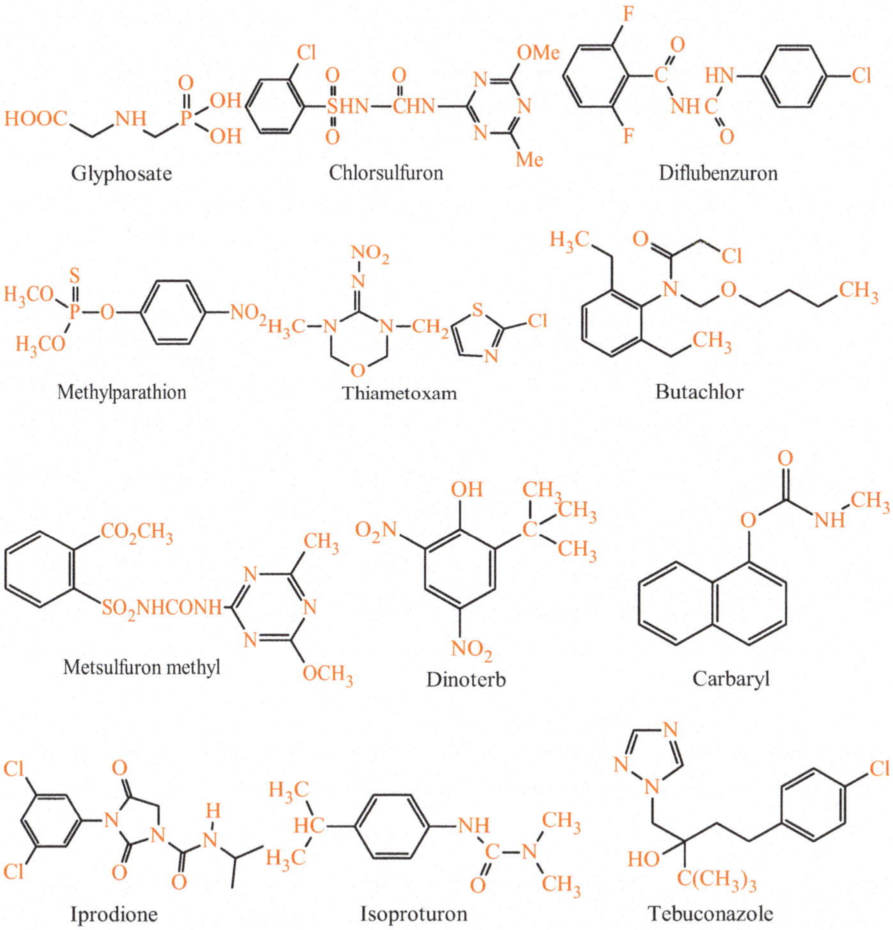

Fig. 2 (continued)

dramatically reduced insect infestations, diseases, and weeds, which can significantly diminish harvestable crop yields and economic returns.

(ii) Pesticide use has contributed to enhanced agricultural production of high-quality food. For instance, the consumption of diets rich in fruits and vegetables outweighs the potential risks associated with consuming low pesticide residues that may remain in some agricultural crops after pesticide application (Pesticide Action Network (Group), 2010).

(iii) Pesticides play a crucial role in preserving wood from destruction by termites and other wood-boring insects. Additionally, they help control the spread of malaria vectors, such as Anopheles mosquitoes (Ross, 2005).

(iv) The adoption of pesticides by farmers for agricultural pest management has motivated scientists and researchers to dedicate significant effort to further

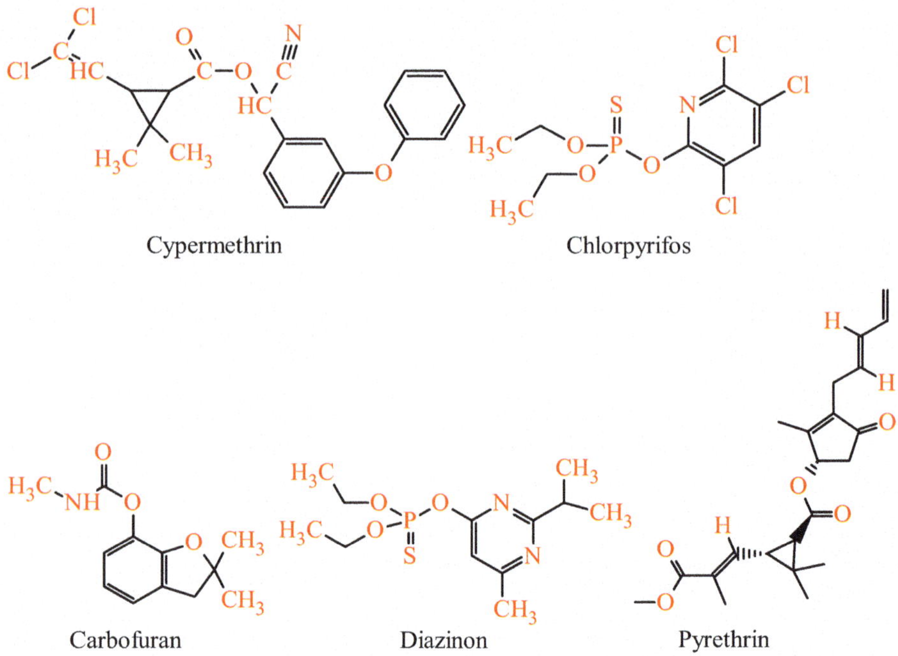

Fig. 2 (continued)

research and development in the pesticide sector. This has led to the discovery of new types of pesticides with diverse modes of action.

8 Disadvantages of Pesticides

(i) The rampant use of pesticides has led to widespread environmental contamination, posing a significant threat to the quality of soil, water, and air. This contamination can have both short-term and long-term consequences, including the contamination of groundwater and drinking water. The accumulation of pesticide residues in non-target organisms, as observed by Struik and Bonciarelli (1997), further exacerbates the environmental impact (Struik & Bonciarelli, 1997).

(ii) Numerous studies have demonstrated the detrimental effects of pesticides on non-target organisms, including beneficial microorganisms, soil quality, soil enzymes, aquatic microorganisms, and algae (Struik & Bonciarelli, 1997). These unintended consequences disrupt the delicate balance of ecosystems and contribute to environmental degradation.

(iii) The continuous reliance on pesticides has led to the development of resistance in target organisms, as well as cross-resistance to other pesticide components. This resistance necessitates the development of more potent pesticides, perpetuating a vicious cycle of increased pesticide use and exacerbated resistance.
(iv) The overreliance on pesticides has discouraged farmers from exploring alternative pest management strategies, such as natural and organic agricultural practices and the use of biopesticides (Zadoks & Waibel, 2000). A broader approach to pest control is essential for reducing the reliance on pesticides and promoting sustainable agricultural practices.
(v) The indiscriminate use of pesticides poses significant health risks, particularly during post-harvest storage of agricultural crops. Exposure to pesticides, such as Gamalin 20 and Malathion, can lead to a range of health problems, including acute and chronic poisoning, neurobehavioral effects, developmental and reproductive disorders, carcinogenic effects, and immunological effects (Casida & Durkin, 2013). The inadequate washing of pesticide-treated produce before consumption, particularly in developing countries, leads to the ingestion of pesticide residues, contributing to food poisoning and even fatalities. Additionally, studies have linked pesticide exposure to increased infertility and birth defects (Smith & Smith, 2014).
(vi) The pervasive use of pesticides has exacerbated environmental pollution, with pesticide applications significantly contributing to the contamination of soil, water, and air. This contamination poses a serious threat to human health, ecosystem health, and the overall quality of the environment.

9 Toxic Effect of Pesticides Pollution to Plant and Animal

(i) *Pesticide Toxicity to Microorganisms*: Pesticides, employed for pest control on plants and soil, inevitably enter the soil environment through direct application and plant residue incorporation. Within the soil matrix, pesticides undergo a range of physical, chemical, and biological transformations. Some of the resulting chemical products exhibit significant toxicity to soil-dwelling microorganisms (Stanley & Preetha, 2016). Numerous studies have documented the detrimental effects of insecticides on soil microbial properties, including alterations in enzymatic activities. For instance, buprofezin has been shown to adversely impact invertase activity in soil (Hashimi et al., 2020).
(ii) *Pesticide Toxicity to Plants*: Herbicides, while targeting specific plant species, can also inadvertently affect non-target plant life. Non-target plant exposure to herbicides can occur through direct spraying, droplet drift, vapor movement, runoff, leaching, erosion, and improper disposal practices (Hashimi et al., 2020; Marshall & Marshall, 2001). These toxicants pose a significant threat to plant biodiversity (Isenring, 2010).

(iii) *Pesticide Toxicity to Animals*: The widespread dispersion of pesticide residues in the environment has led to substantial mortality among non-target organisms, including bees, birds, amphibians, fish, and small mammals (Carvalho, 2017). Pesticides have also been implicated in population declines of marine mammals, alligators, fish-eating birds, and other animal species. A notable example is the association between the accumulation of persistent chlorinated hydrocarbons, such as DDT, polychlorinated biphenyls (PCBs), and dioxins, in the food chain and the deaths of thousands of Arctic seals.

(iv) *Pesticide Toxicity to Pollinators*: Pollinators play a crucial role in maintaining ecosystem health. Cross-pollinated and self-incompatible plants rely on pollination services for seed production and biodiversity conservation. Honeybees, in particular, are not only nectar collectors but also essential pollinators. Pollen grains adhere to specialized hairs on the bodies of bees, enabling the transfer of pollen between flowers as bees move from plant to plant (Hashimi et al., 2020; Martin, 2015). Approximately 84% of commercially cultivated crops require insect pollination, underscoring the immense value of pollinating insects (Stanley & Preetha, 2016). Globally, honeybees contribute over $200 billion in pollination services annually, playing a vital role in the production of two-thirds of global crops and different wild flowering plants. Honey production is a significant economic activity and a valuable source of food worldwide (Hashimi et al., 2020).

10 Improper Disposal of Pesticides and Human Exposure

Improper disposal of pesticides presents a critical environmental hazard, contaminating soil, water, and air, and threatening both human health and natural ecosystems. Farmers in various global regions have reported discarding unused pesticides on land, often due to limited awareness and inadequate waste management systems in rural areas. This practice results in pesticide buildup in soil and water, disrupting ecosystems and harming unintended organisms (Stanley & Preetha, 2016). In addition to direct pesticide disposal, improper handling of pesticide containers exacerbates the problem. Discarded containers left in fields can leak residual pesticides into the environment, and repurposing empty containers for household storage creates risks of accidental exposure (Isenring, 2010). Burning or burying containers further pollutes the soil, water, and air, releasing toxic residues.

Human exposure to pesticides commonly occurs through contaminated food and water or by inhaling pesticide-contaminated air, with pesticides entering the body via ingestion, inhalation, or skin contact. International organizations, including the World Health Organization (WHO), the United States Food and Drug Administration (USFDA), and the Food and Agriculture Organization of the United Nations (FAO), have implemented regulatory frameworks to reduce human exposure. These include setting maximum residue limits (MRLs) for pesticides in food products, taking into account that pesticide effectiveness can be influenced by factors like wind speed and humidity (Davis et al., 2008; Fenik et al., 2011).

Pesticide exposure occurs through two main pathways: occupational and non-occupational. Occupational exposure primarily affects individuals involved in pesticide production, application, or disposal, while non-occupational exposure impacts the general population through contaminated food, water, or air (Davis et al., 2008; Hoh & Hites, 2004; Sharma et al., 2019).

11 Environmental Effect of Pesticides Pollution

Globally, an estimated 40% of agricultural produce is lost each year due to plant diseases, weeds, and pests. Without pesticide use, crop losses would likely increase significantly. These crop-protecting substances not only prevent damage but also improve yields (Aktar et al., 2009; Bajwa et al., 2017; Ross, 2005). Failing to protect crops from pests could reduce production, leading to food shortages and increased food prices, so pesticides indirectly contribute to food price stability. Many crops are also prone to aflatoxin contamination, requiring insect control to prevent the transfer of these carcinogenic toxins from insects to plants. Aflatoxins, known to cause liver and other cancers, can impair immune function and stunt growth in children. Pesticides are often used to reduce aflatoxin contamination by targeting insects responsible for its spread (Boutsalis et al., 2012; Ross, 2005).

The effects of pesticides on populations depend on exposure levels, toxicity, species characteristics, timing, population dynamics, and landscape factors (Schmolke et al., 2010). Some insecticides work by targeting insect nervous systems, such as organophosphates and methyl-carbamates, which inhibit acetylcholinesterase; neonicotinoids, which act on nicotinic acetylcholine receptors; and pyrethroids and DDT, which target sodium channels (Casida & Durkin, 2013). Neonicotinoid use, in particular, is on the rise and has raised concerns about toxicity (van Dijk, 2010). However, the overuse of pesticides poses risks to human health and biodiversity, and their water solubility, stability, and polar properties make it difficult to control their impact on the environment (Agrawal et al., 2010). Pesticides can induce toxicity not only in agriculture but also in industrial and public health sectors, harming the environment, natural flora, fauna, and aquatic ecosystems (Schmolke et al., 2010).

12 Health Effect of Pesticides Pollution

Pesticides have played a critical role in advancing public health by controlling vector-borne diseases. However, their extensive and often indiscriminate use has led to growing concerns about long-term health effects. Infants and young children are particularly vulnerable to pesticide exposure due to the non-specific nature of these chemicals and improper application practices. The surge in pesticide use over the past few decades has increased human exposure, especially in developing countries where the World Health Organization estimates around three million pesticide poisoning cases and 220,000 deaths occur each year (Kaur et al., 2019). Additionally, approximately 2.2 million people, mainly in these regions, are at heightened risk of pesticide exposure (Agrawal et al., 2010).

Certain groups, including young children, agricultural workers, and pesticide applicators, face a higher risk of toxic exposure (Abong'o et al., 2014; Ahmad et al., 2024). Pesticides can enter the human body through ingestion, inhalation, or skin absorption, leading to both acute and chronic health impacts (Pozo et al., 2011), with contaminated food being the primary exposure source. After ingestion, pesticides can cross various biological barriers before reaching tissues or storage sites (van der Werf, 1996). While the human body has detoxification mechanisms, excessive exposure can cause these chemicals to accumulate in the circulatory system, reaching toxic levels (Jabbar & Mallick, 1994). Studies have shown links between cancer risks and certain pesticide residues found in foods like fish (Kaur et al., 2019), water (Buczyńska & Szadkowska-Stańczyk, 2005), seafood (Moon et al., 2009), and dairy products (Pesticide Action Network, 2010). While cancer associations have been noted for specific residues, such as DDT and DDD, no significant link has been established for some other organochlorines, though PCBs (polychlorinated biphenyls) have been found to pose a notable risk (Kaur et al., 2019).

13 Acute Effect of Pesticides on Human Health

Direct exposure to pesticides can trigger a range of immediate health effects, ranging from mild irritation to severe life-threatening conditions. These acute effects can manifest through various routes of exposure, including inhalation, ingestion, dermal contact, and ocular exposure.

Common symptoms associated with acute pesticide exposure include:

- Eye irritation: Stinging, burning, redness, and blurred vision
- Skin irritation: Itching, rashes, blisters, and discomfort
- Respiratory irritation: Coughing, wheezing, shortness of breath, and chest tightness
- Gastrointestinal distress: Nausea, vomiting, abdominal cramps, and diarrhea
- Neurological disturbances: Headache, dizziness, fatigue, and tremors

In severe cases, acute pesticide exposure can lead to more serious health complications, including seizures, convulsions, loss of consciousness, respiratory failure, and death. The severity of acute pesticide poisoning depends on various factors, including the type of pesticide, the amount of exposure, the individual's sensitivity, and the route of exposure (Abong'o et al., 2014; Ahmad et al., 2024; van Dijk, 2010; Hashimi et al., 2020; Moon et al., 2009).

14 Chronic Effect of Pesticides on Human Health

Chronic pesticide exposure poses a serious, often overlooked risk to human health, with adverse effects that may take years to surface. Long-term contact with these toxic substances can disrupt multiple organ systems, leading to a host of serious health problems. Neurological impairment, immune system disruption, heightened cancer risk, reproductive issues, and organ damage are just some of the potential outcomes. Recent studies indicate a possible link between chronic exposure to pesticides and a higher likelihood of developing neurodegenerative conditions like Parkinson's and Alzheimer's diseases (Buczyńska & Szadkowska-Stańczyk, 2005; Moon et al., 2009; Pesticide Action Network, 2010; Rola & Pingali, 1993).

15 Distribution of Pesticides Consumption and Usage Worldwide

A sustained rise in global agricultural pesticide consumption was observed from 1990 to 2021, reaching nearly 3.54 million metric tons in the latter year. This upward trend is likely attributable to the burgeoning demand for crop protection chemicals, driven by the concurrent challenges of climate change, evolving pest resistance, and a growing population. Notably, the Americas held the dominant position in 2021, accounting for over half of global agricultural pesticide use. Asia and Europe followed with considerably lower shares of 27.7% and 14.3%, respectively. In contrast, Africa and Oceania collectively consumed less than 8% of the world's total (Global Pesticide Consumption 1990–2021, n.d.; Pesticide Action Network (Group), 2010; Zhang et al., 2011).

16 Pesticides Usage Worldwide and Their Effect

16.1 The Use of Pesticides in Africa and Its Associated Risks

Agriculture is essential to the African economy, with approximately 59% of the population dependent on farming. Despite this reliance, Africa represents only 2–4% of the global pesticide market and has the world's lowest pesticide usage rate (Abate et al., 2000). However, rising population pressures and food demand are expected to increase pesticide, herbicide, and fungicide use over the next three decades (Snyder et al., 2015). To maintain high yields and profitability, many African farmers are increasingly dependent on pesticides (de Bon et al., 2014).

Government policies have further shaped pesticide use trends. Since the 1970s, many African governments promoted pesticide use, but policy changes in the 1990s reduced input subsidies, leading to less government oversight and a sharp rise in pesticide imports through informal channels. Between 2000 and 2010, the import value of pesticides surged by 261%, largely due to unregulated imports (de Bon et al., 2014).

This lack of regulation has allowed banned pesticides to enter African markets. Inadequate farmer awareness has led to improper pesticide use, and non-compliance with Food and Agriculture Organization (FAO) codes of conduct has resulted in untested, potentially dangerous pesticide applications. Many farmers unknowingly use hazardous pesticides classified under the World Health Organization (WHO) risk system. For instance, the Pesticide Risk Reduction Program (PRRP) in Ethiopia found that 160 of 302 registered pesticides contain WHO class II (moderately hazardous) chemicals (Khan et al., 2010; Joseph Shaba et al., 2019; Kaur et al., 2019). Studies from Botswana, Nigeria, Zambia, and Malawi similarly reveal high usage of WHO-classified hazardous pesticides, including Malathion, cypermethrin, methomyl, and monocrotophos (Obopile et al., 2008; Oluwole & Cheke, 2009; Nyirenda et al., 2011).

Pesticide use in Africa also raises concerns about pest resistance, which threatens crop yields despite pesticide application. For example, pyrethroid resistance has emerged in tomato bollworm (*Heliothis armigera*) and diamondback moth (*Plutella xylostella*) in West Africa (de Bon et al., 2014; Martin, 2015). Resistance in the aphid *Aphis gossypii* to pyrethroids and organophosphates, and in the whitefly *Bemisia tabaci* to multiple pesticide classes, highlights the need for sustainable pest management practices (Carletto et al., 2010; Houndété et al., 2010).

16.2 The Use of Pesticides in Asia and Its Associated Risks

The use of pesticides in agricultural practices has significantly increased in developing countries, particularly in Southeast Asia. According to the World Health Organization (WHO), approximately 20% of global pesticide consumption occurs in these regions. Countries like Cambodia, Laos, and Vietnam have experienced notable annual increases in pesticide imports, with rates of 61%, 55%, and 10%, respectively (Schreinemachers & Tipraqsa, 2012).

In India, pesticide manufacturing began in 1952 with the production of benzene hexachloride, followed by DDT. The industry expanded rapidly, with India producing over 5000 metric tons of pesticides in 1958, reaching 85,000 metric tons by the mid-1990s. By this time, 145 pesticides had been registered, with insecticides being the dominant category (Ahad et al., 2010; Khan et al., 2010).

Similarly, Pakistan's pesticide usage began in 1954, initially importing 250 metric tons (Jabbar & Mallick, 1994). By 2003, annual pesticide consumption in Pakistan had risen dramatically to 78,132 tons (Pozo et al., 2011; Syed & Malik, 2011). During the Green Revolution, Pakistan significantly increased its pesticide imports from Europe and the United States to address pest infestations, control locust outbreaks, and suppress malaria (Ahad et al., 2010).

This surge in pesticide usage across Southeast Asia and South Asia highlights the growing reliance on chemicals to sustain agricultural production, although it also raises concerns regarding environmental and health risks associated with their widespread application.

16.3 The Use of Pesticides in Australia and Its Associated Risks

Australian farmers have shown a willingness to incur higher chemical costs to effectively manage pest infestations, with the agricultural pesticide market valued at A$187 million annually (Llewellyn et al., 2016). This includes herbicides, insecticides, fungicides, and certain growth regulators. The costs of herbicides, in particular, have risen due to the use of alternative herbicides, mixtures, and higher concentration rates, which add approximately A$8 per hectare.

Weed infestations significantly reduce crop productivity and quality, making effective weed control essential for maintaining agricultural productivity. Herbicide application is an effective method for managing weeds, thus making it an attractive choice for farmers. However, concerns arise from the overuse of herbicides, especially those with similar modes of action, which can lead to the development of herbicide-resistant biotypes. This resistance, which has been observed in several weed species, poses a long-term challenge for sustainable agricultural practices and highlights the need for integrated pest management strategies to mitigate resistance development (Walsh & Powles, 2014).

16.4 The Use of Pesticides in Europe and Its Associated Risks

The widespread use of insecticides, herbicides, fungicides, and chemical fertilizers across Europe has contributed to increasing pest infestations in agricultural lands, which, in turn, has caused a decline in natural habitats and reduced the heterogeneity of farmlands and surrounding landscapes (Miraglia et al., 2009; Struik & Bonciarelli, 1997). A joint survey across seven European countries, including Latvia, Denmark, Germany, the Netherlands, Finland, Sweden, and the UK, highlighted the growing use of pesticides in both urban and non-agricultural areas (Walsh & Powles, 2014). Herbicides were identified as the main pesticide contaminants in urban environments, with significant regional variations in political interest, public discourse, and regulations on pesticide use in urban settings.

Countries like Denmark, the Netherlands, Germany, and Sweden have shown strong political and public interest in reducing herbicide use for urban weed control. In contrast, while the UK has implemented stricter regulations, Finland and Latvia still lack specific measures to curb pesticide use. Finland, in particular, has experienced a rise in pesticide use, though it remains relatively low compared to other European regions, with annual pesticide consumption estimated at 5–6 metric tons (Zhang et al., 2011). A 2007 survey of 80 Finnish municipalities revealed that while one-fifth occasionally used pesticides, only about 15% reported regular use.

Additionally, a study in Greece detected pesticides in hot spring water samples, though none exceeded the European Union's Maximum Acceptable Concentration (MAC). Lindane (γ-BHC), a commonly used pesticide, was found in 35% of the samples, with concentrations ranging from 0.005 to 0.01 μg/L, further illustrating the pervasive presence of pesticides in the environment (Sharma et al., 2019).

17 Impacts of Pesticide Pollution and Contamination on Global Ecosystem

The widespread and excessive use of pesticides has raised significant environmental concerns, particularly regarding their safe storage, disposal, and their impact on ecosystems. Pesticide misuse has been linked to soil and water contamination, which adversely affects the delicate balance of soil microflora and microfauna and impairs plants' ability to absorb essential nutrients (van der Werf, 1996). Pesticide leaching exacerbates the problem by contaminating local water bodies, with groundwater contamination posing a serious threat to drinking water supplies (Sharma et al., 2019).

In Australia, pesticide runoff from agricultural lands, especially in the Tasmanian forestry industry, has raised alarms due to its impact on coastal and inshore ecosystems, including the Great Barrier Reef (GBR). The extensive use of triazine herbicides has left residual contamination in agricultural catchments, which affects marine organisms such as seagrass, corals, and algae (Barzman et al., 2015). Herbicides like atrazine and diuron are commonly found in GBR sediments, with seasonal pesticide movement detected in the agricultural floodplains of the lower Burdekin River. The majority of detected pesticides were linked to the sugarcane industry, with runoff being a major contributor to water contamination (Davis et al., 2008).

This contamination, resulting from agricultural waste discharge and runoff, enters the food chain and poses risks to aquatic life. Fish consuming contaminated water may become unsafe for human consumption. In Central America and other regions, the continuous use of pesticides has led to elevated concentrations of hazardous chemicals like DDT, chlordane, p,p′-DDE, and toxaphene in the air (Hoh & Hites, 2004). In South America, pesticides such as chlorpyrifos, endosulfan, and cypermethrin have frequently been detected in water bodies, indicating the ongoing challenge of pesticide pollution in the environment (Ahad et al., 2010; Syed & Malik, 2011). These persistent chemicals threaten not only wildlife but also human health through the contamination of water resources and the food chain.

18 Threats of Pesticide Pollution to Biodiversity

The indiscriminate use of pesticides presents a serious threat to biodiversity, necessitating urgent attention and comprehensive mitigation strategies. Pesticides have far-reaching effects on both aquatic and terrestrial ecosystems, impacting a wide variety of organisms, including plants, animals, and birds. A particularly concerning aspect is the bioaccumulation of pesticides within the food chain, which leads to harmful consequences for predator and raptor populations. Studies have shown that pesticides, when ingested by lower-level organisms, gradually accumulate in the tissues of higher-level predators, affecting their health and reproductive capabilities (van Dijk, 2010; Hashimi et al., 2020; Mahmood et al., 2015).

Furthermore, the widespread application of pesticides can indirectly reduce the abundance of weeds, shrubs, and insects, which are crucial food sources for many species. The loss of these essential food resources can cause declines in the populations of various animals and birds, some of which may be rare or endangered. The indiscriminate spraying of insecticides, herbicides, and fungicides has been linked to a significant decrease in the numbers of certain species, with some experiencing population crashes due to the direct and indirect effects of pesticide exposure. The long-term and frequent use of pesticides exacerbates the issue of bioaccumulation, causing further ecological damage over time (Isenring, 2010; Marshall, 2001).

In addition to harming individual species, the disruption of food webs and ecosystems can lead to cascading effects, undermining ecosystem services that are vital for maintaining biodiversity. These changes can also make ecosystems less resilient to other environmental stresses, such as climate change and habitat loss, creating a feedback loop of ecological degradation. As such, it is crucial to adopt more sustainable agricultural practices and regulations that limit the use of harmful pesticides, promote alternatives, and prioritize ecological balance.

19 Climate Change and Its Impact on Pesticides Pollution and Food Systems

Climate change has become a central issue in the global food system, with far-reaching effects on agricultural production. Rising temperatures, altered rainfall patterns, increased drought frequency, and disrupted seasonal variations all contribute to significant changes in crop yields. Additionally, elevated CO_2 levels and changes in soil quality, as well as the forced relocation of cropping systems due to water scarcity, further exacerbate these challenges (Bailey, 2004; Miraglia et al., 2009; Muriel et al., 2000). These shifts make crops more susceptible to plant diseases and insect infestations, increasing the reliance on pesticides for protection.

The direct impacts of climate change on pesticide efficacy are complex. Water scarcity and degradation due to high atmospheric temperatures reduce soil bioactivity, while elevated temperatures often lead to a reduction in soil organic matter. This can result in greater root uptake of pesticides that would otherwise be bound to the soil's organic matter. Furthermore, higher temperatures may cause pesticides to evaporate more quickly, reducing their effectiveness (Miraglia et al., 2009). In arid conditions, many pesticides are less effective, requiring higher doses or more frequent applications to ensure adequate crop protection (Muriel et al., 2000). Some studies suggest that climate change may also accelerate the degradation of certain pesticides under extreme temperatures, further complicating their use (Bailey, 2004). These effects contribute to food security concerns, as the efficacy and application of pesticides become increasingly unpredictable in the face of climate change.

To address the growing concerns about pesticide use, the concept of "Integrated Pest Management (IPM)" has emerged as a sustainable approach to crop protection. IPM, which began in 1959 as a strategy to control the spotted alfalfa aphid in the United States, integrates various pest management methods based on cost–benefit analysis. It aims to reduce reliance on synthetic pesticides while promoting sustainable agriculture (Kogan, 1998; Stern et al., 1959). The European Union has actively supported the adoption of IPM through the enactment of the "Sustainable Use of Pesticides" Directive (2009/128/EC), which encourages Member States to develop National Action Plans to reduce pesticide use.

IPM programs focus on minimizing pesticide use, enhancing biodiversity, and increasing resilience to pests and diseases. The EU directive promotes the inclusion of IPM as a core component of national pesticide management programs. These initiatives are complemented by measures such as pest forecasting, pesticide resistance monitoring, and training programs for farmers to ensure that IPM strategies are effectively implemented (Pimentel, 2005; Barzman et al., 2015). Globally, the adoption of IPM principles is growing, with farmer field schools engaging millions of farmers in sustainable pest management practices since 1989 (Sharma et al., 2019). This shift toward more sustainable agricultural practices reflects the increasing recognition of the need for systems that reduce pesticide dependency while maintaining food security.

20 Pesticide Legislation and Regulatory Frameworks: A Global Perspective

Pesticide legislation and management practices differ significantly across countries due to various socioeconomic, environmental, and regulatory contexts. The management of pesticides throughout their lifecycle ranging from manufacturing and application to risk reduction, monitoring, and waste disposal requires a multifaceted approach that is often more robust in developed countries compared to developing nations. In many developed countries, stringent pesticide regulations have been implemented, often supported by comprehensive legal frameworks, monitoring programs, and infrastructure. Conversely, developing countries face several challenges, including inadequate legislation, resource constraints, lack of expertise, and insufficient infrastructure for pesticide analysis. These factors contribute to difficulties in enforcing pesticide regulations and managing associated risks effectively (Handford et al., 2015).

A key international organization involved in pesticide management is the Codex Alimentarius Commission (CAC), a joint initiative of the World Health Organization (WHO) and the Food and Agriculture Organization of the United Nations (FAO). The CAC has established standards for pesticide residues in food commodities, offering a global framework to minimize contamination risks and safeguard food safety across the supply chain. However, despite these global standards, a significant gap in pesticide management is the lack of comprehensive guidelines tailored to specific farming systems. Such guidelines would consider the economic, social, and environmental impacts of pesticide use, promoting the production of safe and healthy food through sustainable methods (Leong et al., 2020).

To address the potential health risks posed by pesticides, organizations like the FAO have developed toxicological reference values, such as the Acceptable Daily Intake (ADI) and Acute Reference Dose (ARfD). The ADI represents the amount of a pesticide that can be ingested daily over a lifetime without posing a significant health risk, expressed in milligrams per kilogram of body weight per day. This value is based on scientific data available at the time of evaluation. The European Union (EU) has adopted Maximum Residue Limits (MRLs), which are often more stringent than the standards set by the Codex Alimentarius. The approval process for new pesticides in the EU is rigorous, involving an extensive evaluation by the European Food Safety Authority (EFSA) and EU member states, which can take up to 3 years. Additionally, countries within the EU may establish "national MRLs" to address local dietary patterns and risk profiles, ensuring that pesticide use aligns with regional health and safety needs. Overall, while significant progress has been made in pesticide regulation in developed countries, much work remains to improve pesticide management in developing nations. This includes the need for stronger legal frameworks, better enforcement mechanisms, and more tailored, sustainable approaches to pesticide use that prioritize both public health and environmental protection.

21 Biopesticides: Natural Pesticides an Alternative to Chemical Pesticides

The growing awareness of the detrimental effects of chemical pesticides on human health and the environment has spurred the search for sustainable alternatives. Biopesticides, derived from natural sources such as plants, microbes, and animals, emerged as a promising solution (EU Legislation on MRLs – European Commission, n.d.; Simon et al., 2024). These substances offer a nontoxic approach to pest control, mitigating the environmental and health concerns associated with conventional pesticides. Biopesticides can be broadly classified into two main categories:

biochemical pest control agents and microbial pest control agents. Biochemical agents, including pheromones, hormones, natural plant growth regulators, and enzymes, disrupt the pest's life cycle and reproductive processes without harming non-target organisms. Microbial agents, on the other hand, utilize microorganisms, such as bacteria, fungi, viruses, and protozoa, to directly kill pests or disrupt their development and behavior.

The development of biopesticides with novel modes of action and minimal cross-resistance to conventional pesticides is a subject of ongoing research. Biopesticides offer several advantages over their synthetic counterparts, including:

(i) Environmental Friendliness: Biopesticides are biodegradable and typically have a narrow spectrum of activity, targeting specific pests without harming beneficial organisms or the environment.
(ii) Reduced Human Health Risks: Biopesticides generally exhibit lower toxicity to humans and animals, posing minimal risks to applicators and consumers.
(iii) Organic Production Compatibility: Biopesticides are essential components of certified organic agricultural practices, adhering to strict standards that promote sustainable and environmentally friendly farming methods.
(iv) Reduced Residue Concerns: Biopesticides often have shorter persistence in the environment and leave fewer residues on treated crops, allowing for harvest closer to application.
(v) Rapid Pest Control: Biopesticides can act quickly to disrupt pest feeding, providing immediate relief from pest damage.

Despite their advantages, biopesticides also has various disadvantages such as:

(i) Narrow Spectrum of Activity: Biopesticides often target specific pests, limiting their broad applicability across a range of pests.
(ii) Weather Dependence: Environmental factors, such as temperature, humidity, and sunlight, can affect the efficacy of biopesticides.
(iii) Storage and Shelf Life: Proper storage conditions are crucial to maintain the viability and effectiveness of microbial biopesticides.
(iv) Regulatory Requirements: Extensive testing and registration processes are required to ensure the safety and efficacy of biopesticides before commercialization (Ahmad et al., 2024).

21.1 Types of Natural Pesticides

The realm of natural pesticides encompasses a diverse array of botanicals, microbials, essential oils, and mineral-based agents, many of which are derived from plants, insects, or naturally occurring minerals. These natural alternatives offer a compelling solution to the growing concerns associated with conventional pesticides, which often pose significant risks to human health and the environment. Examples are:

(i) Microbial pesticides, encompassing fungi, bacteria, protozoans, algae, and viruses, exploit the inherent pathogenicity of microorganisms to combat insect populations. Genetically engineered variants further enhance the efficacy of these biopesticides, targeting specific insects with increased potency. The mode of action of microbial pesticides varies, encompassing disease induction, toxin production, and reproductive inhibition. *Bacillus thuringiensis*, commonly known as milky spore, exemplifies a microbial pesticide that disrupts insect development.
(ii) Mineral-based pesticides, such as sulfur and lime-sulfur, have long been employed in pest control strategies. These compounds are typically applied as sprays to combat common insect pests.
(iii) Botanical pesticides harness the insecticidal properties of various plants. Nicotine, neem, rotenone, anabasine, azadirachtin, ryania, essential oils, sabadilla, and pyrethrins are all derived from plant sources. Pyrethrins, extracted from chrysanthemum plants, are particularly effective against flying insects, larvae, and grubs. Botanical pesticides can be classified as constitutive chemicals, naturally present in the plant, or inducible chemicals, activated in response to insect activity. Some botanical pesticides undergo chemical modifications after extraction to enhance their insecticidal properties (Joseph Shaba et al., 2019).

22 Conclusion

The widespread adoption of pesticides in modern agriculture has indeed transformed food production, allowing for higher crop yields and more affordable food prices. This increase in agricultural productivity has significantly contributed to global food security, making nutrient-rich food more accessible to larger populations. Without pesticides, achieving such a feat would have been much more challenging, especially in terms of controlling pests and diseases that could otherwise decimate crops. However, this agricultural success has come with a substantial environmental and health cost. The indiscriminate use of pesticides, often motivated by the pursuit of immediate results and economic affordability, has resulted in widespread environmental contamination. Toxic residues from pesticides persist in the environment, accumulating in soil, water, and air. These chemicals can disrupt ecosystems and harm non-target organisms, including beneficial insects, wildlife, and aquatic life. For example, pesticides can harm pollinators like bees, which are essential for the reproduction of many plants. Moreover, pesticides can lead to the development of resistance in pest populations, prompting the need for even more toxic chemicals, creating a vicious cycle of pesticide use and environmental degradation.

The harmful effects of pesticides also extend to human health. Exposure to these chemicals can lead to a range of health issues, particularly when individuals are exposed to high concentrations or over long periods. Children, pregnant women, and developing organisms are especially vulnerable. Health problems associated with pesticide exposure include neurological disorders, respiratory issues, and an increased risk of cancers and reproductive disorders. Though many harmful pesticides have been banned or restricted in developed countries, they are still widely used in many developing nations where they are seen as cost-effective and efficient solutions for pest control. This continued use in developing regions poses significant health risks to agricultural workers and surrounding communities.

In light of the growing concerns regarding the negative impacts of synthetic pesticides, biopesticides which are pesticides derived from natural sources like plants, microbes, and animals have emerged as a promising alternative. Biopesticides are generally more specific in their action, targeting particular pests without affecting beneficial organisms or the broader ecosystem. They typically break down more quickly in the environment, reducing the risk of long-term environmental contamination. Some biopesticides work by disrupting the growth and development of pests, while others can stimulate natural pest resistance in plants or attract predators that control pest populations. As research into biopesticides continues to advance, they hold the potential to significantly reduce the reliance on synthetic pesticides, offering a more sustainable and eco-friendly solution to pest management in agriculture.

Ultimately, the challenge is to find a balance between maintaining high levels of agricultural productivity and minimizing the negative environmental and health impacts of pesticide use. Shifting toward more sustainable and integrated pest management approaches, including the use of biopesticides, could help address many of these challenges while preserving the benefits of modern agriculture.

References

Abate, T., van Huis, A., & Ampofo, J. K. O. (2000). Pest management strategies in traditional agriculture: An African perspective. *Annual Review of Entomology, 45*, 631–659. https://doi.org/10.1146/annurev.ento.45.1.631

Abong'o, D. A., Wandiga, S. O., Jumba, I. O., Madadi, V. O., & Kylin, H. (2014). *Impacts of pesticides on human health and environment in the River Nyando catchment, Kenya*. http://erepository.uonbi.ac.ke/handle/11295/72981

Abubakar, Y., Tijjani, H., Egbuna, C., Adetunji, C. O., Kala, S., Kryeziu, T. L., Ifemeje, J. C., & Patrick-Iwuanyanwu, K. C. (2020). Pesticides, history, and classification. In *Natural remedies for pest, disease and weed control* (pp. 29–42). Elsevier. https://doi.org/10.1016/B978-0-12-819304-4.00003-8

Agrawal, A., Pandey, R. S., & Sharma, B. (2010). Water pollution with special reference to pesticide contamination in India. *Journal of Water Resource and Protection, 2*(5), Article 5. https://doi.org/10.4236/jwarp.2010.25050

Ahad, K., Mohammad, A., Khan, H., Ahmad, I., & Hayat, Y. (2010). Monitoring results for organochlorine pesticides in soil and water from selected obsolete pesticide stores in Pakistan. *Environmental Monitoring and Assessment, 166*(1–4), 191–199. https://doi.org/10.1007/s10661-009-0995-5

Ahmad, M. F., Ahmad, F. A., Alsayegh, A. A., Zeyaullah, M., AlShahrani, A. M., Muzammil, K., Saati, A. A., Wahab, S., Elbendary, E. Y., Kambal, N., Abdelrahman, M. H., & Hussain,

S. (2024). Pesticides impacts on human health and the environment with their mechanisms of action and possible countermeasures. *Heliyon, 10*(7), e29128. https://doi.org/10.1016/j.heliyon.2024.e29128
Aktar, M. W., Sengupta, D., & Chowdhury, A. (2009). Impact of pesticides use in agriculture: Their benefits and hazards. *Interdisciplinary Toxicology, 2*(1), 1–12. https://doi.org/10.2478/v10102-009-0001-7
Alosaimy, A. M., Abouzied, A. S., Alsaedi, A. M. R., Alafnan, A., Alamri, A., Alamri, M. A., Khaled Bin Break, M., Sabour, R., & Farghaly, T. A. (2023). Discovery of novel indene-based hybrids as breast cancer inhibitors targeting Hsp90: Synthesis, bio-evaluation and molecular docking study. *Arabian Journal of Chemistry, 16*(4), 104569. https://doi.org/10.1016/j.arabjc.2023.104569
Bailey, S. W. (2004). Climate change and decreasing herbicide persistence. *Pest Management Science, 60*(2), 158–162. https://doi.org/10.1002/ps.785
Bajwa, A. A., Walsh, M., & Chauhan, B. S. (2017). Weed management using crop competition in Australia. *Crop Protection, 95*, 8–13. https://doi.org/10.1016/j.cropro.2016.08.021
Barzman, M., Bàrberi, P., Birch, A. N. E., Boonekamp, P. M., Dachbrodt-Saaydeh, S., Graf, B., Hommel, B., Jensen, J. E., Kiss, J., Kudsk, P., Lamichhane, J. R., Messéan, A., Moonen, A. C., Ratnadass, A., Ricci, P., Sarah, J. L., & Sattin, M. (2015). Eight principles of integrated pest management. *Agronomy for Sustainable Development, 35*(4), 1199–1215. https://doi.org/10.1007/s13593-015-0327-9
Boutsalis, P., Gill, G. S., & Preston, C. (2012). Incidence of herbicide resistance in rigid ryegrass (*Lolium rigidum*) across southeastern Australia. *Weed Technology, 26*(3), 391–398. https://doi.org/10.1614/WT-D-11-00150.1
Brown, A. E. (2006). *Mode of action of insecticides and related pest control chemicals for production agriculture, ornamentals, and turf.* https://www.semanticscholar.org/paper/Mode-of-Action-of-Insecticides-and-Related-Pest-for-Brown/9668739012d80c586489ae0588b97fa1a900fbed
Buczyńska, A., & Szadkowska-Stańczyk, I. (2005). Identification of health hazards to rural population living near pesticide dump sites in Poland. *International Journal of Occupational Medicine and Environmental Health, 18*(4), 331–339.
Carletto, J., Martin, T., Vanlerberghe-Masutti, F., & Brévault, T. (2010). Insecticide resistance traits differ among and within host races in Aphis gossypii. *Pest Management Science, 66*(3), 301–307. https://doi.org/10.1002/ps.1874
Carvalho, F. (2017). Pesticides, environment, and food safety. *Food and Energy Security, 6*, 48–60. https://doi.org/10.1002/fes3.108
Casida, J. E., & Durkin, K. A. (2013). Neuroactive insecticides: Targets, selectivity, resistance, and secondary effects. *Annual Review of Entomology, 58*, 99–117. https://doi.org/10.1146/annurev-ento-120811-153645
Claborn, D. M. (2010). The biology and control of leishmaniasis vectors. *Journal of global infectious diseases, 2*(2), 127–134.
Davis, A., Lewis, S., Bainbridge, Z. T., Brodie, J., & Shannon, E. (2008, May 1). *Pesticide residues in waterways of the lower Burdekin region: Challenges in ecotoxicological interpretation of monitoring data.* https://www.semanticscholar.org/paper/Pesticide-residues-in-waterways-of-the-lower-in-of-Davis-Lewis/f2a6b1341acc6fae54566229400347f4ea0f4376
de Bon, H., Huat, J., Parrot, L., Sinzogan, A., Martin, T., Malézieux, E., & Vayssières, J.-F. (2014). Pesticide risks from fruit and vegetable pest management by small farmers in sub-Saharan Africa. A review. *Agronomy for Sustainable Development, 34*(4), 723–736. https://doi.org/10.1007/s13593-014-0216-7
EU legislation on MRLs—European Commission. (n.d.). Retrieved September 23, 2024, from https://food.ec.europa.eu/plants/pesticides/maximum-residue-levels/eu-legislation-mrls_en

Fenik, J., Tankiewicz, M., & Biziuk, M. (2011). Properties and determination of pesticides in fruits and vegetables. *TrAC Trends in Analytical Chemistry, 30*(6), 814–826. https://doi.org/10.1016/j.trac.2011.02.008

Foundation, G. (2018). Pesticides and their applications in agriculture. *Asian Journal of Applied Science and Technology, 2*(2), 894–900.

Global pesticide consumption 1990–2021. (n.d.). *Statista*. Retrieved August 27, 2024, from https://www.statista.com/statistics/1263077/global-pesticide-agricultural-use/

Handford, C. E., Elliott, C. T., & Campbell, K. (2015). A review of the global pesticide legislation and the scale of challenge in reaching the global harmonization of food safety standards. *Integrated Environmental Assessment and Management, 11*(4), 525–536. https://doi.org/10.1002/ieam.1635

Hashimi, M. H., Hashimi, R., & Ryan, Q. (2020). Toxic effects of pesticides on humans, plants, animals, pollinators and beneficial organisms. *Asian Plant Research Journal*, 37–47. https://doi.org/10.9734/aprj/2020/v5i430114

Hatfield, G. (2004). *Encyclopedia of folk medicine: Old world and new world traditions*. ABC-CLIO.

Hoh, E., & Hites, R. A. (2004). Sources of toxaphene and other organochlorine pesticides in North America as determined by air measurements and potential source contribution function analyses. *Environmental Science & Technology, 38*(15), 4187–4194. https://doi.org/10.1021/es0499290

Houndété, T. A., Kétoh, G. K., Hema, O. S. A., Brévault, T., Glitho, I. A., & Martin, T. (2010). Insecticide resistance in field populations of Bemisia tabaci (Hemiptera: Aleyrodidae) in West Africa. *Pest Management Science, 66*(11), 1181–1185. https://doi.org/10.1002/ps.2008

Isenring, R. (2010). Pesticides and the loss of biodiversity: How intensive pesticide use affects wildlife populations and species diversity.

Jabbar, A., & Mallick, S. (1994). *Introduction (Pesticides and environment situation in Pakistan)* (pp. 1–2). Sustainable Development Policy Institute. https://www.jstor.org/stable/resrep00618.4

Joseph Shaba, P., Oguh, C., Ogechi, O., Ubani, C., Okekeaji, U., & Ugochukwu, E. (2019). Natural pesticides (biopesticides) and uses in pest management-A critical review. *Asian Journal of Biotechnology and Genetic Engineering, 2*, 1–18.

Kaur, R., Mavi, G. K., Raghav, S., & Khan, I. (2019). Pesticides classification and its impact on environment. *International Journal of Current Microbiology and Applied Sciences, 8*(03), 1889–1897. https://doi.org/10.20546/ijcmas.2019.803.224

Khan, J., Zia, M., Qasim, M., & s. (2010). Use of pesticides and their role in environmental pollution. *International Scholarly and Scientific Research & Innovation, 4*, 85–91.

Kogan, M. (1998). Integrated pest management: Historical perspectives and contemporary developments. *Annual Review of Entomology, 43*, 243–270. https://doi.org/10.1146/annurev.ento.43.1.243

Kraus, W. (1995). Biologically active ingredients: Section 2.1. In H. Schmutterer (Ed.), *The neem tree* (1st ed., pp. 35–74). Wiley. https://doi.org/10.1002/3527603980.ch2a

Leong, W.-H., Teh, S.-Y., Hossain, M. M., Nadarajaw, T., Zabidi-Hussin, Z., Chin, S.-Y., Lai, K.-S., & Lim, S.-H. E. (2020). Application, monitoring and adverse effects in pesticide use: The importance of reinforcement of Good Agricultural Practices (GAPs). *Journal of Environmental Management, 260*, 109987. https://doi.org/10.1016/j.jenvman.2019.109987

Llewellyn, R., Ronning, D., Clarke, M., Mayfield, A., Walker, S., & Ouzman, J. (2016). Impact of weeds in Australian grain production. Grains Research and Development Corporation, Canberra, ACT, Australia.

Mahmood, I., Imadi, S., Shazadi, K., Gul, A., & Hakeem, K. (2015). Effects of pesticides on environment. https://doi.org/10.1007/978-3-319-27455-3_13

Marshall, E. (2001). Biodiversity, herbicides and non-target plants. In *BCPC conference weeds* (pp. 855–862). BCPC.

Marshall, E. J. P., & Marshall, E. J. P. (2001). Biodiversity, herbicides and non-target plants. In *Pesticide behaviour in soils and water proceedings of a symposium organized by the British*

Crop Protection Council, Brighton, UK, 13–15 November 2001 (pp. 419–426). https://eurekamag.com/research/003/660/003660441.php

Martin, C. (2015). A re-examination of the pollinator crisis. *Current Biology, 25*(19), R811–R815. https://doi.org/10.1016/j.cub.2015.09.022

Miraglia, M., Marvin, H. J. P., Kleter, G. A., Battilani, P., Brera, C., Coni, E., Cubadda, F., Croci, L., De Santis, B., Dekkers, S., Filippi, L., Hutjes, R. W. A., Noordam, M. Y., Pisante, M., Piva, G., Prandini, A., Toti, L., van den Born, G. J., & Vespermann, A. (2009). Climate change and food safety: An emerging issue with special focus on Europe. *Food and Chemical Toxicology: An International Journal Published for the British Industrial Biological Research Association, 47*(5), 1009–1021. https://doi.org/10.1016/j.fct.2009.02.005

Moon, H.-B., Kim, H.-S., Choi, M., Yu, J., & Choi, H.-G. (2009). Human health risk of polychlorinated biphenyls and organochlorine pesticides resulting from seafood consumption in South Korea, 2005–2007. *Food and Chemical Toxicology, 47*(8), 1819–1825. https://doi.org/10.1016/j.fct.2009.04.028

Muriel, P., Downing, T., Hulme, M., Harrington, R., Lawlor, D. W., Wurr, D., Atkinson, C. J., Cockshull, K. E., Taylor, D. R., Richards, A. J., Parsons, D. J., Hillerton, J. E., Parry, M. L., Jarvis, S. C., Weatherhead, K., & Jenkins, G. (2000). *Climate change and agriculture in the United Kingdom [Text]*. Ministry of Agriculture, Fisheries and Food (MAFF). https://repository.rothamsted.ac.uk/item/8846y/climate-change-and-agriculture-in-the-united-kingdom

Nasrollahi, Z., Hashemi, M., Bameri, S., & Mohamad Taghvaee, V. (2020). Environmental pollution, economic growth, population, industrialization, and technology in weak and strong sustainability: Using STIRPAT model. *Environment, Development and Sustainability, 22*(2), 1105–1122. https://doi.org/10.1007/s10668-018-0237-5

Nyirenda, S., Sileshi, G., Belmain, S., Kamanula, J., & Stevenson, P. (2011). *Farmers' ethno-ecological knowledge of vegetable pests and pesticidal plant use in Northern Malawi and Eastern Zambia*. https://www.semanticscholar.org/paper/Farmers'-ethno-ecological-knowledge-of-vegetable-in-Nyirenda-Sileshi/4765812f91baf85b7300f25cf898ec012983902b

Obopile, M., Munthali, D. C., & Matilo, B. (2008). Farmers' knowledge, perceptions and management of vegetable pests and diseases in Botswana. *Crop Protection, 27*(8), 1220–1224. https://doi.org/10.1016/j.cropro.2008.03.003

Oluwole, O., & Cheke, R. A. (2009). Health and environmental impacts of pesticide use practices: A case study of farmers in Ekiti State, Nigeria. *International Journal of Agricultural Sustainability, 7*(3), 153–163. https://doi.org/10.3763/ijas.2009.0431

Pesticide Action Network (Group) (Ed.). (2010). *Communities in peril: Global report on health impacts of pesticide use in agriculture*. Pesticide Action Network Asia Pacific.

Pimentel, D. (2005). Environmental and economic costs of the application of pesticides primarily in the United States. *Environment, Development and Sustainability, 7*(2), 229–252. https://doi.org/10.1007/s10668-005-7314-2

Pozo, K., Harner, T., Lee, S. C., Sinha, R. K., Sengupta, B., Loewen, M., Geethalakshmi, V., Kannan, K., & Volpi, V. (2011). Assessing seasonal and spatial trends of persistent organic pollutants (POPs) in Indian agricultural regions using PUF disk passive air samplers. *Environmental Pollution, 159*(2), 646–653. https://doi.org/10.1016/j.envpol.2010.09.025

Rola, A., & Pingali, P. (1993). *Pesticides, rice productivity, and farmers' health: An economic assessment*. https://www.semanticscholar.org/paper/Pesticides%2C-rice-productivity%2C-and-farmers%27-health%3A-Rola-Pingali/f41ebf60ff1c00c4a2c522faca88d5ab424d10d1

Ross, G. (2005). Risks and benefits of DDT. *Lancet (London, England), 366*(9499), 1771–1772; author reply 1772. https://doi.org/10.1016/S0140-6736(05)67722-7

Schmolke, A., Thorbek, P., Chapman, P., & Grimm, V. (2010). Ecological models and pesticide risk assessment: Current modeling practice. *Environmental Toxicology and Chemistry, 29*(4), 1006–1012. https://doi.org/10.1002/etc.120

Schreinemachers, P., & Tipraqsa, P. (2012). Agricultural pesticides and land use intensification in high, middle and low income countries. *Food Policy, 37*(6), 616–626. https://doi.org/10.1016/j.foodpol.2012.06.003
Sharma, A., Kumar, V., Shahzad, B., Tanveer, M., Sidhu, G. P. S., Handa, N., Kohli, S. K., Yadav, P., Bali, A. S., Parihar, R. D., Dar, O. I., Singh, K., Jasrotia, S., Bakshi, P., Ramakrishnan, M., Kumar, S., Bhardwaj, R., & Thukral, A. K. (2019). Worldwide pesticide usage and its impacts on ecosystem. *SN Applied Sciences, 1*(11), 1446. https://doi.org/10.1007/s42452-019-1485-1
Simon, L. V., Sandhu, D. S., Goyal, A., & Kruse, B. (2024). Japanese encephalitis. In *StatPearls*. StatPearls Publishing. http://www.ncbi.nlm.nih.gov/books/NBK470423/
Smith, T. M., & Smith, R. L. (2014). *Elements of ecology*. Pearson Education.
Snyder, J., Smart, J., Goeb, J., & Tschirley, D. (Eds.). (2015). *Pesticide use in Sub-Saharan Africa: Estimates, projections, and implications in the context of food system transformation*. https://doi.org/10.22004/ag.econ.230980
Stanley, J., & Preetha, G. (2016). *Pesticide toxicity to non-target organisms*. Springer Netherlands. https://doi.org/10.1007/978-94-017-7752-0
Stern, V., Smith, R., van den Bosch, R., & Hagen, K. (1959). The integration of chemical and biological control of the spotted alfalfa aphid: The integrated control concept. *Hilgardia, 29*(2), 81–101. http://hilgardia.ucanr.edu/Abstract/?a=hilg.v29n02p081
Struik, P. C., & Bonciarelli, F. (1997). Resource use at the cropping system level. *European Journal of Agronomy, 7*(1), 133–143. https://doi.org/10.1016/S1161-0301(97)00027-0
Syed, J. H., & Malik, R. N. (2011). Occurrence and source identification of organochlorine pesticides in the surrounding surface soils of the Ittehad Chemical Industries Kalashah Kaku, Pakistan. *Environmental Earth Sciences, 62*(6), 1311–1321. https://doi.org/10.1007/s12665-010-0618-z
Umetsu, N., & Shirai, Y. (2020). Development of novel pesticides in the 21st century. *Journal of Pesticide Science, 45*(2), 54–74. https://doi.org/10.1584/jpestics.D20-201
van der Werf, H. M. G. (1996). Assessing the impact of pesticides on the environment. *Agriculture, Ecosystems & Environment, 60*(2), 81–96. https://doi.org/10.1016/S0167-8809(96)01096-1
van Dijk, T. C. (2010). *Effects of neonicotinoid pesticide pollution of Dutch surface water on non-target species abundance* [Master thesis]. https://studenttheses.uu.nl/handle/20.500.12932/4847
Walsh, M. J., & Powles, S. B. (2014). Management of herbicide resistance in wheat cropping systems: Learning from the Australian experience. *Pest Management Science, 70*(9), 1324–1328. https://doi.org/10.1002/ps.3704
Wang, H., Yuan, X., Wu, Y., Chen, X., Leng, L., Wang, H., Li, H., & Zeng, G. (2015). Facile synthesis of polypyrrole decorated reduced graphene oxide–Fe3O4 magnetic composites and its application for the Cr(VI) removal. *Chemical Engineering Journal, 262*, 597–606. https://doi.org/10.1016/j.cej.2014.10.020
WHO, (2012) Handbook for Integrated Vector Management WHO/HTM/NTD/VEM/2012.3. Geneva: World Health Organization. http://whqlibdoc.who.int/ publications/2012/9789241502801_eng.pdf
Zadoks, J. C., & Waibel, H. (2000). From pesticides to genetically modified plants: History, economics and politics. *NJAS – Wageningen Journal of Life Sciences, 48*(2), 125–149. https://doi.org/10.1016/S1573-5214(00)80010-X
Zhang, W., Jiang, F., & Ou, J.-J. (2011). *Global pesticide consumption and pollution: With China as a focus*. https://www.semanticscholar.org/paper/Global-pesticide-consumption-and-pollution%3A-with-as-Zhang-Jiang/6e8299fcbc98b76637430502c0398d558640c348
Zoccali, C., Catalano, C., & Rastelli, S. (2009). Blood pressure control: Hydrogen sulfide, a new gasotransmitter, takes stage. *Nephrology Dialysis Transplantation, 24*(5), 1394–1396. https://doi.org/10.1093/ndt/gfp053

Adjuvants: Types of Adjuvants and Their Effects on Ecosystems in Relation to Climate Change

Barbara Sawicka, Piotr Pszczółkowski, Piotr Barbaś, and Barbara Krochmal-Marczak

1 Introduction

Definition of Adjuvants: Adjuvants are excipients added to pesticide formulations to enhance their efficiency and effectiveness. They improve the physical, chemical, or biological properties of pesticides, herbicides, or fungicides, helping them to work more effectively by promoting better absorption, adhesion, and distribution on target surfaces. Adjuvants can be divided into different types such as surfactants, wetting agents, penetrants, or emulsifiers, each of which performs specific functions such as reducing surface tension, increasing penetration into plant tissues or stabilizing active ingredients (Hazra et al. 2017; Cui et al., 2024).

The effectiveness of adjuvants depends on the type of chemical and pest species, as well as on environmental conditions such as temperature and vehicles. Key steps include droplet secretion, droplet transfer to form deposits, penetration of the active

B. Sawicka (✉)
Department of Plant Production Technology and Commodities Science, University of Life Sciences in Lublin, Lublin, Poland
e-mail: barbara.sawicka@up.lublin.pl

P. Pszczółkowski
Research Centre for Cultivar Testing in Słupia Wielka, Słupia Wielka, Poland
e-mail: p.pszczolkowski.inspektor@coboru.gov.pl

P. Barbaś
Department Agronomy of Potato, Plant Breeding and Acclimatization Institute—National Research Institute, Serock, Poland
e-mail: p.barbas@ihar.edu.pl

B. Krochmal-Marczak
Department of Plant Production and Food Safety, Carpathian State College in Krosno, Krosno, Poland
e-mail: barbara.marczak@pans.krosno.pl

B. R. Babaniyi, E. E. Babaniyi (eds.), *The Interplay of Pesticides and Climate Change*, https://doi.org/10.1007/978-3-031-81669-7_2

device and its translocation in the plant or pest device. This process can be modified by adjuvants, which use the retention and absorption of the working tool to ensure the effectiveness of the protective tools. Recent studies have shown that plant-derived adjuvants are a source of chemical energy and their biodegradability, which supports sustainable development (Gupta et al., 2023; Angaye et al., 2024).

In the context of agriculture, adjuvants are essential for optimizing pesticide performance, especially under changing environmental conditions, and are increasingly considered for their role in sustainable practices and minimizing environmental impact. They are added to a given active substance as a complement. Due to their properties, they can influence the better spread of liquids on the surface of plants, which increases their effectiveness. Depending on the type, dose and purpose, their effect may be variable. Adjuvants also improve biological effectiveness by modifying the chemical and physical properties of working fluid solutions used in agriculture (Sawicka & Egbuna, 2020; Angaye et al., 2024).

Importance in Modern Agriculture: Discussion on the growing dependence on adjuvants in the agricultural sector to improve pesticide application, especially in the context of increasing environmental stressors. Plant-derived pesticides (PDPs) are more environmentally friendly than synthetic ones, resulting from toxicity and decomposition, which reduces damage to non-target organisms. Synthetic pesticides, effectively counteracting pest resistance and populations of beneficial organisms, which makes PDPs have additional potential in protecting biodiversity (Angaye et al., 2024).

The growing reliance on adjuvants in modern agriculture is driven by increasing environmental pressures such as climate change, pest resistance, and soil degradation, which challenge traditional crop protection methods. Adjuvants, which are additives that enhance the efficacy of pesticides, herbicides, and fertilizers, play a key role in optimizing the use and efficiency of these products. They improve the adhesion, penetration, and spread of chemicals on plant surfaces, ensuring that crops receive maximum benefit while reducing waste and minimizing environmental pollution (Kirkwood et al., 2021; Hazra et al. 2017; Li et al., 2022). With climate change introducing more unpredictable weather patterns, such as heavy rainfall or prolonged droughts, adjuvants can help stabilize pesticide formulations and ensure they remain effective under a variety of extreme conditions. For example, some adjuvants enable better retention of pesticides on crops during heavy rains, reducing the need for repeated applications and saving both resources and labor. This is especially important as farmers seek to minimize chemical use due to environmental and regulatory concerns (Jamar & Lefrancq, 2008; Parker et al., 2023).

Adjuvants also help address the problem of pest resistance; a growing problem due to the excessive use of pesticides. By improving the efficiency of pesticide application, adjuvants reduce the likelihood of pest resistance developing, thus prolonging the effectiveness of existing chemicals (van Zwieten, 2015). Furthermore, the use of precision farming techniques, which involve precise control over inputs such as water and chemicals, offers significant benefits from the use of adjuvants. They enable more targeted and efficient applications, contributing to sustainable agricultural practices by reducing runoff and leaching into water bodies, which is a major environmental problem (Jamar & Lefrancq, 2008; Rout et al., 2023). The importance of adjuvants in the context of climate change: In the context of climate change, it is important to highlight how rising global temperatures, changing rainfall patterns, and the increasing frequency of extreme weather events create more favorable conditions for pest development. These changes are exacerbating the challenges of pest management, leading to larger pest populations, longer growing seasons, and the spread of pests to new regions where they were previously absent (Ranjan et al., 2023; Raj et al., 2022; Zhao et al., 2023).

Adjuvants can play a key role in mitigating these impacts by increasing the efficacy of pest control products. Key considerations include:

- Increased pest pressure: Climate change is altering ecosystems, allowing pests to survive warmer winters and allowing them to reproduce more quickly. This increases pest pressure on crops, making it more difficult to control them using traditional methods alone. Adjuvants can enhance the efficacy of pesticides, herbicides, and fungicides, making them essential tools in addressing this growing challenge (Porter et al., 2014; Noyes et al., 2019; Gupta et al., 2023).
- Spread of invasive species: As temperatures rise, invasive pest species are spreading to new areas, sometimes to other continents. Adjuvants can help ensure that chemical treatments are more effective in these new environments, where pests may exhibit different behaviors or resistance to traditional treatments (Jamieson et al., 2012; Sawicka & Egbuna, 2020; Raj et al., 2022).
- Effectiveness under extreme weather conditions: Climate change is causing more frequent droughts, heavy rainfall, and stronger winds, which can have significant impacts on pesticide use. Adjuvants can improve the stability, spread, and adherence of pesticide formulations, ensuring they remain effective even in adverse weather conditions (Porter et al., 2014; Garcia et al., 2021; Anonymous, 2024; Gupta et al., 2023).
- Sustainable pest management: In a world with growing concerns about environmental sustainability, adjuvants can help reduce the amount of chemicals needed by increasing the bioavailability and efficacy of pesticides. This minimizes environmental impacts while addressing the increased pest challenges caused by climate change (Noyes et al., 2019; Sawicka & Egbuna, 2020).
- Adaptation to changing growth cycles: Changing climate patterns can alter crop growth cycles, which in turn impacts the timing of pest infestations. Adjuvants can support precision agriculture practices by optimizing the timing and effectiveness of pest management strategies under these changing conditions (Porter et al., 2014; Raj et al., 2022).
- Improving biological pest control: As sustainable agricultural practices gain popularity, adjuvants can also be used to improve the effectiveness of biological control agents, helping these natural pest control solutions perform better under changing climatic conditions (Sawicka & Egbuna, 2020; Gupta et al., 2023).

The aim of this chapter is to comprehensively analyze and evaluate the role of adjuvants in modern agriculture, with a special focus on their impact on the effectiveness of plant protection products in the context of changing climatic conditions. The research primarily aims to:

- Characterize the mechanisms of action of adjuvants in the context of improving the physical and chemical properties of pesticides, especially under difficult environmental conditions such as droughts, heavy rainfall, and extreme temperatures.
- Assess the role of adjuvants in pest resistance management. Analyze how adjuvants can support existing plant protection products by reducing the risk of resistance development in pests and ensuring long-term pesticide efficacy.
- Analyze the impact of adjuvants on the environment. Assess the extent to which different types of adjuvants, such as penetrants, surfactants, and wetting agents, contribute to minimizing chemical run-off and environmental contaminants, which is an important element of sustainable agriculture.
- Assess the use of adjuvants in changing crop growth patterns and determine how adjuvants can support the adaptation of agriculture to new climatic conditions, such as changes in crop seasonality and the spread of new pest species.
- Analyze the potential benefits and limitations of precise adjuvant application, especially in the context of reducing chemical inputs and optimizing crop protection management under different climatic conditions.

The results of these analyses will contribute to a better understanding and implementation of adjuvants as a tool supporting sustainable agricultural practices, especially in the face of global challenges related to crop protection and climate change.

2 Types of Adjuvants

2.1 Surfactants Mechanism of Action

How surfactants alter surface tension to improve spray distribution. Surfactants, or surface-active agents, play a key role in the formulation of agricultural sprays, particularly herbicides and insecticides. Their primary function is to modify the physical properties of the spray solution, particularly its surface tension, leading to improved distribution and efficacy (Table 1).

Table 2 highlights the role of different adjuvants (surfactants, oils, and humectants) at different stages of agrochemical leaf penetration. It examines how each type of adjuvant affects key processes involved in the absorption and translocation of active ingredients through plant leaves. The stages of leaf penetration include: (1) Droplet adhesion: All three types of adjuvants (surfactant, oil, and humectant) aid (2) in the initial adhesion of the droplet to the leaf surface.

Below is an explanation of how surfactants achieve this (Stelzer & Schaller, 2021).

1. Understanding Surface Tension Definition: Surface tension is the cohesive force between liquid molecules at the surface of a liquid. It causes the liquid to behave

Table 1 Effect of different types of adjuvants on leaf penetration

Leaf penetration stages surfactant		Oil	Humectant
Droplet adhesion	X	X	X
Changes in sediment	X	X	X
Penetration into cuticle	X	X	X
Diffusion in cuticle	X	0	0
Exit from cuticle	X	0	0
Migration in tissue	0	0	0
Migration in plan	0	0	0

The effect at the time of entry and diffusion in the cuticle depends on the properties of the biologically active substance and the adjuvant: x—effect, 0—no effect

as if its surface were covered by a flexible membrane. Surfactants, or surface-active agents, play a key role in the formulation of agricultural sprays, particularly herbicides and insecticides. Their primary function is to modify the physical properties of the spray solution, particularly its surface tension, leading to improved distribution and efficacy. Below is an explanation of how surfactants achieve this. High surface tension can cause spray droplets to assume spherical shapes, which may not spread effectively on the target surface (e.g., leaves or soil). This can result in poor coverage and less effective pest control.

2. Role of Surfactants Reducing Surface Tension: Surfactants lower the surface tension of a spray solution by disrupting the cohesive forces between liquid molecules.
3. Improved droplet behavior

    - Smaller droplet formation: Lower surface tension leads to smaller droplet formation after spray application. Smaller droplets have a greater surface area relative to volume, which improves coverage (Garrido et al., 2021). Improved wetting: With reduced surface tension, droplets can spread more easily across leaf surfaces, which increases wetting and penetration into plant leaves or soil (Stelzer & Schaller, 2020).

4. Impact on spray distribution

    - Uniform coverage: Improved droplet distribution provides more uniform coverage of the target area, which is crucial for effective pesticide and herbicide application (Sharma et al., 2023).
    - Reduced drift: Surfactants can also help stabilize droplets in flight, reducing the risk of drift and ensuring that more product reaches the intended target (Zhou et al., 2022a, b).
    - Impact on spraying: High surface tension can cause spray droplets to assume spherical shapes that may not spread effectively on the target surface (e.g., leaves or soil). This can result in poor coverage and less effective pest control.

Table 2 Types of adjuvants and their roles

Adjuvants	Role	References
Mineral	Improved retention of results on plants Increased absorption of active application Reduced resultant evaporation Reduction in dry spray Stabilization of emulsion and suspension Protective action against UV radiation Control of release of active mineral application Action enhancing mechanical protection	Jamar and Lefrancq (2008), van Zwieten (2015), Zhou et al. (2022a), and Smith et al. (2023)
Emulsion	Emulsion stabilization Improving the miscibility of components Enhancing droplet retention on plants Improving the absorption of the production solution Reducing evaporation Reducing spray drift Enhancing the bioavailability of the working solution Protecting the industrial product from degradation	Cessna et al. (2023), Nooraei et al. (2023), and Cui et al. (2024)
Polymeric	Improved adhesion and retention Increased rain wash-off load Control of active ingredient release Improved penetration and absorption Reduced spray drift Protection against UV degradation Reduced water consumption Ecological alternative to the supplementary use of adjuvants	Rout et al. (2023), Parker et al. (2023), and Smith et al. (2023)
Saponins	Enhanced retention of spray application Improved penetration through cuticles Emulsifying action Natural alternative to synthetic surfactants Antibacterial and antifungal action Enhanced efficiency of low doses of pesticides	Zhang et al. (2020), Zhang et al. (2023a) and Nooraei et al. (2023)
Derivatives of complements system	Enhancement of response Improvement in antigen uptake Activation of B-cell independent response Function and recruitment of configuration cells Application in new adjuvant technologies	Parker et al. (2023), Smith et al. (2023), Angaye et al. (2024), and Cui et al. (2024)
Taken from bacteria	Activation of toll-like receptors (TLRs) Enhancement of anti-inflammatory potency Enhancement of antigen presentation Induction of humoral and cellular responses Use in cancer vaccines Stability and safety of vaccines	Nooraei et al. (2023), Zhou et al. (2022b), Angaye et al. (2024), and Cui et al. (2024)

Role of Surfactants

- Reduction in surface tension: Surfactants lower the surface tension of the spray solution by disrupting the cohesive forces between liquid molecules.
- Hydrophilic and hydrophobic ends: Surfactants consist of a hydrophilic (water-attracting) head and a hydrophobic (water-repelling) tail. When added to a liquid, surfactants orient themselves at the air–water interface, reducing surface tension and allowing the liquid to spread more easily (Wang et al., 2022).

Improved Droplet Behavior

- Formation of smaller droplets: Lower surface tension leads to the formation of smaller droplets after spray application. Smaller droplets have a greater surface area relative to volume, which improves coverage (Garrido et al., 2021).
- Improved wetting: With reduced surface tension, droplets can spread more easily across the leaf surface, which increases wetting and penetration into plant leaves or soil (Stelzer & Schaller, 2020).

Impact on Spray Distribution

- Uniform coverage: Improved droplet distribution provides more uniform coverage of the target area, which is key for effective pesticide and herbicide application (Sharma et al., 2023).
- Reduced drift: Surfactants can also help stabilize droplets in flight, reducing the risk of drift and ensuring that more product reaches the intended target (Zhou et al., 2022a, b).

Types of Surfactants

- Non-ionic surfactants: They are commonly used in agricultural applications due to their compatibility with various formulations and reduced toxicity (Sharma et al., 2023).
- Anionic and cationic surfactants: Although effective, these types may be less commonly used due to potential interactions with active ingredients and environmental concerns (Wang et al., 2022).

The incorporation of surfactants into spray formulations is essential for optimizing the performance of pesticides and herbicides. By reducing surface tension, surfactants increase droplet size, shape, and behavior, leading to better spray distribution and efficiency. Ultimately, this contributes to more effective pest control practices, minimizing waste, and increasing crop protection.

Applications

Formulation development: Adjuvant knowledge can be used to develop more effective pesticide formulations by selecting appropriate surfactants.

Application techniques: Farmers can also optimize spraying techniques based on the properties of the surfactants used, leading to better pest control results.

Types of Surfactants

- Non-ionic surfactants: They are commonly used in agricultural applications due to their compatibility with various formulations and reduced toxicity (Sharma et al., 2023).

- Anionic and cationic surfactants: Although effective, these types may be less commonly used due to potential interactions with active ingredients and environmental concerns (Wang et al., 2022).

2.2 Penetrants

Penetrants are a key class of surfactants that enhance pesticide absorption by improving their ability to pass through the protective plant cuticle. The plant cuticle acts as a barrier, and many pesticides have difficulty penetrating deep enough to reach target pests or pathogens. Penetrants work by softening or dissolving the waxy layer of the cuticle, allowing active ingredients to move more efficiently into plant tissue. Their functionality is primarily to improve the absorption of pesticides through the plant cuticle (Sharma et al., 2022). Mechanism of action: Penetrants reduce the surface tension between the pesticide solution and the cuticle, breaking down the waxy layer. This increases the permeability of the cuticle, allowing pesticides to be more easily absorbed. Key examples: Methylated seed oils (MSO): Methylated seed oils are commonly used as penetrants in pesticide formulations. They improve the movement of systemic pesticides into plant tissues by increasing cuticle penetration. These oils are derived from fatty acids found in seeds such as soybean or rapeseed. Other vegetable oils, such as rapeseed or cottonseed oil, are also effective penetrants. They offer a sustainable alternative to petroleum-based oils and are biodegradable, reducing the environmental impact compared to synthetic chemicals (Sharma et al., 2022; Gupta et al., 2023).

In general, vegetable oils are considered more environmentally friendly because they degrade faster than their synthetic counterparts, leaving less residue in soil or water. However, they still need to be used with caution, as overuse can cause oil to accumulate in the soil, affecting its microbial life. Their important feature is functionality, mainly retaining moisture on plant surfaces to prolong the action of pesticides (Sharma et al., 2022).

Humectants and wetting agents are used in pesticide formulations to retain moisture on plant surfaces, ensuring that active ingredients remain in solution longer and are absorbed more efficiently. This is especially important in low humidity or when water availability is limited (Sharma et al., 2022).

Mechanism of action: Humectants absorb water from the atmosphere, keeping the pesticide solution moist on the leaf surface for longer periods of time. Wetting agents, in turn, reduce the surface tension of water, allowing the pesticide to spread more evenly across the plant surface (Stelzer & Schaller, 2021; Gupta et al., 2023).

Importance of humectants in water-stressed environments: In areas where drought or water stress is common, humectants and wetting agents play a key role. They ensure that pesticides do not dry out too quickly, which would reduce their effectiveness. Given the increasing frequency of droughts due to climate change, these agents are becoming essential for crop protection in arid regions.

Thus, mineral adjuvants are designed to:

1. Improve retention of results on plants: Mineral adjuvants, such as clays or silica, are supplied in increased amounts that remain on the surface of the energy. They limit access to the network, allowing the spray application to appear on other working surfaces.
2. Increase absorption of the active application: Biological adjuvants can help the penetration of the active tool through the leaf cuticle. Substances such as silicon oxide improve the efficiency of the action to penetrate the wax controller, which increases the effectiveness of herbicides, fungi, and insects.
3. Reduce evaporation: Excipients located on the surface of the plants that limit the evaporation of spray droplets. This allows the active module to have more time to penetrate the plants, which increases the effect of the pesticide.
4. Reduction in dry spray: Minerals such as aluminosilicates or bentonite can increase the viscosity of the spray liquid, which helps to reduce the drift phenomenon—unintentional transport of droplets outside the spray area. To minimize pesticide losses and reduce environmental losses.
5. Stabilization of emulsions and suspensions: Mineral adjuvants such as kaolinite or talc, used to stabilize emulsions and suspensions in spray liquids. To eliminate application separation, providing a solution based on application.
6. Control of active application release.
7. Enhancement of mechanical protection (Jamar & Lefrancq, 2008; van Zwieten, 2015; Zhou et al., 2022a, b; Smith et al., 2023; Nooraei et al., 2023; Rout et al., 2023; Cessna et al., 2023; Nooraei et al., 2023; Cui et al., 2024).

Adjuvants used in emulsions play a key role as a switch in the enhancement of agrochemicals such as pesticides, fungi, and herbicides. The most important role of adjuvants in emulsions is as follows:

1. Stabilization of emulsion: The use of adjuvants in emulsions ensures the stability of the mixture. Adjuvants prevent phase separation, which is the separation of oily (e.g., oils) from water. Stabilization of emulsion is crucial to prevent an event from occurring after spraying.
2. Improved miscibility of components: Adjuvants in emulsions facilitate the mixing of components that are normally difficult to mix (e.g., oils with water). Thanks to this, the active function is more widespread on the online platform, which increases the effectiveness of the action.
3. Strengthening the retention of droplets on plants.
4. Improving the absorption of the production solution: Adjuvants in emulsions can facilitate the penetration of the tool through the leaf cuticle, obtaining their effectiveness. Thanks to this, more active substance gets into the interior of animals or pest animals, which leads to more effective control of diseases, pests or weeds.
5. Reduction in evaporation: Thanks to adjuvants, emulsions can be launched on the surface water of plants, which slow down the evaporation of results. It gives more time for the absorption of the active application, which increases the effectiveness of plant protection.

6. Reduction in spray drift: Adjuvants in emulsions can help in the application of the spray drift phenomenon (lifting of spray drops with the wind outside the target area). Improves droplet size to prevent pesticide loss and minimize the risk of drift (spraying).
7. Enhancement of the bioavailability of the working solution: In oil emulsions, adjuvants can increase the bioavailability of pesticides by better dissociating the active function, which increases their effectiveness in terms of pest control and weeds.
8. Protection of the industrial product from degradation: Some adjuvants in emulsions can protect against chronic activation and against degradation caused by UV radiation or moisture.

Examples of adjuvants in emulsions: Surfactants (emulsifiers); Vegetable or mineral oils—Silicone; Polymers.

Adjuvants in emulsions have multiple functions, they are tools for stabilizing aggregates, through the retention and absorption solution of the industrial device, up to the application of spray drift. Thanks to them, spraying is more extended and secondary, which can cause plant protection effects and negative impact on the environment.

Polymer adjuvants cause a number of beneficial activities such as:

1. Improved adhesion and retention: Polymer adjuvants as switches on the surfaces of control devices, which help to stop the spray droplet on leaves and other plants. Thanks to this, more active application reaches the target directly, allows the use of losses related to evaporation or flow of the spray. Improved retention allows the use of available doses of herbicides, fungi, or insecticides, and still to the accessible device.
2. Increased load on rain wash-off: Polymers, primarily of a hydrophobic nature, can cause consequences from rain wash-off. This means that protection methods can be more advanced, even under unfavorable conditions, which are often used in traditional plant protection products from the surface of leaves.
3. Control of the release of the active substance.
4. Improved penetration and absorption: Polymeric adjuvants are tools for penetrating the active agent through the waxy control of the leaf cuticle. Polymers can generate application data that emerges from the spray, allowing access to the leaf surface and penetration of the test tool into the system. Extended use of additional absorption allows for the application of lower pesticide doses.
5. Reduced spray drift: Polymeric adjuvants have anti-drift properties, meaning that spray droplets are dispersed off-target by wind. Reducing dry spray reduces the risk of pesticide effects and exposure in nearby applications, which is particularly important for protecting biodiversity and human health.
6. UV degradation protection: There are additional polymer modules that can protect against UV radiation. This allows herbicides, insecticides, and fungi to influence the effects, allowing for less frequent application and use of lower amounts of applied chemicals.
7. Reduction in water consumption: By enhancing retention and extension, polymeric adjuvants facilitate the availability of water needed for pesticide applica-

tion. This is important in the context of water resource management, especially in regions with hydrological drought.

8. Ecological alternative to adjuvant supplementation: Biodegradable adjuvants of polymeric origin, which are similar to synthetic polymers but more environmentally friendly, are becoming increasingly important. Such additives can be used with plant sources, which reduces their impact on ecosystems and enables more sustainable agricultural management (Santos et al. 2021, Gupta et al., 2023).

Thus, polymeric adjuvants connected to a radio module, improving pesticide application, reducing environmental impact and ensuring the use of lower chemical doses. Due to their properties, there is a possibility of retention, controlled release, protection against wash-off and degradation, as polymeric adjuvants provide an accessible element of the adjuvant range.

Saponins, natural compounds in plants, equipped with an increasingly large device in the formulation of adjuvants in agriculture. As natural surfactants, saponins have several properties that can lead to the use of plant protection products and other agrochemicals. Here are the key roles of saponins in adjuvants:

1. Enhancement of spray retention: Saponins, due to their surface-active properties, reduce access to surface water, which allows for better application of spray drops on working surfaces. In this way, adjuvants containing saponins spread better when sprayed on the leaves, which translates into greater effectiveness of action and reduces losses of chemicals.
2. Improvement in penetration through the cuticles—application with a barrier for the absorbed tool by cuticular technology—a waxy protective layer on the leaves. Saponins can increase the effectiveness of this function, using the research tool to the interior of the exit. Due to this, adjuvants with saponins can improve the absorption of herbicides, fungi, and insecticides.
3. Emulsifying action: Saponins also have a strong emulsifying action, which allows them to stabilize water–oil emulsions. Therefore, this function is possible because it allows the mixing of different ingredients that apply and improves the effectiveness of the formulation. Saponins can be used in sprays.
4. Natural alternative to synthetic surfactants: Saponins, as plant compounds, are an ecological alternative to synthetic surfactants in many commercial adjuvants. Replacing synthetic effects with saponins can cause toxicity of the adjuvants and their impact on the environment, in turn influencing the development in agriculture.
5. Antibacterial and antifungal action: There are some natural antibacterial and antifungal properties. Their action is not possible to protect against pathogens on their own, in the article with other plant protection elements that can protect against infections, reducing the risk of developing plant diseases.
6. Enhancement of the efficiency of low-dose pesticides: The addition of saponins to adjuvants can be effectively used for pesticides when using reduced doses. By using adhesion and absorption, it is possible that even with this small amount of chemicals, which reduces costs and environmental burden.

7. Biodegradability and risk-bearing: As a natural hazard, saponins are biodegradable and do not accumulate in the environment. This makes adjuvants containing saponins more environmentally friendly in the common adjuvants on synthetic surfactants, which may have a general degradation time and cause negative ecological effects (Gupta et al., 2023).

Thus, saponins have a number of advantages, such as: correct retention of spray appearance, penetration through the cuticle, emulsifying action and natural antibacterial and antifungal properties. Due to their biodegradability and ecological after application, saponins can be approved by agriculture, distributed to more diverse and environmentally friendly use of pesticides.

Derivatives of the Complement System cause:

- Enhancement of response: The complement system is a virus, and its activation leads to stimulation of cells of the physical system, such as macrophages and dendritic cells. Complement derivatives can be devices as adjuvants that enhance antigen presentation and activate these cells, which leads to adaptive effects, that is, the use of T-cell functions and cells. In this way, derivatives to the complement enhance the action of vaccines, leading to protection against the pathogen.
- Improvement in antigen uptake. Complement derivatives, such as C3d (C3 fragment), can occur with antigens, which improves their reception by the occurrence of antigens. This process increases the uptake and transmission of antigens through the configuration network, which in turn leads to effective stimulation of T and B lymphocytes.
- Activation of a B-cell-independent response: Complement derivatives can also directly remove B lymphocytes, governing their function and proliferation. It acts as a "second use" in B lymphocyte supply. Complement derivatives such as C3d spread to CR2 receptors on the surface of B lymphocytes, which amplifies the signal from the antigen. In the rest to the rest of the production, which may be important.
- Function and recruitment of configuration cells: Some complement derivatives, such as C5a, act as a combination of chemoattractant that recruit to sites where immune cells such as neutrophils and macrophages are formed. To accelerate the local inflammatory response, which promotes dendritic cell maturation and antigen presentation, thus enhancing the immune response.
- Control: The complement system, also connected to the driver, is important in the context of vaccine adjuvants. Complement derivatives can modulate the response of the system in such a way that another, equivalently acting response can be used to generate a protective response. This balance is crucial to minimize the effects of vaccines.
- Application in new adjuvant technologies: Complement derivatives for the storage of the target in studies on the types of adjuvants in the supply. There is potential to connect complement derivatives to other technologies, plug-ins such as nanoparticles or lipid adjuvants, to ensure precision and delivery of antigens and to apply immunological function (Angaye et al., 2024; Cui et al., 2024).

- Complement derivatives in adjuvants are therefore crucial in enhancing the response to adjuvants, improving antigen presentation, activating control modules and control functions. Through these mechanisms, adjuvants derived from complement derivatives can increase vaccine efficacy while minimizing losses.

In general, humectants help cope with drought stress, enabling plants to grow from pesticides in areas where water is not readily available, reducing the need for supplemental irrigation, and helping to maintain crop health under difficult conditions.

Adjuvants are an additional source of power in stimulating the response, especially in the context of vaccines and immunotherapy. Bacterial products such as wall fragments or lipopolysaccharides (LPS) can act as potent stimulators of the reinforcement structure, enhancing the efficacy of vaccines. Key roles of bacterial adjuvants include:

(i) Activation of Toll-like receptors (TLRs): Bacterial adjuvants often activate the immune response via Toll-like receptors (TLRs), which recognize pathogens. For example, bacterial lipopolysaccharides (LPS) are associated with the TLR4 receptor, which is associated with inflammation. Such activation leads to the production of cytokines, which in turn activate T and B lymphocytes, enhancing the adaptive response (Fig. 1).
(ii) Enhancing anti-inflammatory power: Bacterial products such as peptidoglycan fragment or flagellin can induce a strong source of energy. Adjuvants that bind to antigens (APCs), such as macrophages and dendritic cells.

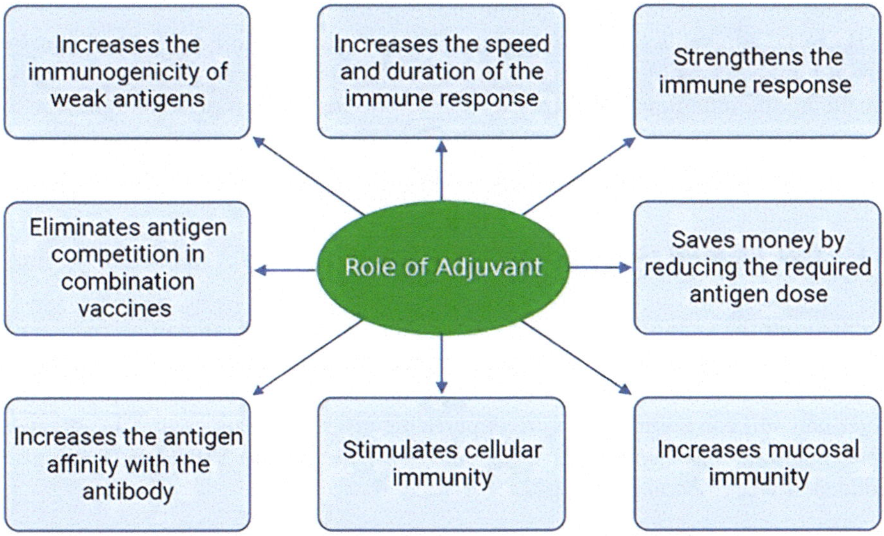

Fig. 1 Ways in which adjuvants can play a role in improving the immune response

(iii) Enhancement of antigen presentation: Bacterial adjuvants connected to the antigen presentation process through the dendritic nest. For example, adjuvants containing fragments of immune DNA, such as CpG, are recognized by TLR9 and can promote the original presentation of antigens by these cells. This in turn results in a stronger activation of lymphocytes.

Induction of humoral and cellular responses

1. Bacterial products can be used for both humoral (antibody) and cellular (cytotoxic T lymphocytes) responses. For example, bacterial fragments such as MPL (monophosphoryl lipid A), which is a derivative of LPS, are used in some vaccines to obtain a T cell response and product testing.
2. Use in cancer vaccines: Adjuvants used in immune therapy are also used in cancer immunotherapy. Bacterial products can activate dendritic cells, which is crucial for inducing an antitumor response. In this way, bacteria and their sources can assist the viral system in detecting and destroying cancer cells.
3. Stability and safety: Bacterial adjuvants are valued for their fairly good chemical solution. However, their use requires the use of a strong bacterial adjuvant, such as LPS, which can cause side effects. Distinctive forms of these bacterial products are found in adjuvants to take advantage of their toxicity while maintaining the ability to configure stimulation. Examples of bacterial adjuvants include: Monophosphorylate lipid A (MPL); Lipoteichoic acid (LTA); Flagellin—DNA, CpG—.

Figure 1 illustrates the various mechanisms by which adjuvants enhance the host immune response. Key roles of adjuvants include: (1) Enhancement of antigen presentation: Adjuvants promote the uptake of antigens by antigen-presenting cells (APCs), such as dendritic cells, which process and present these antigens to T lymphocytes, initiating a more potent immune response.

Bacterial adjuvants are key to differentiating vaccines and immunotherapies by activating the extensible system via innate mechanisms such as Toll-like receptors, enhancing the inflammatory response, and improving antigen presentation. Due to these properties, bacterial adjuvants act on the principle of an ongoing response.

2.3 *Buffers and pH Modifiers*

Maintaining the stability of pesticide formulations under changing environmental conditions is very important for humans and the environment. Buffers and pH modifiers are crucial to ensure the stability of pesticide formulations, as the pH of both water and soil can significantly affect pesticide efficacy. Changes in pH can cause faster degradation of some active ingredients or their lower solubility, leading to reduced efficacy (Zhou et al., 2022a, b).

Mechanism Buffers help maintain a constant pH in the spray tank, ensuring that the pesticide remains stable and active. pH modifiers can lower or raise the pH of a solution to the optimal range for a given pesticide. For example, many insecticides and herbicides are most effective at slightly acidic pH (5–6), while others, such as some fungicides, perform better in neutral or slightly alkaline conditions (Zhou et al., 2022a, b). Case studies:

Effects of soil and water pH Climatic stress, such as excessive rainfall or drought, can alter soil and irrigation water pH. For example, in regions with high rainfall, increased leaching of soil minerals can lead to a decrease in pH, making the environment more acidic. This can reduce the effectiveness of some herbicides, which are less effective at low pH conditions (Garrido et al., 2021).

Farmers are increasingly using pH modifiers in areas where climate-induced changes in soil and water chemistry are becoming common. This ensures that pesticides remain effective even when environmental conditions fluctuate (Sharma et al., 2023).

3 Impact of Adjuvants on Ecosystems

3.1 Toxicity to Non-target Species

Adjuvants, although used in agriculture to help increase the effectiveness of pesticides, can have adverse effects on non-target species such as pollinators (e.g., bees), birds and aquatic organisms. Recent studies have shown that some surfactants, such as siloxanes and ethoxylated fats, can disrupt the biological systems of insects, affecting their ability to pollinate and leading to reduced pollinator populations. For example, studies on honeybees suggest that exposure to some types of adjuvants can weaken the immune system of these insects and impair their orientation in the field (Chagnon et al., 2015; Sanchez-Bayo & Tennekes, 2021). In aquatic ecosystems, adjuvants can increase the toxicity of pesticides to aquatic organisms such as fish, amphibians, and invertebrates. For example, nonylphenol ethoxylates (NPEs) are commonly used as adjuvants because they are known to have endocrine-disrupting properties and affect the reproductive performance of fish and amphibians (Sass & Colangelo, 2006).

Case studies: Studies by Chagnon et al. (2015) have shown that adjuvants used in agriculture can increase the toxicity of neonicotinoids, causing increased mortality and orientation disruption in bees (*Apis mellifera*).

Aquatic ecosystems There are known and described cases of mass fish kills in Canadian lakes where higher concentrations of adjuvants were found to sensitize aquatic organisms to pesticides. Analyses of adjuvant accumulation in food chains have shown that adjuvants, like many pesticides, can bioaccumulate in organisms and enter food chains. In particular, surfactants and emulsifiers can bind to lipids in aquatic organisms, which increases their presence in higher trophic levels. The phenomenon of bioaccumulation means that organisms higher in the food chain, such as predatory fish, birds, and mammals, are exposed to higher concentrations of toxic substances (Sass & Colangelo, 2006; Malaj et al., 2014). Bioaccumulation potential: Some adjuvants, such as polyethylene glycols, are resistant to biodegradation, which makes them persistent in the environment and increases the risk of accumulation in the food chain (Wang et al., 2023). Case examples include the accumulation of surfactants in planktonic organisms, which in turn are consumed by larger aquatic organisms, transferring toxins to higher levels of aquatic ecosystems.

Soil health and microbial activity Adjuvants' effects on beneficial soil organisms Adjuvants, when used in conjunction with pesticides, can affect the microbial community of the application, including organisms responsible for the degradation of substances and the cycling of nutrients. This effect depends on the type of adjuvant and its interactions with the active chemicals. Some adjuvants, such as surfactants or emulsifiers, can enable data transmission across cells, leading to the transmission of cells and their competence to contents such as nitrogen compounds or the decomposition of outgoing matter.

Nitrogen fixation Herbicide adjuvants can be associated with nitrogen-fixing bacteria, such as Rhizobium, which are essential for nitrogen fixation in soil, primarily in symbiosis with legumes. Studies on some surfactants reduce the populations of sensitive microorganisms, which affects soil nitrogen availability (Jamieson et al., 2012).

Microbial diversity The impact of adjuvants on microbial application, a key health effect. Excessive exposure to chemicals in adjuvants can act on microbial diversity, impairing soil ecosystems and their capacity to function and respond to electrical stress. Reductions in microbial biodiversity can limit the ability to benefit from the principles of biogeochemical cycling (Smith et al., 2023).

Consequences of soil structure and water retention on soil health and erosion: Adjuvants can be revealed in residues, through their effects on soil particle aggregation and through available water. This has an impact on changing weather conditions, where extreme weather events such as droughts and heavy rains are becoming more common.

Soil aggregation Some adjuvants can reduce soil aggregation problems, which leads to the establishment of erosion risk. The weakening of this structure occurs through the use of adjuvants, which can cause losses due to water activity (Six et al., 2002). In use, this leads to a decrease in the ability to support plant growth and a decrease in water resources.

Water retention Adjuvants can also control the ability of the soil to retain water. On the one hand, wetting agents can increase water retention, which is a consequence in drought periods, improving water absorption by plants. On the other hand, the drainage of some adjuvants can be through drainage, reducing its water storage capacity, which ensures the separate action of agricultural ecosystems on climate change (Arif et al., 2022).

Implications for climate change Climate changes that are visible can be mitigated by adjuvants via a driver—they can act depending on their composition and application. In high rainfall conditions, the soil structure adapted to erosion is weakened, while in drought periods, the soil may have difficulty retaining water if adjuvants are not appropriately selected for climatic conditions. Empirical studies that lead to the use of adjuvants as part of sustainable agricultural practices, applied as conservation agriculture, can prevent these effects by controlling and adhering to safety rules in compliance with regulations. Soil degradation and food security: adjuvants can cause health effects on humans and ecosystems. Non-representational studies focus on their long-term effects and ways to minimize negative impacts on ecosystems, especially in the context of changing climate (Arif et al., 2022).

3.2 Changing Precipitation Patterns and Adjuvant Efficiency

Extreme weather events Changing global precipitation patterns, including increased storm frequency and erratic rainfall, are significantly affecting pesticide retention in crops. Heavy rains can wash away pesticides, reducing their effectiveness, while prolonged periods of drought can impede their uptake. This changing climate requires specialized adjuvants—additives that enhance pesticide performance—designed to enhance retention in wet conditions and improve uptake in dry conditions. Adjuvants with properties such as improved wetting agents or sticking agents are particularly valuable because they help pesticides adhere better to plants, even during storms. In regions experiencing increased drought, more systemic formulations that work through uptake by plants may be beneficial (Jamieson et al., 2012; Arif et al., 2022; Okonkwo et al., 2023). Recent studies (Anonymous, 2024) highlight how extreme weather events and droughts affect pesticide retention and efficacy. Climate change, such as increased rainfall intensity or irregular rainfall, can significantly reduce pesticide retention on crops due to wash-off, making precise formulations of adjuvants essential to enhance their adhesion and efficacy under such conditions. On the other hand, during droughts, the role of adjuvants becomes critical to optimize pesticide uptake, as water deficit can stress plants, altering their uptake rate and reducing the efficacy of traditional pesticide applications (Jamieson et al., 2012; Arif et al., 2022). Studies highlight, for example, how extreme weather patterns in Europe, including droughts and intense storms, disrupt the stability of agricultural ecosystems. These factors require a more sophisticated approach to adjuvant use, focusing on formulations that enhance pesticide penetration and efficacy despite unpredictable environmental conditions. Furthermore, with declining soil moisture levels in many regions, drought-tolerant adjuvant formulations are required to mitigate reduced water availability and its impact on pesticide efficacy (Okonkwo et al., 2023; Anonymous, 2024).

Drought conditions Drought conditions pose unique challenges to pesticide efficacy. Adjuvants play a key role in maintaining pesticide efficacy when water is scarce. Drought conditions reduce soil moisture, which affects the rate of herbicide degradation, often slowing it down. Some adjuvants can optimize pesticide performance by improving penetration through plant waxes, a barrier that thickens during water stress. Humectants in adjuvants, which help retain moisture, are also useful in such environments because they improve the distribution of pesticides on the plant surface (Jamieson et al., 2012). Recent studies have shown that drought not only affects pesticide efficacy but also changes pest behavior, further complicating control efforts. As a result, adjuvants designed for use in drought conditions are becoming increasingly important in maintaining pesticide efficacy (Okonkwo et al., 2023). Okonkwo et al. (2023) examined how climate-change-induced plant disease lesions are mediated by insecticides and adjuvants. This has important implications for developing adaptive agricultural practices and integrating climate-related disease control strategies. It explores the ecological implications of pest control to mitigate health effects that may occur with pesticide use. Also valuable are insights into sustainable agricultural practices and new technologies that are being used to control pests under changing climatic conditions (Okonkwo et al., 2023).

3.3 Soil Health and Microbial Activity

Impact on soil microbiota Adjuvants, commonly used in agricultural practices, can have both positive and negative effects on soil microbiota, which plays a key role in maintaining soil health and fertility. Different types of adjuvants, such as surfactants, oils, and polymers, can affect the activity and diversity of beneficial microorganisms in the soil, thus influencing processes such as nitrogen fixation and nutrient cycling (Bardgett & van der Putten, 2014). Beneficial soil organisms, such as various Rhizobium species, participate in symbiotic nitrogen fixation in plants of the *Papilionaceae* family, and are essential for converting atmospheric nitrogen into a form usable by plants. Some adjuvants can enhance or inhibit the growth of these microorganisms depending on their chemical composition. For example, biodegradable adjuvants are less likely to harm microbial populations and can even stimulate beneficial organisms by providing additional sources of carbon. In contrast, adjuvants containing synthetic chemicals can suppress microbial activity, leading to reduced nitrogen fixation and soil fertility (Bardgett & van der Putten, 2014; Bhardwaj et al., 2014).

Microbial diversity is the diversity of the soil microbiome, which is a key indicator of soil health because it supports various other functions, such as organic matter decomposition, disease suppression, and nutrient availability. Adjuvants can alter the microbial balance by selectively promoting or inhibiting specific groups of microorganisms (Doran & Zeiss, 2000; Bardgett & van der Putten, 2014). Recent studies (Zhang et al., 2020) suggest that organic adjuvants (e.g., vegetable oils) tend to support greater microbial diversity, compared to synthetic options, which may suppress sensitive microbial taxa. Maintaining or increasing microbial diversity is crucial, especially in agroecosystems stressed by climate change, soil degradation, or intensive agriculture (Rillig & Mummey, 2006).

Soil structure and water retention Adjuvants can also have a strong effect on soil structure and water retention, which are essential for maintaining soil health in the face of climate change. Soil structure, defined by the arrangement of soil particles and pore spaces, directly affects water infiltration, aeration, and root penetration. Proper use of adjuvants can help improve these physical properties, especially in degraded soils (Haynes, 2005; Singh & Reddy, 2011).

Soil aggregation Some adjuvants, particularly humic substances or polymer-based products, can increase soil aggregation by binding soil particles, thereby improving soil porosity and reducing erosion. Improved soil structure allows for better water infiltration and retention, reducing the risk of runoff and erosion during heavy rainfall events. This is especially important in a changing climate where extreme weather events such as droughts or heavy rainfall are becoming more common (Liang et al., 2014; Haynes, 2005; Zhang et al., 2020).

Water retention In arid and semi-arid regions where water conservation is critical, adjuvants can increase the soil's ability to hold water. For example, superabsorbent polymers (SAPs) have been shown to retain large amounts of water, which can be gradually released to plants during droughts. This ability to retain moisture in the soil not only improves plant growth but also helps reduce the need for frequent irrigation, making agriculture more sustainable in water-limited environments (Zhang et al., 2020).

Erosion control Soil erosion is a growing problem, especially in areas exposed to heavy rainfall or wind. Adjuvants that improve soil aggregation and increase organic matter content can help mitigate erosion by stabilizing the soil surface. By reducing soil movement, these adjuvants protect the topsoil, which is rich in nutrients and organic matter, thereby preserving long-term soil fertility (Six et al., 2002; Zhao et al., 2016). Thus, the use of adjuvants has the potential to significantly impact soil health through their effects on microbial activity, soil structure, and water retention. Selecting the right type of adjuvant is essential to supporting beneficial microbial communities and increasing soil resilience in the face of climate change. While some adjuvants can improve nitrogen fixation and water retention, others can have detrimental effects on soil microbiota and ecosystem balance.

Thus, the use of adjuvants has the potential to significantly impact soil health through their effects on microbial activity, soil structure, and water retention. Selecting the right type of adjuvant is essential to supporting beneficial microbial communities and increasing soil resilience in the face of climate change. While some adjuvants can improve nitrogen fixation and water retention, others can have detrimental effects on soil microbiota and ecosystem balance. Therefore, incorporating environmentally friendly, biodegradable, or organic adjuvants into agricultural practices can play a key role in promoting sustainable soil health management strategies.

4 Water Contamination

Runoff and Leaching Adjuvants and Water Pollution through Agricultural Runoff

Adjuvants, often used in agriculture to effectively increase the efficiency of crop protection products and fertilizers, can inadvertently contribute to surface and groundwater contamination. Recent studies show that adjuvants affect the mobility of active substances, which can lead to increased leaching of pesticides, herbicides and other chemicals into aquatic systems. For example, Smith et al. (2023) emphasize that hydrophilic adjuvants can increase the desorption and migration of chemicals in soil, which causes them to enter groundwater and surface water more quickly via runoff and leaching. One of the key problems is the increased solubility of some active substances in water in the presence of adjuvants, which is confirmed by the studies of Li et al. (2022) which show that adjuvants can increase the rate of transport of agrochemicals in soil by 30–50%, depending on the type of chemicals used. In turn, Garcia et al. [2021] draw attention to differences in adjuvant behavior depending on soil type and climate, suggesting that areas with high rainfall and light soils are particularly susceptible to increased groundwater contamination.

Impact on Aquatic Ecosystems Changes in Toxicity Profiles under Climate Change

Adjuvants, in addition to directly affecting soil migration processes, can alter the toxicity of aquatic pollutants. Increased temperatures and changes in precipitation resulting from climate change can lead to intensified impacts on aquatic ecosystems. Studies by Zhao et al. (2023) have shown that under conditions of elevated temperature, adjuvants can increase the toxicity of chemicals by accelerating their degradation to more toxic intermediates. As a result, aquatic microorganisms and planktonic organisms become more susceptible to the effects of chemicals, leading to a decrease in biodiversity.

Thompson et al. (2022) show that changing precipitation patterns—especially the intensification of extreme precipitation events—can lead to increased runoff of chemicals from fields into rivers and lakes, with direct impacts on fish, invertebrates, and aquatic vegetation. Moreover, these studies suggest that these impacts will be even greater in regions with decreasing soil water retention, where more frequent drought episodes can increase the concentration of chemicals in residual water bodies.

Anderson et al. (2021) also note that adjuvants can increase the accumulation of chemicals in bottom sediments, leading to long-term toxic effects that are difficult to reverse. As climate change continues, rising water temperatures can change the chemical properties of adjuvants and chemicals, leading to unpredictable ecological effects, such as increased biomagnification in food chains. New studies have shown that adjuvants play a significant role in water pollution processes, and their interactions with pesticides and climate change can lead to increased toxicity in aquatic ecosystems (Gupta et al., 2023). With climate change, predictions indicate that this problem may increase, hence it is necessary to further study this phenomenon and introduce preventive measures, such as the development of more ecological adjuvants.

4.1 Adjuvants and Climate Change: Interactions and Implications

4.1.1 Increased Pest Resistance

Rising temperatures and pest spread: Climate change, characterized by rising temperatures, altered precipitation patterns, and increased frequency of extreme weather events, is significantly affecting agricultural ecosystems. One of the most pressing challenges posed by these changes is the spread of pests and pathogens that threaten crop yields and food security. Warmer temperatures can lead to accelerated life cycles of many pests, resulting in increased population densities and more frequent disease outbreaks (Bebber et al., 2022). For example, studies have shown that pests such as the Colorado potato beetle and various aphid species exhibit increased reproductive rates under warmer conditions, leading to greater crop damage (Ranjan et al., 2023). As a result, the agricultural sector is being forced to adopt stronger pesticide formulations and increase application rates to effectively manage these growing pest populations. However, increased reliance on chemical control methods can lead to a cycle of increased pest resistance, where pests develop mechanisms that allow them to survive even the most potent pesticides (O'Neill et al., 2022).

4.1.2 Role of Adjuvants in Delaying or Accelerating Resistance

Adjuvants are substances added to pesticide formulations to enhance their efficacy, stability, and application characteristics. Their role in pest resistance dynamics is complex; they can contribute to or help delay the development of resistance. Some studies indicate that certain adjuvants can enhance pesticide efficacy by improving penetration and adhesion to plant surfaces, thereby increasing the likelihood of pest mortality and reducing the chances of resistance development (Rout et al., 2023). On the other hand, there is concern that the increased efficacy provided by some adjuvants may lead to overreliance on specific classes of pesticides, accelerating selection pressure on resistant pest populations. For example, a case study of glyphosate combined with adjuvants showed that although efficacy increased, it also contributed to the rapid emergence of glyphosate-resistant weed populations (Kirkwood et al., 2021).

In the context of climate change, the interplay of adjuvants and pest resistance is crucial. As pests adapt more rapidly to changing environmental conditions, strategies for managing their populations must also evolve. Predictive models suggest that integrating adjuvants with alternative pest management strategies, such as biopesticides and crop rotation, can help delay the development of resistance while maintaining effective pest control (Parker et al., 2023).

5 Sustainable Adjuvants: Innovations and Future Directions

5.1 Biodegradable and Eco-friendly Adjuvants

The development of biodegradable adjuvant options and the drive toward sustainable agriculture have led to an increasing interest in biodegradable and plant-based adjuvants as alternatives to conventional synthetic chemicals. Researchers are now focusing on bio-based adjuvants, often derived from vegetable oils, chitosan, and other natural polymers, which offer eco-friendly options with reduced toxicity and environmental persistence. For example, adjuvants derived from soy lecithin, coconut oil, and other renewable sources are being investigated for their ability to enhance pesticide efficacy while minimizing environmental impact (Porter et al., 2014; Bebber et al., 2022). These biodegradable alternatives break down into harmless substances, reducing the accumulation of these harmful residues in the ecosystem. *Reduced environmental footprint:* Innovations in biodegradable adjuvants aim to mitigate the negative impacts of synthetic chemicals on soil health, water quality, and biodiversity. Current research emphasizes the design of adjuvants that degrade more rapidly in the environment, leaving minimal residues. Techniques such as encapsulation of pesticides in natural carriers (e.g., starch or cellulosic) can improve pesticide delivery while ensuring that the carriers degrade innocuously. These solutions address concerns about the impact of traditional adjuvants on non-target organisms such as pollinators and beneficial insects (Anderson et al., 2021).

Policy and regulatory frameworks Global Adjuvant Regulation: The regulatory landscape for adjuvants is evolving, with major agricultural regions such as the European Union (EU) (Bebber et al., 2022), the United States, and Brazil increasingly imposing restrictions on the use of synthetic adjuvants due to environmental and public health sustainability concerns (EPA, 2024). The EU Green Deal and its Farm to Fork Strategy aim to reduce the use of harmful chemicals in agriculture, including adjuvants, and promote the adoption of sustainable alternatives. Similarly, the U.S. Environmental Protection Agency (EPA) has tightened regulations on adjuvants that pose a risk to human health or contribute to environmental degradation. Its Office of Emergency Management is working with federal partners to prevent accidents and maintain a well-adapted incident response capacity (EPA, 2024). International Initiatives for Sustainable Agriculture International initiatives such as the Sustainable Development Goals (SDGs) and programs led by the Food and

Agriculture Organization (FAO, 2023) that are driving the transition to sustainable agricultural practices (Zhao et al., 2016). Adjuvant regulations are increasingly aligned with these global SDGs, with a focus on reducing pesticide use and promoting integrated pest management (IPM) strategies. Countries are encouraging the development of adjuvants that support climate-resilient agriculture, encouraging the adoption of products that increase the efficacy of biopesticides and reduce the amount of chemicals (Parker et al., 2023).

5.2 Future Trends

Nanotechnology is playing a key role in agriculture and food production, enabling new tools and mechanisms for delivering agrochemicals such as pesticides and fertilizers. Nanoformulations improve this effect by reducing the effectiveness of chemicals, and nanosensors detect diseases and agrochemicals. Nanotechnology can also support gene transfer in plants (Bonny, 2016), the development of insect-resistant varieties (Bebber et al., 2022), animal health, and the occurrence of food product durability. Examples of applications also include intelligent delivery systems, access technologies (Zdyb, 2018; Sawicka et al., 2022). Nanotechnology has found application in many branches of industry and production, including agriculture. Nanomaterials serve there as carriers of chemical plant protection products, as well as active substances of pesticides (Bebber et al., 2022). However, we know little about the effects of human exposure to nanopesticides. New directions of the use of nanoparticles in agriculture should inform people about the routes of exposure of farmers to their effects and the possible toxicity of nanomaterials to mammalian cells (Matysiak et al., 2017). This is related to techniques for detecting nanoparticles in the work environment and biomarkers, which should be used to assess exposure and the effects of exposure. The use of nanotechnology achievements in agriculture can bring tangible benefits in the form of reducing the amount of chemical protection agents used (Sawicka & Egbuna, 2020; Sawicka et al., 2022). However, there are no clinical and in vitro studies determining whether the use of nanoparticles as carriers does not increase the effects of the harmful effects of pesticides on the human body. The results of studies, both on cell lines and animal models, indicate that nanoparticles used as active substances may be toxic to mammalian cells. There are also no epidemiological studies on this aspect. In the near future, the effects of human and animal exposure to nanopesticides may require special attention not only from scientists, but also from doctors taking care of food producers, farmers, and their families (Sekhon, 2014; Bonny, 2016; Matysiak et al., 2017).

Integration with precision agriculture: Integration of adjuvants with precision agriculture technologies—such as artificial intelligence (AI) and the Internet of Things (IoT)—is expected to revolutionize the way adjuvants are applied (Kirkwood et al., 2021; Parker et al., 2023; Rout et al., 2023; Smith et al., 2023; Cui et al., 2024). Real-time environmental data collected from sensors, drones, and other similar devices can optimize adjuvant application, ensuring that they are applied only when and where needed (Kirkwood et al., 2021; Cessna et al., 2023). By adjusting adjuvant use based on weather conditions, soil moisture, and pest pressure, these technologies have the potential to significantly reduce overuse of chemicals and improve the sustainability of agricultural practices (Gaskin et al. 2005; Hazra et al., 2017; Li et al., 2022; Raj et al., 2022). The use of smart sprayers and other precision tools will enable farmers to make more informed decisions about the use of adjuvants, further reducing the environmental impact of pesticide use (Xu et al., 2021; Cui et al., 2024). Nanopesticides are generally divided into two types: (1) organic molecular active ingredients, nanocarrier complexes, and (2) nanoscale active ingredients without carriers (Zdyb, 2018; Sawicka et al., 2022). Nanostructures are objects with a size of up to 100 nm, where one dimension not exceeding this value is required. They include, among others, nanoparticles, nanowires, and nanolayers. Due to their small size, nanostructures, quantum effects, which are available on their unique conditions, distinguishing them from materials with a larger number. Nanomaterials can be quasi-zero-wear (e.g., nanoparticles), quasi-one-dimensional (e.g., nanofibers), or quasi-two-dimensional (e.g., nanolayers). They can also be created in matrices of another material, and their structure can be both amorphous and quasi-crystalline. Nanoparticles as pesticide carriers can pose an active risk (Zdyb, 2018). Nanotechnology in agriculture offers solutions to problems with pesticide harmfulness, reducing their amount and effects of use. Nanoparticles, used as carriers of plant protection products, improve the solubility of pesticides, control their release, release and protect against degradation, which allow to reduce their effects at lower doses. Nanopesticides include both carriers and active protections. For example, "silver nanoparticles" have strong antibacterial, antifungal, and antiviral effects, and "copper nanoparticles," "titanium dioxide," and "zinc oxide" effectively combat plant pathogens. These particles are used outside traditional plant protection products due to the lack of resistance acquisition by pests. In the case of classic pesticides, the acquisition of resistance to nanoparticles is harmful, which causes disorders related to enzymatic changes in many diseases. Studies by Bonny (2016), Matysiak et al. (2017) show that nanomaterials can be a threat as insecticides and herbicides, which limits their potential in agriculture.

6 The Future of Adjuvants

Nanotechnology presents a promising path for the future of adjuvants, with the development of nano adjuvants that offer more precise and efficient delivery of pesticides and nutrients. These nano-sized carriers improve the solubility, stability, and absorption of active ingredients, ensuring that pesticides are delivered in a more controlled manner.

Nano-adjuvants could potentially reduce the total volume of chemicals needed, leading to less environmental pollution. For example, nano-encapsulation techniques are being explored to ensure that pesticides are released only when needed, minimizing their impact on non-target organisms. Integration with precision agriculture: The integration of adjuvants with precision agriculture technologies—such as artificial intelligence (AI) and the Internet of Things (IoT)—is expected to revolutionize the way adjuvants are applied. Real-time environmental data collected from sensors, drones, and satellite imagery can optimize the use of adjuvants, ensuring they are applied only when and where needed. By adjusting adjuvant use based on weather conditions, soil moisture, and pest pressure, these technologies have the potential to significantly reduce overuse of chemicals and improve the sustainability of agricultural practices. The use of smart sprayers and other precision tools will enable farmers to make more informed decisions about adjuvant use, further reducing the environmental impact of pesticide use (van Zwieten, 2015; Zhang et al., 2020; Zhang et al. 2023b; Zhou et al. 2022a, b; Cessna et al., 2023; Smith et al., 2023; Rout et al., 2023).

Current status of adjuvant sustainability, regulatory trends and future innovations: Combining findings from the latest research and regulatory frameworks can provide a future perspective on the role of nano-encapsulation techniques. They are being studied to ensure that pesticides are released only when necessary, minimizing their impact on non-target organisms.

6.1 Sustainable Adjuvant Development Using Plant-Based Materials

Plant-based materials are the basis for the use of natural resources to reduce the negative impact on the environment and the efficiency of agrochemicals such as pesticides and fertilizers. Adjuvants are accessories added to agricultural formulations for their separation, stability and plant uptake capacity. Traditionally, it was one of the synthesized utility devices, but for the application in the sustainable agricultural process in the development of plant-based adjuvants. Main benefits of plant-based adjuvants: Biodegradability: Plant-based materials such as vegetable oils, waxes or extracts are readily biodegradable, which limits their accumulation in the environment and to minimize the risk of complications and groundwater (Gupta et al., 2023).

6.2 Innovations in Bio-Based Adjuvants: Current Progress

Bio-based adjuvants are becoming increasingly popular as an alternative to synthetic additives in agriculture, especially in the context of applying practical knowledge and applications of negative environmental restrictions. Innovations in this field focus on natural organic sources that can use pests, fertilizers, and other plant protection products while minimizing their toxicity to ecosystems and humans.

The use of plant extracts is one of the key trends in the field of bio-based adjuvants is an extract from plants, from sources such as oils, saponins, and other sources. For example, saponins, which are naturally occurring compounds in plants, characteristic of surfactants, improving droplet retention and penetrating the tool through the leaf cuticle. They have antibacterial effects, which can additionally protect against pathogens. Natural polymers, polymers of plant origin, such as natural gums, are used as adjuvants due to their hydrophilic properties, which help increase retention of safety on the leaf surface. The use of such solutions can result in spraying while reducing losses resulting from spraying.

6.3 Probiotic Microorganisms

An innovation in the field of biological adjuvants are also probiotic microorganisms, which support the health of the action and the plant, improving the knowledge of ingredients and restrictions on abiotic and biotic stress. Bacteria such as *Bacillus subtilis* can take full advantage of adjuvants, supporting biological control of diseases and pests. Additionally, some microorganisms can cause natural side effects on plant growth, which can increase the effectiveness of fertilizer application.

6.4 Sustainable Adjuvants from Plant Waste

Other applications are the use of plant waste, such as hemp raw materials, cellulose from plant branches or algae extracts, as a component for the production of adjuvants. This action reduces waste but also reduces the release of substances. The use of biodegradable materials with an impact on ecosystems that affect the environment of agriculture.

Biopolymers and nanotechnology. Recent advances include biopolymers developed from starting materials that can be combined with nanotechnology, creating delivery systems, and solutions of the results. Thanks to this, biopolymer adjuvants can provide controlled release of the device, leading to their further use and reduction in pesticide use.

6.5 Benefits and Challenges

The benefits of using innovative adjuvants are: Reduction in toxicity: Adjuvants meaning universal are less toxic in the sense of synthetic suitable, which translates into a greater risk to human and animal health and less medical treatment and groundwater.

Sustainability: Repurposing plants, microorganisms, or plant wastes, contributing to the spread of the carbon footprint and supporting the development of more ecologically friendly agricultural practices.

6.6 *Improving Biodiversity*

Plant-based adjuvants support health and ecosystems by introducing natural components that do not disrupt international sustainability.

6.7 *Special Event*

Due to the natural origin of adjuvants, their effects may be less common and more alternative to specific environmental conditions than synthetic adjuvants.

6.8 *Production Costs*

Although adjuvants have many advantages, their production can be more expensive, especially in the case of biopolymers and nanotechnologies, which poses a challenge for their commercialization at high speed.

6.9 *Regulations*

Introduction of new adjuvants to the market requires compliance with the issuance of an environmental and health protection release, which may delay the commercialization process.

6.10 *Future Outlook*

Considering further actions in adjuvant research resulting from the notification, which will be included in the application. Research on their exclusion and exclusion of ecosystems will be necessary. Introduction of innovations such as biopolymers with nanotechnology may lead to the development of more ecological and alternative crop technologies or processing technologies that provide effective crop protection. From global perspectives such as climate change and environmental degradation, adjuvants may be available and will introduce key effects in case of spreading in more sustainable and resilient technologies in food production systems.

Although adjuvants based on natural plants and products have many advantages, their production can be more expensive, especially in the case of biopolymers and nanotechnology, which poses a challenge for their commercialization.

FAO's Role in Promoting the Sustainable Use of Pesticides and Adjuvants
The status of adjuvant sustainability, regulatory trends and future innovations, integrating findings from the latest research and regulatory frameworks to provide a future perspective on the role of nano-encapsulation techniques are being studied to ensure that pesticides are released only when necessary, minimizing their impact on non-target organisms.

Nanotechnology in Agricultural Adjuvants: New Frontiers
Nanotechnology in agriculture is developing rapidly and its application in adjuvants is one of the most widely used fields. Adjuvants such as surfactants, application-related compounds, wetting agents, or buffering threats are a key plug in the power supply protecting against pests, herbicides, growth regulators, and fertilizers. In the publication of software, there is a problem with undisclosed, known solution and their analyzer in the environment. With nanotechnology, it is possible to solve these problems by:

- Enhancing the bioavailability of bioactive substances (Adetunji & Sawicka, 2021)
- Pesticide reduction strategy (Xu et al., 2021)
- Optimizing the penetration of nanopesticides (Adetunji & Sawicka, 2021; Sawicka et al., 2022, 2023)
- Nanoemulsions (Sawicka et al., 2022, 2023)
- Nanocapsules (Xu et al., 2021; Adetunji & Sawicka, 2021)
- Metal nanoparticles (Gupta et al., 2023).

7 Applications

Research on nanoformulations of pesticides, for example, zinc and copper oxide nanoparticles, may translate into their use. An additional advantage of nanotechnology in adjuvants is the possibility of lower chemical consumption. Smaller doses are required, which reduce environmental contamination, groundwater contamination, and side effects. Additionally, controlled-release nanomaterials reduce their consumption (Xia et al., 2021; Sass & Colangelo, 2006; Sawicka et al., 2023).

Although nanotechnology in agricultural adjuvants opens up new perspectives, there are also some challenges. Their impulsive action affects human and environmental health (Xu et al., 2021). This aspect should be better studied.

8 The Role of AI and IoT in Optimizing the Use of Adjuvants in Precision Agriculture

It is expected that adjuvants will be integrated with precision agriculture technologies such as artificial intelligence (AI) and the Internet of Things (IoT), which will ultimately revolutionize the way adjuvants are used. This will allow you to gain control and insight into every aspect of production, from soil condition to animal health. It will be possible to optimize processes, save resources, and make better decisions based on hard data (Sawicka et al., 2023). Real-time environmental data collected from various sensors, drones, and satellite imagery can optimize adjuvant application, ensuring that they are applied only when and where needed. By adjusting adjuvant application based on weather conditions, soil moisture, and pest pressure, these technologies can significantly reduce overuse of chemicals and improve the sustainability of agricultural practices. The use of smart sprayers and other precision tools will enable farmers to make more informed decisions about adjuvant use, further reducing the environmental impact of pesticide use (Bhardwaj et al., 2014; Hazra et al., 2017; Sharma et al., 2022; Gupta et al., 2023). Thus, the current state of adjuvant sustainability, regulatory trends, and future innovations integrate findings from the latest research and regulatory frameworks to provide a future perspective on adjuvant use (Gupta et al., 2023).

9 Conclusions

Conclusions can be drawn from two perspectives: a sustainable future for agriculture and environmental impacts, and lessons for the future.

Balancing agricultural efficiency and environmental impacts Adjuvants are key to integrated agriculture. They improve the efficacy of pesticides and fertilizers, leading to reduced equipment use. Acting as surfactants, oils or humectants, adjuvants improve retention, penetration, and dissemination of the equipment, which can reduce the doses of chemicals that provide effective crop protection. On the other hand, inappropriate use of adjuvants can lead to additional chemicalization of agriculture and the environment, especially in the context of climate change, where resource conservation is becoming a priority. This is necessary because it maximizes the effects while minimizing the effects on ecosystems.

Toward the Future Sustainable availability will require rigorous research on adjuvants to better understand their impact on the environment and health. The development of more ecological alternative directions, such as plant-based adjuvants or biodegradable formulations, is crucial for the future. This is necessary to apply the regulatory elements and effects that are applied later, adjuvants on biodiverse and auxiliary function. As a result of the changing environment and pressures on energy consumption, it is necessary for adjuvants to be an integral part of control strategies, electricity distribution and food and general security.

References

Adetunji, C. O., & Sawicka, B. H. (2021). Chapter 33: Application of nanoceutical technology for fast and efficient control of illness. Part III: Management, protocols, regulation, and situation indicators of phytochemicals the use of nanoceutical technology for treatment and treatment of diseases. In A. G. Mtewa & C. Egbuna (Eds.), *Phytochemistry, the military and health phytotoxins and natural defenses mtewa_phytochemistry*. Elsevier. https://doi.org/10.1016/B978-0-12-821556-2.00014-1. https://editorial.elsevier.com/app/book?execution=e1s3

Anderson, R. J., Lee, S. Y., & Patel, V. K. (2021). Accumulation of agricultural chemicals in sediments and its ecological implications. *Environmental Pollution, 268*, 115647. https://doi.org/10.1016/j.envpol.2020.115647

Angaye, T. C. N., Ukoroije, R. B., & Angaye, G. P. (2024). Plant-derived pesticides (PDPs) versus synthetic pesticides: An eco-friendly approach to biodiversity conservation. *International Journal of Innovative Environmental Studies Research, 12*(1), 67–75. www.seahipaj.org; ISSN: 2354-2918.

Anonymous. (2024). *Extreme weather events: Floods, droughts and heatwaves*. https://www.eea.europa.eu/en/topics/in-depth/extreme-weather-floods-droughts-and-heatwaves

Arif, M. et al. (2022). Antibacterial and antibiofilm activity of mannose-modified chitosan/PMLA nanoparticles against multidrug-resistant Helicobacter pylori. *International Journal of Biological Macromolecules 223*(Pt A), 418–432. https://doi.org/10.1016/j.ijbiomac.2022.10.265

Bardgett, R. D., & van der Putten, W. H. (2014). Belowground biodiversity and ecosystem functioning. *Nature, 515*(7528), 505–511.

Bebber, D. P., Holmes, T. J., & Gurr, S. J. (2022). The global spread of crop pests and pathogens. *Nature Climate Change, 12*(1), 1–6. https://doi.org/10.1038/s41558-021-01197-6

Bhardwaj, D., Ansari, M. W., Sahoo, R. K., & Tuteja, N. (2014). Biofertilizers function as key player in sustainable agriculture by improving soil fertility, plant tolerance and crop productivity. *Microbial Cell Factories, 13*(1), 66.

Bonny, S. (2016). Genetically modified herbicide-tolerant crops, weeds, and herbicides: Overview and impact. *Environmental Management, 57*(1), 31–48. Available at SSRN: https://ssrn.com/abstract=2914952. Accessed October 12, 2024.

Cessna, A. J., Lander, A. W., & Lutz, R. S. (2023). Volatilization of pesticide formulations: The role of adjuvants. *Environmental Toxicology and Chemistry, 42*(5), 1234–1245. https://doi.org/10.1002/etc.5564

Chagnon, M., Kreutzweiser, D., Mitchell, E. A., Morrissey, C. A., Noome, D. A., & Van der Sluijs, J. P. (2015). Risks of large-scale use of systemic insecticides to ecosystem functioning and services. *Environmental Science and Pollution Research, 22*(1), 119–134.

Cui, Y., Ho, M., Hu, Y., & Shi, Y. (2024). Vaccine adjuvants: Current status, research and development, licensing, and future opportunities. *Journal of Materials Chemistry B, 12*(17), 4118–4137. https://doi.org/10.1039/d3tb02861e

Doran, J. W., & Zeiss, M. R. (2000). Soil health and sustainability: Managing the biotic component of soil quality. *Applied Soil Ecology, 15*(1), 3–11.

EPA (2024) Annual Report For the year ended 30 June 2024, chromeextension://efaidnbmnnnibpcajpcglclefindmkaj/https://www.epa.govt.nz/assets/RecordsAPI/EPA-Annual-Report-2024.pdf (accessed 10.01.2025)

FAO (2023). The state of food and agriculture revealing the true cost of food to transform agrifood systems. Food and Agriculture Organization of the United Nations Rome, 2023

Garcia, M. S., Black, R. J., & Thompson, D. A. (2021). The role of soil type and climate in the transport of agricultural pollutants. *Environmental Science & Technology, 55*(18), 12345–12353. https://doi.org/10.1021/acs.est.1c03001

Garrido, A., Gonçalves, F., & Vale, C. (2021). Surfactants in pesticide formulations: Mechanisms of action and recent advances. *Journal of Agricultural and Food Chemistry, 69*(12), 3547–3560. https://doi.org/10.1021/acs.jafc.0c08012

Gaskin, R. E., Steele, K. D., & Forster, W. A. (2005). Characterising plant surfaces for spray adhesion and retention. *New Zealand Plant Protection Society, 58*, 179–183.

Gupta, S., Bhagat, M., & Singh, R. (2023). Climate-induced shifts in pest management: Emerging trends and solutions in agriculture. *Journal of Agricultural Science and Climate Change, 45*(2), 109–123. https://doi.org/10.1007/s12356-023-02113

Haynes, R. J. (2005). Labile organic matter fractions as central components of the quality of agricultural soils: An overview. *Advances in Agronomy, 85*, 221–268.

Hazra, D. K., Karmakar, R., Poi, R., Bhattacharya, S., Mondal, S. (2017). Recent advances in pesticide formulations for eco-friendly and sustainable vegetable pest management: a review. *Archives of Agriculture and Environmental Science, 2*(3), 232–237.

Jamar, L., & Lefrancq, B. (2008). The impact of adjuvants on rain resistance and effectiveness of spore suspension formulations of biocontrol agents against apple scab. *Crop Protection, 27*(2), 366–371. https://doi.org/10.1016/j.cropro.2007.05.011

Jamieson, M. A., Arowbridge, A. M., Raffa, K. F., & Lindroth, R. F. (2012). Consequences of climate warming and altered precipitation patterns for plant-insect and multitrophic interactions. *Plant Physiology, 160*(4), 1719–1727. https://doi.org/10.1104/pp.112.206524

Kirkwood, R. C., Taylor, A., & Morrison, C. (2021). Glyphosate and adjuvants: The role of formulation in herbicide resistance. *Weed Science, 69*(4), 407–416. https://doi.org/10.1017/wsc.2021.25

Li, X., Chen, Y., & Wang, T. (2022). Effects of adjuvants on the leaching of pesticides in different soil types. *Chemosphere, 302*, 134749. https://doi.org/10.1016/j.chemosphere.2022.134749

Liang, Y., He, X., Chen, C., & Zhang, X. (2014). Adjuvants' impact on the efficacy and environmental behavior of pesticides. *Journal of Agricultural and Food Chemistry, 62*(46), 10871–10882.

Malaj, E., Peter, C., Grote, M., Kühne, R., Mondy, C. P., & Scheringer, M. (2014). Aquatic ecotoxicity of pesticides in European surface waters: Refined threshold screening approach. *Environmental Toxicology and Chemistry, 33*(6), 1207–1216.

Matysiak, M., Kruszewski, M., & Kapka-Skrzypczak, L. (2017). Nanopesticides – Light or dark side of the force/Nanopestycydy – Jasna czy ciemna strona mocy. *Medycyna Pracy, 68*(3), 423–432. (in Polish).

Nooraei, S., Sarkar Lotfabadi, A., Akbarzadehmoallemkolaei, M., & Rezaei, N. (2023). Immunogenicity of different types of adjuvants and nano-adjuvants in veterinary vaccines: A comprehensive review. *Vaccine, 11*, 453. https://doi.org/10.3390/vaccines11020453

Noyes, P. D., McElwee, M. K., Miller, H. D., Clark, B. W., Van Tiem, L. A., Walcott, K. C., Erwin, K. N., & Levin, E. D. (2019). The interaction between climate change and pest dynamics: Implications for food security. *Environmental Science and Technology, 53*(7), 4016–4034. https://doi.org/10.1021/acs.est.9b00182

O'Neill, P. M., Swanson, L. D., & Jayakumar, A. (2022). Evolution of pest resistance to pesticides: Implications for future pest management strategies. *Agricultural Systems, 195*, 103312. https://doi.org/10.1016/j.agsy.2022.103312

Okonkwo, C., Hilary, E. A., & Essien, S. N. (2023). Ameliorating the impact of climate variability on pesticide dynamics and efficacy, 2023, preprint. https://doi.org/10.32388/IK3NC2

Parker, R. A., McLellan, J. W., & Leach, J. E. (2023). Integrated pest management in the age of climate change: The role of adjuvants and biopesticides. *Frontiers in Plant Science, 14*, 203234. https://doi.org/10.3389/fpls.2023.203234

Porter, J. R., Xie, L., & Challinor, A. J. (2014). Food security and food production systems. In *Climate change 2014: Impacts, adaptation, and vulnerability. Part A: Global and sectoral aspects. Contribution of working group II to the fifth assessment report of the Intergovernmental Panel on Climate Change* (pp. 485–533). Cambridge University Press.

Raj, S., Kumar, M., & Mehta, V. (2022). The role of adjuvants in improving pesticide efficiency under varying climatic conditions: A review. *Journal of Pesticide Science, 47*(3), 215–228. https://doi.org/10.1007/s00048-022-02017

Ranjan, A., Garg, S., & Singh, S. P. (2023). Climate change and its effects on the population dynamics of agricultural pests: A review. *Pest Management Science, 79*(3), 500–515. https://doi.org/10.1002/ps.6765

Rillig, M. C., & Mummey, D. L. (2006). Mycorrhizas and soil structure. *New Phytologist, 171*(1), 41–53.

Rout, P. K., Raghavendra, A. S., & Kumar, N. (2023). Adjuvants as a tool for enhancing pesticide efficiency and reducing resistance. *Journal of Agricultural and Food Chemistry, 71*(10), 4003–4011. https://doi.org/10.1021/acs.jafc.2c08211

Sanchez-Bayo, F., & Tennekes, H. A. (2021). Neonics and their impact on pollinators. *Pesticide Biochemistry and Physiology, 169*, 104641.

Sass, J. B., Colangelo, A. (2006). European Union bans atrazine, while the United States negotiates continued use. *International journal of occupational and environmental health, 12*(3), 260–267. https://doi.org/10.1179/oeh.2006.12.3.260

Sawicka, B., Pszczółkowski, P., Barbaś, P., Skiba, D., & Bienia, B. (2022). Chapter 7. The role and importance of microorganisms in environmental sustainability. In P. Chowdhary, S. Mani, & P. Chaturvedi (Eds.), (pp. 107–134). John Wiley & Sons, Inc.. https://doi.org/10.1002/9781119834489.ch7. First published: 09 September 2022, ISBN: 9781119834458.

Sawicka, B., Barbaś, P., Skiba, D., Pszczółkowski, P., & Krochmal-Marczak, B. (2023). Chapter 9. Nanotechnology for enhancement of food value. In A. Kumari, R. Garg, & R. Garg (Eds.), *Nanotechnology for sustainable agriculture, food and environment* (1st ed., pp. 109–123). An imprint of CRC Press. https://doi.org/10.1201/9781003397861-9. First published 2023; e-book ISBN: 9781003397861.

Sawicka B., Egbuna Ch. (2020). Chapter 1 - Pests of Agricultural Crops and Control Measures, Editor(s): Chukwuebuka Egbuna, Barbara Sawicka, Natural Remedies for Pest, Disease and Weed Control, Academic Press, 2020, 1–16, ISBN 9780128193044, https://doi.org/10.1016/B978-0-12-819304-4.00001-4

Sekhon, B. S. (2014). Nanotechnology in agro-food production. An overview. *Nanotechnology, Science and Applications, 7*(2), 31–53. https://doi.org/10.2147/NSA.S39406

Sharma, A., Singh, M., & Nair, R. (2022). Improving pesticide penetration using plant-based oils: Environmental benefits and crop safety. *Agriculture, 12*(4), 478. https://doi.org/10.3390/agriculture12040478

Sharma, P., Sharma, K., & Nair, P. (2023). Effect of surfactants on the efficacy of herbicides: A review. *Pest Management Science, 79*(3), 1245–1258. https://doi.org/10.1002/ps.6834

Singh, B. K., & Reddy, M. S. (2011). Microbial biotechnology and sustainable agriculture. *Microbial Biotechnology, 4*(4), 415–420.

Six, J., Conant, R. T., Paul, E. A., & Paustian, K. (2002). Stabilization mechanisms of soil organic matter: Implications for C-saturation of soils. *Plant and Soil, 241*(2), 155–176.

Smith, J. A., Johnson, L. R., & Martinez, E. P. (2023). Impact of adjuvants on the mobility of agricultural chemicals in soil. *Journal of Environmental Quality, 52*(3), 540–550. https://doi.org/10.2134/jeq2023.03.0010

Stelzer, R. S., & Schaller, C. (2020). Influence of surfactants on droplet behavior in spray applications. *International Journal of Pest Management, 66*(4), 1–10. https://doi.org/10.1080/09670874.2020.1747923

Stelzer, R. S., & Schaller, C. (2021). Influence of surfactants on pesticide adherence and rainfastness in variable weather conditions. *Journal of Pest Science, 94*(4), 1190–1205. https://doi.org/10.1007/s10340-021-01365-4

Thompson, R. E., Nguyen, T. H., & Li, K. (2022). Extreme weather events and their impact on agricultural runoff in river systems. *Water Research, 207*, 117854. https://doi.org/10.1016/j.watres.2021.117854

van Zwieten, L. (2015). Adjuvants in agriculture: The role of adjuvants in improving pesticide and fertilizer efficiency. In *Agricultural systems research*. Springer.

Wang, J., Zhao, Q., & Zhang, Y. (2022). The role of surfactants in enhancing the efficacy of pesticides: A comprehensive review. *Agricultural Sciences, 13*(1), 15–28. https://doi.org/10.4236/as.2022.131002

Wang, Y., Chen, D., Huang, X., & Yu, Y. (2023). Environmental impacts of surfactants in pesticides: From field applications to aquatic ecosystems. *Science of the Total Environment, 850*, 158280.

Xia, J., Wang, Z., & Yu, X. (2021). Drift potential of pesticide applications: Impacts of droplet size and wind conditions. *Agricultural and Forest Meteorology, 296*, 108200. https://doi.org/10.1016/j.agrformet.2021.108200

Xu, X., Sharma, P., Shu, S. et al. (2021) Global greenhouse gas emissions from animal-based foods are twice those of plant-based foods. *Nature Food 2*, 724–732. https://doi.org/10.1038/s43016-021-00358-x

Zdyb, A. (2018). Nanostructures in energy. In M. Maciąg & K. Maciąg (Eds.), *Latest reports on nanotechnology*. Copyright by Wydawnictwo Naukowe TYGIEL sp. z o.o. ISBN 978-83-65932-26-6, 7-14. (in Polish).

Zhang, Z., Liu, X., Li, X., & Chen, J. (2020). Effect of adjuvants on pesticide droplet spread and penetration in relation to pest management. *Pest Management Science, 76*(3), 905–913.

Zhang, L., Liu, X., & Zhang, J. (2023a). Enhancing pesticide adherence and efficacy under variable environmental conditions: The role of modern surfactants. *Crop Protection, 157*, 106945. https://doi.org/10.1016/j.cropro.2023.106945

Zhang, T., Wang, X., & Li, J. (2023b). The role of methylated seed oils in enhancing pesticide efficacy and sustainability. *Pesticide Biochemistry and Physiology, 188*, 105321. https://doi.org/10.1016/j.pestbp.2023.105321

Zhao, J., Zeng, Z., He, X., & Wu, Z. (2016). Effects of agricultural practices on soil microbial activity and community composition. *Science of the Total Environment, 565*, 463–473.

Zhao, H., Wu, G., & Zhang, Y. (2023). Climate change impacts on the toxicity of agrochemicals in aquatic ecosystems. *Aquatic Toxicology, 266*, 105946. https://doi.org/10.1016/j.aquatox.2023.105946

Zhou, Q., Zhang, Y., & Liu, T. (2022a). Effects of adjuvant type on spray droplet size and drift potential. *Pest Management Science, 78*(10), 3456–3464. https://doi.org/10.1002/ps.6710

Zhou, X., Liu, Y., & Wang, L. (2022b). Reducing pesticide drift with surfactants: Mechanisms and applications. *Crop Protection, 150*, 105785. https://doi.org/10.1016/j.cropro.2021.105785

Resilience of Ecosystems: Navigating the Challenges of Pesticide Exposure and Climate Change

Mukul Machhindra Barwant, Shreya Singh, Shivendra Singh, B. Sri Sai Siddartha Naik, and Odangowei Inetiminebi Ogidi

1 Introduction

The presence of numerous land-based pesticides in waterways through runoff or spray drift poses a hazard to fish and other nontarget animals (Ogidi et al., 2023). Pesticides exacerbate and interact with other ecological processes, in addition to surface-water diversions, nonnative species, toxic algal outbreaks, and habitat loss for spawning and rearing. Evaluation of the cumulative effects of pesticides on ecosystem services or species has been challenging due to philosophical, legal, pragmatic, and historical constraints. In order to examine these issues, it is necessary to consider land-based nutrients, sediment, pathogens, and refuse sources as pollutants. Furthermore, aquatic ecosystems, including lakes, rivers, estuaries, and near-shore marine ecosystems, are frequently exposed to a variety of substances, including metals, pharmaceuticals, personal care products, plasticizers, and

M. M. Barwant (✉)
Department of Botany, Sanjivani Rural Education Society (SRES), Sanjivani Arts, Commerce and Science College, Kopargaon, Ahilyanagar, Maharashtra, India

S. Singh
Department of Agriculture, Ramlalit Singh Mahavidyalaya, Kailhat, Chunar, Mirzapur, Uttar Pradesh, India

S. Singh
Department of Agriculture, Udai Pratap College (An Autonomous Institution), Bhojubir, Varanasi, Uttar Pradesh, India

B. Sri Sai Siddartha Naik
Department of Agronomy, Acharya N.G Ranga Agricultural University, Guntur, Andhra Pradesh, India

O. I. Ogidi
Department of Biochemistry, Faculty of Basic Medical Sciences, Bayelsa Medical University, Yenagoa, Bayelsa State, Nigeria

B. R. Babaniyi, E. E. Babaniyi (eds.), *The Interplay of Pesticides and Climate Change*, https://doi.org/10.1007/978-3-031-81669-7_3

industrial agents that are still in the initial stages of development, such as engineered nanoparticles. It is possible that one or more of these chemical contaminants will cause injury (Ogidi & Akpan, 2023).

Scholz et al. conducted a study in 2012. The long-term health and viability of Earth are contingent upon the resilience of its ecosystems. However, the combination of pesticide exposure and climate change is increasingly causing a double burden for ecosystems worldwide. Although pesticides are essential for modern farming, they pose a threat to numerous non-target species and ecological processes. Climate change is also hastening the arrival of environmental stresses like temperature swings, changed precipitation patterns, and extreme weather events. The interplay of these factors is not just reducing biodiversity but also endangering pollination, water purification, and soil fertility, among other ecosystem services upon which human societies depend. Resilient ecosystems are crucial for human survival and community adaptation to climate change, as is becoming more and more obvious. In order to better adapt to climate change, the governing bodies of the UNFCCC and the CBD urge all parties to adopt "ecosystem-based adaptation" (EbA) methods that make use of biodiversity and ecosystem services. The restoration of mangroves as a buffer against storm surges, the control of floods and droughts via watershed management, the halting of desertification through rangeland management, and the assurance of food security through sustainable forest and fisheries management are just a few of the many themes covered by these methods (Chong, 2014).

This is to provide a comprehensive overview of how pesticide exposure and climate change interact to affect ecosystem resilience. By understanding the mechanisms through which ecosystems are influenced by these factors, we can better assess how to preserve their functionality and resilience in the face of escalating environmental challenges. Additionally, this chapter will review the strategies that have been developed to mitigate these impacts, including integrated pest management, habitat restoration, and the promotion of biodiversity conservation practices. In contrast, toxic chemical pollutants are far more prevalent, diverse, challenging to identify, and costly to monitor (Ogidi et al., 2024b). Because of their intricate migration across aquatic habitats, fate-and-exposure modeling frequently produces difficult predictions. Examples of recent, intensive, and widely distributed activities in the US include the National Water Quality Assessment Program of the US Geological Survey. Additionally, the ability of an organism or other component of an ecosystem to endure or recover from environmental toxicity has traditionally been evaluated in the domains of ecotoxicology (for impacts on communities and species) or medicine (for impacts on human health) (Ogidi et al., 2024a).

Therefore, in order to include toxics in an ecosystem-based framework, a basic knowledge of toxicology is required. Thus, familiarity with toxicology is necessary to incorporate toxics into an ecosystem-based framework, even though there have been previous demands for combining toxicity and conservation. Humans are the principal cause of climate change, which is a serious and urgent problem, because we release greenhouse gases into the air. We need to do more than just increase agricultural output to address this issue and ensure long-term food security; we must also work to mitigate and adapt to the consequences of climate change. A rise in the concentration of greenhouse gases like methane (CH_4) and carbon dioxide (CO_2) is caused by the combustion of fossil fuels, industrialization, and deforestation. These gases are responsible for the greenhouse effect, which raises global temperatures (also known as warming) by absorbing heat from the sun. Heatwaves, storms, and heavy rains are becoming more often and more destructive as a result of this warming trend's impact on weather patterns.

The organochlorine family of insecticides was the primary concern when individuals began to express concern about the effects of pesticides on the environment and human health in the mid-twentieth century. This category included chlordane, dieldrin, endrin, heptachlor, toxaphene, and DDT (dichlorodiphenyltrichloroethane). The organochlorine insecticides were notoriously harmful to aquatic life and frequently resulted in mass fish deaths upon their discharge into the water, despite their low frequency of use. The pesticides' movements can be traced to the present day, as they were and continue to be detectable in sediments, tissues, and, to a lesser extent, water (Ogidi et al., 2024c; Iyiola et al., 2023).

Additionally, they accumulate in animals at higher trophic levels and are lipophilic and extremely persistent. In the 1970s, the majority of DDT and other organochlorine applications were outlawed in the USA, partly due to pesticides (Scholz et al., 2012). Pesticides, primarily used to control pests in agriculture, have become ubiquitous in natural environments due to runoff, aerial drift, and leaching into soil and water systems. Although they help protect crops, their indiscriminate use can harm non-target species, reduce biodiversity, and degrade ecosystem health (Pimentel et al., 1992). According to Geiger et al. (2010), keystone species are vital for sustaining the structure of ecosystems. When pesticides are used excessively, they can disrupt food webs, decrease pollinator abundance, and hasten the loss of these species. According to Holling (1973), ecosystem resilience is the capacity of an ecosystem to endure perturbations while preserving its fundamental functions and structure.

It is a measure of an ecosystem's capacity to adapt to changes, resist degradation, and recover from stressors, both natural and anthropogenic. In recent decades, two of the most pervasive and interconnected challenges impacting ecosystem resilience are pesticide use and climate change. Organochlorine bans decreased the amount of well-known pesticides that were introduced into aquatic environments, such as the San Francisco Estuary. Yet, the number and variety of pesticides (including herbicides, insecticides, and fungicides) used today have increased dramatically during the previous half-century. Even though they are becoming less common, fish deaths do not always represent the whole impact of pollutants on ecosystems. Instead, sublethal health impacts such as endocrine disruption (Ogidi & Tawariowei, 2023), lower growth, altered behaviors, aberrant development, immune system impairment, and reproductive impairment are the focus of cutting-edge toxicological research. Sublethal toxicity can affect interspecific relationships (e.g., disease transmission, predator-prey dynamics) as well as individual fitness (Ogidi & Tawariowei, 2023). However, a more involved experimental design is needed to measure the sublethal physiological effects of pesticides, especially in the field, as opposed to acute death in a controlled laboratory environment (Scholz et al., 2012). Staudt et al.

(2013) looked into how pesticides could modify their environmental fate due to changes in temperature and precipitation patterns caused by climate change.

They found that increased temperatures accelerated the breakdown of some pesticides, but also increased their volatilization, which allowed them to spread over greater distances. This process poses new risks for ecosystems that were previously less exposed to pesticide contamination. Given the intertwined nature of pesticide use and climate change, holistic approaches are necessary to preserve ecosystem resilience. Biological control, crop rotation, and reduced pesticide use are all parts of integrated pest management (IPM), a promising strategy for sustainable pest population management (Pretty & Bharucha, 2015). IPM helps reduce the reliance on chemical pesticides, thus lowering the risk of ecological harm while maintaining agricultural productivity. Additionally, promoting biodiversity in agricultural landscapes can enhance ecosystem resilience to both pesticides and climate change. Ecosystem services like pollination and pest regulation are enhanced when habitat complexity increases, which in turn supports a broader variety of species that can buffer against pest outbreaks (Tscharntke et al., 2012).

Climate change is increasingly recognized as a major factor influencing ecosystem dynamics and resilience. As temperatures rise, ecosystems are experiencing shifts in species distributions, phenology, and interactions. Some species may thrive in new conditions, while others face extinction due to the inability to adapt quickly enough (Root et al., 2003). Variations in precipitation and temperature also have an effect on pest populations, which can lead to more intense and frequent outbreaks.

This leads to an increased use of pesticides, which further compounds the pressures on ecosystems. For example, warmer temperatures have been linked to an increased incidence of pest infestations in agricultural landscapes, particularly in temperate regions (Rosenzweig et al., 2001). The need to control these pests has led to greater pesticide use, which in turn affects surrounding ecosystems, as pesticides can drift beyond their intended areas and accumulate in non-target habitats. According to Parmesan and Yohe (2003), climate change is altering the natural rhythms of ecosystems, causing mismatches between species interactions, such as the timing of pollination and the flowering of plants. These mismatches can disrupt food webs and impair ecosystem services, making ecosystems less resilient to additional stressors such as pesticide exposure. Recent research highlights the synergistic effects of pesticide exposure and climate change on ecosystems. These two stressors often interact in ways that amplify their individual impacts. Some pesticides, for instance, can become more volatile and persistent as a result of climate change, which can make them more harmful to species that are not intended targets (Noyes et al., 2009). Precipitation patterns around the world are being significantly impacted by the main human-caused drivers of climate change, which include deforestation and the combustion of fossil fuels. Weather extremes, such as protracted droughts, torrential rainfall, and shifting temperature regimes, are becoming more often and severe as a consequence of this process.

An industry that depends heavily on steady weather conditions, agriculture is seriously threatened by these shifts. Due to these extreme weather events and altering growing seasons, crop yields are becoming more and more unpredictable, which has an impact on global food security and lives. Climate resilience agriculture has become a key strategy to solve these issues. This strategy entails implementing agricultural techniques and innovations intended to endure and adjust to shifting weather patterns (Sarma et al., 2024). In order to withstand unpredictable weather, it will be essential to practice intercropping, which involves strategically planting different crop species together to create a more diverse field and environment. Some examples of systems and approaches that integrate animals and crops to use applied inputs and natural resources more efficiently are permaculture, agroforestry, and alley cropping. With an emphasis on spatial variation, future systems will be customized to specific fields, soil types, and unique agroecozones. Planting a variety of species combinations in the same field will increase production resilience, and creating a variety of goods that can be sold through various channels would increase economic resilience. In the future, the development of local food webs will be more appropriate than the domination of global food chains (Francis, 2019).

These effects are made worse by the fact that climate change is making things worse for ecosystems. Deutsch et al. (2018) discovered that many species are seeing changes in their habitats as a result of increased pests and illnesses, changing precipitation patterns, and more frequent extreme weather events. This, in turn, can lead to greater reliance on pesticides, creating a vicious cycle of ecosystem degradation. Pesticide exposure has long been recognized as a significant threat to ecosystem health. According to studies, pesticides can alter species composition by eliminating sensitive organisms, leading to an imbalance in ecosystems. For instance, aquatic ecosystems are particularly vulnerable to pesticide contamination, as these chemicals often find their way into streams and rivers through agricultural runoff (Schulz, 2004). Pesticides such as neonicotinoids and organophosphates have been implicated in the decline of pollinators like bees, which are critical for maintaining plant biodiversity and crop production (Goulson et al., 2015; Ogidi, 2023). Researchers Beketov et al. (2013) discovered that pesticides, even at low concentrations, can drastically lower biodiversity in freshwater ecosystems, making these places less resilient to other stresses. In addition to lowering species diversity, these chemicals hinder ecological processes including primary production and nutrient cycling, which are crucial to the stability of ecosystems. Also, ecosystems can get even more stressed out from pesticide abuse, which might cause pests to become resistant and necessitate even stronger chemicals (Pimentel, 2005).

2 Pesticide Exposure and Ecosystem Health

Pesticides, which are synthetic mixtures, are employed to eradicate vegetation, insects, rodents, and other species (Sawyer et al., 2023). Various pesticides and herbicides, as well as nematicides, fungicides, molluscicides, rodenticides, and plant development controllers, are employed in agriculture, as outlined by Zhan et al. (2020), Bhatt et al. (2021a), and Zhang et al. (2021). Its primary function is to prevent vector-borne diseases, which encompasses agricultural security, food conservation, and numerous business and food-related practices (Mieldazys et al., 2015; Sharma et al., 2019; Ogidi & Emaikwu, 2024). The term "vermin" is employed to denote any living organism that poses a hazard to humans or other animals. Pesticides are compounds that are employed to either eliminate or impede the proliferation of undesirable organisms.

An estimated 4.19 million metric tons of insecticides were used worldwide in 2019. Following the USA and Brazil were Argentina and China, with the USA contributing 408,000 metric tons, Brazil 377 thousand metric tons, and China 1.76 million metric tons (Fernández, 2021). Cambodia, Laos, and Vietnam are among the 20% of agricultural nations that buy pesticides, according to the World Health Organisation (Schreinemachers & Tipraqsa, 2012; Schreinemachers et al., 2015). According to Khan et al. (2010) and Pozo et al. (2011), India is one of the main Asian nations that produces pesticides. Specifically, 90 thousand tons of organochlorine pesticides, like DDT and benzene hexachloride, are produced annually. From 2010 to 2014, the average yearly usage of total pesticides per kilogram of produce was 0.645 g. There was a benefit-cost ratio of 2.784 kg ha^{-1}. China (10.45 kg ha^{-1}), Mexico (7.87 kg ha^{-1}), Brazil (6.16 kg ha^{-1}), Germany (5.12 kg ha^{-1}), France (4.85 kg ha^{-1}), Joined Realm (4.03 kg ha^{-1}), Joined Status (3.88 kg ha^{-1}), and India (0.26 kg ha^{-1}) had the highest average pesticide usage from 2010 to 2014. Japan came in second. Zhang (2018) is where we got this data. Figure 1 shows that herbicides make up 47.5% of pesticide commitments, while insect sprays, fungicides, and other types of insect poisons account for 29%, 17%, and 5.5%, respectively (Gill & Garg, 2014; Zhang, 2018; Sharma et al., 2019). There are several different ways in which pesticides are categorized. Common criteria for classifying pesticides include their mode of transportation, the makeup of the component, and the target it aims to achieve. Toxic or harmful effects were the basis for pesticide classification by the World Health Organisation (WHO) and the Global Health System (GHS), with a focus on public health.

Fig. 1 Parentage of pesticide distribution

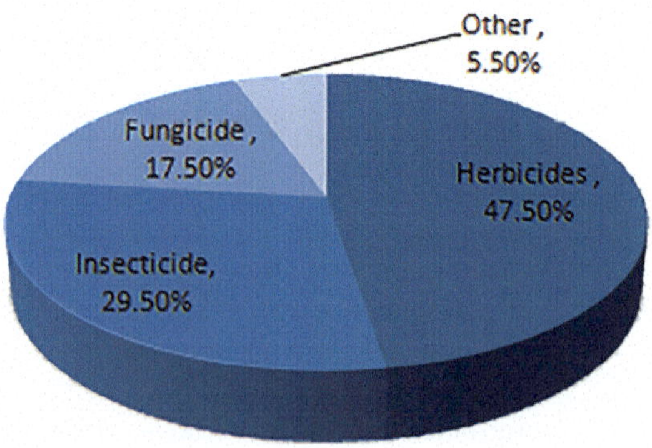

2.1 *Classification of Pesticides*

Pesticides' synthetic plans and amounts reveal their harmfulness to living systems. The term "pesticide" refers to a broad category that includes many different types of chemicals, including those that kill germs, sanitizers, fungi, rodents, and herbicides (Garcia et al., 2012). Based on their chemical and physical characteristics, pesticides are broadly divided into two groups. Typical pesticide classifications (engineered and non-engineered) and follow-up on annoyance kind are outlined. The majority of developed pesticides were composed of natural synthetic substances. Organophosphates, organochlorines, carbamates, and pyrethroids were the four main groups into which they were placed. Quite a few pesticides are ubiquitous. Biopesticides, or naturally occurring pesticides, are shaped by living organisms including plants, bacteria, and growth, according to Abubakar et al. (2020) and Bhatt et al. (2020, 2021b) (Fig. 2).

2.2 *Behavior of Pesticides in the Ecosystem*

The potential for pesticides to disperse and decompose in the environment is present when they are administered locally by ranchers, as a result of the use of natural microbial strains and physicochemical components. Upon accessing the biological system, they produce a diverse array of effects on both designated plants and non-designated animals (Tudi et al., 2021). Pesticides break down in our bodies through a combination of physical and microbiological processes, including exposure to light, heat, humidity, oxygen, and microorganisms. The byproducts of this breakdown process are called metabolites, and the toxic or non-toxic properties of these compounds depend on their synthetic structure (Liu et al., 2015; Marie et al., 2017). Both the pesticides and their metabolites are transported from one place to another through various processes, including adsorption, filtration, volatilization, and surface overflow (Tudi et al., 2021). The sorption framework attraction between soil particles and pesticides, which is affected by soil organic matter and the surface of the soil, has a negative effect on biology and a long half-life on Earth (Qin et al., 2014) (Table 1).

3 Climate Change and Ecosystem Dynamics

The biosphere, the thin layer of life on Earth that is extremely dependent on the hydrosphere and climate and sustains human social orders, is vulnerable to environmental and oceanic changes. Accordingly, degrading or restoring portions of the biosphere will likely result in consequences on a regional or global scale. Outflows of anthropogenic ozone-depleting substances promote climate change and ocean

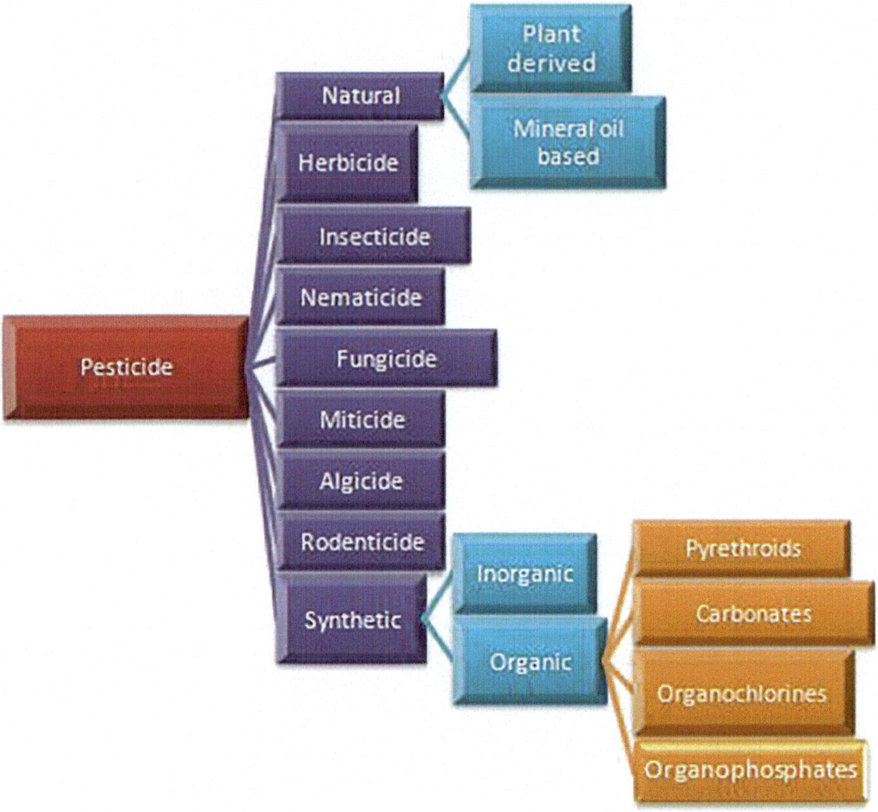

Fig. 2 Classification of pesticides

acidification, which in turn weakens the resilience and adaptability of typically occurring biological systems and the social orders that depend on them (Bennett et al., 2016). These threats have recently become more noticeable, and their effects can be substantial. Earth is currently experiencing a significantly warmer climate and is projected to continue warming in the future unless there is a dramatic shift in the direction of fossil fuel byproducts (Ogidi et al., 2024d).

The structure of Earth's environment and its relationship with the biosphere are being better understood through logical investigation. Environmental change is likely to have negative consequences for both biological systems and people, according to most predictions. Environmental factors are undeniably becoming more noticeable and seem to be increasing in both severity and frequency. Several Earth-based environmental tipping points are now showing signs of activation. The 2018 World Board on Environmental Change Extraordinary Report on 1.5 °C warns that if the planet warms beyond 1.5 °C, it will cause harmful environmental changes for humans and biodiversity, such as droughts, floods, heat waves, and rising ocean

Table 1 Impact of pesticides on ecosystems

Impact area	Physical/chemical factors	Effects	Mechanism
Soil	Molecular weight, ionizability, adsorption coefficients, pH	Adsorption, degradation, reduced microbial diversity	Physical forces (van der Waals, hydrogen bonding), microbial metabolism leading to breakdown
Water	Volatility, lipophilicity, solubility, temperature	Contamination of groundwater, eutrophication	Leaching through soil layers, surface runoff into water bodies causing nutrient loading
Aquatic ecosystems	Bioaccumulation potential, persistence, toxicity levels	Toxic effects on fish, amphibians, and invertebrates	Uptake via gills or skin, incorporation into the food web leading to biomagnification
Human health	Exposure duration, concentration, route of exposure (inhalation, dermal)	Acute and chronic health issues, endocrine disruption	Neurotoxicity (affecting nervous system), respiratory problems, long-term health conditions like cancer
Plant growth	Herbicide mechanism (photosystem II inhibitors), oxidative stress, nutrient uptake interference	Growth inhibition, metabolic disorders, reduced yield	Disruption of photosynthesis, interference with hormone balance, root absorption issues
Biodiversity	Species sensitivity, habitat alteration, food chain dynamics	Decline in non-target species, habitat loss	Disruption of food webs, loss of habitat through direct toxicity or alteration of ecosystems
Soil health	Soil organic matter, microbial activity, soil structure	Reduced fertility, erosion susceptibility	Loss of beneficial microorganisms, changes in soil composition affecting nutrient cycling
Pollinator health	Exposure levels, habitat availability, pesticide formulation	Decline in pollinator populations (e.g., bees)	Neurotoxic effects leading to disorientation, decreased reproduction, habitat loss
Climate impact	Interactions with greenhouse gases, land-use changes	Increased carbon emissions, altered climate feedbacks	Altered land use due to pesticide application can affect carbon sequestration and contribute to climate change

levels. Although 2 °C was formerly the internationally agreed-upon goal, this half-degree difference could reduce the risk of widespread degradation of cold and coral reef ecosystems. The world has about 12 years to cut global net fossil fuel emissions in half to avoid the main effects of a 1.5 °C maximum warming goal, but even if this goal is achieved, the expected effects of warming will likely continue for a very long time, maybe even hundreds of years (Hoegh-Guldberg et al., 2018).

In this topical issue, we present commitments that finish from conversations held at the 2018 Illustrious Society-Public Foundation of Sciences Discussion on Environmental Change and Biological Systems (National Academy of Sciences,

2019). The points of the gathering, mutually coordinated by the two social orders, were to fabricate new open doors for worldwide joint effort, feature the most recent examination discoveries on the central subject, distinguish research holes and future exploration needs, and talk about how research in this field might illuminate global strategy. The gathering analyzed the most recent science on what environmental change can mean for earthly, amphibian, and marine biological systems, frequently in communication with different elements (Lenton et al., 2019). Specifically, it tended to explore wildernesses like the impacts of changes in environment fluctuation and limits; connections among various stressors; edges and the potential for unexpected change; and multi-trophic cooperations, across a scope of earthly, sea-going, and marine biological systems. The discussion likewise considered chances to help and oversee environments to upgrade both their versatility and cultural flexibility to environmental change by investigating a scope of science and strategy aspects (Nicholls et al., 2018). This included how environments can best be figured out how to upgrade their versatility to environmental change, their capacity to change under environmental change, and how biological system the board can be a technique for more broad transformation to change. Subsequently, a focal center was to consider how environment the executives and reclamation can possibly contribute "nature-based arrangements" (NbS) to handle both the causes and outcomes of environmental change. In any case, the adequacy, adaptability, and extent of various nature-based methodologies should be investigated, better comprehended, and assessed.

Environmental change poses threats to biological systems, opportunities exist to make environments more adaptable to these changes, and the rebuilding of biological systems and their environments can aid in the alleviation and variation of these changes; these are the topics that will be covered in the next topical issue, which is also my first encounter with it (Turner et al., 2020) (Table 2).

4 Synergistic Effects of Pesticide Exposure and Climate Change

The environment is facing a significant challenge as a consequence of human-caused global warming. This investigation examines one of the most conspicuous recent outcomes of environmental change: the influence of climate change on the toxicity and ecological transit of synthetic pollutants. An emphasis was placed on the linkages between contaminants and ecological boundaries, temperature, precipitation, and salinity, and how these are influenced by climate change. Unique examination papers, surveys, reports from governments and worldwide organizations, and more were reviewed in a logical fashion. Persistent organic pollutants (POPs), air contaminants, and several types of pesticides were the three most concentrated forms of chemical toxicity on a global scale. The POPs do not include organ chlorine insecticides. In general, as temperatures increase, pollutants will become more

Table 2 Socio-economic consequences on climate change

Aspect	Description	Impacts on agriculture	Impacts on biodiversity	Human health implications
Natural disasters	Extreme events like earthquakes, tsunamis, and cyclones cause significant death tolls and damage.	Reduces crop productivity due to weather extremes and disrupted seasons.	Alters species' habitats, leading to biodiversity loss and species extinction.	Increases vulnerability to Climate-related health issues like heat strokes, respiratory issues, and vector-borne diseases.
Climate change and greenhouse gases	Agriculture contributes 30–40% of all global greenhouse emissions.	Rising temperatures reduce yields, impact water resources, and increase pest invasions.	Disrupts ecosystems, causing habitat loss and affecting food chains.	Promotes the spread of antimicrobial resistance and other diseases.
Decline in crop productivity	Cereals, rice, wheat, and maize are vulnerable to rising temperatures and extreme weather.	Yields of major crops like wheat, rice, and maize expected to decline significantly by 2050.	Climate-driven habitat changes may disrupt species reliant on agricultural landscapes.	Loss of food security can exacerbate malnutrition and poverty-related health issues.
Forestry sector	Climate change affects forest ecosystems, causing fires, droughts, and pest outbreaks.	Reduces forest-based agricultural practices like agroforestry, increasing reliance on non-sustainable practices.	Degradation of forest ecosystems causes loss of species and reduced carbon sequestration.	Deforestation and forest degradation increase exposure to diseases like malaria.
Economic impact of climate change	Developing countries suffer the most due to lack of infrastructure and resilience.	Economic losses due to crop failures and higher costs for adaptation.	Loss of biodiversity impacts tourism, fishing, and ecosystem services.	Healthcare costs rise due to increased diseases, malnutrition, and climate-related health risks.

hazardous and regional convergences of troposphere ozone will increase. Nevertheless, it is logical to assume that synthetic pollution will increase at the same rate (Ogidi & Izah, 2024).

Future research should be guided by apprehensions regarding the potential adverse effects of climate change and air pollution on the health of individuals in contaminated metropolitan areas. A variety of natural processes, such as the melting of ice and snow, alterations in food webs and lipid components, and other climatic changes, could contribute to an increase in POP concentrations (Ogidi & Akpan, 2024). Furthermore, there is irrefutable evidence that wildlife that is exposed to contaminants may experience adverse effects as a consequence of rising temperatures. For example, the biotransformation of contaminants to other beneficial molecules may be altered if the water is too hot, which could disrupt homeostasis. Species that are already on the brink of their physiological resistance range may encounter an even greater challenge in traversing the complex web of interactions between environmental change and pollutants due to the limited space for adaptation. It is anticipated that climate change will result in an increase in global temperatures and modifications to regional precipitation patterns (Noyes et al., 2009).

Pesticide and POP volatilization into the environment may be more prevalent in regions with lower precipitation. The reduction in precipitation will result in an increase in air pollution in urban areas, which has already had a negative impact on health and could become even more severe as temperatures continue to rise. Air pollution levels are generally lower in regions with higher rainfall; however, there is a greater probability of surface deposition of airborne POPs and pesticide run-off. The occurrence of increasingly powerful storms induced by climate change may also lead to a rise in the frequency and severity of synthetic contamination of water bodies and their associated watersheds. As a free stressor, modifications to salinity have the potential to affect marine life. In certain cases, they can even modify the bioavailability or toxicity of synthetic compounds. The identification of the species and communities that are most vulnerable to environmental pollution and the numerous other physical, chemical, and biological pressures that may arise as a consequence of climate change should be a top priority. Additionally, it is imperative to identify inflection points that can initiate or accelerate synergistic relationships between environmental change and pollution openings (Ogidi & Richard-Nwachukwu, 2024).

5 Assessing Ecosystem Vulnerability and Resilience

Versatility and resilience share normal targets. Both locations how biological systems answer worldwide change (e.g., aggravations, dangers, dangers, or stressors (Folke, 2006). The two ideas are worried about the support of specific properties, works, or administrations, characterized by a scope of values or pointers that might address a reference or target state (Eakin & Luers, 2006). A shared characteristic between them is that the two ideas are often utilized in administration and strategy settings. Strength is ordinarily pushed as an objective of woodland the executives to expand the limit of backwoods to adapt to worldwide change Thus, the resilience of timberlands to worldwide change has been broadly surveyed across the board units to distinguish areas of high need for variation measures (Weise et al., 2020). Disregarding the shared characteristics in their goals, flexibility and resilience contrast in their starting point and history, technique, parts, and spatiotemporal concentration (Forzieri et al., 2021). All the more as of late, it has likewise become noticeable in friendly biological appraisals, by combining natural and social aspects and explicitly taking into account transformation (Turner et al., 2020). Paradoxically, the resilience idea grew basically in a more financial setting (at first tending to common security and chance administration against catastrophes), yet its utilization is very different (e.g., inside ecological, geophysical, and political-monetary settings (Lenton et al., 2019). In the woodland natural examination, concentrates connected with flexibility are more normal than those connected with resilience (Folke et al., 2010).

An expansive scope of understandings and strategies of backwoods strength have arisen, remembering the concentration for recuperation from unsettling influence (named designing versatility and the unique upkeep of the framework inside a given attractor or balance state (named biological flexibility) (Adger, 2006). In spite of (or in light of) a wide range of ways of surveying flexibility, the idea has frequently been scrutinized for its absence of versatility and materialness, and various endeavors have been made to unequivocally versatility it (Greiner et al., 2020). Conversely, the resilience idea has been imagined as a functional way to deal with illuminating strategy choices and arranging all along, not least on account of its focal job in IPCC evaluations beginning around 2001 (Standish et al., 2014). Be that as it may, resilience evaluations have a more limited history of utilization with regard to backwoods social-biological frameworks (Holling, 1973).

Versatility and resilience likewise contrast in the systems applied for their evaluation. Strength portrays a unique reaction (ability to endure and recuperate through time from unsettling influences), while resilience is normally centered around depicting a static condition (the capacity to adapt to and adjust to unsettling influence at a given time). Strength concentrates commonly expect to distinguish the component's fundamental versatility—see later—and the related indicators (Mina et al., 2020). Likewise, flexibility is in many cases zeroed in on distinguishing limits and tipping focuses past which a framework moves unexpectedly from a reference state to an elective one (Ogden & Innes, 2009). Rather than recognizing instruments, resilience every now and again involves pointers as heuristic apparatuses for evaluation, and edges are seldom surveyed. Moreover, the principal parts of strength and resilience are unmistakable (Oliver et al., 2015). Obstruction and recuperation depict strength to unsettling influence and follow an express worldly succession (with the opposition being significant during aggravation, and recuperation after aggravation, while resilience centers around vulnerability and versatile limit with transience often being just verifiable. Versatile limit is viewed as seldom in flexibility appraisals and provided that this is true, it is with regard to social-biological strength (normally thought to be a significant component to improve recuperation) (Messier et al., 2021).

Likewise, the consideration of unsettling influence contrasts relying upon the idea. In flexibility, aggravations are recognized as the sort of tension in the face of which we need to keep up with the framework (strength to what?, and they are in many cases not unequivocally remembered for the appraisal but rather will be somewhat thought to be as an outside constraining of the framework) (be that as it may, see (McLaughlin & Dietz, 2008). Resilience thus is many times implanted in a more extensive setting of hazard evaluation, which requires the express measurement of unsettling influence greatness and openness (Spears et al., 2015). These distinctions are additionally clear in the spatiotemporal focal point of the two ideas, as versatility principally centers around the fleeting aspect and resilience on the spatial aspect notwithstanding, see moreover (McCarthy & Intergovernmental Panel on Climate Change, 2001). Resilience will in general be planned, regularly alluding to dangers or risks that poor person yet happened, though versatility is more review, alluding to aggravations and stressors that have proactively happened (yet advancing future flexibility is a shared objective, which as a rule is construed by distinguishing significant strength related processes).

6 Future Directions of Pesticide Exposure and Climate Change

Pesticide accessibility and dangers can be impacted by environmental conditions, as our review shows. Pesticide treatments in this example study caused more localized precipitation occurrences than expected future temperature increases. That is why the Mediterranean area's future pesticide risk assessment models need to account for shifting precipitation patterns, especially for the most severe rainfall occurrences.

The risk assessment concluded that in order to reduce the pesticide risks to frogs, it will be necessary to use natural security measures in the near future, such as the portion reduction strategy advocated by the "Homestead to-Fork" technique. However, even with the 50% reduction, there is still a risk connected with certain pesticides. The results of studies by Alexoaei et al. (2022), Martín et al. (2020), Pavlidis and Karasali (2020), and Silva et al. (2022) suggest that integrated pest management practices, the use of less dangerous substances, or the construction of artificial wetlands can decrease the amount of pesticides released into nearby bodies of water. A recent study by Rodrigo et al. (2022) showed that Albufera Lake can reduce metal loading and pesticide runoff when two artificial wetlands are located near rice-growing areas. To fully comprehend the pesticide transport in the ANP seepage trenches and the potential risks to Albufera Lake's sea-going networks under different climate and pesticide board scenarios, further research is necessary.

There are some limitations to the presented display method, which, from one angle, fails to account for the spatial-temporal arrangement of potentially harmful pesticide metabolites and change items (Li, 2021). Nevertheless, it does provide promising opportunities for pesticide risk prediction in a complex ecological context. Performing the computations for parent mixtures and metabolites and accounting for the impacts of temperature on pesticide change rates in each scenario would have helped to reduce the disparities between the standard and future gamble scenarios. Despite including a few cycles to represent metabolite development, we discovered that the RICEWQ model did not provide enough data to explain how temperature affects the arrangement rate and ecotoxicological dangers of metabolites. As a result, this perspective was disregarded throughout the evaluation.

The fact that we only used concentrated and sustained SSDs to show how the ANP's ocean biological systems responded to a small number of (standard) test species—species that do not reflect the diversity of Mediterranean wetland habitats—is another major restriction. Furthermore, it was expected that the maintenance of these aquatic habitats would remain unchanged for both the environmental scenarios of 2050 and 2100 (i.e., future status assessments will use similar SSD limits). Research by Camp and Buchwalter (2016), Roth et al. (2022), and Vilas-Boas et al. (2021) suggests that pesticide exposure can be mitigated by raising water temperatures by 5 °C or higher. As a result, there may be some real, unacknowledged environmental hazards linked to our gambling estimates. Nevertheless, environmental change and associated extreme weather events are expected to impact the growth of maritime networks (Polazzo et al., 2022). Species that may be more or less vulnerable to particular pesticides will be identified through this screening. Future pesticide risk assessments for the Mediterranean area should look into these viewpoints and maybe include them.

7 Conclusion

The resilience of ecosystems is being tested by the dual challenges of pesticide exposure and climate change. The interaction between these stressors has the potential to undermine ecosystem health and reduce the capacity of ecosystems to provide essential services. Understanding the mechanisms through which pesticides and climate change affect ecosystems is crucial for developing effective mitigation strategies. Integrated approaches, such as IPM and biodiversity conservation, offer promising pathways for reducing these impacts and supporting ecosystem resilience in a changing world.

References

Abubakar, M. S., Zulkifli, R. A., Kamarudin, N. H. A., & Salleh, F. H. M. (2020). Biopesticides: An eco-friendly approach for sustainable agriculture. *Egyptian Journal Of Biological Pest Control, 30*, 76. https://doi.org/10.1186/s41938-020-00268-6

Adger, W. N. (2006). Vulnerability. *Global Environmental Change, 16*(3), 268–281. https://doi.org/10.1016/j.gloenvcha.2006.02.006

Alexoaei, A., Silva, E., Martín, E., Pavlidis, K., & Karasali, H. (2022). Environmental safety and pesticide reduction measures for sustainable agriculture: Integrating climate change predictions. *Environmental Research, 204*, 112126. https://doi.org/10.1016/j.envres.2022.112126

Beketov, M. A., Kefford, B. J., Schafer, R. B., & Liess, M. (2013). Pesticides reduce regional biodiversity of stream invertebrates. *Proceedings of the National Academy of Sciences, 110*(27), 11039–11043. https://doi.org/10.1073/pnas.1305618110

Bennett, E. M., Carpenter, S. R., Gordon, L. J., Ramankutty, N., Balvanera, P., Campbell, B. M., Cramer, W., Foley, J. A., Folke, C., Karlberg, L., & Liu, J. (2016). Bright spots: Seeds of a

good Anthropocene. *Frontiers in Ecology and the Environment, 14*(8), 441–448. https://doi.org/10.1002/fee.1309
Bhatt, P., Bhatt, K., Sharma, A., & Gangola, S. (2020). Enzymatic degradation of organophosphate pesticides by different bacterial strains: An eco-friendly approach. *Ecotoxicology and Environmental Safety, 197*, 110603. https://doi.org/10.1016/j.ecoenv.2020.110603
Bhatt, P., Kumar, A. J., Gangola, S., & Bhandari, G. (2021a). Novel approaches in degradation of organophosphate pesticides through microbial degradation, enzymatic degradation, and photocatalysis. *Frontiers in Microbiology, 12*, 687853. https://doi.org/10.3389/fmicb.2021.687853
Bhatt, P., Huang, Y., Zhang, W., Sharma, A., Chen, S., & Gangola, S. (2021b). Biodegradation of toxic pesticides: Transforming harmful chemicals into harmless molecules. *Journal of Hazardous Materials, 411*, 125152. https://doi.org/10.1016/j.jhazmat.2021.125152
Camp, A. A., & Buchwalter, D. B. (2016). Elevated temperatures increase sensitivity to pesticides in aquatic insects. *Environmental Toxicology and Chemistry, 35*(11), 2806–2814. https://doi.org/10.1002/etc.3469
Chong, J. (2014). Ecosystem-based approaches to climate change adaptation: Progress and challenges. *International Environmental Agreements: Politics, Law and Economics, 14*, 391–405.
Deutsch, C. A., Tewksbury, J. J., Tigchelaar, M., Battisti, D. S., Merrill, S. C., Huey, R. B., & Naylor, R. L. (2018). Increase in crop losses to insect pests in a warming climate. *Science, 361*(6405), 916–919. https://doi.org/10.1126/science.aat3466
Eakin, H., & Luers, A. L. (2006). Assessing the vulnerability of social-environmental systems. *Annual Review of Environment and Resources, 31*, 365–394. https://doi.org/10.1146/annurev.energy.30.050504.144352
Fernández, C. (2021). Global use of pesticides: Trends and future prospects. *Environmental Science & Policy, 122*, 107–117. https://doi.org/10.1016/j.envsci.2021.04.001
Folke, C. (2006). Resilience: The emergence of a perspective for social–ecological systems analyses. *Global Environmental Change, 16*(3), 253–267. https://doi.org/10.1016/j.gloenvcha.2006.04.002
Folke, C., Carpenter, S. R., Walker, B., Scheffer, M., Chapin, T., & Rockström, J. (2010). Resilience thinking: Integrating resilience, adaptability and transformability. *Ecology and Society, 15*(4), 20. https://doi.org/10.5751/ES-03610-150420
Forzieri, G., Dakos, V., McDowell, N. G., Ricotta, C., & Cescatti, A. (2021). Emergent vulnerability to climate-driven disturbances in European forests. *Nature Communications, 12*(1), 1081. https://doi.org/10.1038/s41467-021-21399-7
Francis, C. A. (2019). Crop production resilience through biodiversity for adaptation to climate change. In *Oxford research encyclopedia of environmental science*. Oxford University Press.
Garcia, F. P., Ascencio, S. D., Oyarzun, J. C. G., Hernandez, A. C., & Alavarado, P. V. (2012). Pesticides: Classification, uses, and toxicity. *Measures of Toxicity and Hazard Assessment of Pesticides, 17*, 1–35.
Geiger, F., Bengtsson, J., Berendse, F., Weisser, W. W., Emmerson, M., Morales, M. B., et al. (2010). Persistent negative effects of pesticides on biodiversity and biological control potential on European farmland. *Basic and Applied Ecology, 11*(2), 97–105. https://doi.org/10.1016/j.baae.2009.12.001
Gill, H. K., & Garg, H. (2014). Pesticide: Environmental impacts and management strategies. In M. L. Larramendy & S. Soloneski (Eds.), *Pesticides – Toxic aspects* (pp. 187–230). InTech. https://doi.org/10.5772/57399
Goulson, D., Nicholls, E., Botías, C., & Rotheray, E. L. (2015). Bee declines driven by combined stress from parasites, pesticides, and lack of flowers. *Science, 347*(6229). https://doi.org/10.1126/science.1255957
Greiner, S. M., Bradford, J. B., D'Amato, A. W., & Palik, B. J. (2020). Managing for resilience? Examining management implications of resilience in southwestern national forests. *Journal of Forestry, 118*(4), 433–443. https://doi.org/10.1093/jofore/fvaa015
Hoegh-Guldberg, O., Jacob, D., Taylor, M., Bindi, M., Brown, S., Camilloni, I., Diedhiou, A., Djalante, R., Ebi, K., Engelbrecht, F., Guiot, J., Hijioka, Y., Mehrotra, S., Payne, A., Seneviratne,

S. I., Thomas, A., Warren, R., & Zhou, G. (2018). *Global warming of 1.5°C: An IPCC special report on the impacts of global warming of 1.5°C above pre-industrial levels and related global greenhouse gas emission pathways, in the context of strengthening the global response to the threat of climate change, sustainable development, and efforts to eradicate poverty*. IPCC.

Holling, C. S. (1973). Resilience and stability of ecological systems. *Annual Review of Ecology and Systematics, 4*(1), 1–23. https://doi.org/10.1146/annurev.es.04.110173.000245

Iyiola, A. O., Kolawole, A. S., Ajayi, F. O., Ogidi, O. I., & Ogwu, M. C. (2023). Sustainable water use and management for agricultural transformation in Africa. In *Water resources management for rural development: Challenges and mitigation* (Vol. 1, pp. 287–300). Elsevier. https://doi.org/10.1016/B978-0-443-18778-0.00001-5

Khan, H., Sharma, C., Kathuria, V., & Bhardwaj, P. (2010). Environmental risks associated with pesticide residues in India. *Journal of Environmental Science and Health, Part B, 45*(3), 192–199. https://doi.org/10.1080/03601230903471969

Lenton, T. M., Rockström, J., Gaffney, O., Rahmstorf, S., Richardson, K., Steffen, W., & Schellnhuber, H. J. (2019). Climate tipping points—Too risky to bet against. *Nature, 575*(7784), 592–595. https://doi.org/10.1038/d41586-019-03595-0

Li, Y. (2021). Influence of temperature on pesticide transformation and degradation products: Implications for risk assessment. *Journal of Environmental Management, 297*, 113204. https://doi.org/10.1016/j.jenvman.2021.113204

Liu, Y., Chen, W., Li, Y., & Zhao, X. (2015). Environmental fate of pesticides in the aquatic system. *Journal of Environmental Management, 160*, 332–340. https://doi.org/10.1016/j.jenvman.2015.05.020

Marie, C., Céline, A., Benoit, M., & André, T. (2017). Metabolites of pesticides in surface waters and their risks to humans and aquatic systems: A case study of the Rhône River basin. *Environmental Science and Pollution Research, 24*(14), 12994–13008. https://doi.org/10.1007/s11356-017-8862-3

Martín, E., Pavlidis, K., & Karasali, H. (2020). Effects of climate change on pesticide risk and the need for integrated management in Mediterranean wetlands. *Ecological Indicators, 115*, 106382. https://doi.org/10.1016/j.ecolind.2020.106382

McCarthy, J. J., & Intergovernmental Panel on Climate Change (Eds.). (2001). *Climate change 2001: Impacts, adaptation, and vulnerability: Contribution of Working Group II to the third assessment report of the Intergovernmental Panel on Climate Change*. Cambridge University Press.

McLaughlin, P., & Dietz, T. (2008). Structure, agency and environment: Toward an integrated perspective on vulnerability. *Global Environmental Change, 18*(1), 99–111. https://doi.org/10.1016/j.gloenvcha.2007.05.003

Messier, C., Puettmann, K. J., Chazdon, R., Andersson, K., Angers, V. A., Brotons, L., Filotas, E., Tittler, R., Parrott, L., & Levin, S. A. (2021). For the sake of resilience and multifunctionality, let's diversify planted forests! *Conservation Letters, 15*(1), e12829. https://doi.org/10.1111/conl.12829

Mieldazys, A., Tamkutė, L., Maikštėnienė, S., & Povilaitis, V. (2015). Residue levels of selected pesticides in fruits and vegetables grown in Lithuania. *Food Additives & Contaminants: Part B, 8*(4), 268–275. https://doi.org/10.1080/19393210.2015.1059730

Mina, M., Lloret, F., Martínez-Vilalta, J., & Retana, J. (2020). Network analysis can guide resilience-based management in forest landscapes under global change. *Ecological Applications, 31*(1), e02221. https://doi.org/10.1002/eap.2221

National Academy of Sciences. (2019). *Climate change and ecosystems*. The National Academies Press. Retrieved from https://www.ipcc.ch/report/srccl/

Nicholls, R. J., Hinkel, J., Lincke, D., & van der Pol, T. (2018). Stabilization of global temperature at 1.5°C and 2.0°C: Implications for coastal areas. *Philosophical Transactions of the Royal Society A: Mathematical, Physical and Engineering Sciences, 376*(2119), 20160448. https://doi.org/10.1098/rsta.2016.0448

Noyes, P. D., McElwee, M. K., Miller, H. D., Clark, B. W., Van Tiem, L. A., Walcott, K. C., Erwin, K. N., & Levin, E. D. (2009). The toxicology of climate change: Environmental contaminants in a warming world. *Environment International, 35*(6), 971–986. https://doi.org/10.1016/j.envint.2009.02.006

Ogden, A. E., & Innes, J. L. (2009). Application of structured decision making to an assessment of climate change vulnerabilities and adaptation options for sustainable forest management. *Ecology and Society, 14*(1), 11. https://doi.org/10.5751/ES-02715-140111

Ogidi, O. I. (2023). Sustainable utilization of important medicinal plants in Africa. In *Sustainable utilization and conservation of Africa's biological resources and environment. Sustainable development and biodiversity*. Springer Nature Singapore Pte Ltd. https://doi.org/10.1007/978-981-19-6974-4_12

Ogidi, O. I., & Akpan, U. M. (2023). Impacts of chemical use in agricultural practices: Perspectives of soil microorganism and vegetation. In *One Health implications of agrochemicals and their sustainable alternative*. Springer Nature Singapore Pte Ltd.

Ogidi, O. I., & Akpan, U. M. (2024). Microbial and phyto-remediation of crude oil contaminated soil. In *Eco-restoration of the environment*. CRC Press, Routledge Taylor and Francis Group.

Ogidi, O. I., & Emaikwu, N. G. (2024). Plant phenolic compound isolation and its bioinformatics approaches of molecular mechanisms in antimicrobial activities and resistance. In *Computational approaches in biomaterials, bioinformatics and biomedical engineering applications*. CRC Press, Taylor and Francis Group.

Ogidi, O. I., & Izah, S. C. (2024). Emerging technologies in the Global South food industry: Prospects and challenges. In *Food safety and quality in Global South*. Springer Nature Singapore Pte Ltd.

Ogidi, O. I., & Richard-Nwachukwu, N. (2024). Microbe assisted remediation of pesticide residues from soil and water. In *Eco-restoration of the environment*. CRC Press, Routledge Taylor and Francis Group.

Ogidi, O. I., & Tawariowei, A. M. (2023). Endocrine disrupting chemicals and their role in cancer-a review. *Molecular Sciences and Applications, 3*, 5–14. https://doi.org/10.37394/232023.2023.3.2

Ogidi, O. I., Izah, S. C., & Akpan, U. M. (2023). Food quality and agrochemical use: Integrated monitoring, assessment, and management policies. In *One health implications of agrochemicals and their sustainable alternative*. Springer Nature Singapore Pte Ltd.

Ogidi, O. I., Ajoko, I. T., & Tawariowei, A. M. (2024a). Sustainable application of genetic ecotoxicological techniques. In *Biomonitoring for environmental sustainability. Biomonitoring of pollutants in the Global South*. Springer Nature Singapore Pte Ltd.

Ogidi, O. I., Angaye, T. C. N., & Carbom, H. E. (2024b). Microorganisms in the monitoring of terrestrial and aquatic ecosystems: Applications, challenges and future prospects. In *Biomonitoring of pollutants in the Global South*. Springer Nature Singapore Pte Ltd.

Ogidi, O. I., Izah, S. C., & Tobia, P. S. (2024c). Biosurfactant production from cassava processing wastes. In *Sustainable cassava: Strategies from production through waste management*. Elsevier.

Ogidi, O. I., Onwuagba, C. G., & Richard-Nwachukwu, N. (2024d). Biomonitoring tools, techniques and approaches for environmental assessments. In *Biomonitoring of pollutants in the Global South*. Springer Nature Singapore Pte Ltd.

Oliver, T. H., Heard, M. S., Isaac, N. J., Roy, D. B., Procter, D., Eigenbrod, F., et al. (2015). Biodiversity and resilience of ecosystem functions. *Trends in Ecology & Evolution, 30*(11), 673–684. https://doi.org/10.1016/j.tree.2015.08.009

Parmesan, C., & Yohe, G. (2003). A globally coherent fingerprint of climate change impacts across natural systems. *Nature, 421*(6918), 37–42. https://doi.org/10.1038/nature01286

Pavlidis, K., & Karasali, H. (2020). Climate variability and the importance of adaptive pest management practices for sustainable agricultural landscapes. *Agricultural Systems, 183*, 102870. https://doi.org/10.1016/j.agsy.2020.102870

Pimentel, D., Acquay, H., Biltonen, M., Rice, P., Silva, M., Nelson, J., et al. (1992). Environmental and economic costs of pesticide use. *BioScience, 42*(10), 750–760. https://doi.org/10.2307/131199

Pimentel, D., Zuniga, R., & Morrison, D. (2005). Update on the environmental and economic costs associated with alien-invasive species in the United States. *Ecological economics, 52*(3), 273–288.

Polazzo, F., Vilas-Boas, P., & Roth, N. (2022). Resilience of aquatic ecosystems under extreme weather conditions: Impacts on species sensitivity to pesticides. *Science of the Total Environment, 827*, 154276. https://doi.org/10.1016/j.scitotenv.2022.154276

Pozo, K., Harner, T., Lee, S. C., Wania, F., Muir, D. C., & Jones, K. C. (2011). Assessing seasonal and spatial trends of persistent organic pollutants (POPs) in the global atmosphere using passive samplers. *Environmental Science & Technology, 45*(6), 1452–1459. https://doi.org/10.1021/es303744q

Pretty, J., & Pervez Bharucha, Z. (2015). Integrated pest management for sustainable intensification of agriculture in Asia and Africa. *Insects, 6*(1), 152–182.

Qin, G., Gong, J., Fan, M., & Yu, H. (2014). Sorption and degradation of pesticides in soils and the environmental fate of pesticide residues: A review. *Journal of Agricultural and Food Chemistry, 62*(27), 6634–6648. https://doi.org/10.1021/jf5009157

Rodrigo, A., Vilas-Boas, P., & Silva, E. (2022). Role of constructed wetlands in reducing pesticide and metal loads from agricultural drainage: A case study in Mediterranean rice fields. *Journal of Environmental Quality, 51*(3), 721–732. https://doi.org/10.1002/jeq2.20318

Root, T. L., Price, J. T., Hall, K. R., Schneider, S. H., Rosenzweig, C., & Pounds, J. A. (2003). Fingerprints of global warming on wild animals and plants. *Nature, 421*(6918), 57–60.

Rosenzweig, C., Iglesius, A., Yang, X. B., Epstein, P. R., & Chivian, E. (2001). Climate change and extreme weather events-Implications for food production, plant diseases, and pests.

Roth, N., Polazzo, F., & Vilas-Boas, P. (2022). Aquatic species sensitivity to temperature-induced pesticide exposure under climate change scenarios. *Environmental Pollution, 292*, 118333. https://doi.org/10.1016/j.envpol.2021.118333

Sarma, H. H., Borah, S. K., Dutta, N., Sultana, N., Nath, H., & Das, B. C. (2024). Innovative approaches for climate-resilient farming: Strategies against environmental shifts and climate change. *International Journal of Environment and Climate Change, 14*(9), 217–241.

Sawyer, W. E., Jacob, D. E., Ogidi, O. I., Izah, S. C., & Popoola, A. M. (2023). Threats and conservation strategies of common edible vegetables that possess pharmacological potentials in Nigeria. In *Reference series in phytochemistry. Herbal medicine phytochemistry*. Springer Nature Singapore Pte Ltd.

Scholz, N. L., Fleishman, E., Brown, L., Werner, I., Johnson, M. L., Brooks, M. L., & Schlenk, D. (2012). A perspective on modern pesticides, pelagic fish declines, and unknown ecological resilience in highly managed ecosystems. *BioScience, 62*(4), 428–434.

Schreinemachers, P., & Tipraqsa, P. (2012). Agricultural pesticides and land use intensification in high, middle, and low income countries. *Food Policy, 37*(6), 616–626. https://doi.org/10.1016/j.foodpol.2012.07.002

Schreinemachers, P., Schad, I., Tipraqsa, P., Williams, P. M., Neef, A., Riwthong, S., Sangchan, W., & Grovermann, C. (2015). Can public GAP standards reduce agricultural pesticide use? The case of fruit and vegetable farming in northern Thailand. *Agriculture and Human Values, 32*(4), 603–619. https://doi.org/10.1007/s10460-014-9571-2

Schulz, W. (2004). Reconstructing mediatization as an analytical concept. *European journal of communication, 19*(1), 87–101.

Sharma, A., Kumar, V., Shahzad, B., Tanveer, M., Sidhu, G. P. S., Handa, N., Kohli, S. K., Yadav, P., Bali, A. S., Parihar, R. D., & Thukral, A. K. (2019). Worldwide pesticide usage and its impacts on ecosystem. *SN Applied Sciences, 1*(11), 1446. https://doi.org/10.1007/s42452-019-1485-1

Silva, E., Rodrigo, A., & Pavlidis, K. (2022). The impact of pesticide use and extreme weather events on Mediterranean aquatic ecosystems: Integrating climate predictions in risk assessments. *Science of the Total Environment, 828*, 154254. https://doi.org/10.1016/j.scitotenv.2022.154254

Spears, B. M., Adriaenssens, V., Burthe, S. J., Davidson, T. A., Gunn, I. D., Hawkes, J. A., et al. (2015). FORUM: Effective management of ecological resilience—Are we there yet? *Journal of Applied Ecology, 52*(5), 1311–1315. https://doi.org/10.1111/1365-2664.12497

Standish, R. J., Hobbs, R. J., Mayfield, M. M., Bestelmeyer, B. T., Suding, K. N., Battaglia, L. L., et al. (2014). Resilience in ecology: Abstraction, distraction, or where the action is? *Biological Conservation, 177*, 43–51. https://doi.org/10.1016/j.biocon.2014.06.008

Staudt, A., Leidner, A. K., Howard, J., Brauman, K. A., Dukes, J. S., Hansen, L. J., ... & Solórzano, L. A. (2013). The added complications of climate change: understanding and managing biodiversity and ecosystems. *Frontiers in Ecology and the Environment, 11*(9), 494–501.

Tscharntke, T., Tylianakis, J. M., Rand, T. A., Didham, R. K., Fahrig, L., Batáry, P., & Westphal, C. (2012). Landscape moderation of biodiversity patterns and processes-eight hypotheses. *Biological reviews, 87*(3), 661–685.

Tudi, M., Ruan, H. D., Wang, L., Lyu, J., Sadler, R., Connell, D., & Phung, D. T. (2021). Agriculture development, pesticide application and its impact on the environment. *International Journal of Environmental Research and Public Health, 18*(3), 1112. https://doi.org/10.3390/ijerph18031112

Turner, M. G., Donato, D. C., & Romme, W. H. (2020). Climate change, ecosystems, and abrupt change: Science priorities. *Philosophical Transactions of the Royal Society B: Biological Sciences, 375*(1794), 20190105. https://doi.org/10.1098/rstb.2019.0105

Vilas-Boas, P., Roth, N., & Polazzo, F. (2021). Climate change and pesticide risk in aquatic environments: Exploring temperature-induced shifts in sensitivity. *Environmental Toxicology and Chemistry, 40*(6), 1605–1612. https://doi.org/10.1002/etc.5014

Weise, H., Ostergard, H., & Heilmann-Clausen, J. (2020). Resilience trinity: Safeguarding ecosystem functioning and services across three different time horizons and decision contexts. *Oikos, 129*(3), 445–456. https://doi.org/10.1111/oik.06993

Zhan, H., Sun, Y., Wang, Y., & Zhu, L. (2020). Selective removal of organophosphate pesticides from aqueous solution by magnetic molecularly imprinted polymers. *Journal of Hazardous Materials, 384*, 121323. https://doi.org/10.1016/j.jhazmat.2019.121323

Zhang, W. (2018). Global pesticide use: Profile, trend, cost/benefit, and more. *Proceedings of the International Academy of Ecology and Environmental Sciences, 8*(1), 1–27. https://doi.org/10.13140/RG.2.2.18329.06248

Zhang, X., Xu, X., Yin, Y., Liu, Z., & Cheng, H. (2021). Ecological risk assessment of pesticide residues in agricultural products: A case study in the Pearl River Delta, China. *Ecotoxicology and Environmental Safety, 208*, 111678. https://doi.org/10.1016/j.ecoenv.2020.111678

Pesticide Runoff and Its Impact on Water Quality

Sayed Rashad, Ghadir A. El-Chaghaby, and Muhammad Abdul Moneem

Abbreviations

AOPs	Advanced oxidation processes
ChF	Chlorpyrifos
DDT	Dichlorodiphenyl trichloroethane
EPA	Environmental Protection Agency
EPF	Entomopathogenic fungi
IPM	Insect pest management
NIPs	National implementation plans
OCPs	Organochlorine pesticides
pg/L to ng/L	Picogram/liter to nanogram/liter
UAV	Unmanned aerial vehicles
USDA	United States Department of Agriculture
WHO	World Health Organization

1 Introduction

Pesticide use in aquatic environments is a pervasive issue that is causing concern throughout the world because it can build up in the bodies of aquatic organisms and sediment soil, endangering human health. Due to the widespread use of pesticides in both agricultural and non-agricultural sectors, the global pesticide market has

S. Rashad (✉) · G. A. El-Chaghaby
Regional Center for Food and Feed, Agricultural Research Center, Giza, Egypt

M. Abdul Moneem
El-Fostat Laboratory, Cairo Water Company, Cairo, Egypt

B. R. Babaniyi, E. E. Babaniyi (eds.), *The Interplay of Pesticides and Climate Change*, https://doi.org/10.1007/978-3-031-81669-7_4

grown rapidly, leading to the introduction of numerous pesticide formulations. Pesticides end up in the water body because of runoff from industrial wastewater and agricultural fields. Particularly during the precipitation event, soluble pesticides are carried away by water molecules, which eventually seeped into the soil layers, surface, and groundwater (Islam et al., 2022). Pesticides are essential tools for protecting crops against harmful pests and diseases, contributing to the global population's standard of living. An average of two million tons of pesticides are used annually, with herbicides and insecticides being the most common types. The pest control revolution began in the 1970s with the development of toxic heavy metal-based pesticides, followed by the discovery of dichlorodiphenyl trichloroethane (DDT) during World War II. DDT's effectiveness against almost all pest species at low dosages led to its widespread use, but its adverse impact on the environment and mankind became apparent. Synthetic pesticides, such as organophosphate and pyrethroid, have been produced but are still toxic to the environment (Tudi et al., 2021). The continuous and excessive use of pesticides harms non-target species and causes pesticide residues to appear at unexpected sites. Under constant chemical pressure, pesticides have led to the development of resistant strains, causing pests and insects to become immune to the pesticide. Pesticides have benefits such as increased food quality and reduced insect-borne diseases, but they also pose potential environmental and water resource threats. The dissolution of pesticides depends on the nature of the compound, application techniques, and climatic factors. Nondegrading pesticides accumulate in soils or move from one site to another, posing unknown toxicity to human health (Syafrudin et al., 2021). The impact of climate change on precipitation patterns and water availability and quality will be covered in this chapter. Examine how aquatic ecosystems and human health are impacted by pesticide drift, leaching, and runoff, especially in light of changing hydrological cycles.

2 Pesticides, Definition, and History

Pesticides are chemicals that are intended to kill pests. A chemical or biological agent that deters, neutralizes, or eradicates pests such as bacteria, viruses, antimicrobials, or disinfectants is commonly referred to as a pesticide. The terms "plant protection product" and "pesticide" are often used interchangeably due to the widespread use of pesticides. Various agricultural pests that can damage crops and livestock and reduce farm productivity are commonly managed or eliminated with their help. Typically, pesticides are used to control mold, mildew, and fungi; they are also used to kill insects, weeds, and rodents (Tudi et al., 2021).

Pesticides are defined as toxic chemicals, chemical mixtures, or biological agents that are intentionally released into the environment to prevent, discourage, control, and/or eradicate populations of insects, weeds, rodents, fungi, or other unwanted pests. The way pesticides work is by attracting, enticing, and finally killing or controlling the pests. Pests are generally defined as "the plants or animals that jeopardize our food, health, and/or comfort (Hassaan & El Nemr, 2020)."

The invention of pesticides is not new. To keep insects and other pests away from their crops, numerous ancient societies employed pesticides. Elemental sulfur was used by the ancient Sumerians to keep insects away from their crops. In contrast, farmers in the middle Ages experimented with chemicals on common crops by using lead and arsenic.

To get rid of body lice and other pests, the Chinese used compounds containing arsenic and mercury, whereas the Greeks and Romans employed substances like oil, ash, and sulfur, among others, to keep pests away from their crops, livestock, and themselves. In the meantime, nineteenth-century researchers concentrated more on natural methods utilizing compounds derived from chrysanthemum and tropical vegetable roots (Oberemok et al., 2015).

3 Classification of Pesticides

Pesticides are crucial for averting crop losses and diseases that affect humans. They belong to the categories of mitigating, repelling, and destroying agents. Unfortunately, excessive use of commercial pesticides has made insects and other pests immune to them. In addition to improving integrated pest management and providing thorough preventive and corrective treatments, chemical pesticides and insecticides are increasingly being used as the primary means of eradicating pests when paired with potent natural enemies (Pathak et al., 2022). Based on several criteria, pesticides have been categorized into several classifications (Fig. 1).

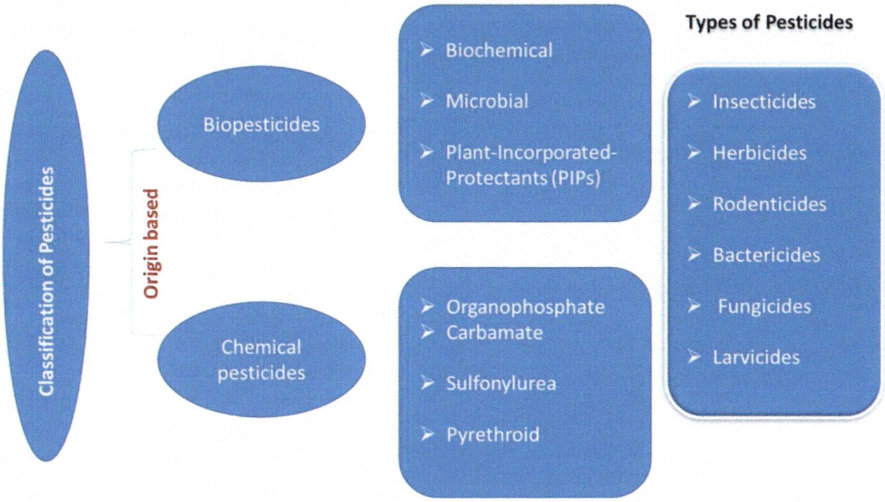

Fig. 1 Pesticide classification

Pesticide effects are contingent upon exposure, toxicity, and other variables such as life history, characteristics, application timing, population density, and landscape architecture. Neuroactive insecticides specifically target nerve targets, including voltage-gated sodium channels for dichlorodiphenyltrichloroethane and pyrethroids, nicotinic acetylcholine receptors for neonicotinoids, gamma-aminobutyric acid receptor channels for polychlorocyclohexanes and fiproles, and acetylcholinesterase for organophosphates and methylcarbamates (Fayaz et al., 2024).

Table 1 Types of pesticides

Types of pesticides	
Grouped by target pest species	Avicide—birds Bactericides—bacteria Fungicide—fungi Herbicide—plants and weeds Insecticide—insects Larvicides—larvae Miticide—mites Molluscicide—snails, slugs Nematicide—nematodes Rodenticides—rodents (rats and mice)
Grouped by their biodegradability	*Biodegradable*: Biodegradable materials are those that microorganisms and other living things can convert into innocuous compounds *Persistent*: On the other hand, persistent ones might take years or even months to decompose
Grouped by their chemistry	*Organophosphate*: Insecticides make up the majority of organophosphates; they act on the nervous system by interfering with the enzyme that controls a neurotransmitter *Carbamate*: By interfering with an enzyme that controls neurotransmitter function, carbamate pesticides act on the nervous system similarly to organophosphorus pesticides. The effects of enzymes can, however, usually be reversed *Organochlorine insecticides*: Previously widely used, these pesticides—such as DDT, chlordane, and toxaphene—have been banned from sale in many nations due to their persistent negative effects on human health and the environment These are synthetic forms of pyrethrin, a pesticide that is naturally present in chrysanthemum flowers. To maximize their stability in the environment, they were developed *Sulfonylurea herbicides*: Products like pyrithiobac-sodium, cyclosulfamuron, bispyribac-sodium, terbacil, and sulfometuron-methyl have been commercially available for the control of weeds. The following sulfur compounds are sulfosulfuron, rimsulfuron, pyrazosulfuron-ethyl, imazosulfuron, nicosulfuron, oxasulfuron, nicosulfuron, flazasulfuron, primisulfuron-methyl, halosulfuron-methyl, flupyrsulfuron-methyl-sodium, ethoxysulfuron, chlorimuron-ethyl, bensulfuron-methyl, azimsulfuron, and amidosulfuron *Biopesticides*: These are specific kinds of insecticides made from naturally occurring substances like bacteria, plants, animals, and minerals.

Pesticides are mostly divided into herbicides, insecticides, fungicides, bactericides, and rodenticides according to their intended use (Table 1). Pesticides can also be classified into pyrethroids, carbamates, dithiocarbamates, amides, coumadin compounds, phenoxyl, triazine, and organochlorines (OC) based on their chemical properties.

Pesticides used worldwide can be classified as herbicides, insecticides, fungicides, rodenticides, molluscicides, nematicides, or regulators of plant growth. Pesticide use that is not controlled has resulted in environmental catastrophes, endangering biodiversity and human health. Reducing the lethal nature of pesticides is challenging because of their increased water solubility, heat stability, and polarity. Depending on the target species, they can be toxic to aquatic life, natural flora, and wildlife (Benbrook et al., 2021).

4 The Role of Pesticides

Pesticides have shown to be extremely helpful in a number of areas, including agriculture and public health. Pesticides are used in public health to eradicate pests like rats, mice, ticks, and mosquitoes, which lowers the burden of diseases carried by these vectors. Additionally essential to agricultural productivity are insecticides, which suppress weeds and insects and significantly boost the yield of agricultural products (Kaur et al., 2024).

Since the start of the twentieth century, the rate of population growth worldwide has increased dramatically; by 2050, an additional 9.4–10 billion people are expected to live on the planet. The use of pesticides, which lower harvest losses brought on by weeds, illnesses, and insect pests, is primarily to blame for this growth. Pesticides are used in the production of about one-third of agricultural products; fruit, vegetable, and cereal production would all suffer greatly in their absence. Around the world, pesticides are essential for lowering disease rates and increasing crop yields, which helps to end hunger and increase access to a plentiful supply of high-quality food. These advantages have been made possible by the use of pesticides, which will help ensure a more sustainable future for all people (Tudi et al., 2021).

In the long run, using pesticides can improve food safety, nutrition and health, increase farm and agribusiness revenue, improve quality of life, and increase the range of crops that can be grown. It also lowers medical and veterinary expenses, encourages a population that is more fit, and lowers greenhouse gas emissions. Eliminating pests from pastures can also boost livestock productivity. Because affordable, safe, and nutrient-rich food improves health and life expectancy, farming families can also make more money by using pesticides properly (Garrigou et al., 2020).

5 Environmental Behavior of Pesticides

Pesticides can release new chemicals into the environment during their transfer and degradation processes. They can disperse from the target site to other environmental media or non-target plants by adsorption, leaching, volatilization, spray drift, and runoff (Pérez-Lucas et al., 2018). Figure 2 summarizes the dispersion routes of pesticides into the environment.

5.1 *Pesticide Degradation*

Pesticides are compounds that are sprayed on living things and broken down by light, chemical reactions, or bacteria. The rate at which the pesticide degrades can vary greatly, ranging from hours to days or even years, contingent upon the surrounding circumstances and the chemical makeup of the agent (Ray et al., 2024). It produces various metabolites and regulates the persistence of pesticides in soils, giving the pesticide a half-life in the environment. For instance, the metabolite of chlorpyrifos is more mobile and toxic than that of its parent. Groundwater, sediments, and soils are common places to find chlorpyrifos and its breakdown products, which are regarded as endocrine disruptors and may be harmful to human health (Wołejko et al., 2022).

Fig. 2 Pesticides dispersion routes into the surrounding environment

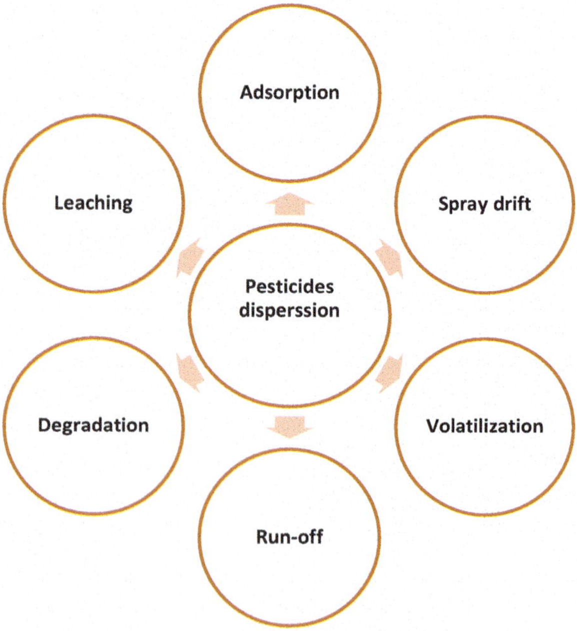

Pesticides can degrade in three different ways: chemically, biologically, or via photodegradation. Pesticides can be broken down by microorganisms through a process known as microbial degradation, or by chemical reactions that occur in the soil. A number of variables, including soil temperature, pH, moisture content, and insecticide binding to the soil, affect the type and rate of chemical degradation. In contrast, photodegradation occurs when pesticides are exposed to sunlight, and their rate of degradation varies based on the light's intensity, length of exposure, and insecticide characteristics (Ortiz-Hernández et al., 2013).

5.2 *Pesticide Sorption*

Pesticide use can severely contaminate soil, despite the fact that it is necessary in the fight against plant diseases. Through a process known as sorption, which is impacted by a number of factors including pH, organic matter, and soil amendment, pesticides are bound to soil particles. Soils rich in organic matter or clay are more adsorbent to pesticides than coarse, sandy soils because of their larger particle surface area or more sites for insecticide binding. The adsorption and desorption capacities of endosulfan may be affected by the amount of clay and organic matter present in the soil (Rasool et al., 2022).

Numerous soil parameters, such as pH, organic matter content, and cationic exchange capacity, have been found to influence how well fungicides adsorb and desorb. Moisture has an impact on pesticide adsorption as well; dry soils absorb insecticides more quickly than wet soils. Temperature affects ammonium nitrogen adsorption as well, but humic acid colloid affects DDT adsorption in sediments. Some pesticides can be absorbed by plants during growth and stay in the soil for a long time, which could damage crops or leave residues. Simple binding is facilitated by the attraction of positively charged pesticide molecules to negatively charged clay particles (Cara et al., 2022).

5.3 *Pesticide Leaching*

Despite being extensively used all over the world, some pesticides have the ability to contaminate water by seeping into groundwater. Among the factors influencing leaching are solubility, soil permeability, adsorption coefficient (Koc), and half-life in aerobic soil (DT50). The duration of an insecticide's presence in the environment has an effect on leaching as well. For instance, imidacloprid is persistent and exhibits strong environmental fate characteristics (Taidi & El laghdach, 2024). The leaching properties of pesticides are also influenced by meteorological factors, such as average annual rainfall and temperature. While temperature affects soil evapotranspiration and thus the behavior of pesticide leaching, precipitation affects the flux of downward leaching to groundwater. Water percolation and the movement of pesticides into groundwater are influenced by soil properties, including soil texture and organic content. The primary factor influencing pesticide and water movement transport is soil texture (Steffens et al., 2013).

Phenazines break down under specific circumstances determined by the pH, the amount of organic matter in the soil, and the presence of anaerobic microorganisms. There is little to no relationship between pesticide adsorption and the pH, C, and N contents of soil organic matter, and the organo-mineral complex of the soil. Despite a multidimensional spatial association, adsorption is not correlated with particle sizes or cation exchange capacity. It is also necessary to consider the possibility of evaporation when considering the effects of crop residues on soil water and temperature regimes (Ou et al., 2020).

5.4 Pesticide Spray Drift

Pesticides can contaminate food and the surrounding environment through spray drift, or the movement of the spray droplets. Aquatic environments are especially vulnerable to pesticide residues like chlorpyrifos (ChF), which can be toxic to aquatic life. It has been shown through research that tilapia tissues exposed to sublethal levels of ChF can undergo histological changes and oxidative stress. Application of unmanned aerial vehicles (UAV) for low-volume droplet delivery has become more common; however, the associated spray drift has not received much attention, particularly for fine droplets generated by spinning disk nozzles. The environment and regulations are affected by this (Ray & Shaju, 2023).

5.5 Pesticide Volatilization

The process by which a solid or liquid turns into a gas is known as volatilization. The term volatilization refers to this procedure. The elements that influence volatilization are variations in temperature, humidity, airflow, and soil characteristics. The insecticide's increased volatility is caused by low relative humidity, air movement, high temperatures, and high vapor pressure, respectively. The spread of organochlorine pesticides (OCPs) is more likely in tropical regions than in temperate ones. Tightly adsorbed insecticides on soil particles reduce their volatilization. Contaminated surface waters may be the main source of human exposure to OCPs through volatilization.

5.6 *Pesticide Surface Runoff*

Pesticides traverse in runoff or water over a sloping surface. Pesticides can travel in the form of substances that dissolve in water or cling to soil particles that are eroding. This is closely related to a number of variables, such as the area's grade or slope, edibility, soil texture and moisture content, rainfall quantity and timing, and irrigation. When water is added to a field at a speed that prevents the soil from absorbing it, runoff results. When surface water levels are too high due to over-irrigation, insecticide runoff occurs. Pesticide pollution in streams, ponds, lakes, and wells is caused by pesticide runoff, and pesticide contamination may have detrimental effects on people, animals, and plants.

6 Pesticides Ecotoxicology and Environmental Pollution

Pesticides not only accumulate in crops but may be carried by air, soil, and water over far distances, establishing an enormous pollution source in ecosystems (Fig. 3).

Pesticide use is causing direct or indirect contamination of the agroecosystem. Pesticides have the ability to kill a variety of non-target organisms, such as beneficial insects, birds, fish, and non-target plants, in addition to reducing harmful insects, pathogens, or weeds. Herbicides can also pose a risk to living things that are not the intended target, although insecticides are typically the pesticide class that is most acutely toxic. Pesticide residues cause disruption to agricultural systems when they contaminate the soil, surface, and ground waters, and enter the ecosystem (Ali et al., 2021). In a similar vein, pesticide transformation products build up in soils and have the potential to disrupt the beneficial soil microbial life for our crop system. The presence of heavy pesticide deposits in agricultural soils and their environs

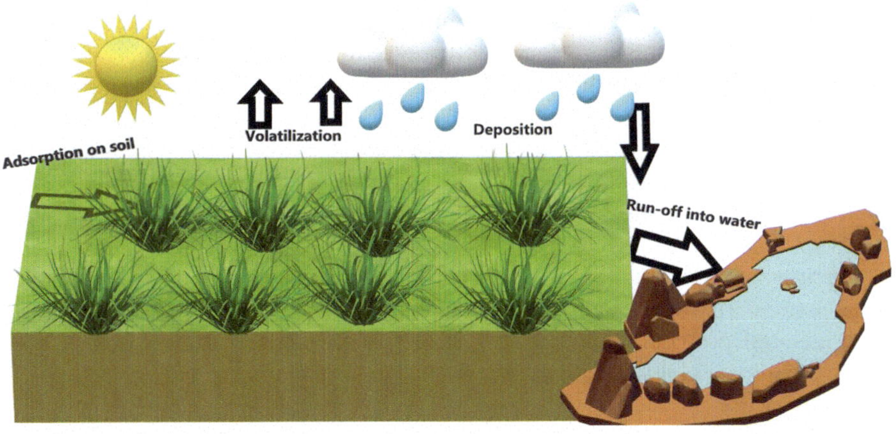

Fig. 3 Pesticides in the environment

is causing a decline in beneficial microfauna. For instance, heavy soil herbicide and insecticide applications disrupt the bacteria that fix nitrogen. Mycorrhizal fungi have also been shown to have similar effects. Future crop productivity may be affected by decreased populations of beneficial microbes brought on by prolonged pesticide exposure (Mustapha et al., 2018).

Due to their frequent and unselective application, pesticides build up in the soil and alter its microbiome and properties through adsorption–desorption courses, transformation products, and degradation processes. Degrading pesticides changes the microorganisms and soil textures, which affects the soil ecosystem and fertility. Pesticides work by modifying the growth, colonization capacity, and metabolic processes of soil microorganisms that invade roots. Pseudomonas strains in agricultural soil are reduced by herbicides such as chlorsulfuron, sulfonyl-urea, metsulfuron, and thifensulfur on methyl, which is essential for soil fertility. Additionally, by influencing specific soil microbial organisms or their enzyme activities, pesticide residues disrupt soil biochemical reactions such as nitrogen-fixation, ammonification, and nitrification. Pesticides have an impact on organic matter and soil mineralization, which control soil productivity. To reduce their harmful effects, it is essential to evaluate how soil microbes and enzyme activities react to pesticide depositions (Pathak et al., 2022).

Pesticides have benefits, but they can also be harmful to people and the environment. Numerous substances are dangerous, bioaccumulative, and stable in the environment. Pesticides can linger in the environment for years due to their persistence. Pesticides are employed in the management of pests and weeds. Large-scale pesticide losses are caused by spray drift, off-target deposition, runoff, and photodegradation. These losses can affect particular species, communities, ecosystems, and people. Another issue is that many medications may not affect organisms in a way that is immediately apparent at low dosages. Nevertheless, they might result in serious consequences, like genetic defects and physiological alterations that eventually shorten lifespans. Due to their high toxicity and potentially harmful effects at low concentrations (pg/L to ng/L), many pesticide types found in natural waters and wastewater resources pose a serious environmental risk. Pesticides can bring on burns, nausea, or convulsions. Shortly after exposure, acute effects appear and are probably fairly severe (Ahmad et al., 2024).

Pesticides are substances, either synthetic or natural, that are used to prevent, control, and get rid of weeds, insects, and other pests that hinder the growth of plants. The demand for pesticides has surged globally since the mid-1940s as a result of commercial farming. It is present in fruits, vegetables, processed foods, water, air, and soil. Excessive and uncontrolled use has led to food contamination and environmental, agricultural, and aquatic pollution. Serious public health concerns, particularly in developing nations, are the acute and long-term health effects of agricultural pesticide exposure and dietary consumption. Chemical pesticides may cause mutagenicity, cytotoxicity, and cancer (Pathak et al., 2022).

Pesticides can produce reactive oxygen species, which reduce antioxidant levels and their ability to protect cells from oxidative damage. Reactive oxygen species and oxidative stress have long-term effects on health. The frequent and negligent use of pesticides has detrimental effects on human health, raising the risk of illnesses such as diabetes, leukemia, Alzheimer's, Parkinson's, neurotoxicity, and various cancers. Contrarily, pesticides provide a host of advantages for increased agricultural output. They safeguard the amount and caliber of harvested crops, boosting yields and lowering losses due to pests, illnesses, and weeds. Food security is aided by this. Pesticides are used in public health to manage disease-carrying insects such as ticks, mosquitoes, and fleas, thereby preventing the spread of illnesses like Lyme disease, dengue fever, and malaria (Sule et al., 2022).

7 Pesticides and Aquatic Ecosystem

Pesticide runoff from treated soil and plants can contaminate surface water bodies, potentially disrupting aquatic life over time. The US Geological Survey claims that pesticide pollution in urban water resources is greater than that in agricultural water bodies. The groundwater system becomes accumulated with pesticides that seep from soil surfaces. In a US Geological Survey study, 90% of the well water samples were found to be contaminated with pesticides (Parvin & Tareq, 2021).

Pesticides can enter water resources through unintentional spills; waste effluents; surface runoff and shifting from pesticide-applied soils; cleaning of spray machinery after spraying; leaching into water bodies; and aerial sprays to kill water-dwelling pests. Afterward, pesticides disrupt the food webs in addition to affecting the fish fauna. There have been numerous persistent pesticides found in the main food webs in the Arctic Ocean, including organochlorines and polychlorinated biphenyls (Pradhan et al., 2022). Pesticides' solubility in water and ability to penetrate an organism determines their residual effects on the aquatic environment. Since the pesticide's residual mass is hydrophilic or lipophobic, it is less accessible in aquatic organisms' fatty tissues, which eventually reduces its toxicity. The agroecosystem will be seriously threatened if pesticide use is not stopped, so it is imperative that we learn as much as we can about this problem and develop environmentally responsible alternatives (Kalyabina et al., 2021).

The aquatic environment can be described by the way sediment, water, and biota work together to produce food for humans. Hazardous inorganic and organic pollutants have accumulated, leading to a significant decline in aquatic environments that may have long-term negative effects on human health. Among the most hazardous substances when it comes to toxic pollutants are pesticides. Many pests and animals that spread disease are eradicated from human habitats and surrounding areas by the widespread use of pesticides. These animals include ticks, mice, rats, and mosquitoes. It also aids in lessening insect infestation in crops used in agriculture (Bashir et al., 2020).

Aquatic resources and fisheries, which include ponds, lakes, rivers, streams, and oceans, are particularly valuable natural resources. Concern over the effects of an increasing human population and their activities on the aquatic environment has grown in tandem with the recognition of aquatic systems and fisheries. One important class of toxic substances that have been connected to human exposure and have a significant negative influence on aquatic environments is pesticides. Pesticides are any material or combination of materials used to prevent, limit, eradicate, or disperse pests like weeds, insects, rats, nematodes, and mites. The increasing population that requires more food is the reason behind the steady increase in pesticide applications worldwide (Shefali et al., 2021).

Pesticide use affects about one-third of agricultural products worldwide, and it is a major cause of environmental and health problems. Fruits, vegetables, and cereals could lose production if they are not used. The World Health Organization (WHO) reports on efficient pesticide control methods for malaria-carrying insects. On the other hand, incorrect application can have extremely harmful effects on both people and the environment. The widespread use and release of pesticides are frequent causes of pesticide metabolites in rivers, sediment, and the atmosphere, endangering both human health and ecosystems (Sule et al., 2022).

Pollution is mostly caused by organochlorine pesticides (OCPs), which are most prevalent in the Middle East, Asia, Africa, Eastern Europe, and Latin America. OCPs are highly soluble in lipids, less soluble in water, and less polar, which allows them to bioaccumulate in food chains and pose serious risks to both human health and the environment. Long after they are first applied, they can linger in food, water, and soil sediments for several decades (Ohoro & Wepener, 2023).

OCPs can enter the environment in a number of ways, such as by spraying or dumping contaminated waste into landfills or industrial effluents. Wind can carry airborne OCPs into isolated locations, and some of them can stick to dirt or airborne particles. After adsorbing into food chains, DDT and dieldrin metabolites can linger in soil for years. Animals and humans are both exposed to OCP absorption and accumulation when they consume contaminated food. Food has been found to contain over 90% of OCP adulterants, especially in fish. It is critical to assess the distribution and contamination of organic pesticides in water, sediment, and aquatic biota, as well as their long-term impacts on human health, in order to minimize their use and maintain environmentally friendly aquatic ecosystems (Arisekar et al., 2019).

Environmentally friendly pesticide development and production have grown significantly in the advanced world's pesticide trade. However, to ensure sustainable and successful pest control strategies, pesticide use must be balanced with considerations for the environment and human health. Responsible pesticide practices, such as integrated pest management, should be promoted (Verma et al., 2023).

Every government needs to regulate pesticides both nationally and regionally. The "Interregional Research Project Number 4" (IR-4 Project) of the United States Department of Agriculture (USDA) was initiated in 1963 and has been producing data to ascertain pesticide tolerance. Additionally, the project has been assisting the Environmental Protection Agency (EPA) in the United States with safe registration and safe control measures.

While product authorization and usage guidelines are set at the national level, "safe uses" of pesticides are decided upon at the EU level to safeguard human health, animals, and the aquatic environment (Aparicio & De Gerónimo, 2024; Lindahl et al., 2024). Several recent case studies have focused on assessing and monitoring the relevance of different pesticide residues in water resources, some are summarized in Table 2.

Table 2 Case studies monitoring pesticide residues in different water types

Study region/ country	Water type	Pesticide	Conclusion	Reference
Doñana Natural Park (south-west of Spain)	Freshwater	Organophosphate pesticides, pyrethroids dinitroanilines, diphenyl ethers, oxadiazoles, and carboxamides	High pesticide levels were discovered indicating possible threats to freshwater and sediment-dwelling creatures in the investigated locations	Peris et al. (2024)
Coastal area of northwestern Mexico	Seawater	Chlorfenapyr, chlorpyrifos, cypermethrin, endosulfan, fipronil, myclobutanil, pendimethalin, PCNB, PCA, pirimicarb, and pyriproxyfen	18 pesticides were detected with cypermethrin and chlorpyrifos having the highest quantities. These two pesticides have been linked to acute toxic effects on algae, crustaceans, and fish.	Leyva-Morales et al. (2024)
Pampas region-Argentina	Drinking water	Atrazine, metolachlor, imidacloprid, hydroxyatrazine, imazethapyr, and 2.4D	The pesticide content in drinking water surpassed the European standard on an individual basis.	Aparicio and De Gerónimo (2024)
11 European countries	Freshwater	67 fungicides, 62 herbicides, 63 insecticides, and 1 synergist	Nine out of the 11 case study locations had one or more pesticides that were beyond the threshold values set out in the European Water Framework Directive for surface water.	Navarro et al. (2024)
China	Agricultural water	Carbendazim, chlorothalonil, difenoconazole, mancozeb, thiophanate-methyl, and thiram (fungicides); acetochlor, atrazine, butachlor, glyphosate, pendimethalin, and propisochlor (herbicides); and abamectin, bisultap, chlorpyrifos, cypermethrin, imidacloprid, and phoxim (insecticides)	The water was dominated by a few number of pesticides used on particular crops. In the category of fungicides, mancozeb has the largest contribution (19%), followed by acetochlor (19%) in the herbicides category, and cypermethrin (43%) in the insecticides category.	Yi et al. (2024)

(continued)

Table 2 (continued)

Study region/ country	Water type	Pesticide	Conclusion	Reference
Mekong River, Cambodia	Surface-water	64 pesticides	Tebufenozide was the pesticide with the highest observed concentration in surface water, followed by two other insecticides, bromophos-ethyl and dichlorvos, and a very high risk fungicide, iprobenfos.	Ngin et al. (2024)
Kenya	River water	Organochlorine pesticides (OCPs)	19 pesticide residues were found in the river's water. Furthermore, OCPs exceeded maximum residue levels.	Ogola et al. (2024)
Yangtze River Delta, East China	Drinking water	Organochlorine pesticides (OCPs)	All of the drinking water from the tap included OPPs. Many prohibited OPPs had residual levels that were still more than 1 ng/ L.	Zhang et al. (2023)
Paraná state—Brazil	Drinking water	11 proven, probable, or potentially carcinogenic pesticides	Drinking water was found to be heavily contaminated with legacy pesticides. The majority of the towns were well over the maximum amounts permitted for each of the pesticides that were already in use.	Panis et al. (2022)
Sri Lanka	River water	Pretilachlor, oxyfluorfen, thiamethoxam, chlorantraniliprole, fenobucarb, fipronil, diazinon, etofenprox, tebuconazole, and captan	Concentrations of all determined pesticides exceeded national regulatory threshold limits	Jayasiri et al. (2022)
Vineyard area of La Rioja (Spain)	Groundwater	15 fungicides +2 DP, 7 insecticides +1 DP, and 3 herbicides +2 DP	The fungicides metalaxyl, tebuconazole, and boscalid were the chemicals most commonly found in water samples.	Manjarres-López et al. (2021)

(continued)

Table 2 (continued)

Study region/ country	Water type	Pesticide	Conclusion	Reference
Lake Ziway, Ethiopia	Freshwater	Insecticides and fungicides	Malathion, dimethoate, metalaxyl, diazinon, chlorpyrifos, fenitrothion and endosulfan were detected in water	Merga et al. (2021)
Maputo Bay, Mozambique	Estuaries	37 different pesticides	In three of the four nursery sites that were evaluated, the risk assessment revealed that many herbicides had environmental thresholds that were exceeded.	Sturve et al. (2021)
South Africa	Watersheds	96 sprayed pesticides	The insecticides imidacloprid, thiacloprid, chlorpyrifos and acetamiprid and the herbicide terbuthylazine exceeded the limits	Curchod et al. (2020)
Kashmir valley	Drinking water	26 commonly used pesticides	Water tests revealed the presence of pesticide residues from ten of the twenty-six regularly used pesticides. Of the pesticides found, difenoconazole had the greatest concentration.	Ganaie et al. (2023)
Gujarat, India	Surface water	Chlorpyrifos, methyl parathion, hexachlorocyclohexane (HCH), dichloro diphenyl trichloroethane (DDT), and endosulfan	Surface water contained methyl parathion and chlorpyrifos indicating that extremely dangerous pesticides are still being used in the neighborhood	Hashmi et al. (2020)

8 Impact of Climate Change on Pesticide Runoff and Water Quality

There is indisputable scientific proof that climate change is occurring. The effects of climate change are seen in a variety of environmental elements. In agriculture, pesticides are essential for increasing both the amount and quality of crops. The way that pesticides work on the intended pests will also be impacted by climate change. Any pesticide used in agricultural areas has a fate mostly determined by biotic and abiotic variables. The method of pesticide dissipation and movement is largely influenced by abiotic variables, including the quantity and intensity of rainfall. The pesticides that are weakly bound to soil particles will be removed because of the quick flow paths and episodic preferential flow that occurs during irregular rainfall. Surface runoff and leaching from pesticides onto agricultural fields are likely to spread to other environmental compartments, contaminating ground and surface water supplies (Pandey et al., 2020).

Studies have indicated that the decomposition of pesticides is contingent upon various abiotic factors such as soil pH, temperature, humidity, and climate, as well as variations in plant species and/or microbial communities, pesticide properties like hydrophobicity, pH, and all associated biochemical processes (Farha et al., 2016).

8.1 *Impacts of Increased CO_2 on Pesticide*

The majority of pesticides used on crops are primarily absorbed by the soil, and one important metabolic pathway that helps to detoxify these chemicals is microbial degradation. Pesticide degradation is indirectly impacted by elevated CO_2 because it influences soil microbial activity indirectly. Pesticide fate and behavior in the environment may be affected by changes in the volatility, deposition, and degradation characteristics of pesticides due to climate change. On the other hand, little is known about how higher CO_2 affects pesticide fate. Only modeling has been used in studies to forecast pesticide fate in scenarios where atmospheric CO_2 levels rise in the future (Mustapha et al., 2018).

8.2 *Increased Temperature and Pesticide*

A substantial reduction in freshwater resources and an increase in global temperature are expected as a result of the predicted 1.1–6.4 °C increase in global temperatures. Pesticides and other xenobiotics are predicted to be impacted by this shift both directly and indirectly. Different soil temperatures have been reported in different parts of the world, including China, Turkey, Canada, Alaska, Siberia, and Russia, as a result of global warming. Pesticides volatilize more readily in an atmosphere with higher temperatures, which speeds up their degradation. Temperature affects both biotic and abiotic factors that contribute to pesticide degradation. It can speed up or slow down degradation reactions and alter the environment's active degradation channels. Because of the high temperatures, pesticides such as gamma-HCH, dieldrin, and heptachlor epoxide have been reemitted from previously contaminated soil or snow surfaces (Davamani et al., 2024).

Variations in climate parameters, such as temperature and precipitation, have a direct impact on the fate of pesticides, and one variable can offset another. For example, higher temperatures or soil moisture contents will accelerate the rate of pesticidal degradation, whereas higher rainfall will accelerate leaching. There are reports regarding the proportionate relationship between temperature and pesticide toxicity, which indicates that temperature also affects the ecotoxicity of pesticides. Nonetheless, in colder climates, pyrethroids and DDT exhibit greater toxicity (Kaka et al., 2021).

8.3 Climate Change Associating with Leaching of Pesticides

Climate change is expected to have a significant impact on global agriculture, making it more challenging to guarantee that food is available to all people. Because of shifts in the micro- and macroclimate, crop–pest interactions will shift and favor pests over crops. Due to this, more pesticides will likely be used in developing countries; potentially, some banned or restricted pesticides may even be reintroduced (Skendžić et al., 2021).

Only 10–15% of pesticides that are used to control targeted pests actually reach the target site; the remaining 70–90% reach nontargeted sites. According to climate change projections, wind circulation is likely to change, which will have an impact on drift loss when applying pesticides. By verifying the ideal wind conditions for dusting or spraying, farmers can take preventative measures (Kannan, 2023).

It seems likely that climate change will result in more frequent storms and floods, which will have an immediate impact on pesticide losses through runoff. Pesticides that makeup 5–15% of runoff have the potential to contaminate surface water bodies. As a result of washing loss brought on by increased precipitation due to climate change, frequent pest control will be required. A significant amount of pesticides end up in the soil, where they serve as a sink for additional pesticides. Pesticides can be absorbed by soil and interact with the soil environment in a number of ways. Temperature and moisture content of the soil have a direct relationship with pesticide volatilization and photodegradation. Elevated temperatures, increased sunshine, and increased precipitation levels increase the likelihood of volatilization (Bolan et al., 2024).

Pesticides are also applied to the soil, and pesticide contamination contaminates irrigation water. Three main processes control the fate of pesticides in soil: leaching, degradation, and sorption. The amount of pesticide in the soil that is available for subsequent interactions like degradation and leaching is determined by sorption.

The inverse relationship between pesticide adsorption onto soil and climate change, which indicates warmer temperatures, will increase the concentration of pesticides available in the soil. The most significant process following sorption and degradation is pesticide leaching, which is the downward movement of pesticides through various soil profiles and groundwater. Pesticide leaching is significantly influenced by soil characteristics, weather, and pesticide characteristics. Pesticides are often rated according to their leaching potential using the groundwater ubiquity score (GUS), which divides them into extremely low to moderate leachers and high to very high leachers (Araya et al., 2024).

8.4 Climate Change's Effects on Pesticides' Bioavailability and Toxicity

Pesticides frequently contaminate groundwater and open water bodies; the primary causes are both direct and indirect applications. Runoff, wash-off, spillage loss, drainage from farmlands and urban areas, public health applications, and accidental spills are the main ways that indirect roots cause contamination. Pesticides can infiltrate the food chain by contaminating the lowest trophic level and then using bioaccumulation and biomagnification to reach the highest trophic level when they are present in water. The frequency of storms, rain, and floods is predicted to increase due to climate change, which will increase the amount of pesticides in water bodies (Kaur et al., 2024).

Water's physical characteristics, pesticide molecules' structural characteristics, and the physiology of each aquatic species all influence how ecotoxic pesticides are to aquatic life. Since bioaccumulation and toxicity data are necessary for these pesticides prior to registration and marketing, the log Kow (octanol–water partition coefficient) is frequently used as a measure of bioaccumulation potential. Aquatic organisms' physiological behavior, the pH of the water, the characteristics of dissolved pesticides, and the presence of suspended particles all have an impact on how toxic pesticides are to them (Pereira et al., 2016).

9 Methods for Treating Water Contaminated with Pesticides

Because pesticides are used excessively in today's agriculture sector, the amount of pesticides in water has grown. The kind of pesticide and the effectiveness of the treatment procedure must be considered when selecting an appropriate water treatment technique for pesticide removal (Saleh et al., 2020). Water pollutants pose obstacles to natural breakdown, prompting the construction of wastewater treatment facilities using chemical, physical, and biological approaches (Fig. 4.).

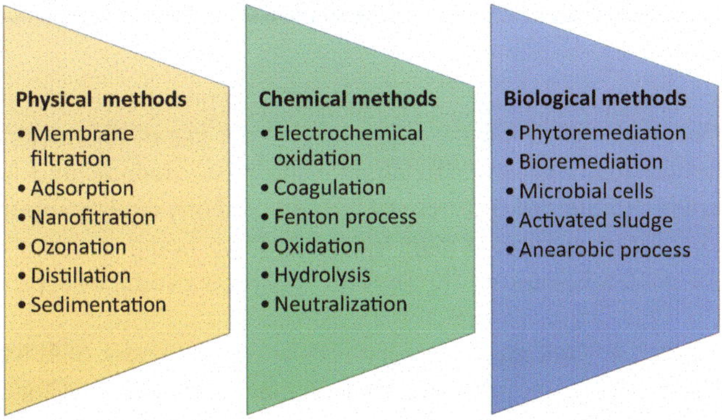

Fig. 4 Methods for treating water

Chemical wastewater treatment is made up of a series of chemical processes that help to hydrolyze pollutants into safer compounds. The primary chemical approaches include coagulation and advanced oxidation processes (AOPs), such as ozonation and Fenton treatment. Chemical approaches often entail the use of chemical agents such as radicals or reagents to transform pesticides into harmless or less toxic forms. Chemical treatments are typically employed in conjunction with photo-catalysis and membrane methods (Saleh et al., 2020).

Biological treatment is another way of removing pesticides from water and wastewater. The biological process is distinguished by biodegradation. This process transforms organic contaminants into less hazardous compounds like CO_2 and H_2O. There are three forms of biological treatments using microorganisms: bioaugmentation, natural attenuation, and biostimulation. Bioremediation can utilize fungal and bacterial strains to break down pesticides. Biological treatment is divided into aerobic and anaerobic components. The mix of aerobic and anaerobic conditions improves absorption efficiency. Pesticide elimination is most effectively accomplished by combining aerobic and anaerobic conditions (Nabil et al., 2023).

Physical approaches use physical separation to remove pollutants. The key physical processes in this category are membrane and adsorption. Membranes provide excellent selectivity by separating based on pore size. However, in the case of membranes with hole sizes bigger than a pesticide, the affinity of the component for the membrane influences the percentage rejection. Adsorption procedures are quite straightforward. The efficacy of adsorption separation is strongly dependent on the adsorbent and adsorbate, their interactions, and the operating circumstances. Adsorption procedures are quite straightforward. The efficacy of adsorption separation is strongly dependent on the adsorbent and adsorbate, their interactions, and the operating circumstances (Ponnuchamy et al., 2021).

According to Saleh et al. (2020), the adopted treatment techniques should also be countries' appropriate, in a way that they suit each country's environmental conditions and rely on the available resources (adsorbents and others), which will increase the cost-effectiveness of the treatment plant.

10 Prospects and Necessities for the Future

In order to prevent pest infestations in the pre- and post-harvest food production processes, appropriate solutions are required, and the use of chemical pesticides is a widely acknowledged concern. To change the pest's habitat and lessen stress, physical control techniques like structural barriers and physical manipulation can be used. Treatments for insect pest management (IPM) are being sought after more and more to lessen pest infestations throughout the food production process (Berhe et al., 2022).

Multidisciplinary efforts are necessary for the effectiveness of IPM in unfavorable climate conditions, compositions that enhance permanence, longer lifespan, and ease of use, as well as pathogenic aggressiveness and range of effect. Because entomopathogenic fungi EPF actions have a significant impact on the sustainability of pest abundance and distribution in ecological systems, research into the potential application of infectious fungi in insect pest control is being conducted.

The environment benefits from biological treatment, and farmers must be trained to correctly apply cultural approaches and incorporate them into other strategies. To get the best results, it is essential to look at different preparations of microbiological products. Every stage of the product lifecycle, from producers to consumers, should support and encourage biological control methods (Deka et al., 2021).

It is possible to carry out promotional campaigns, exhibitions, ads, and apprenticeship programs to raise consumer awareness of biostimulants. It is important to keep looking for novel microbial inoculants that can be used in a range of temperatures and environments. To increase cost-effectiveness, usability, and effectiveness, better surface treatments, conceptualization, manufacturing, and preservation techniques must be developed. Important agricultural plants can be conventionally bred with the inclusion of microbiological genomes. To guarantee their long-term use and comprehend the epidemiology of insect diseases, more research is required (Katsenios et al., 2023).

The use of various chemicals in agriculture, including pesticides, has increased crop yield and reduced postharvest losses. However, this has led to concerns about the potential adverse effects on human health. Exposure to pesticides is most likely among farm workers and less controlled uses. Limited epidemiological data exists to evaluate the health effects of pesticides on humans. Only a small proportion of a population is likely to receive a pesticide dose high enough to cause acute severe effects, but many more may be at risk of developing chronic effects depending on the type of pesticide they are exposed to. Pesticides currently in use include a wide variety with varying modes of action, uptake by the body, metabolism, elimination from the body, and toxicity to humans. The main hazard lies in acute, short-term exposures with highly acute toxicity, while long-term exposure, even to small doses, is associated with long-term exposure. Active ingredients, impurities, solvents, carriers, emulsifiers, and other constituents of the formulated product may cause adverse effects (Anaduaka et al., 2023).

Pesticide use has long been associated with agricultural development and has a substantial impact on both agricultural practices and public health. Pesticides are used in daily life to eradicate pests like rats, mice, ticks, and mosquitoes. They are categorized based on chemical classes, functional groups, modes of action, and toxicity. Farmers have also employed them to manage insects and weeds in agricultural cultivation, which has resulted in a notable rise in agricultural output. On the other hand, inappropriate pesticide application and management may result in contamination of food, water, soil, and air (Taidi & El laghdach, 2024).

Socioeconomic and environmental factors, such as soil conditions, crop growth, and the prevalence of insect pests, weeds, and diseases associated with pesticide use, are influenced by climate change factors. Changes in soil organic matter, soil's ability to store and cycle carbon, and the size and frequency of soil cracks are all impacted by climate change, which has an impact on how pesticides are applied. The geographic distribution of crops, crop productivity, pest abundance, pest outbreaks, vector dispersal, weed evolution, and disease stimulation are all impacted by climate change, which leads to a rise in the usage of a greater range of pesticides (Skendžić et al., 2021).

Additionally, pesticide behaviors such as volatilization, runoff, leaching processes, and pesticide degradation are accelerated by climate change. Controlling pesticide contamination and its detrimental effects on the environment and non-target organisms is therefore necessary. To gain a better understanding of pesticide use and management in the future, more research should concentrate on occupational and environmental exposures and the health risks associated with pesticide use. To lessen the harmful effects of pesticides, new scientific technologies and methods like integrated pest management (IPM), legislation prohibiting high-risk pesticides, and the creation of national implementation plans (NIPs) should be put into place (Pathak et al., 2022).

11 Conclusion

The use of pesticides in agriculture to boost crop yield and lower postharvest losses has sparked worries about possible negative health effects on people. Pesticide exposure is most common among farm laborers and in less regulated settings. There is a paucity of epidemiological data to assess the impact of pesticides on human health. Pesticides are used to control weeds, insects, and pests, which increases agricultural productivity. Food, water, soil, and air contamination can result from improper application and management, though. The effects of climate change on crop growth, soil properties, and pesticide usage all contribute to an increase in pesticide use. It is imperative to control pesticide contamination to prevent negative impacts on the environment and non-target organisms. Future studies should concentrate on the health risks connected to pesticide use as well as exposures in the workplace and environment. Legislation, national implementation plans, integrated pest management, and other new scientific technologies and methods ought to be put into practice.

References

Ahmad, M. F., Ahmad, F. A., Alsayegh, A. A., Zeyaullah, M., AlShahrani, A. M., Muzammil, K., Saati, A. A., Wahab, S., Elbendary, E. Y., Kambal, N., Abdelrahman, M. H., & Hussain, S. (2024). Pesticides impacts on human health and the environment with their mechanisms of action and possible countermeasures. *Heliyon, 10*(7), e29128. https://doi.org/10.1016/j.heliyon.2024.e29128

Ali, S., Ullah, M. I., Sajjad, A., Shakeel, Q., & Hussain, A. (2021). Environmental and health effects of pesticide residues. In M. Inamuddin, I. Ahamed, & E. Lichtfouse (Eds.), *Sustainable agriculture reviews 48: Pesticide occurrence, analysis and remediation Vol. 2 Analysis* (pp. 311–336). Springer International Publishing. https://doi.org/10.1007/978-3-030-54719-6_8

Anaduaka, E. G., Uchendu, N. O., Asomadu, R. O., Ezugwu, A. L., Okeke, E. S., & Chidike Ezeorba, T. P. (2023). Widespread use of toxic agrochemicals and pesticides for agricultural products storage in Africa and developing countries: Possible panacea for ecotoxicology and health implications. *Heliyon, 9*(4), e15173. https://doi.org/10.1016/j.heliyon.2023.e15173

Aparicio, V., & De Gerónimo, E. (2024). Pesticide pollution in Argentine drinking water: A call to ensure safe access. *Environmental Challenges, 14*, 100808. https://doi.org/10.1016/j.envc.2023.100808

Araya, G., Perfetti-Bolaño, A., Sandoval, M., Araneda, A., & Barra, R. O. (2024). Groundwater leaching potential of pesticides: A historic review and critical analysis. *Environmental Toxicology and Chemistry*, 1–14. https://doi.org/10.1002/etc.5869

Arisekar, U., Shakila, R. J., Jeyasekaran, G., Shalini, R., Kumar, P., Malani, A. H., & Rani, V. (2019). Accumulation of organochlorine and pyrethroid pesticide residues in fish, water, and sediments in the Thamirabarani river system of southern peninsular India. *Environmental Nanotechnology, Monitoring & Management, 11*, 100194. https://doi.org/10.1016/j.enmm.2018.11.003

Bashir, I., Lone, F. A., Bhat, R. A., Mir, S. A., Dar, Z. A., & Dar, S. A. (2020). Concerns and threats of contamination on aquatic ecosystems. In *Bioremediation and biotechnology: Sustainable approaches to pollution degradation* (pp. 1–26). https://doi.org/10.1007/978-3-030-35691-0_1

Benbrook, C., Kegley, S., & Baker, B. (2021). Organic farming lessens reliance on pesticides and promotes public health by lowering dietary risks. *Agronomy, 11*(7), 1–36. https://doi.org/10.3390/agronomy11071266

Berhe, M., Subramanyam, B., Chichaybelu, M., Demissie, G., Abay, F., & Harvey, J. (2022). Post-harvest insect pests and their management practices for major food and export crops in East Africa: An Ethiopian case study. *Insects, 13*(11). https://doi.org/10.3390/insects13111068

Bolan, S., Padhye, L. P., Jasemizad, T., Govarthanan, M., Karmegam, N., Wijesekara, H., Amarasiri, D., Hou, D., Zhou, P., Biswal, B. K., Balasubramanian, R., Wang, H., Siddique, K. H. M., Rinklebe, J., Kirkham, M. B., & Bolan, N. (2024). Impacts of climate change on the fate of contaminants through extreme weather events. *Science of the Total Environment, 909*, 168388. https://doi.org/10.1016/j.scitotenv.2023.168388

Cara, I. G., Ţopa, D., Puiu, I., & Jităreanu, G. (2022). Biochar a promising strategy for pesticide-contaminated soils. *Agriculture, 12*(10). https://doi.org/10.3390/agriculture12101579

Curchod, L., Oltramare, C., Junghans, M., Stamm, C., Dalvie, M. A., Röösli, M., & Fuhrimann, S. (2020). Temporal variation of pesticide mixtures in rivers of three agricultural watersheds during a major drought in the Western Cape, South Africa. *Water Research X, 6*, 100039. https://doi.org/10.1016/j.wroa.2019.100039

Davamani, V., John, J. E., Poornachandhra, C., Gopalakrishnan, B., Arulmani, S., Parameswari, E., Santhosh, A., Srinivasulu, A., Lal, A., & Naidu, R. (2024). A critical review of climate change impacts on groundwater resources: A focus on the current status, future possibilities, and role of simulation models. *Atmosphere, 15*(1). https://doi.org/10.3390/atmos15010122

Deka, B., Baruah, C., & Babu, A. (2021). Entomopathogenic microorganisms: Their role in insect pest management. *Egyptian Journal of Biological Pest Control, 31*(1), 121. https://doi.org/10.1186/s41938-021-00466-7

Farha, W., Abd El-Aty, A. M., Rahman, M. M., Shin, H. C., & Shim, J. H. (2016). An overview on common aspects influencing the dissipation pattern of pesticides: A review. *Environmental Monitoring and Assessment, 188*(12), 693. https://doi.org/10.1007/s10661-016-5709-1

Fayaz, T., Rana, S. S., Goyal, E., Ratha, S. K., & Renuka, N. (2024). Harnessing the potential of microalgae-based systems for mitigating pesticide pollution and its impact on their metabolism. *Journal of Environmental Management, 357*, 120723. https://doi.org/10.1016/j.jenvman.2024.120723

Ganaie, M. I., Jan, I., Mayer, A. N., Dar, A. A., Mayer, I. A., Ahmed, P., & Sofi, J. A. (2023). Health risk assessment of pesticide residues in drinking water of upper Jhelum region in Kashmir Valley-India by GC-MS/MS. *International Journal of Analytical Chemistry, 2023*, 6802782. https://doi.org/10.1155/2023/6802782

Garrigou, A., Laurent, C., Berthet, A., Colosio, C., Jas, N., Daubas-Letourneux, V., Jackson Filho, J.-M., Jouzel, J.-N., Samuel, O., Baldi, I., Lebailly, P., Galey, L., Goutille, F., & Judon, N. (2020). Critical review of the role of PPE in the prevention of risks related to agricultural pesticide use. *Safety Science, 123*, 104527. https://doi.org/10.1016/j.ssci.2019.104527

Hashmi, T. A., Qureshi, R., Tipre, D., & Menon, S. (2020). Investigation of pesticide residues in water, sediments and fish samples from Tapi River, India as a case study and its forensic significance. *Environmental Forensics, 21*(1), 1–10. https://doi.org/10.1080/15275922.2019.1693441

Hassaan, M. A., & El Nemr, A. (2020). Pesticides pollution: Classifications, human health impact, extraction and treatment techniques. *Egyptian Journal of Aquatic Research, 46*(3), 207–220. https://doi.org/10.1016/j.ejar.2020.08.007

Islam, M. A., Amin, S. M. N., Rahman, M. A., Juraimi, A. S., Uddin, M. K., Brown, C. L., & Arshad, A. (2022). Chronic effects of organic pesticides on the aquatic environment and human health: A review. *Environmental Nanotechnology, Monitoring & Management, 18*, 100740. https://doi.org/10.1016/j.enmm.2022.100740

Jayasiri, M. M. J. G. C. N., Yadav, S., Dayawansa, N. D. K., Propper, C. R., Kumar, V., & Singleton, G. R. (2022). Spatio-temporal analysis of water quality for pesticides and other agricultural pollutants in Deduru Oya river basin of Sri Lanka. *Journal of Cleaner Production, 330*, 129897. https://doi.org/10.1016/j.jclepro.2021.129897

Kaka, H., Opute, P. A., & Maboeta, M. S. (2021). Potential impacts of climate change on the toxicity of pesticides towards earthworms. *Journal of Toxicology, 2021*, 8527991. https://doi.org/10.1155/2021/8527991

Kalyabina, V. P., Esimbekova, E. N., Kopylova, K. V., & Kratasyuk, V. A. (2021). Pesticides: Formulants, distribution pathways and effects on human health – A review. *Toxicology Reports, 8*, 1179–1192. https://doi.org/10.1016/j.toxrep.2021.06.004

Kannan, N. (2023). An analysis of the climate change effects on pesticide vapor drift from ground-based pesticide applications to cotton. *Scientific Reports, 13*(1), 9740. https://doi.org/10.1038/s41598-023-36941-4

Katsenios, N., Sparangis, P., Vitsa, S., Leonidakis, D., & Efthimiadou, A. (2023). Application of biostimulants and herbicides as a promising co-implementation: The incorporation of a new cultivation practice. *Agronomy, 13*(10). https://doi.org/10.3390/agronomy13102634

Kaur, R., Choudhary, D., Bali, S., Bandral, S. S., Singh, V., Ahmad, M. A., Rani, N., Singh, T. G., & Chandrasekaran, B. (2024). Pesticides: An alarming detrimental to health and environment. *Science of the Total Environment, 915*, 170113. https://doi.org/10.1016/j.scitotenv.2024.170113

Leyva-Morales, J. B., del Bastidas-Bastidas, P. J., del Salvatierra-Stamp, V. C., González-Márquez, L. C., Davizon, Y. A., Márquez-Pacheco, H., Amillano-Cisneros, J. M., Cruz-Acevedo, E., Amézquita-López, B. A., Romo Gómez, C., De la Torre Espinosa, Z. Y., & Rodríguez-Aguilar, B. A. (2024). Temporal distribution and ecological risk assessment for pesticides in water from the north-central coastal zone of Sinaloa, Mexico. *Heliyon, 10*(15), e35207. https://doi.org/10.1016/j.heliyon.2024.e35207

Lindahl, A., Reichenberger, S., Pohlert, T., Multsch, S., Boström, G., Gönczi, M., Stenemo, F., Kreuger, J., Markensten, H., & Jarvis, N. (2024). A web-based pesticide risk assessment tool

for drinking water protection zones in Sweden. *Journal of Environmental Management, 357*, 120700. https://doi.org/10.1016/j.jenvman.2024.120700
Manjarres-López, D. P., Andrades, M. S., Sánchez-González, S., Rodríguez-Cruz, M. S., Sánchez-Martín, M. J., & Herrero-Hernández, E. (2021). Assessment of pesticide residues in waters and soils of a vineyard region and its temporal evolution. *Environmental Pollution, 284*, 117463. https://doi.org/10.1016/j.envpol.2021.117463
Merga, L. B., Mengistie, A. A., Alemu, M. T., & Van den Brink, P. J. (2021). Biological and chemical monitoring of the ecological risks of pesticides in Lake Ziway, Ethiopia. *Chemosphere, 266*, 129214. https://doi.org/10.1016/j.chemosphere.2020.129214
Mustapha, M., Halimoon, N., Wan Johari, W. L., & Abd Shakur, M. (2018). Soil microorganisms and their potential in pesticide biodegradation: A review article. *Journal of Sustainable Agricultural Sciences*. https://doi.org/10.21608/jsas.2018.3465.1061
Nabil, M., Abdel-Gawad, H., Elantabli, F. M., El-Medani, S. M., & Abdelhameed, R. M. (2023). Removal methods of pesticides and their improvement using metal-organic frameworks. *Egyptian Journal of Chemistry, 66*(13), 39–75. https://doi.org/10.21608/EJCHEM.2023.233038.8537
Navarro, I., de la Torre, A., Sanz, P., Abrantes, N., Campos, I., Alaoui, A., Christ, F., Alcon, F., Contreras, J., Glavan, M., Pasković, I., Pasković, M. P., Nørgaard, T., Mandrioli, D., Sgargi, D., Hofman, J., Aparicio, V., Baldi, I., Bureau, M., et al. (2024). Assessing pesticide residues occurrence and risks in water systems: A Pan-European and Argentina perspective. *Water Research, 254*, 121419. https://doi.org/10.1016/j.watres.2024.121419
Ngin, P., Haglund, P., Proum, S., & Fick, J. (2024). Pesticide screening of surface water and soil along the Mekong River in Cambodia. *Science of the Total Environment, 912*, 169312. https://doi.org/10.1016/j.scitotenv.2023.169312
Oberemok, V. V., Laikova, K. V., Gninenko, Y. I., Zaitsev, A. S., Nyadar, P. M., & Adeyemi, T. A. (2015). A short history of insecticides. *Journal of Plant Protection Research, 55*(3), 221–226. https://doi.org/10.1515/jppr-2015-0033
Ogola, J. O., Olale, K., Mogwasi, R., & Mainya, O. (2024). Organochlorine pesticide residues in water and sediments in river Kibos-Nyamasaria in Kisumu County: An inlet river of Lake Victoria, Kenya. *Scientific African, 23*, e02094. https://doi.org/10.1016/j.sciaf.2024.e02094
Ohoro, C. R., & Wepener, V. (2023). Review of scientific literature on available methods of assessing organochlorine pesticides in the environment. *Heliyon, 9*(11), e22142. https://doi.org/10.1016/j.heliyon.2023.e22142
Ortiz-Hernández, M. L., Sánchez-Salinas, E., Dantán-González, E., & Castrejón-Godínez, M. L. (2013). Pesticide biodegradation: Mechanisms, genetics and strategies to enhance the process. In R. Chamy & F. Rosenkranz (Eds.), *Biodegradation*. IntechOpen. https://doi.org/10.5772/56098
Ou, J., Li, H., Ou, X., Yang, Z., Chen, M., Liu, K., Teng, Y., & Xing, B. (2020). Degradation, adsorption and leaching of phenazine-1-carboxamide in agricultural soils. *Ecotoxicology and Environmental Safety, 205*, 111374. https://doi.org/10.1016/j.ecoenv.2020.111374
Pandey, N., Rana, D., Chandrakar, G., Gowda, G. B., Patil, N. B., Pandi G, G. P., Annamalai, M., Pokhare, S. S., Rath, P. C., & Adak, T. (2020). Role of climate change variables (standing water and rainfall) on dissipation of chlorantraniliprole from a simulated rice ecosystem. *Ecotoxicology and Environmental Safety, 205*, 111324. https://doi.org/10.1016/j.ecoenv.2020.111324
Panis, C., Candiotto, L. Z. P., Gaboardi, S. C., Gurzenda, S., Cruz, J., Castro, M., & Lemos, B. (2022). Widespread pesticide contamination of drinking water and impact on cancer risk in Brazil. *Environment International, 165*, 107321. https://doi.org/10.1016/j.envint.2022.107321
Parvin, F., & Tareq, S. M. (2021). Impact of landfill leachate contamination on surface and groundwater of Bangladesh: A systematic review and possible public health risks assessment. *Applied Water Science, 11*(6), 100. https://doi.org/10.1007/s13201-021-01431-3
Pathak, V. M., Verma, V. K., Rawat, B. S., Kaur, B., Babu, N., Sharma, A., Dewali, S., Yadav, M., Kumari, R., Singh, S., Mohapatra, A., Pandey, V., Rana, N., & Cunill, J. M. (2022). Current

status of pesticide effects on environment, human health and it's eco-friendly management as bioremediation: A comprehensive review. *Frontiers in Microbiology, 13*, 962619. https://doi.org/10.3389/fmicb.2022.962619

Pereira, V. J., Cunha, J. P. A. R., Morais, T. P., Ribeiro-Oliveira, J., & Morais, J. B. (2016). Physical-chemical properties of pesticides: Concepts, applications, and interactions with the environment. *Propriedades Físico-Químicas Dos Agrotóxicos: Bioscience Journal, 32*(3), 627–641.

Pérez-Lucas, G., Vela, N., El Aatik, A., & Navarro, S. (2018). Environmental risk of groundwater pollution by pesticide leaching through the soil profile. In M. Larramendy & S. Soloneski (Eds.), *Pesticides*. IntechOpen. https://doi.org/10.5772/intechopen.82418

Peris, A., Soriano, Y., Picó, Y., Bravo, M. A., Blanco, G., & Eljarrat, E. (2024). Pesticides in water and sediments from natural protected areas of Spain and their associated ecological risk. *Chemosphere, 362*, 142628. https://doi.org/10.1016/j.chemosphere.2024.142628

Ponnuchamy, M., Kapoor, A., Senthil Kumar, P., Vo, D.-V. N., Balakrishnan, A., Mariam Jacob, M., & Sivaraman, P. (2021). Sustainable adsorbents for the removal of pesticides from water: A review. *Environmental Chemistry Letters, 19*(3), 2425–2463. https://doi.org/10.1007/s10311-021-01183-1

Pradhan, S. S., Gowda, G. B., Adak, T., Guru-Pirasanna-Pandi, G., Patil, N. B., Annamalai, M., & Rath, P. C. (2022). Pesticides occurrence in water sources and decontamination techniques. In M. L. Larramendy & S. Soloneski (Eds.), *Pesticides*. IntechOpen. https://doi.org/10.5772/intechopen.103812

Rasool, S., Rasool, T., & Gani, K. M. (2022). A review of interactions of pesticides within various interfaces of intrinsic and organic residue amended soil environment. *Chemical Engineering Journal Advances, 11*, 100301. https://doi.org/10.1016/j.ceja.2022.100301

Ray, S., & Shaju, S. T. (2023). Bioaccumulation of pesticides in fish resulting toxicities in humans through food chain and forensic aspects. *Environmental Analysis, Health and Toxicology, 38*(3), e2023017-0. https://doi.org/10.5620/eaht.2023017

Ray, S. S., Parihar, K., Goyal, N., & Mahapatra, D. M. (2024). Synergistic insights into pesticide persistence and microbial dynamics for bioremediation. *Environmental Research, 257*, 119290. https://doi.org/10.1016/j.envres.2024.119290

Saleh, I. A., Zouari, N., & Al-Ghouti, M. A. (2020). Removal of pesticides from water and wastewater: Chemical, physical and biological treatment approaches. *Environmental Technology & Innovation, 19*, 101026. https://doi.org/10.1016/j.eti.2020.101026

Shefali, Kumar, R., Sankhla, M. S., Kumar, R., & Sonone, S. S. (2021). Impact of pesticide toxicity in aquatic environment. *Biointerface Research in Applied Chemistry, 11*(3), 10131–10140. https://doi.org/10.33263/BRIAC113.1013110140

Skendžić, S., Zovko, M., Živković, I. P., Lešić, V., & Lemić, D. (2021). The impact of climate change on agricultural insect pests. *Insects, 12*(5). https://doi.org/10.3390/insects12050440

Steffens, K., Larsbo, M., Moeys, J., Jarvis, N., & Lewan, E. (2013). Predicting pesticide leaching under climate change: Importance of model structure and parameter uncertainty. *Agriculture, Ecosystems & Environment, 172*, 24–34. https://doi.org/10.1016/j.agee.2013.03.018

Sturve, J., Gustavsson, M., Moksnes, P.-O., & de Abreu, D. C. (2021). Effects of pesticides and metals on penaeid shrimps in Maputo Bay, Mozambique – A field study. *Marine Pollution Bulletin, 173*, 112964. https://doi.org/10.1016/j.marpolbul.2021.112964

Sule, R. O., Condon, L., & Gomes, A. V. (2022). A common feature of pesticides: Oxidative stress-the role of oxidative stress in pesticide-induced toxicity. *Oxidative Medicine and Cellular Longevity, 2022*, 5563759. https://doi.org/10.1155/2022/5563759

Syafrudin, M., Kristanti, R. A., Yuniarto, A., Hadibarata, T., Rhee, J., Al-Onazi, W. A., Algarni, T. S., Almarri, A. H., & Al-Mohaimeed, A. M. (2021). Pesticides in Drinking Water-A Review. *International Journal of Environmental Research and Public Health, 18*(2), 468. https://doi.org/10.3390/ijerph18020468

Taidi, Y., & El Laghdach, A. (2024). Pesticides: Environmental fate, human health impacts, and analytical techniques from extraction to removal. *Moroccan Journal of Chemistry, 12*(1), 89–129. https://doi.org/10.48317/IMIST.PRSM/morjchem-v12i1.40056

Tudi, M., Daniel Ruan, H., Wang, L., Lyu, J., Sadler, R., Connell, D., Chu, C., & Phung, D. T. (2021). Agriculture development, pesticide application and its impact on the environment. *International Journal of Environmental Research and Public Health, 18*(3), 89. https://doi.org/10.3390/ijerph18031112

Verma, N. S., Kuldeep, D. K., Chouhan, M., Prajapati, R., & Singh, S. K. (2023). A review on eco-friendly pesticides and their rising importance in sustainable plant protection practices. *International Journal of Plant & Soil Science, 35*(22), 200–214. https://doi.org/10.9734/ijpss/2023/v35i224126

Wołejko, E., Łozowicka, B., Jabłońska-Trypuć, A., Pietruszyńska, M., & Wydro, U. (2022). Chlorpyrifos occurrence and toxicological risk assessment: A review. *International Journal of Environmental Research and Public Health, 19*(19). https://doi.org/10.3390/ijerph191912209

Yi, J., Gerbens-Leenes, P. W., & Aldaya, M. M. (2024). Crop grey water footprints in China: The impact of pesticides on water pollution. *Science of the Total Environment, 935*, 173464. https://doi.org/10.1016/j.scitotenv.2024.173464

Zhang, S. W., Wang, R., Wang, F., & Cai, M. (2023). Assessment of currently used and restricted organophosphorus pesticides and their degradation products in urban drinking water: An investigation of eight cities in Yangtze River Delta urban agglomeration, East China. *Journal of Hazardous Materials Advances, 9*, 100211. https://doi.org/10.1016/j.hazadv.2022.100211

Global Trade and Pesticide Use

Alia Syafiqah Abdul Hamed and Nur Hidayah

1 Introduction

The demand for agricultural products in industrialized nations fuels a system where production is increasingly outsourced to developing countries. Global supply chains have transformed food production, making it more efficient but also more reliant on scarce resources like water and fertile soil (Gómez & Lee, 2023). The internationalization of production has led to a shift from domestic to global networks, significantly altering agricultural practices and trade patterns (Singh et al., 2023). As these countries expand their agricultural sectors to meet global demand, intensive farming techniques have been used, including the widespread use of pesticides. Pesticides encompass various types, including herbicides, insecticides, fungicides, rodenticides and nematicides. These chemicals are frequently used in modern agriculture as an effective and cost-efficient way to enhance both the quality and quantity of crop yields, helping to ensure food security for the growing global population. The dependency on agrochemicals to boost yields highlights the interconnectedness between global trade and environmental impacts, particularly through the lens of pesticide use.

One of the most concerning outcomes of this dynamic is the shifting of environmental burdens. High-income nations, while consuming imported goods, inadvertently drive land-use change, deforestation and pollution in lower-income countries where these goods are produced. The result is a form of environmental decoupling,

A. S. A. Hamed (✉)
Faculty of Ocean Engineering Technology, Universiti Malaysia Terengganu,
Kuala Nerus, Terengganu, Malaysia
e-mail: p5334@pps.umt.edu.my

N. Hidayah
Centre for Earth Sciences and Environment, Faculty of Science and Technology, Universiti Kebangsaan Malaysia, Bangi, Selangor, Malaysia

B. R. Babaniyi, E. E. Babaniyi (eds.), *The Interplay of Pesticides and Climate Change*, https://doi.org/10.1007/978-3-031-81669-7_5

where the negative impacts of agricultural intensification, such as biodiversity loss and pesticide contamination, are displaced from the consumption regions to the production regions (Ortiz et al., 2021). This chapter explores the complex relationship between global trade and the environmental consequences of pesticide use, drawing attention to how these pressures manifest in countries like Brazil, Indonesia and Ethiopia.

Weak governance and regulatory frameworks in many developing countries further worsen the environmental and health risks associated with pesticide use. In countries like Ethiopia, where pesticide supply chains are poorly regulated, international market pressures lead to aggressive marketing and the proliferation of unsafe chemicals (Mengistie et al., 2016). This results in significant environmental and health risks, underlining how weak governance in supply chains can exacerbate the adverse effects of pesticides. The environmental and social costs of such practices are often handled most acutely in these vulnerable regions, where small-scale farmers bear the brunt of harmful exposure and environmental degradation, all while trying to meet the demands of foreign markets.

Through the lens of environmental economics and policy, this chapter will also address the global spillover effects of pesticide use, as well as the responsibility of wealthier nations in driving these patterns. Bibliometric data on pesticide's trade connects academic research with real-world applications, enabling informed decision-making in research evaluation, funding and strategic planning. A bibliometric analysis using the Scopus database search query TITLE-ABS-KEY (pesticide* AND trade*), between 1994 and 2023, shows the total of 1773 research articles related to pesticides and trade were published. As presented in Fig. 1, the cumulative number of publications grew exponentially over time ($R^2 = 0.9285$). During the first two decades of this period, fewer than 60 papers were published annually on average, indicating that the field was in its early stages. Interestingly, publication numbers began to rise sharply from 2015 onwards, reaching a peak in 2022. This trend reflects the increasing recognition of the need to understand the environmental costs associated with pesticide use, as developed countries, through their consumption, indirectly drive pesticide use and pollution in producer countries (Tang et al., 2022).

This analysis of the Scopus data also revealed that the United States, China and India contribute the most research on this topic, with 466, 140 and 136 articles, respectively. The French National Research Institute for Agriculture, Food, and Environment (INRAE) leads as a most productive affiliations with 51 publications. By linking consumption patterns in developed countries with production practices in developing nations, this chapter emphasizes the need for more equitable and sustainable approaches to managing pesticide use in a globalized world. As global supply chains continue to evolve, understanding the environmental costs embedded in international trade will be crucial to mitigating future harm (Malik et al., 2023; Reis et al., 2023).

This chapter will be organized as follows: First, it will examine the influence of international trade on pesticide use patterns, exploring how market demands shape agricultural practices. Next, it will analyse the environmental effects of pesticide

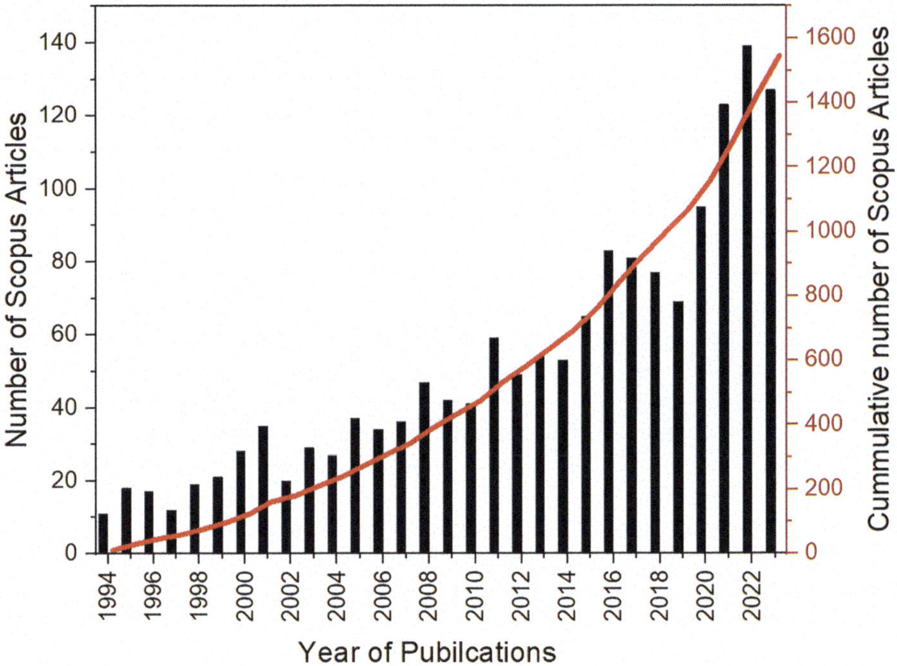

Fig. 1 Annual publication and cumulative number of Scopus articles on pesticides trade

use, including impacts on biodiversity and ecosystem health. The discussion will then turn to the interplay between climate change and trade dynamics, highlighting how shifting climatic conditions affect agricultural practices and pesticide reliance. Following this, this chapter will outline collaborative efforts addressing pesticide issues, focusing on policy initiatives and international agreements aimed at reducing the adverse effects of pesticides. Finally, this chapter will conclude with reflections on the need for a more sustainable and equitable approach to pesticide management within the context of global trade, emphasizing the importance of addressing both environmental and social dimensions.

2 Influence of International Trade on Pesticide Use Patterns

The increase in research interest for trade of pesticides coincides with significant trends in global pesticide use, which reached 3.7 million tonnes of active ingredients in 2022, representing a 50% increase over the past three decades since 1990 (FAO, 2024). International trade has a profound impact on global pesticide use patterns, shaping both consumption levels and regulatory frameworks across different regions. A key insight from various studies is that countries with higher levels of agricultural exports tend to use more pesticides. This observation aligns with the

International Political Economy (IPE) perspective, which suggests that trade liberalization in agricultural products leads to increased pesticide consumption (Longo and York 2008). As agricultural products move across borders more freely, the demand for higher yields and pest-free produce increases, fuelling pesticide usage.

Table 1 provides a breakdown of global agricultural pesticide use in metric tonnes across different pesticide categories. Herbicides lead agricultural pesticide use, with 1,942,604.1 metric tonnes. The percentage of 52.6% reflects the dominance of herbicides in global pesticide use, followed by fungicides/bactericides and insecticides, while rodenticides and other types of pesticides represent only a small fraction of total usage. This substantial amount reflects the extensive reliance on herbicides for pest control in large-scale agricultural operations. The significant increase in pesticide use, especially due to the expansion of international trade, highlights the pressing need to explore how global supply chains influence where and how pesticides are used. This increased reliance on herbicides has been linked to the growing adoption of herbicide-tolerant genetically modified (GM) crops, which further intensifies the volume of herbicide usage worldwide (Brookes & Barfoot, 2018). As agricultural trade continues to expand globally, there is concern that the intensification of pesticide use in export-oriented countries could increase environmental and health risks associated with chemical exposure.

The total pesticide uses in agriculture, as indicated by the data, reach nearly 3.7 million metric tonnes. Insecticides and fungicides/bactericides, while second to herbicides, also contribute significantly, with insecticides at 772,457.4 metric tonnes and fungicides/bactericides at 793,922.6 metric tonnes, which cover 20.9% and 21.5%, respectively. This indicates the widespread need for pest and disease control measures in various crops, particularly high-value fruits and vegetables. However, excessive pesticide use can lead to environmental contamination, pesticide resistance and human health risks. Numerous studies emphasize the need for integrated pest management (IPM) practices that reduce the dependency on chemical pesticides and focus on sustainable alternatives. As global trade and agricultural production continue to rise, the implementation of stricter pesticide regulations and better risk assessment frameworks will be critical for minimizing the adverse impacts of pesticide use.

As illustrated in Fig. 2, the global landscape of pesticide consumption in 2022 was marked by significant disparities across continents, with Brazil leading the list with the used of 800,650 metric tonnes of pesticides. Brazil's extensive agricultural

Table 1 Global agricultural pesticide use by type (metric tonnes)

Type of pesticide	Total for agriculture use (tonnes)	Pesticide percentage (%)
Herbicides	1,942,604.1	52.6
Fungicides and bactericides	793,922.6	21.5
Insecticides	772,457.4	20.9
Rodenticides	17,761.6	0.5
Others	164,189.0	4.5
Total	3,690,934.7	100

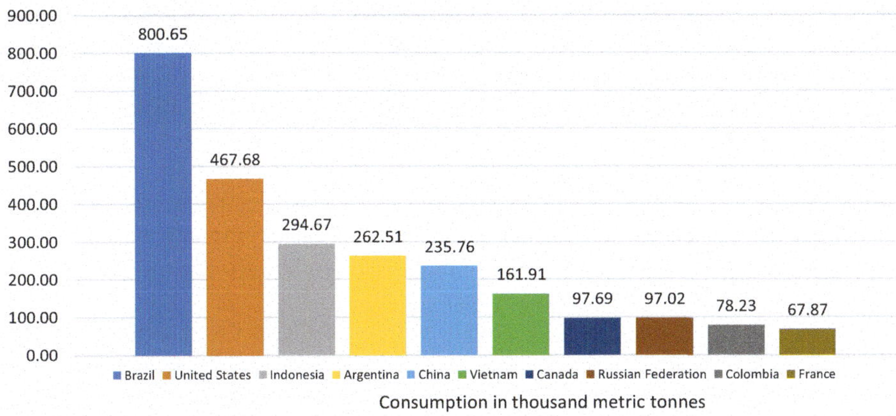

Fig. 2 Leading countries in agricultural consumption of pesticides worldwide in 2022

sector, particularly focused on soybean and sugarcane cultivation, has been a driving force behind its heavy reliance on pesticides. This makes Brazil not only the largest consumer in South America but also worldwide, reflecting its intensive farming practices that depend on agrochemical inputs to control pests and maximize yield in these export crops (Lopes-Ferreira et al., 2022). In addition, United States followed as the second-largest pesticide consumer, utilizing around 467,680 metric tonnes in 2022. The country's diverse agricultural landscape, including corn, soybeans, and cotton, depends heavily on herbicides, insecticides and fungicides to sustain high productivity levels. Such large-scale pesticide use is driven by the need to maintain competitive crop yields while managing weeds and pests that threaten agriculture across vast farmlands.

In Asia, Indonesia and China stand out as significant pesticide consumers. Indonesia, with a consumption of around 294,670 metric tonnes, demonstrates its reliance on pesticides for rice, palm oil and rubber plantations. In contrast, China, although one of the largest agricultural producers, consumed approximately 235,760 metric tonnes in 2022. Despite China's massive production scale, the use of IPM strategies in some regions is helping to moderate pesticide use to some extent (Lengai et al., 2020; Das et al., 2024). Argentina's consumption of 262,510 metric tonnes underscores its agricultural reliance on chemical inputs, particularly for crops like soybeans, maize and wheat, similar to Brazil. This extensive pesticide use highlights the region's dependence on agrochemicals to maintain crop health and yield under challenging environmental conditions. Other contributors included Russia and Canada in Europe and North America, respectively, along with Spain, France and Australia, which further illustrate the global reliance on pesticides across different agricultural contexts. This data underscores critical environmental and health concerns linked to such high levels of chemical inputs, emphasizing the urgent need for sustainable agricultural practices worldwide.

International trade accounts for approximately 32% of the global pesticide footprint, with high-income countries, such as China, Germany and the United Kingdom, being the largest net importers of pesticide hazards (Tang et al., 2022). The globalization of pesticide trade has further intensified these patterns by creating denser international trade networks. The size, density and reciprocity of these networks have grown, facilitating the widespread distribution and use of pesticides across the globe. This interconnectedness highlights the role of global markets in spreading pesticide products, making their use more prevalent worldwide. Furthermore, economic factors also play a significant role in determining pesticide use. Countries with differing levels of economic development exhibit varying consumption patterns. For instance, lower-income countries often underutilize pesticides, while higher-income nations tend to overuse them.

Additionally, economic variables such as foreign direct investment (FDI) are closely linked to pesticide consumption trends, reflecting broader economic influences on agricultural practices. Trade liberalization often boosts the productivity of the agricultural activities, by allowing the farmers a greater access to the market. This distance often hides the environmental impacts of pesticide use, especially in low-income regions where pesticide application has soared. For instance, in low-income countries, pesticide use has increased to 153% in the past decade, largely because of agricultural growth linked to trade (Shattuck et al., 2023). The relationship between trade and pesticide regulation is also notable. Trade liberalization can prompt stricter pesticide regulations as countries shift towards importing more food and relying less on domestic agricultural production. However, these regulatory measures can affect the competitiveness of agricultural sectors in international markets and contribute to increased food price volatility (Assouto et al., 2020). This trade-off between regulation and market competitiveness underscores the complex dynamics at play in the global pesticide market.

Moreover, the international trade of pesticides has led to a homogenization of pesticide use worldwide. Commonly used chemicals such as glyphosate, chlorpyrifos and mancozeb are now prevalent across many countries, reflecting a standardized global market shaped by international trade (Valbuena et al., 2021). This trend towards uniformity indicates how trade influences not only the volume but also the types of pesticides used globally. Effective international cooperation and the development of sustainable agricultural policies are critical to mitigate the negative impacts of pesticide use driven by global trade. While international trade can improve the food security, it also necessitates a re-evaluation of pesticide management strategies to prevent environmental degradation and reduce the health risks associated with the rise of pesticide application in trade-intensive agricultural systems (Tang et al., 2022; Wisnujati, 2023).

Therefore, international trade exerts a significant influence on pesticide use patterns by promoting higher consumption in export-heavy countries, fostering dense trade networks and creating economic disparities in pesticide application. Trade liberalization can also result in stricter regulations, affecting domestic agricultural competitiveness. Finally, the global market shows a clear trend towards the homogenization of pesticide use, driven by these international trade dynamics.

3 Environmental Effects of Pesticide Use

International trade and global supply chains significantly impact the trend of pesticide use, which thus have a various environmental impact. The globalization of food trade often conceals the ecological significances of pesticide application because the geographical separation between production and consumption masks its effects. Pesticides are known to be significant pollutants, endangering biodiversity and disrupting the human ecosystem. Recent research has also shown that these pollutants persist in the environment, leading to long-term contamination of soil and water systems, which exacerbates the degradation of ecosystem services vital for food security (Silva et al., 2023). Figure 3 illustrated the environmental effects of pesticides usage. The pesticides application in agriculture has significant environmental impacts, affecting various ecosystems, biodiversity and human health (Santosh, 2022). Even though these toxic chemicals are designed to control pest populations, it also often has unintended consequences on non-target species and the environment as a whole.

Pesticides can significantly reduce soil quality by killing beneficial soil organisms, such as earthworms and microbes, which are crucial for nutrient cycling, decomposition, and soil structure. This disrupts soil ecosystems, decreases fertility and leads to reduced biodiversity. Neonicotinoids and fipronil have been shown to negatively affect soil invertebrates and microorganisms that maintain ecosystem

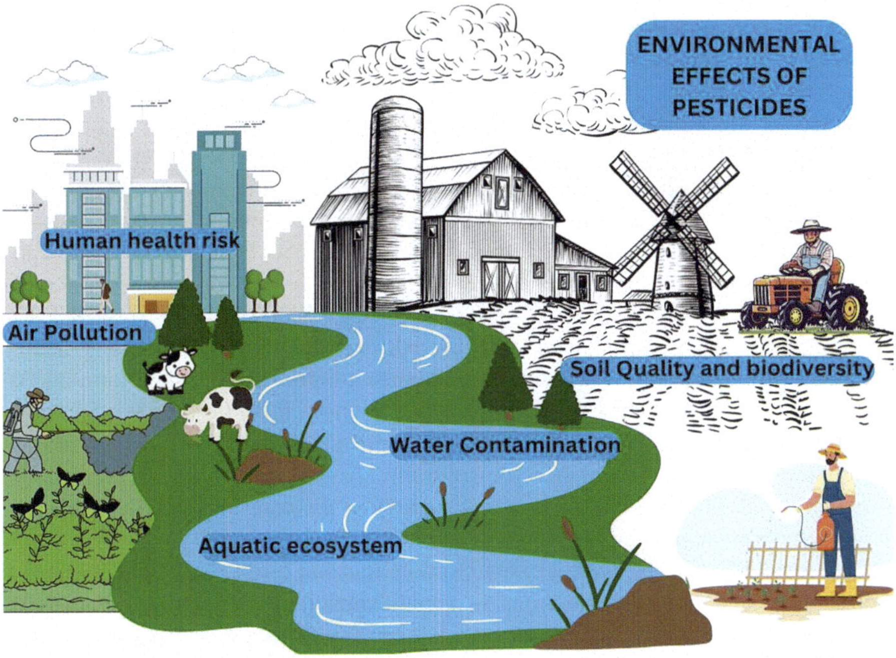

Fig. 3 Environmental effect of pesticides

services like decomposition and soil respiration (Pisa et al., 2015). Moreover, herbicides like glyphosate can alter microbial biomass and nutrient availability in the soil, negatively impacting the soil's ability to support plant growth (Castrejón-Godínez et al., 2021). Moreover, agricultural runoff is a major contributor to water pollution, introducing pesticides into rivers, lakes and groundwater. Pesticides such as insecticides are found at levels exceeding regulatory thresholds, threatening aquatic biodiversity. More than 50% of the detected insecticide concentrations in surface waters globally exceed safe levels, which drastically reduces aquatic biodiversity, including fish and macroinvertebrates (Stehle & Schulz, 2015).

In addition, pesticides have severe effects on aquatic ecosystems, impacting the health of fish and other aquatic organisms. Insecticides like organophosphates are highly toxic to fish, affecting their growth, reproduction and behaviour. Fish exposed to pesticide-laden water show signs of immune suppression, reduced growth rates and reproductive disorders. Over time, pesticides can bioaccumulate in aquatic ecosystems, leading to long-term damage to food webs (Banaee, 2012). Pesticides can also contribute to air pollution through volatilization and spray drift during application. Neonicotinoids, for example, can be dispersed into the air in dust particles during crop sowing, which then settle in distant ecosystems, contaminating non-target species and environments. This drift not only reduces air quality but also poses risks to pollinators such as bees, which are crucial for biodiversity and food production (Chagnon et al., 2015). These effects show that pesticides significantly impact soil, water and air ecosystems, posing a threat to biodiversity, ecosystem services and public health.

The trade of commodities related to the high pesticide usage contributes considerably to global biodiversity loss, with an estimated 30% of this decline attributed to such trade practices. In particular, insecticides such as neonicotinoids have been shown to severely harm pollinators like bees, which are crucial for maintaining biodiversity and supporting agricultural productivity (Woodcock et al., 2017). Moreover, the rise in pesticide use, particularly in low-income countries, poses severe health risks to agricultural workers and local communities (Sarkar et al., 2021). The global soybean trade exemplifies this issue, with 64% of traded soybeans linked to significant environmental and health risks, largely due to pesticide use concentrated in a few countries (Wang et al., 2024). Studies have also found that pesticide residues on crops can be transferred through the food chain, posing dietary risks to consumers in both exporting and importing countries (Ray & Shaju, 2023).

A study by Bhat et al. (2019) highlights the environmental effects of various pesticides on agriculture ecosystems. The findings show that Chlorpyrifos and trifluralin are notably toxic to fish, with trifluralin also causing vertebral deformities. Meanwhile, glyphosate leads to erratic behaviour in fish, and 2,4-D herbicides induce physiological stress in salmon and reduce food-gathering in rainbow trout. The adverse effects of pesticides extend to dolphins and birds, with DDT impairing reproductive and immune functions in aquatic mammals. The contamination of freshwater ecosystems by such chemicals has cascading effects, potentially disrupting aquatic food webs and leading to reduced fish populations, which are essential for both biodiversity and local fisheries (Bashir et al., 2020).

Furthermore, in the soil ecosystem, triclopyr suppresses beneficial bacteria and mycorrhizal fungi, while glyphosate and 2,4-D suppress nitrogen-fixing bacteria and blue-green algae, hindering nutrient cycling. This disruption of the soil microbial community reduces not only the fertility of the soil but also the capacity for carbon sequestration in the soil, thereby enhancing climate change. In general, these pesticides pose a high level of risk to aquatic ecosystems and soil health, affecting biodiversity and proper ecosystem functioning. The other factor affecting the demand of the control products in exporting countries is dictated by export market requirements; this usually leads to an increase in the application of pesticides to meet quality standards demanded by international markets.

Lack of awareness on safe use of pesticides among farmers leads to increased risk of poisoning due to pesticides, particularly in areas where education relating to proper handling and use is limited (Bhandari, 2019). In regions where regulatory frameworks are weak or poorly enforced, pesticide overuse is rampant, further intensifying the environmental and health challenges associated with these chemicals. This "export effect" results in heightened pesticide use, even when domestic consumption patterns do not justify such practices. As the global demand for agricultural products continues to rise, it is essential for policymakers to implement strategies that minimize the environmental impacts of pesticide use.

In order to reduce the reliance on pesticides uses, stricter regulations on pesticide application in exporting countries, combined with the promotion of sustainable agricultural practice can be used. The adoption of agroecological approaches, such as crop rotation, biological pest control and organic farming, offers promising alternatives to chemical pesticides and can significantly reduce environmental harm while maintaining crop yields. By addressing these problems comprehensively, the environment and the general public's health can be protected while fostering a more sustainable global food system.

4 Climate Change and Trade Dynamics

Climate change introduces another layer of intricacy to global trade dynamics and pesticide usage. As climate patterns shift, crop pest distributions and intensities change, requiring adjustments in pest management strategies. Agricultural production regions, facing these new climate pressures, often rely more heavily on pesticides to manage emerging pest problems, which can increase the use of chemical inputs and heighten associated environmental risks (Kumar et al., 2024). Climate change has also been linked to the accelerated development of pest resistance to pesticides, further increasing the reliance on chemical solutions.

Yang et al. (2024) highlight how climate change could amplify agriculture's environmental impacts by increasing greenhouse gas emissions, water use, soil erosion and agrochemical pollution. Figure 4 shows the impact climate change to the pesticide use. Rising temperatures and changing climate conditions are causing more pests and plant diseases, forcing farmers to use more pesticides and creating a

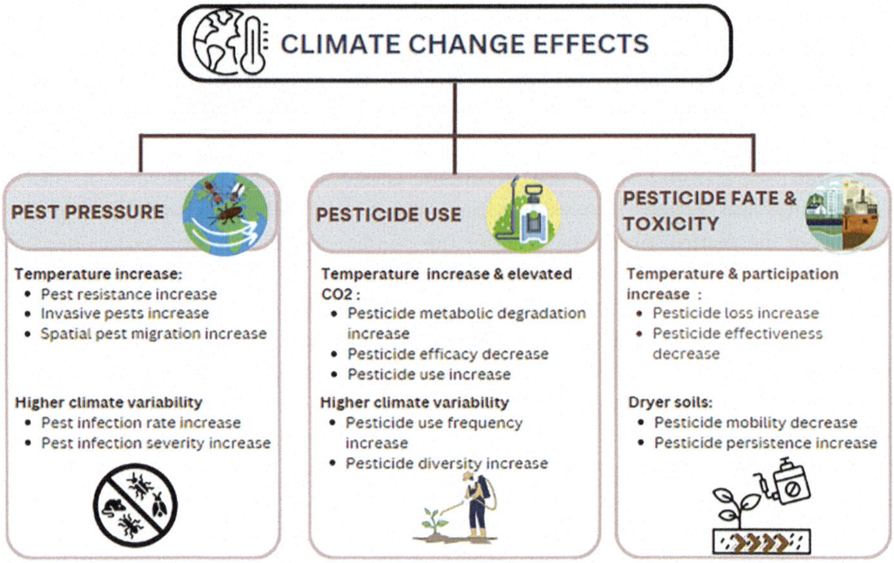

Fig. 4 Climate change effect on pest and pesticide usage

growing reliance on agrochemicals. Climate change significantly impacts pest pressure by altering environmental conditions in ways that increase pest populations and spread. Rising temperatures facilitate the development of pest resistance and the migration of invasive species into new regions. Warmer climates extend the growing season for many pests, leading to higher infestation rates and more widespread crop damage (Subedi et al., 2023). Additionally, increased climate variability, such as extreme weather events and fluctuating humidity levels, further amplifies pest infection rates and the severity of their impact on crops. These changes demand more intensive pest control strategies, which may exacerbate the use of chemical interventions.

Furthermore, climate change also affects pesticide use and efficacy. Higher temperatures and elevated CO_2 levels accelerate pesticide degradation, reducing their effectiveness and requiring farmers to apply more chemicals or use them more frequently (Subedi et al., 2023). As climate variability intensifies, more diverse and frequent pesticide applications are necessary to address a broader range of pests and diseases, leading to increased overall pesticide use. Moreover, rising temperatures and changing precipitation patterns alter the fate and toxicity of pesticides, with higher temperatures increasing pesticide runoff and loss, while drier soils prolong pesticide persistence, which can lead to long-term environmental contamination.

Notably, extreme weather events such as heavy precipitation, prolonged droughts, tornadoes and floods are anticipated to disrupt pest lifecycles, which may lead to unpredictable outbreaks, increasing pesticide use unpredictably across different regions. Climate-induced disruptions in global supply chains, such as extreme weather events affecting transportation or climate policy-driven changes to trade agreements, can further alter pesticide availability and use patterns. These disruptions reveal the intricate links between climate change, trade and pesticide management. For instance, in 2023, droughts can severely affect the main trade routes, such as the Panama Canal, causing significant disruptions to global shipping and trade volumes (Karl Plume, 2023).

The study by Bibi and Rahman (2023) focuses on the damaging impact of climate change on soil properties in agriculture, emphasizing how abiotic stresses like salinity, drought and temperature fluctuations are negatively affecting crop physiology, productivity and yield. These challenges threaten global food security and the stability of agroecosystems. Besides, the extensive use of chemical fertilizers and pesticides further worsens climate change and soil degradation. To address these issues, it is crucial to adopt more eco-friendly and sustainable practices that mitigate the adverse effects of climate change on agriculture. These also highlight the urgent need for sustainable solutions to protect soil, agriculture and the environment from the accelerating impacts of climate change.

Additionally, higher temperatures and increased rainfall may also enhance export values for some countries, further influencing trade routes and performances. Bozzola et al. (2023) discover that rising temperatures and rainfall can boost the value of agri-food exports in some regions, signifying that climate change may have positive trade effects in certain areas. However, while some regions may benefit from these shifts, others may experience declining crop yields and export capacities due to extreme weather conditions or degraded soil quality, leading to increased reliance on food imports. The impact of climate change on crop yields varies by region, altering comparative advantages and trade patterns (Hedlund et al., 2022).

Countries often increase trade to encounter the food shortages, relying on international trade to enhance food security. However, the stability of global food trade networks is declining, particularly for agricultural crops, making food security systems more vulnerable. The combination of supply chain disruptions, increased demand for pesticides and unpredictable weather patterns threatens long-term food security and underscores the need for global cooperation in trade and sustainable agriculture practices (Tchonkouang et al., 2024). This situation highlights the need for sustainable agricultural practices to improve resilience to climate change and other disruptions.

5 Collaborative Efforts on Pesticide Issues

Collaborative efforts on pesticide issues are crucial for addressing the environmental and health risks related with their use. The necessity for global cooperation is emphasized, particularly in managing pesticide risks that arise from international trade, as food consumption in one country can lead to extensive pesticide use in another, often with less stringent regulations (Wang et al., 2024). IPM strategies, supported by global research initiatives, are crucial for building resilience against these challenges (Kumar et al., 2024). Policymakers should promote sustainable practices and develop regulatory frameworks that encourage the adoption of effective pest control measures across national borders. Collaboration at the intergovernmental level is also key to harmonizing regulatory frameworks for pesticide usage, ensuring a shared responsibility for environmental protection across the global supply chain.

Gong et al. (2023) state the interaction between three key stakeholders in the pesticide industry: the government, pesticide operating enterprises and farmers. The relationships between these entities indicate a system of regulation, supervision and participation in social oversight. The government plays a central role by regulating and inspecting pesticide sales to ensure enterprises comply with marketing practices. In parallel, the government supervises the behaviour of pesticide trading enterprises to ensure they provide high-quality, qualified products. This framework ensures that the pesticide operating enterprises act within legal and quality standards. On the other hand, farmers are also involved in this supervision process by participating in social oversight. The government encourages farmers' participation to ensure a broader societal check on the pesticide market.

Besides, farmers rely on pesticide operating enterprises to supply safe and effective products. The joint relationship between the pesticide operating enterprises and the farmers demonstrates the mutual dependence: while enterprises provide qualified pesticides, farmers participate in the social supervision of these companies' practices. Overall, this highlights a structured system of checks and balances to ensure the responsible marketing and use of pesticides, protecting both consumer interests and public safety.

Despite the challenges posed by climate change, there are also opportunities for countries to collaborate on sustainable agricultural practices and trade strategies, fostering resilience in the face of environmental changes. The adoption of methods such as IPM, organic farming, and the use of biopesticides can help reduce reliance on harmful chemicals. For instance, the promotion of agroecological practices in Latin America has shown success in reducing chemical inputs while maintaining yields, demonstrating how region-specific collaborations can yield effective results (Altieri & Toledo, 2011). Although global pesticide use has steadily increased, especially in low- and lower-middle-income countries, improving the accuracy of pesticide use data is essential for achieving international targets, such as the COP 15 goal to reduce pesticide risks by 50% by 2030 (Shattuck et al., 2023).

Addressing cross-border pesticide issues requires collaboration at various levels. Harmonizing pesticide regulations and promoting sustainable agricultural practices can be achieved through international agreements, such as "Towards a Chemical Pesticide-Free Agriculture", an initiative signed in 2020 by 24 research organizations from 16 European countries (Jacquet et al., 2022). In addition to governmental collaboration, non-governmental organizations (NGOs) and private sectors can play a crucial role in advocating for pesticide-free certification and supply chain monitoring, helping to raise consumer awareness and demand for sustainable agricultural products. This alliance can enhance global effort includes developing preventive approaches based on agroecological principles, which emphasize biodiversity and ecosystem health.

In addition, governments play an essential role in reducing pesticide use through legislation and policy frameworks. For example, the European Union's Green Deal, which aims to reduce pesticide use by 50% by 2030, provides a model for integrating environmental objectives with agricultural policy (Schneider et al., 2023). Integrating climate policies with trade regulations can further strengthen efforts to manage pesticide use effectively. Countries can adopt standards that promote environmentally friendly practices in agricultural exports, ensuring that products entering their markets meet sustainability criteria. This not only helps reduce pesticide usage but also supports the development of greener agricultural technologies.

At a regional level, initiatives promoting legal policy and sustainable agricultural practices can help reduce pesticide reliance. This also can support in maintaining crop productivity and environmental integrity. Regional collaborations, such as the Association of Southeast Asian Nations (ASEAN) network for pesticide management, have demonstrated that collective efforts can lead to more effective control over pesticide use and enforcement of environmental standards. ASEAN prioritizes agricultural market integration to tackle food security and rural poverty, but challenges like cross-border supply chains and farm diversification require careful policy planning (Roy et al., 2023). Collaborative research and knowledge-sharing platforms allow countries to exchange best practices and technologies, enabling them to implement context-specific pesticide management strategies aligned with broader environmental sustainability goals.

6 Conclusion

In conclusion, this chapter has underscored the intricate connections between international trade, pesticide use and environmental sustainability. It reveals how global market demands significantly influence agricultural practices, often leading to increased pesticide reliance, which in turn has serious environmental repercussions, including the degradation of biodiversity and ecosystem health. The interplay between climate change and trade dynamics further exacerbates these issues, as shifting climatic conditions push farmers towards greater pesticide use to maintain crop yields in a more challenging environment. However, addressing these challenges requires more than just reactive measures. Collaborative efforts through policy initiatives and international agreements are critical in reducing the negative impacts of pesticide use. Such initiatives must not only address the environmental harms but also consider the social and economic dimensions of pesticide management, ensuring that vulnerable farming communities are not disproportionately affected. Ultimately, this chapter emphasizes the need for a more sustainable and equitable approach to pesticide management within the global trade framework. Moving forward, it is crucial to adopt strategies that balance agricultural productivity with environmental protection, fostering a system that promotes long-term resilience for both ecosystems and communities. This holistic approach is essential to achieving a truly sustainable agricultural future.

References

Altieri, M. A., & Toledo, V. M. (2011). The agroecological revolution in Latin America: Rescuing nature, ensuring food sovereignty and empowering peasants. *Journal of Peasant Studies, 38*, 587–612. https://doi.org/10.1080/03066150.2011.582947

Assouto, A. B., Houensou, D. A., & Semedo, G. (2020). Price risk and farmers' decisions: A case study from Benin. *Scientific African, 8*, e00311. https://doi.org/10.1016/J.SCIAF.2020.E00311

Banaee, M. (2012). Adverse effect of insecticides on various aspects of Fish's biology and physiology. In *Insecticides – Basic and other applications*. IntechOpen. https://doi.org/10.5772/30545

Bashir, I., Lone, F. A., Bhat, R. A., et al. (2020). Concerns and threats of contamination on aquatic ecosystems. In *Bioremediation and biotechnology: Sustainable approaches to pollution degradation* (pp. 1–26). Springer International Publishing.

Bhandari, L. R. (2019). Increasing trend of pesticides' use in vegetable farming and its impact for human health. *Nepal Journal of Multidisciplinary Research, 2*, 50–67. https://doi.org/10.3126/njmr.v2i3.26975

Bhat, R., Khajuria, M., & Mansotra, D. K. (2019). A systematic review on global environmental risks associated with pesticide application in agriculture. In *Contaminants in agriculture and environment: Health risks and remediation* (pp. 96–110). Agro Environ Media – Agriculture and Environmental Science Academy.

Bibi, F., & Rahman, A. (2023). An overview of climate change impacts on agriculture and their mitigation strategies. *Agriculture (Switzerland), 13*, 1508. https://doi.org/10.3390/AGRICULTURE13081508/S1

Bozzola, M., Lamonaca, E., & Santeramo, F. G. (2023). Impacts of climate change on global agri-food trade. *Ecological Indicators, 154*, 110680. https://doi.org/10.1016/j.ecolind.2023.110680

Brookes, G., & Barfoot, P. (2018). Environmental impacts of genetically modified (GM) crop use 1996-2016: Impacts on pesticide use and carbon emissions. *GM Crops & Food, 9*, 109–139. https://doi.org/10.1080/21645698.2018.1476792

Castrejón-Godínez, M. L., Tovar-Sánchez, E., Valencia-Cuevas, L., et al. (2021). Glyphosate pollution treatment and microbial degradation alternatives, a review. *Microorganisms, 9*, 2322.

Chagnon, M., Kreutzweiser, D., Mitchell, E. A. D., et al. (2015). Risks of large-scale use of systemic insecticides to ecosystem functioning and services. *Environmental Science and Pollution Research International, 22*, 119–134. https://doi.org/10.1007/S11356-014-3277-X

Das, N., Sujatha, G., Teja, K. S. S., et al. (2024). Integrated pest management: Success stories and key takeaways. *Uttar Pradesh Journal of Zoology, 45*, 229–244. https://doi.org/10.56557/UPJOZ/2024/V45I164304

FAO. (2024). *Pesticides use and trade, 1990–2022*. https://doi.org/10.4060/CD1486EN

Gómez, M. I., & Lee, D. (2023). Transforming food supply chains for sustainability. *Journal of Supply Chain Management, 59*, 79–92. https://doi.org/10.1111/JSCM.12310

Gong, J., Du, H., & Sun, Y. (2023). Collaboration among governments, pesticide operators, and farmers in regulating pesticide operations for agricultural product safety. *Agriculture (Switzerland), 13*, 2288. https://doi.org/10.3390/agriculture13122288

Hedlund, J., Carlsen, H., Croft, S., et al. (2022). Impacts of climate change on global food trade networks. *Environmental Research Letters, 17*, 124040. https://doi.org/10.1088/1748-9326/aca68b

Jacquet, F., Jeuffroy, M. H., Jouan, J., et al. (2022). Pesticide-free agriculture as a new paradigm for research. *Agronomy for Sustainable Development, 42*, 8.

Karl Plume. (2023). Panama Canal drought to delay grain ships well into 2024. *Reuters*. https://www.reuters.com/markets/commodities/panama-canal-drought-delay-grain-ships-well-into-2024-2023-12-11/. Accessed 24 Sept 2024.

Kumar, S., Choudhary, M., Reddy, K. J., et al. (2024). A review on the impact of climate change on plant pathogen interactions. *Journal of Advances in Microbiology, 24*, 11–27. https://doi.org/10.9734/JAMB/2024/V24I8843

Lengai, G. M. W., Muthomi, J. W., & Mbega, E. R. (2020). Phytochemical activity and role of botanical pesticides in pest management for sustainable agricultural crop production. *Scientific African, 7*, e00239. https://doi.org/10.1016/J.SCIAF.2019.E00239

Longo S, York R (2008). Agricultural exports and the environment: A cross-national study of fertilizer and pesticide consumption. *Rural Sociology, 73*:82–104. https://doi.org/10.1526/003601108783575853

Lopes-Ferreira, M., Maleski, A. L. A., Balan-Lima, L., et al. (2022). Impact of pesticides on human health in the last six years in Brazil. *International Journal of Environmental Research and Public Health, 19*, 3198. https://doi.org/10.3390/IJERPH19063198

Malik A, Lafortune G, Dahir S, et al (2023) Global environmental and social spillover effects of EU's food trade. *Global Sustainability, 6*. https://doi.org/10.1017/sus.2023.4

Mengistie, B. T., Mol, A. P. J., & Oosterveer, P. (2016). Private environmental governance in the Ethiopian pesticide supply chain: Importation, distribution and use. *NJAS: Wageningen Journal of Life Sciences, 76*, 65–73. https://doi.org/10.1016/J.NJAS.2015.11.005

Ortiz, A. M. D., Outhwaite, C. L., Dalin, C., & Newbold, T. (2021). A review of the interactions between biodiversity, agriculture, climate change, and international trade: Research and policy priorities. *One Earth, 4*, 88–101. https://doi.org/10.1016/J.ONEEAR.2020.12.008

Pisa, L. W., Amaral-Rogers, V., Belzunces, L. P., et al. (2015). Effects of neonicotinoids and fipronil on non-target invertebrates. *Environmental Science and Pollution Research International, 22*, 68. https://doi.org/10.1007/S11356-014-3471-X

Ray, S., & Shaju, S. T. (2023). Bioaccumulation of pesticides in fish resulting toxicities in humans through food chain and forensic aspects. *Environmental Analysis Health and Toxicology, 38*, e2023017. https://doi.org/10.5620/EAHT.2023017

Reis TNP dos, Ribeiro V, Garrett RD, et al (2023). Explaining the stickiness of supply chain relations in the Brazilian soybean trade. *Global Environmental Change, 78*. https://doi.org/10.1016/j.gloenvcha.2022.102633

Roy, D., Kamar, A., Pradhan, M., et al. (2023). *Agricultural transformation and market integration in the ASEAN region: Responding to food security and inclusiveness concerns* (IFPRI Discussion Papers). IFPRI.

Santosh, P. S. (2022). Environmental impacts of pesticides – A review. *International Journal of Zoological Investigations, 8*, 847–856. https://doi.org/10.33745/ijzi.2022.v08i02.102

Sarkar, S., Gil, J. D. B., Keeley, J., et al. (2021). *The use of pesticides in developing countries and their impact on health and the right to food: Study*. European Parliament.

Schneider, K., Barreiro-Hurle, J., & Rodriguez-Cerezo, E. (2023). Pesticide reduction amidst food and feed security concerns in Europe. *Nature Food, 4*, 746–750. https://doi.org/10.1038/s43016-023-00834-6

Shattuck, A., Werner, M., Mempel, F., et al. (2023). Global pesticide use and trade database (GloPUT): New estimates show pesticide use trends in low-income countries substantially underestimated. *Global Environmental Change, 81*, 102693. https://doi.org/10.1016/J.GLOENVCHA.2023.102693

Silva, V., Gai, L., Harkes, P., et al. (2023). Pesticide residues with hazard classifications relevant to non-target species including humans are omnipresent in the environment and farmer residences. *Environment International, 181*, 108280. https://doi.org/10.1016/J.ENVINT.2023.108280

Singh, K., Kaur, S., & Shalu, K. (2023). Changing pattern of world trade and role of global supply chains: Evidences from recent literature. *Gyan Management Journal, 17*, 16–22. https://doi.org/10.48165/GMJ.2022.17.2.3

Stehle, S., & Schulz, R. (2015). Agricultural insecticides threaten surface waters at the global scale. *Proceedings of the National Academy of Sciences of the United States of America, 112*, 5750–5755. https://doi.org/10.1073/PNAS.1500232112/-/DCSUPPLEMENTAL

Subedi, B., Poudel, A., & Aryal, S. (2023). The impact of climate change on insect pest biology and ecology: Implications for pest management strategies, crop production, and food security. *Journal of Agriculture and Food Research, 14*, 100733. https://doi.org/10.1016/J.JAFR.2023.100733

Tang, F. H. M., Malik, A., Li, M., et al. (2022). International demand for food and services drives environmental footprints of pesticide use. *Communications Earth & Environment, 3*, 272. https://doi.org/10.1038/s43247-022-00601-8

Tchonkouang, R. D., Onyeaka, H., & Nkoutchou, H. (2024). Assessing the vulnerability of food supply chains to climate change-induced disruptions. *Science of the Total Environment, 920*, 171047. https://doi.org/10.1016/J.SCITOTENV.2024.171047

Valbuena, D., Cely-Santos, M., & Obregón, D. (2021). Agrochemical pesticide production, trade, and hazard: Narrowing the information gap in Colombia. *Journal of Environmental Management, 286*, 112141. https://doi.org/10.1016/J.JENVMAN.2021.112141

Wang, J., Geng, X., Wang, P., et al. (2024). Pesticide-related risks embodied in global soybean trade. *Cell Reports Sustainability, 1*, 100055. https://doi.org/10.1016/j.crsus.2024.100055

Wisnujati, N. S. (2023). Unveiling the epoch: Exploring pesticide utilization and trade trends globally and regionally from 1990 to 2020. *SCIENTIA: Journal of Multi Disciplinary Science, 2*, 108–121. https://doi.org/10.62394/SCIENTIA.V2I2.16

Woodcock, B. A., Bullock, J. M., Shore, R. F., et al. (2017). Country-specific effects of neonicotinoid pesticides on honey bees and wild bees. *Science, 356*, 1393–1395. https://doi.org/10.1126/SCIENCE.AAA1190

Yang, Y., Tilman, D., Jin, Z., et al. (2024). Climate change exacerbates the environmental impacts of agriculture. *Science, 385*, eadn3747. https://doi.org/10.1126/SCIENCE.ADN3747/ASSET/57517331-FB84-4DF6-A764-12D6FEA01871/ASSETS/IMAGES/LARGE/SCIENCE.ADN3747-FA.JPG

Part II
Dynamics

This part details the complex interplay between pesticides and climate change, addresses the intricate feedback loops that impact pest populations, and modifies the efficacy of pesticides. It scrutinizes the contribution of pesticides to greenhouse gas emissions, evaluates the ecotoxicological repercussions on non-target organisms, and investigates the emerging technologies for sustainable pest management.

Pesticides and Climate Change Feedback Loop

Katherine Georgina Menon, Kondakindi Venkateshwar Reddy, B. H. Boje Gowd, P. Paul Vijay, and Routhu Jhansi

1 Introduction

Pesticides are compounds, either naturally derived or chemically synthesized, used to control various pests across sectors such as food production, forestry, agriculture, and aquaculture. These chemicals exhibit toxicity in living organisms. These include insecticides, herbicides, fungicides, molluscicides, rodenticides, plant growth regulators, and various other compounds Among these, most technologically advanced nations outlawed or severely limited the use of organochlorine (OC) pesticides after the 1960s, despite their great effectiveness in combating diseases like typhus and malaria (Schiliro et al., 2020). The subsequent introduction of other synthetic insecticides—such as organophosphates (OP) in the 1960s, carbamates in the 1970s, and pyrethroids in the 1980s—along with the development of herbicides and fungicides during the 1970s and 1980s significantly advanced pest control and boosted agricultural productivity (Bhatt et al., 2021). Ideally, pesticides should be lethal only to target pests without harming non-target species, including humans. However, this is often not the case, leading to growing concerns over the misuse and overuse of

K. Georgina Menon (✉)
Department of Microbiology, School of Allied and Healthcare Sciences Malla Reddy University, Hyderabad, Telangana, India

K. Venkateshwar Reddy · B. H. Boje Gowd
Center for Biotechnology, University College of Engineering Science and Technology, Jawaharlal Nehru Technological University Hyderabad, Hyderabad, Telangana, India

P. Paul Vijay · R. Jhansi
Department of Microbiology, School of Allied and Healthcare Sciences Malla Reddy University, Hyderabad, Telangana, India

Center for Biotechnology, University College of Engineering Science and Technology, Jawaharlal Nehru Technological University Hyderabad, Hyderabad, Telangana, India

B. R. Babaniyi, E. E. Babaniyi (eds.), *The Interplay of Pesticides and Climate Change*, https://doi.org/10.1007/978-3-031-81669-7_6

pesticides. The widespread belief that "if a little is good, more must be better" has caused serious harm to human health and other life forms (Schiliro et al., 2020).

Over the past two decades, considerable efforts have been made to understand the fate and movement of pesticides in surface and groundwater, with the goal of using this knowledge to shape environmental policies and regulations. This research aids in exploring the relationship between pesticides, climate change, and their effects on the environment. While various types of pesticides can help prevent diseases and pests, thereby enhancing agricultural productivity, they also contribute to increased health and environmental risks and can disrupt the balance of ecosystems.

2 Understanding the Pesticide–Climate Nexus

Pesticides are harmful chemicals intentionally introduced into the environment to control specific pests. However, a significant portion of these chemicals ends up in places other than their intended targets (Menegaki et al., 2022). Pesticides can easily pollute the air, soil, and water through runoff from fields, leakage from storage containers, improper disposal, and particularly through aerial spraying (Fang et al., 2020).

Water
Pesticides are commonly detected in rain, groundwater, streams, rivers, lakes, and oceans. There are four primary pathways through which pesticides can enter water sources:

- Drift outside of the application area
- Leaching through soil
- Runoff from treated areas
- Accidental spills

UK government studies reveal that, in some cases, pesticide concentrations in river and groundwater samples exceed safe limits for drinking water (Faust & Aly, 1964).

Soil
Pesticides reduce soil biodiversity, leading to lower soil quality. Soil that remains free of chemicals retains more water, which is crucial for plant growth (Singh, 2012).

Plants
Pesticides disrupt nitrogen fixation, a vital process for the growth of many plants, potentially causing significant reductions in crop yields. Additionally, pesticide use on blooming crops can kill honeybees, essential pollinators, further reducing plant reproduction and crop production.

Animals
Pesticide residues on food can poison animals, and applications in certain regions can deplete vital food supplies, forcing animals to move, change their diets, or go

hungry. Additionally, pesticide poisoning can affect creatures higher up the food chain that eat polluted prey, such as birds that eat worms and insects exposed to pesticides (Whitehead & Peaslee, 2023).

Birds
There is evidence that pesticides are negatively affecting bird populations. The accumulation of pesticides in the bodies of certain bird species has led to their decrease, as noted in Rachel Carson's "Silent Spring." Certain fungicides may destroy earthworms, an essential food source, which has an indirect effect on birds and mammals even if they may only have a little effect on them (Gibbons et al., 2015). Another danger from granular insecticides is that birds may confuse them for food, and a single granule can be fatal. By decreasing their habitats, herbicides pose a hazard to birds as well.

Aquatic Life
Pesticide-contaminated water can severely impact fish and other aquatic organisms. Herbicide applications to water bodies can kill plants, depleting oxygen and suffocating fish (Jiang et al., 2022). Prolonged exposure to certain pesticides can lead to physiological and behavioral changes in fish, such as abandoning nests, weakened immunity, and an increased inability to avoid predators, ultimately reducing their populations (Singh, 2012).

3 Pesticides Impact on Climate Change

The environmental cost of species invasions has generally gone unrecognized, but it is represented by greenhouse gas (GHG) emissions from the use of pesticides to suppress invasive species. In this investigation, we determine the greenhouse gas emissions associated with the soybean aphid (Aphis glycines) invasion of North America from Asia (Hodgson et al., 2012). The soybean aphid was first found in North America in 2000, and since then, insecticide use in soybean crops has significantly increased. According to our calculations, every hectare of soybeans treated with pesticides results in the emission of about 10.6 kg of greenhouse gases equal to carbon dioxide (CO_2) during manufacture, transportation, and application (Hill et al., 2012). Depending on the pest population, this has led to annual emissions ranging between 6 and 40 million kg of CO_2 equivalents in the United States since the aphid's arrival.

If farmers were not encouraged not to spray below an economic threshold for aphid density, these emissions would have been far greater. Many farmers spray preventively in the absence of this threshold, but the threshold enables them to rely on the aphids' naturally occurring biological control, which can be quite successful. Adopting this level, in our estimation, will cut annual emissions in the United States by about 300 million kg of CO_2 equivalents. Previous studies have demonstrated that over more than half of the soybean land in the United States, biological control agents like lady beetles may lower aphid concentrations below this threshold.

Therefore, annual emission reductions in more than 200 million kg of CO_2 equivalents may be attributed to biological control alone. These results demonstrate how invasive species' interactions with their natural predators can have a major impact on greenhouse gas emissions.

3.1 *Climate Change Impacts on Pesticide Use and Efficacy*

Pesticides and Climate Change

Few studies assess the greenhouse gas (GHG) emissions from pesticide use throughout their entire life cycle—covering production, storage, transportation, application, and breakdown—which likely leads to an underestimation of their true impact (Berntsen et al., 2007). Remarkably, fossil fuels are the source of 99% of synthetic chemicals, including insecticides. They nevertheless get a lot less attention than nitrogen fertilizers, which are another important agricultural input that raises greenhouse gas emissions. According to research, it usually takes ten times as much energy to produce 1 kg of pesticide as it does nitrogen fertilizer.

Pesticides may also produce greenhouse gas emissions after they are applied. Fumigant insecticides, for instance, have been demonstrated to dramatically raise the synthesis of nitrous oxide in soils. Furthermore, a lot of pesticides contribute to the creation of ground-level ozone, a greenhouse gas that is bad for plants and people's health. Figure 1 in this book chapter represents the effects of pesticides on climate change.

Climate Change and Pesticide Use

As pesticide use contributes to climate change, research indicates that the effects of a changing climate are likely to lead to increased reliance on synthetic pesticides Several factors are at play:

- Rising temperatures, heat stress, and altered rainfall patterns are reducing crop resilience. For instance, drought conditions can weaken plants' natural defenses and modify their biology, making them more susceptible to pests
- Warmer temperatures are expected to enhance insect population growth in certain areas and shift their geographic distribution, allowing them to survive winters better.
- Weeds, with their greater genetic diversity and adaptability, are more resilient to climate change than cultivated crops. This increased resilience may allow weeds to outcompete agricultural crops, resulting in lower yields (Guo et al., 2021).
- Climate change accelerates the degradation of pesticides, reducing their effectiveness over time and prompting farmers to apply them more frequently.

An Environmental Justice Issue

Experts warn that climate change will facilitate the movement of pesticides away from their intended targets, further contaminating the environment and threatening

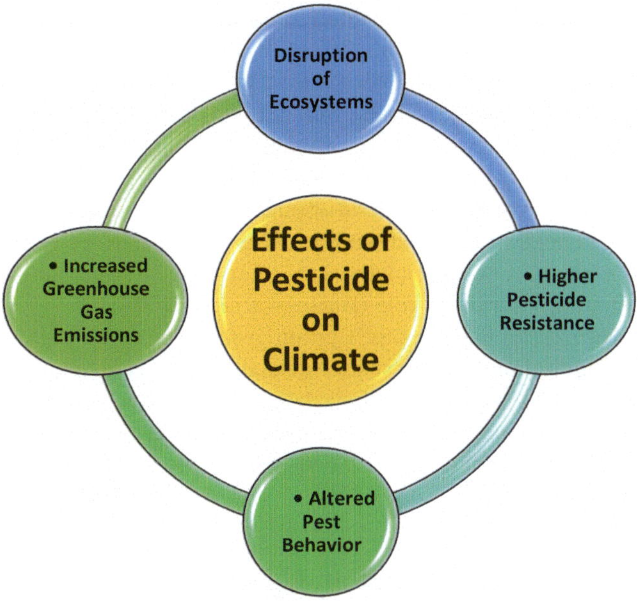

Fig. 1 Effect of pesticides on climate change

public health. Higher temperatures are likely to increase pesticide volatilization—when pesticides turn into gas—resulting in more pesticide exposure in the air.

In the United States, those most exposed to pesticides—such as agricultural workers, rural communities, and residents near pesticide production sites—are often low-income individuals and people of color. The anticipated rise in pesticide use will exacerbate the impacts of climate change on these communities, including extreme heat and increased wildfire smoke (Arcury et al., 2010).

4 Consequences for Human Health and Environment

The use of pesticides has considerable implications for health, stemming from various exposure pathways and mechanisms:

1. *Acute Health Effects*: Symptoms that arise from direct contact with pesticides include headaches, nausea, dizziness, skin irritations, and respiratory issues. When applying pesticides or coming into contact with surfaces that have been treated, several acute reactions may happen.
2. *Chronic Health Effects*: Long-term exposure to pesticides is associated with serious health issues. Studies have linked chronic exposure to various cancers, neurological disorders such as Parkinson's disease, reproductive problems, and

endocrine disruption, which can interfere with hormone function and lead to developmental issues (Perrin et al., 2021).
3. *Contamination of Water and Soil*: Pesticides can leach into groundwater and contaminate drinking water sources. Soil contamination affects agricultural productivity and can lead to indirect health risks for communities that rely on these resources for their food and drinking water (Morais, 2016).
4. *Loss of Biodiversity*: The widespread application of pesticides contributes to the decline of beneficial insects and wildlife. This loss disrupts ecosystems and can lead to an increase in pest populations, necessitating even greater pesticide use, thus perpetuating a harmful cycle (Bengtsson et al., 2005).
5. *Impact on Vulnerable Populations*: Certain groups, including children, pregnant women, and agricultural workers, are especially vulnerable to the effects of pesticide exposure. Children, due to their developing bodies, may experience more severe health consequences, while agricultural workers face heightened risks from direct contact and inhalation during application.

5 Impact of Climate Change and Variability on Agricultural Productivity

A considerable decline in crop yield in many Asian regions is linked to irregular and heavy rainfall patterns during the past few decades, as well as a loss in timely water and rainfall availability (Aryal et al., 2020). Climate change would make it challenging to maintain output and improve food security for Asia's impoverished rural areas, even with the gain in agricultural productivity brought about by the Green Revolution. In the least developed countries, damage from climatic fluctuations may endanger food security and national economic productivity. Different regions see different yield declines in different crops (such rice and wheat) due to variances in climate patterns (Ahmad et al., 2019). While CO_2 fertilization cannot mitigate the effects of rising temperatures, it can counteract the severe effects of rising temperatures on C_3 plants and increase crop yields. The unpredictable nature of rainfall combined with rising temperatures has hampered crop development. In Asia, rice and wheat play a crucial role in guaranteeing food security (Arunrat et al., 2018). It will be difficult to meet growing food demands if wheat production is not increased by 60% by 2050. Rising temperatures at lower latitudes are associated with decreased crop productivity in arid to semi-arid regions. In China, yields of rice, wheat, and maize have all declined due to drought and flooding, and it is expected that these issues will have a bigger effect on crop production in the future. Rice can withstand a gradual rise in temperature at night; yield and biomass will drop by 16–52% if the temperature climbs by 2 °C above the 24 °C threshold (Rezaei et al., 2018; Chen et al., 2018). Asia's semi-arid to dry regions are at risk of becoming much more so, as low productivity and drought stress are already problems. Rising temperatures and heavy, unpredictable rainfall have had a severe impact on wheat crop quality (protein content, sugars, and starch) and grain yield. Climate fluctuation has caused a significant decrease in wheat yield in China, India, and Egypt's North Nile Delta (up to 17.6%). This drop is associated with a rise in temperature, inconsistent rainfall, and an increase in insect pest infestation (Aryal et al., 2020; Kheir et al., 2019). By 2080, rice yield in rain-fed regions of South Asia might drop by 10% under the RCP 8.5 scenario and 14% under the RCP 4.5 scenario (Chun et al., 2016). Due to their detrimental effects on the booting and anthesis stage, high temperatures and dryness have reduced rice yields throughout Asia, particularly in China and Pakistan (Zafar et al., 2018; Ahmad et al., 2019). Similar to this, heat stress poses a serious risk to rice since it reduces the number of productive tillers, causes grains to shrivel, and eventually lowers rice yield (Wang et al., 2019).

6 Climate Variability and Change Impact Assessment

Agriculture will be unpredictable in the future due to its complex structure and interactions with many components, posing a significant danger to regional food security. Thus, it is necessary to assess how climate change is reducing agricultural productivity and develop climate change adaptation measures. The detrimental consequences of climate change on agriculture are being quantified globally using simulation models such as general circulation models (GCMs) and representative concentration pathways (RCPs). These models can also be used to evaluate the negative effects of climate change on crop productivity, which is crucial for food availability and access (Rahman et al., 2018).

By taking into account the rise in temperature and increased CO2 at crucial stages of crop growth, crop models have the potential to explain the internal workings of plants (Challinor et al., 2018). Aside from the modelling approach, there are no advanced tools or technologies available to establish how climatic variability and change effect agricultural and livestock productivity. Furthermore, modelling approaches are available and used extensively to analyze the effects of climatic variability and change on agricultural and livestock output (Asseng et al., 2015). To evaluate the implications of future climate on farmers' livelihoods, it is recommended to examine the entire agricultural system using a complete methodology of climate, crop, and economic modelling (RAPs) techniques (Hoogenboom et al., 2015; Rahman et al., 2019). We discovered that the agricultural model intercomparison and improvement project (AgMIP) was the most effective approach.

7 Mitigating the Impacts of Climate Change on Agriculture/ Crop Production

1. *Integrated Pest Management*: In theory, integrated pest management (IPM) is a flexible and comprehensive system. This approach views the agro-ecosystem as an interconnected whole, employing a variety of biological, cultural, genetic, physical, and chemical approaches to keep pest populations below economically destructive levels while causing the least amount of disruption to the environment and cropping ecosystems. Food and Agriculture Organization describes integrated pest management (IPM) as an ecosystem-based strategy to agricultural production and protection that combines multiple management approaches and tactics to support healthy crop growth while lowering pesticide use. Compared to alternative solutions in terms of crop productivity, profit, and safety, integrated pest management (IPM) applies the most effective combination of control measures for a specific pest problem. The United States Environmental Protection Agency defines integrated pest management (IPM) as a pest control method that is both environmentally responsible and effective. It is based on a variety of commonsense techniques. Pesticides and herbicides are part of a management system that involves a variety of actions aimed at economically feasible pest control. Integrated pest management (IPM) is a comprehensive ecological approach to structural and agricultural pest management. IPM also tries to prevent infestations from starting, observe infestation patterns as they arise, and respond appropriately (without employing poisons) as needed. In essence, IPM is the intentional selection and execution of pest control strategies that achieve positive ecological, socioeconomic, and economic outcomes.
2. *Conservation Agriculture*: A farming method that promotes "minimum soil disturbance, diversified crop rotations, and maintenance of organic soil cover" as key components of agricultural techniques (Nyanga et al., 2020). One of the best management practices (BMPs) tools, it uses three crop management concepts to support ecosystems: crop rotation, permanent soil cover (crop residues or cover crops), and direct crop planting (i.e., no-till) (Pittelkow et al., 2015). BMPs include social initiatives, various management strategies, and methods for conserving water and soil that have been proven to be successful in a certain area and useful instruments for safeguarding the environment. It is rare for a single practice or activity to resolve the issue; instead, a variety of strategies are frequently employed. BMPs can be as simple as not applying manure ahead of predicted rainfall or as complex as changing farming practices such as conservation tillage and crop rotation, crop nutrient management, pest management, conservation buffers, irrigation management, grazing management, animal feeding operations management, erosion and sediment control.
3. *Good Agricultural Practices (GAP)*: A "collection of principles to apply for on-farm production and postproduction processes, resulting in safe and healthy food and nonfood agricultural products, while taking into account economic,

social, and environmental sustainability," according to the FAO (Sareen, 2016). His main focus is on "preventing problems before they happen." Its three primary goals are to (1) guarantee the safety and quality of produce across the food chain; (2) exploit new market opportunities by altering the governance of the supply chain; and (3) enhance the utilization of natural resources, the health of employees, and working conditions. Preventing problems before they arise is a crucial component of GAP.

4. *Organic Farming*: It is described as "a production system that sustains the health of soils, ecosystems, and people" by the International Federation of Organic Agriculture Movements (IFOAM). It is dependent on biodiversity, biological processes, and cycles that are tailored to the local environment rather than harmful inputs. Organic agriculture integrates science, tradition, and innovation to provide high-quality food, improve the shared environment, and foster equitable relationships and a good quality of life for all parties involved (Miśniakiewicz et al., 2021): tradition, innovation, and science. It is covered by the certification, which is the third party's declaration. When feasible, agronomic, biological, and mechanical techniques are used in place of synthetic materials to achieve this.
5. *Sustainable Intensification (SI)*: It was founded in order to deal with the problem of widespread agriculture. It focuses on maximizing the productivity of agricultural resources to yield more food on the same acreage of land while mitigating adverse effects on the environment and society (Petersen & Snapp, 2015). The use of methods like reforestation, shallow bowls filled with organic materials, and stone bunds along contours are a few examples of practices that are promoted to meet the SI objectives (Manzano Lepe, 2016).
6. *Permaculture*: It is the intentional creation and upkeep of productive agricultural ecosystems that reflect the diversity, stability, and resilience of natural ecosystems. The sustainable provision of food, energy, housing, and other material and non-material requirements is achieved through the harmonious coexistence of people and the natural environment (Holmgren, 2020). For instance, you may restore the soil, conserve water, and free up area for permaculture waste streams. In essence, the idea is to create a system for producing plants that take into account the roles played by each plant and how they interact with one another so that the components of each plant support one another. This method is inspired by everyday connections found in nature.
7. *Research and Development*: Many regional farming systems imitate natural processes, guaranteeing that natural resources (land, nutrients, groundwater, radiation, and rainfall) are used as efficiently as possible. This highlights the significance of ecological insights, like those discussed above, in gradually creating technological advancements in current systems. Farmers must modify their methods as the environment and climate change, as well as population density, rises, to guarantee long-term sustainability and higher field output. As was already noted, smallholder systems depend on a complex and wide range of alternatives that allow them to adapt to the whims of farming. These (local) solutions have helped the population to survive in frequently difficult environ-

ments as they have developed and been proven over generations. One of the authors of a study being done in South Africa observed that when farmers in a vegetable garden project encountered drought circumstances following the theft of their irrigation pump, they turned to customary techniques and "traditional" crops. Increased productivity and sustainability are two goals that can be achieved by farmers through a wider choice of technology solutions that are made available to them by national and international agricultural research and development (R&D). A broad range of options is necessary due to the extreme diversity (and variability) in the agroecological and social conditions of farming, even at the micro-field level. This suggests that scientists need to be more flexible. Moreover, farmers will need to modify their farm management practices more frequently as the demand to increase and commercialize their output mounts. This pertains specifically to the scheduling of field operations like as weeding and planting, but it also covers the professional use (the informed use) of agrochemicals, crop rotations that are altered, intercropping, and companion planting techniques. Close cooperation between farmers and researchers is necessary in this setting, but doing so necessitates a number of rather fundamental changes to agricultural specialists' training programs as well as to the way agricultural research is currently conducted and organized, including institutional arrangements (Stoop & Hart, 2005).

8. *Education and Training*: Family farms are the foundation of the farming industry in many regions of the world. They are distinguished by their special qualities by the high level of interaction between the reproductive and productive domains. According to Dawson and Mussolino, the following factors contribute to the distinctiveness of family businesses: (a) socioemotional wealth (affective endowment); (b) the business's fundamental component (family engagement); and (c) familiness (unique bundle of resources) (Dawson & Mussolino, 2014). This is especially true for farming activities, where decision-making takes place within communal entrepreneurial processes and entrepreneurial processes can be explained by a family business viewpoint based on the idea that the family assumes business risk (Jervell, 2011; Errington & Gasson, 1993; Davidova & Thomson, 2014). By offering a "setting where normative systems (affect, altruism, tradition) and utilitarian systems (economic rationality) are combined," this method gives an intriguing viewpoint on entrepreneurship (Alsos et al., 2014; Marcello et al., 2017). As a result, the farmer's disposition toward receiving family criticism is mentioned as a strategic talent (de Wolf et al., 2007; De Rosa et al., 2019). In order to account for the great diversity across family farm enterprises, including family composition and life cycle stage, the study of these firms entails high degrees of variability (Bartoli & de Rosa, 2013; Offutt, 2002). Conversely, a typical feature of family businesses is their high levels of persistence, which can be attributed to a combination of collective entrepreneurial skills and the F-connection (family, friend, firm), which increases resilience (Ben-Porath, 1980; Pollak, 1985; Darnhofer et al., 2016). The literature on both emerging and wealthy nations has long acknowledged the importance of education as a crucial component of human capital in farming activities. An

extensive body of literature, beginning with Becker's groundbreaking work (Becker, 1964), emphasizes the value of human capital primarily in relation to increased agricultural productivity and innovation (Zubovic et al., 2009; World Bank, 2006; Ninh, 2021; Paltasingh & Goyari, 2018). Several studies emphasize education as a specific competency in the transition to sustainable farming, as exemplified by the agroecological transition (Carlisle et al., 2019). Furthermore, a substantial corpus of research highlights the importance of education in its various forms for accelerating the shift to sustainable farming, addressing the normative implications of educational initiatives to assist farmers (Ahmed et al., 2017; Valderrama-Hernández et al., 2019; Banks et al., 2002).

9. *Carbon Farming*: A collection of agricultural techniques and land-use plans known as "carbon farming" is intended to absorb carbon dioxide (CO2) from the atmosphere and slow down global warming. Enhancing soil and vegetation carbon uptake while advancing sustainable and regenerative farming practices is the main objective. Agricultural landscapes operate as a natural carbon sink, storing carbon to help offset anthropogenic CO2 emissions. Plants store carbon in the soil and biomass as they take up CO_2 during photosynthesis, preventing it from causing the greenhouse effect. This procedure helps restore ecosystems and lowers the overall carbon impact (Gajjela Indira et al., 2023). Furthermore, there are several other advantages to using carbon farming techniques, including better soil fertility, enhanced water retention, and increased resistance to harsh weather. Carbon farming strengthens environmental sustainability and improves ecosystems' overall capacity to adapt to the effects of climate change by supporting sustainable agriculture. Farmers are economically motivated to engage in carbon farming by financial incentives such as subsidies for adopting sustainable techniques and participation in carbon markets. This in turn promotes broad adoption and helps carbon farming efforts become more scalable. For carbon farming operations to be supported and regulated on a larger scale, policy frameworks and international collaboration are essential. To ensure a concerted effort in the battle against climate change, governments, organizations, and global initiatives can offer policies, structures, and incentives that encourage the adoption of sustainable farming methods (Gajjela Indira et al., 2023).
10. *Bio-Energy*: A review of the generation of liquid fuels from certain energy crops, such grains and oilseeds, is necessary in light of worries about the long-term sustainability of relying too heavily on fossil fuels. However, the kind of production methods used to make and cultivate biofuels has a significant impact on their ability to lower carbon emissions (Alawa et al., 2014).
11. *Avoidance of Deforestation*: Deforestation occurs when trees are taken down for urban development or building construction throws off the natural equilibrium of carbon and oxygen in the atmosphere. Due to this imbalance, a significant amount of CO_2 remains in the atmosphere, adding to the global GHG stock that exacerbates climate change issues that have an impact on crop productivity. By improving the carbon-oxygen balance and raising awareness, this practice can be changed, which will lower the amount of CO_2 in the atmosphere.

Reduced CO_2 emissions lessen the impact of climate change on the world's agriculture productivity (Alawa et al., 2014).

12. *Fertilizer Management*: N_2O emissions and the amount of nutrient load can be reduced by moving to organic farming or optimizing fertilizer application techniques. The possible impact on yield would need to be balanced against the total benefits, though (IPPC, 2007). Furthermore, in many developing nations, obtaining food security is hampered by a lack of availability to soil nutrients necessary for restoring the condition of degraded soils. When applying fertilizers or using organic manure, the food security index should be properly taken into account to make sure the method does not worsen the issue of national food insecurity (Alawa et al., 2014).
13. *Livestock and Manure Management*: Enhancing digestion with better feeding techniques and dietary supplements is one way to lower methane emissions from enteric fermentation. The quality of the feeds, the age and type of the animals, and whether they are grazing or stall-fed all affect how effective these strategies are. The livestock in developing nations receive feed of inferior quality, which causes their emission rate per animal to be higher than that of herds in rich nations (Mark et al., 2008). Techniques such as chilling and employing solid covers for solid tanks and lagoons, separating solid from slurry, and securing the methane released are pertinent to the management of livestock manure.
14. *Design and Implementation of Good Overall Development Policies and Programs*: Good development policies and programs rank among the best investments for climate change adaptation given the existing uncertainties regarding the precise location and effects of climate change. In addition to supporting agricultural sustainability and pro-growth, a proper development strategy also helps the globe adapt to climate change and ensure food security. When people have greater resources and live in a responsive and flexible economic context, they are better able to adapt to climate change (Alawa et al., 2014).
15. *Increased Investment in Agricultural Productivity*: To address these expectations, agricultural science and technological solutions are crucial. A new and more difficult demand on agricultural productivity is brought about by climate change. Research on biotechnology and other crop productivity-enhancing techniques will be crucial to overcoming the strains brought on by climate change. Rather than performing very well in a limited range of climatic circumstances, crops operate rather well in a variety of producing environments. In order to lower CH4 emissions, research on modifications to irrigation management techniques is required, and government funding at all levels should support this research. One of the most important lessons from the Green Revolution is that increased agricultural output may be a potent tool for indirectly reducing poverty by raising food prices and creating jobs, even if it is not directed towards the poorest of the poor. Productivity upgrades that strengthen farmers' resilience in the face of climate change pressures will likely have similar poverty-reducing impacts. In order for farmers to benefit from improved crop varieties and management practices, rural infrastructure is crucial. Maintaining and

expanding the density of rural road networks are necessary to boost yields and increase planted areas, as it facilitates market access and lowers transaction costs. Infrastructure for irrigation needs to be invested in as well, particularly to increase water usage efficiency (Alawa et al., 2014).

8 Breaking the Feedback Loops

Climate-smart pest management (CSPM) is a relatively new idea that combines a variety of interdisciplinary approaches and strategies to aid primary produce in adapting to changing climatic circumstances. This encompasses supporting functions such as extension services and research, as well as an enabling environment that includes regulations, infrastructure, and human capital. These elements, taken together, create the geographical boundaries for the distribution of species such as crops, pests, and natural enemies.

Implementing CSPM provides for more effective control of the mounting risks from both new and existing crop pests, improving farmer livelihood resilience and strengthening local and national food security in the face of climate change. Furthermore, CSPM contributes to climate change mitigation through enhancing the overall greenhouse gas (GHG) balance. For example, lowering pest-related yield losses reduces GHG emissions per unit of food produced (Altieri et al., 2015). For CSPM to be effective, it must be considered as part of a larger climate-smart agriculture (CSA) plan in which pest management is an important component. CSPM provides farmers with the information and tools they need to proactively apply pest control techniques, such as crop variety, the construction of natural habitats, and careful water management, which improve farm health and minimize vulnerability to pest disturbances (Neufeldt et al., 2013). Farmers can identify and execute specific pest-mitigation methods before they escalate by combining climate and pest monitoring with risk forecasting. When pest numbers approach economic harm levels, CSPM allows farmers to make timely, educated decisions about effective reactive pest control strategies.

While pest control measures are implemented on a farm level, CSPM must be handled more broadly, necessitating science-based political leadership. The immediate implications of increased insect pressure and crop losses affect farmer incomes and livelihoods, particularly for smallholder farmers. However, the broader repercussions of using chemical pesticides incorrectly and experiencing lower yields can disturb ecosystems and undermine the critical services they provide. This, in turn, may jeopardize local and national food security, economic stability, and employment, potentially leading to a variety of socioeconomic challenges, including migration. CSPM is a shared responsibility that requires coordinated assistance from extension services and research institutes to ensure that the services given are relevant, regionally adapted, and accessible to all farmers, especially marginalized groups like women, the elderly, and ethnic minorities.

For example, universities and national agricultural research centers can perform field studies to evaluate and quantify the effects of climate change on crop-pest-natural enemy interactions. This research will help to build tailored adaptive responses, which are now lacking. Furthermore, analyzing historical weather and climate data and developing niche models to anticipate pest species distributions under different climatic scenarios will allow for improved pest risk forecasting and proactive management techniques.

CSPM also emphasizes the importance of creating an enabling environment to help people embrace its techniques. Public and private sector investment is critical for assisting institutions in carrying out their missions, particularly in developing countries, and improving the infrastructure required for effective information and knowledge exchange across stakeholders. It is suggested to develop appropriate regulations, such as incentive-based systems that promote CSPM use, as well as offer financial services to assist farmers in overcoming adoption costs and embracing new techniques. A significant example is China's "Green Control Policy," which offers a subsidy system for non-chemical and least hazardous pest management solutions, undertaken in collaboration with the plant smart development program sponsored by CABI. (Centre for Agriculture and Bioscience International).

9 Conclusion

Climate change poses a significant threat to agricultural productivity in Asia, jeopardizing food security for millions. Erratic rainfall patterns, rising temperatures, and increased frequency of extreme weather events are all contributing to declining crop yields, particularly for staples like rice and wheat. This conclusion highlights the urgency of the situation and proposes various adaptation strategies.

The negative impacts of climate change on agriculture are well documented. Studies using sophisticated modelling techniques confirm a decrease in crop yields across the region. Rising temperatures, coupled with irregular rainfall, are detrimental to crop growth and development. This is particularly concerning for rice and wheat, which are vital for ensuring food security in Asia. Countries in dry and semi-arid areas are especially vulnerable, as rising temperatures exacerbate drought stress and decrease overall productivity.

To mitigate these threats, a multipronged approach is necessary. Implementing Integrated Pest Management (IPM) can minimize reliance on harmful pesticides while effectively controlling pests. Conservation agriculture practices promote minimal soil disturbance, diversified crop rotations, and permanent soil cover, all of which contribute to a healthier and more resilient agricultural system. Good Agricultural Practices (GAP) emphasize preventative measures and responsible resource management, ensuring the safety and quality of agricultural products.

Organic farming offers a long-term solution by focusing on ecological principles and fostering biodiversity. Sustainable intensification (SI) aims to maximize agricultural output while minimizing environmental damage. Permaculture practices encourage the creation of self-sustaining ecosystems that mimic natural systems.

Investing in research and development is crucial for developing climate-resilient crops and innovative farming techniques. Additionally, education and training programs empower farmers with the knowledge and skills necessary to adapt to changing conditions. Policies promoting carbon farming practices can incentivize sustainable agriculture and mitigate climate change.

Beyond on-farm practices, broader strategies are essential. Reducing deforestation helps maintain the natural carbon balance in the atmosphere. Optimizing fertilizer application techniques minimizes environmental impact while ensuring food security. Improved livestock and manure management practices can decrease methane emissions. Implementing effective development policies fosters economic resilience, allowing farmers to better adapt to climate change.

Finally, increased investment in agricultural productivity research and infrastructure development is critical. New technologies and improved irrigation systems can enhance resource use efficiency and crop yields. By working together, governments, farmers, researchers, and international organizations can implement these adaptation strategies, ensuring a more sustainable and food-secure future.

References

Ahmad, S., Abbas, G., Ahmed, M., Fatima, Z., Anjum, M. A., Rasul, G., Khan, M. A., & Hoogenboom, G. (2019). Climate warming and management impact on the change of phenology of the rice-wheat cropping system in Punjab, Pakistan. *Field Crops Research, 230*, 46–61. https://doi.org/10.1016/j.fcr.2018.10.008

Ahmed, S., Sclafani, A., Aquino, E., Kala, S., Barias, L., & Eeg, J. (2017). Building student capacity to lead sustainability transitions in the food system through farm-based authentic research modules in sustainability sciences (FARMS). *Elementa: Science of the Anthropocene, 5*, 46. https://doi.org/10.1525/elementa.239

Alawa, D. A., Asogwa, V. C., & Ikelusi, C. O. (2014). Measures for mitigating the effects of climate change on crop production in Nigeria. *American Journal of Climate Change, 3*(2), 161–168. https://doi.org/10.4236/ajcc.2014.32015

Alsos, G. A., Carter, S., & Ljunggren, E. (2014). Kinship and business: How entrepreneurial households facilitate business growth. *Entrepreneurship & Regional Development, 26*(1–2), 97–122. https://doi.org/10.1080/08985626.2013.870235

Altieri, M. A., Nicholls, C. I., Henao, A., & Lana, M. A. (2015). Agroecology and the design of climate change-resilient farming systems. *Agronomy for Sustainable Development, 35*(3), 869–890. Available at: https://link.springer.com/content/pdf/10.1007/s13593-015-0285-2.pdf

Arcury, T. A., Grzywacz, J. G., Talton, J. W., Chen, H., Vallejos, Q. M., Galván, L., et al. (2010). Repeated pesticide exposure among North Carolina migrant and seasonal farmworkers. *American Journal of Industrial Medicine, 53*, 802–813. https://doi.org/10.1002/ajim.20856

Arunrat, N., Pumijumnong, N., & Hatano, R. (2018). Predicting local-scale impact of climate change on rice yield and soil organic carbon sequestration: A case study in Roi Et Province, Northeast Thailand. *Agricultural Systems, 164*, 58–70. https://doi.org/10.1016/j.agsy.2018.04.001

Aryal, J. P., Sapkota, T. B., Khurana, R., Khatri-Chhetri, A., Rahut, D. B., & Jat, M. L. (2020). Climate change and agriculture in South Asia: Adaptation options in smallholder production systems. *Environment, Development and Sustainability, 22*(6), 5045–5075. https://doi.org/10.1007/s10668-019-00414-4

Asseng, S., Ewert, F., Martre, P., Rötter, R. P., Lobell, D. B., Cammarano, D., Kimball, B. A., Ottman, M. J., Wall, G. W., White, J. W., Reynolds, M. P., Alderman, P. D., Prasad, P. V. V., Aggarwal, P. K., Anothai, J., Basso, B., Biernath, C., Challinor, A. J., de Sanctis, G., et al. (2015). Rising temperatures reduce global wheat production. *Nature Climate Change, 5*(2), 143–147. https://doi.org/10.1038/nclimate2470

Banks, J., Long, A., & van der Ploeg, J. D. (2002). *Living countrysides: Rural development processes in Europe: The state of the art*. Elsevier.

Bartoli, L., & de Rosa, M. (2013). Family farm business and access to rural development polices: A demographic perspective. *Agricultural and Food Economics, 1*(1), 12. https://doi.org/10.1186/2193-7532-1-12

Becker, G. S. (1964). *Human capital: A theoretical and empirical analysis, with special reference to education*. University of Chicago Press.

Bengtsson, J., et al. (2005). The effects of organic agriculture on biodiversity and abundance: A meta-analysis. *Journal of Applied Ecology, 42*(2), 261. http://www3.interscience.wiley.com/cgibin/fulltext/118735203/HTMLSTART

Ben-Porath, Y. (1980). The F-connection: Families, friends, and firms and the organization of exchange. *Population and Development Review, 6*(1), 1. https://doi.org/10.2307/1972655

Berntsen, N. L., Solomon, S., Qin, D., Alley, R. B., Bindoff, Z., et al. (2007). Climate change 2007: The physical science basis. In *Contribution of Working Group I to the fourth assessment report of the Intergovernmental Panel on Climate Change*. Cambridge University Press.

Bhatt, P., Zhou, X., Huang, Y., Zhang, W., & Chen, S. (2021). Characterization of the role of esterases in the biodegradation of organophosphate, carbamate, and pyrethroid pesticides. *Journal of Hazardous Materials, 411*, 125026. https://doi.org/10.1016/j.jhazmat.2020.125026

Carlisle, L., Montenegro de Wit, M., DeLonge, M. S., Iles, A., Calo, A., Getz, C., Ory, J., Munden-Dixon, K., Galt, R., Melone, B., Knox, R., & Press, D. (2019). Transitioning to sustainable agriculture requires growing and sustaining an ecologically skilled workforce. *Frontiers in Sustainable Food Systems, 3*, 96. https://doi.org/10.3389/fsufs.2019.00096

Challinor, A. J., Müller, C., Asseng, S., Deva, C., Nicklin, K. J., Wallach, D., Vanuytrecht, E., Whitfield, S., Ramirez-Villegas, J., & Koehler, A.-K. (2018). Improving the use of crop models for risk assessment and climate change adaptation. *Agricultural Systems, 159*, 296–306. https://doi.org/10.1016/j.agsy.2017.07.010

Chen, H., Liang, Z., Liu, Y., Jiang, Q., & Xie, S. (2018). Effects of drought and flood on crop production in China across 1949–2015: Spatial heterogeneity analysis with Bayesian hierarchical modeling. *Natural Hazards, 92*(1), 525–541. https://doi.org/10.1007/s11069-018-3216-0

Chun, J. A., Li, S., Wang, Q., Lee, W.-S., Lee, E.-J., Horstmann, N., Park, H., Veasna, T., Vanndy, L., Pros, K., & Vang, S. (2016). Assessing rice productivity and adaptation strategies for Southeast Asia under climate change through multi-scale crop modeling. *Agricultural Systems, 143*, 14–21. https://doi.org/10.1016/j.agsy.2015.12.001

Darnhofer, I., Lamine, C., Strauss, A., & Navarrete, M. (2016). The resilience of family farms: Towards a relational approach. *Journal of Rural Studies, 44*, 111–122. https://doi.org/10.1016/j.jrurstud.2016.01.013

Davidova, S., & Thomson, K. (2014). *Family farming: A Europe and Central Asia perspective; Background Report for Regional Dialogue on Family Farming: Working towards a strategic approach to promote food security and nutrition*. Food and Agriculture Organization of the United Nations (FAO).

Dawson, A., & Mussolino, D. (2014). Exploring what makes family firms different: Discrete or overlapping constructs in the literature? *Journal of Family Business Strategy, 5*(2), 169–183. https://doi.org/10.1016/j.jfbs.2013.11.004

de Rosa, M., McElwee, G., & Smith, R. (2019). Farm diversification strategies in response to rural policy: A case from rural Italy. *Land Use Policy, 81*, 291–301. https://doi.org/10.1016/j.landusepol.2018.11.006

de Wolf, P., Schoorlemmer, H., & Rudmann, C. (2007). Important trends and required skills: An international synthesis. In *Exploring the significance of entrepreneurship in agriculture*. Research Institute of Organic Agriculture.

Errington, A., & Gasson, R. (1993). *The farm family business*. CAB International.

Fang, W., Ahsan, F., & Chandio, A. A. (2020). Climate change impacts on cereal crops production in Pakistan. *International Journal of Climate Change Strategies and Management, 12*(2), 257.

Faust, S. D., & Aly, O. M. (1964). Water pollution by organic pesticides. *Journal of American Water Works Association, 56*(3), 267–279.

Gajjela Indira, Alladi Chandrakanth, Anjali, Suneel Kumar, & Tarun Tomar. (2023). Carbon farming: A sustainable approach for mitigating climate change. *The Agriculture Magazine, 3*, 503–506.

Gibbons, D., Morrissey, C., & Mineau, P. (2015). A review of the direct and indirect effects of neonicotinoids and fipronil on vertebrate wildlife. *Environmental Science and Pollution Research, 22*(1), 103–118. https://doi.org/10.1007/s11356-014-3180-5

Guo, L., Li, H., Cao, X., Cao, A., & Huang, M. (2021). Effect of agricultural subsidies on the use of chemical fertilizer. *Journal of Environmental Management, 299*, 113621.

Hill, C. B., Chirumamilla, A., & Hartman, G. L. (2012). Resistance and virulence in the soybean-Aphis glycines interaction. *Euphytica, 186*, 635–646. https://doi.org/10.1007/s10681-012-0695-z

Hodgson, E. W., McCornack, B. P., Tilmon, K., & Knodel, J. (2012). Management recommendations for soybean aphid (Hemiptera: Aphididae) in the United States. *Journal of Integrated Pest Management, 3*, E1–E10.

Holmgren, D. (2020). *Essence of permaculture*. Melliodora.

Hoogenboom, G., Jones, J. W., Wilkens, P. W., Porter, C. H., & Uryasev, O. (2015). *Decision support system for agrotechnology transfer, version 4.6*. DSSAT Foundation.

IPPC. (2007). Impact adaptation and vulnerability: Contributions of Working Group 11 to the fourth assessment report of the IPCC.

Jervell, A. M. (2011). The family farm as a premise for entrepreneurship. In *The handbook of research on entrepreneurship in agriculture and rural development*. Edward Elgar Publishing. https://doi.org/10.4337/9780857933249.00009

Jiang, J., et al. (2022). Health risks of sulfentrazone exposure during zebrafish embryo-larvae development at environmental concentration. *Chemosphere, 288*, 132632. Available at: https://www.sciencedirect.com/science/article/abs/pii/S0045653521031040

Kheir, A. M. S., el Baroudy, A., Aiad, M. A., Zoghdan, M. G., Abd El-Aziz, M. A., Ali, M. G. M., & Fullen, M. A. (2019). Impacts of rising temperature, carbon dioxide concentration and sea level on wheat production in North Nile delta. *Science of the Total Environment, 651*, 3161–3173. https://doi.org/10.1016/j.scitotenv.2018.10.209

Manzano Lepe, B. (2016). *The role of agriculture cooperatives and farmer organizations on the sustainable agricultural practices adoption in Uganda*. Wageningen University.

Marcello, D. R., Felice, A., & Yari, V. (2017). Building up collective actions to qualify GIs. *Land Use Policy, 66*, 340–345. https://doi.org/10.1016/j.landusepol.2017.05.007

Mark, W. S., Rowena, V., Mandy, E., Gray, Y., Lan, B., & Saleemul, H. (2008). *Climate change and agriculture: Treats and opportunities*. Technische Zusammenarbeit (GTZ) GmbH, Deutsche Gesellsechaft Fur.

Menegaki, A. N., Tsounis, N., & Agiomirgianakis, G. M. (2022). The economic impact of climate change (CC) on the Greek economy. *Environment, Development and Sustainability, 24*(6), 8145–8161.

Miśniakiewicz, M., Łuczak, J., & Maruszewska, N. (2021). Improvement of organic farm assessment procedures on the example of organic farming in Poland—Recommendations for organic farming in Poland. *Agronomy, 11*(8), 1560. https://doi.org/10.3390/agronomy11081560

Morais, J. B. (2016). Physical-chemical properties of pesticides: Concepts, applications, and interactions with the environment. *Bioscience Journal, 32*, 627–641.

Neufeldt, H., Jahn, M., Campbell, B. M., Beddington, J. R., DeClerck, F., De Pinto, A., Gulledge, J., Hellin, J., Herrero, M., & Jarvis, A. (2013). Beyond climate-smart agriculture: Toward safe operating spaces for global food systems. *Agriculture & Food Security, 2*, 12.

Ninh, L. K. (2021). Economic role of education in agriculture: Evidence from rural Vietnam. *Journal of Economics and Development, 23*, 47–58.

Nyanga, P. H., Umar, B. B., Chibamba, D., Mubanga, K. H., Kunda-Wamuwi, C. F., & Mushili, B. M. (2020). Reinforcing ecosystem services through conservation agriculture in sustainable food systems. In *The role of ecosystem services in sustainable food systems* (pp. 119–133). Elsevier. https://doi.org/10.1016/B978-0-12-816436-5.00006-8

Offutt, S. (2002). The future of farm policy analysis: A household perspective. *American Journal of Agricultural Economics, 84*, 1189–1200.

Paltasingh, K. R., & Goyari, P. (2018). Impact of farmer education on farm productivity under varying technologies: Case of paddy growers in India. *Agricultural and Food Economics, 6*(1), 7. https://doi.org/10.1186/s40100-018-0101-9

Perrin, L., Spinosi, J., Chaperon, L., Kab, S., Moisan, F., & Ebaz, A. (2021). Pesticides expenditures by farming type and incidence of Parkinson disease in farmers: A French nationwide study. *Environmental Research, 197*, 111161.

Petersen, B., & Snapp, S. (2015). What is sustainable intensification? Views from experts. *Land Use Policy, 46*, 1–10. https://doi.org/10.1016/j.landusepol.2015.02.002

Pittelkow, C. M., Liang, X., Linquist, B. A., van Groenigen, K. J., Lee, J., Lundy, M. E., van Gestel, N., Six, J., Venterea, R. T., & van Kessel, C. (2015). Productivity limits and potentials of the principles of conservation agriculture. *Nature, 517*(7534), 365–368. https://doi.org/10.1038/nature13809

Pollak, R. A. (1985). A transaction cost approach to families and households. *Journal of Economic Literature, 23*, 581–608.

Rahman, M. H., Ahmad, A., Wang, X., Wajid, A., Nasim, W., Hussain, M., Ahmad, B., Ahmad, I., Ali, Z., Ishaque, W., Awais, M., Shelia, V., Ahmad, S., Fahd, S., Alam, M., Ullah, H., & Hoogenboom, G. (2018). Multi-model projections of future climate and climate change impacts uncertainty assessment for cotton production in Pakistan. *Agricultural and Forest Meteorology, 253–254*, 94–113. https://doi.org/10.1016/j.agrformet.2018.02.008

Rahman, M. H., Ahmad, A., Wajid, A., Hussain, M., Rasul, F., Ishaque, W., Islam, M. A., Shelia, V., Awais, M., Ullah, A., Wahid, A., Sultana, S. R., Saud, S., Khan, S., Fahad, S., Hussain, M., Hussain, S., & Nasim, W. (2019). Application of CSM-CROPGRO-Cotton model for cultivars and optimum planting dates: Evaluation in changing semi-arid climate. *Field Crops Research, 238*, 139–152. https://doi.org/10.1016/j.fcr.2017.07.007

Rezaei, E. E., Siebert, S., Hüging, H., & Ewert, F. (2018). Climate change effect on wheat phenology depends on cultivar change. *Scientific Reports, 8*(1), 4891. https://doi.org/10.1038/s41598-018-23101-2

Sareen, S. A. (2016). *Scheme and training manual on good agricultural practices*. FAO.

Schiliro, T., Zhang, C., Gea, M., Bianchi, S., Spinello, A., Magistrato, A., et al. (2020). Molecular basis for endocrine disruption by pesticides targeting aromatase and estrogen receptor. *International Journal of Environmental Research and Public Health, 17*, 5664. https://doi.org/10.3390/ijerph17165664

Singh, D. K. (2012). Pesticides and environment. *Pesticide Chemistry and Toxicology, 1*, 114–122.

Stoop, W. A., & Hart, T. (2005). Research and development towards sustainable agriculture by resource-poor farmers in sub-Saharan Africa: Some strategic and organisational considerations in linking farmer practical needs with policies and scientific theories. *International Journal of Agricultural Sustainability, 3*(3), 206–216. https://doi.org/10.1080/14735903.2005.9684757

Valderrama-Hernández, R., Sánchez-Carracedo, F., Alcántara Rubio, L., & Limón-Domínguez, D. (2019). Methodology to analyze the effectiveness of ESD in a higher degree in education. A case study. *Sustainability, 12*(1), 222. https://doi.org/10.3390/su12010222

Wang, Y., Wang, L., Zhou, J., Hu, S., Chen, H., Xiang, J., Zhang, Y., Zeng, Y., Shi, Q., Zhu, D., & Zhang, Y. (2019). Research progress on heat stress of rice at flowering stage. *Rice Science, 26*(1), 1–10. https://doi.org/10.1016/j.rsci.2018.06.009

Whitehead, H. D., & Peaslee, G. F. (2023). Directly fluorinated containers as a source of perfluoroalkyl carboxylic acids. *Environmental Science & Technology Letters, 10*(4), 350–355.

World Bank. (2006). *Enhancing agricultural innovation: How to go beyond the strengthening of research systems*. World Bank.

Zafar, S. A., Hameed, A., Nawaz, M. A., Wei, M. A., Noor, M. A., Hussain, M., & Mehboob-ur-Rahman. (2018). Mechanisms and molecular approaches for heat tolerance in rice (Oryza sativa L.) under climate change scenario. *Journal of Integrative Agriculture, 17*(4), 726–738. https://doi.org/10.1016/S2095-3119(17)61718-0

Zubovic, J., Domazet, I., & Stosic, I. (2009). *Development of human capital as a tool for improving productivity of agricultural sector – Case of Serbia*. Paper prepared for presentation at the 113th EAAE Seminar, 451–459.

Climate Change and Pesticide Dynamics

Mukul Machhindra Barwant, Odangowei Inetiminebi Ogidi, Balwant Singh, Varsha Sadashiv Patil, and Shalagha A. B. Sharma

1 Introduction

Plant growth hormones, fertilizers, and pesticides are all examples of agrochemicals, which are synthetic commodities used in agriculture. Agrochemicals are man-made substances that are included into farming practices in an effort to enhance crop quality, yield, and overall agricultural production. If agrochemicals can help make agricultural production systems more financially viable and alleviate food security worries, which would be great (Ogidi & Izah, 2024). The fast increase in food production, particularly since the beginning of the Green Revolution in the 1930s, is believed to have been caused by them. Agrochemicals provide nutrients, protect crops from pests and diseases, and promote effective farming methods; without them, modern agriculture would not be possible. The first herbicide, or weed killer, was 2,4-D, which was developed in 1945. The 1980s saw the development of glyphosate and the 1970s saw the creation of atrazine, thanks to this (Shafiq

M. M. Barwant (✉)
Department of Botany, Sanjivani Rural Education Society's, Sanjivani Arts Commerce and Science College Kopargon, Ahmednagar, Maharashtra, India

O. I. Ogidi
Department of Biochemistry, Faculty of Basic Medical Sciences, Bayelsa Medical University, Yenagoa, Bayelsa State, Nigeria

B. Singh
Department of Botany, K. S. Saket PG College, Dr. Ram Manohar Lohia Avadh University, Ayodhya, Uttar Pradesh, India

V. S. Patil
Department of Botany, Shirdi Sai Rural Institute's, Arts, Science and Commerce College, Rahata, Maharashtra, India

S. A. B. Sharma
School of Biological Engineering, SIET, Meerut, Uttar Pradesh, India

B. R. Babaniyi, E. E. Babaniyi (eds.), *The Interplay of Pesticides and Climate Change*, https://doi.org/10.1007/978-3-031-81669-7_7

et al., 2024). For billions of people around the world, agriculture provides a steady supply of food. Nevertheless, this industry is highly dependent on chemical inputs such as pesticides to manage crop-threatening pests, diseases, and weeds. In order to increase agricultural output, pesticide use has skyrocketed since the middle of the twentieth century.

Concurrently, human activities are causing the climate to change in ways that have never been seen before, resulting in higher average temperatures, different patterns of precipitation, and more frequent and severe weather events. Ecosystems and agricultural systems across the globe are already feeling the effects of these changes. Many environmental problems are brought about by intensive agriculture, such as greenhouse gas emissions, chemical pesticide use, excess nutrient runoff, and biodiversity loss. It is already extremely difficult to reduce these detrimental impacts of agriculture while feeding the world's expanding population, and climate change is probably going to make things worse. It is generally predicted that as the world's population grows and becomes wealthier, there will be a greater demand for agricultural products such as foods, feeds, fibers, and biofuels, which will have an increasing influence on the environment. Although it has not gotten as much focus, climate change is another element that could amplify the environmental impacts of agriculture. Rising temperatures are disrupting the normally stable weather patterns that have supported the growth and success of modern agriculture for decades, and they also pose a significant threat to agricultural output (Yang et al., 2024). In order to prepare for and deal with the consequences of climate change, it is essential to understand how this phenomenon will influence various plant types, ecosystem services, and crops (Ogidi, 2023). But many of the parts are likely to be far too complex for simple models to provide reliable estimates. As we can see from the present financial crisis, there might be disastrous results from not having enough system awareness. A common premise in economic models is the persistence of existing regulatory mechanisms and the existence of compensating forces to maintain desired equilibrium. As the near-global economic recession that started in 2008 shows, terrible things can happen when people cannot predict or handle the complexity of systems properly (Garrett et al., 2011).

By definition, feedback is any kind of impact or communication that relays details regarding an action's or process's outcomes back to the person or thing that initiated it. As a result, the changes it makes might have an indirect impact on some part of the system. Feedback loops are a kind of control mechanism that can be found in both ecological and socioeconomic systems. One use of negative loops that stabilizes change is the regulation of core body temperature, whereas positive feedback is used in other contexts. The transition from a regional to a global food production system has improved economic efficiency, but it has come at the expense of numerous social and environmental concerns. In the section that follows, we will explore the current trends in the market. These tendencies pertain to the food system and how feedback is received, understood, and communicated. We conclude that the food system's feedback loops need to be strengthened based on the limitations of existing feedback management. If people want society to be better at reading ecosystem signals, responding to change, and adjusting management methods and consumer habits appropriately, then feedback loops need to be tightened. One way to improve is to learn from small-scale resource management based on local ecological principles; another is to rely more on local ecosystems (Sundkvist et al., 2005).

Pesticides and climate change are part of a complex feedback loop. On the one side, climate change is reshaping pest dynamics, making some species more resilient and leading to longer growing seasons for pests. This increases the demand for pesticides. On the other side, the increased use of pesticides contributes to environmental pollution, biodiversity loss, and greenhouse gas emissions (Sharma et al., 2019). These interactions create a vicious cycle where the very tools used to mitigate crop losses from climate-induced pest pressures contribute to accelerating climate change. Agriculture systems that support biodiversity and rely less on chemical inputs and more on ecosystem services are becoming more and more necessary. In this research, we concentrate on crop protection techniques that don't include using pesticides. Plant diversity in agricultural areas over different time and space scales is a powerful socioeconomic and agroecological tool for pest management in the long run. We conducted a literature review on the ecological effects of plant diversity on pests and their natural enemies and a social science literature review on the requirements for farmers to adopt the corresponding practices and behaviors in order to assess the potential benefits of plant diversification (Vialatte et al., 2021). The introduction of new crops to a region often prompts changes in farming methods and cropping practices, which might cause inexperience with the cultivation of these crops and, in turn, increase the likelihood of plant disease epidemics. These kinds of climate change have a profound effect on agricultural systems (Ogidi & Emaikwu, 2024).

A big challenge is separating the direct consequences of climate change from its indirect effects through adaptation in agriculture. Alternative agricultural systems can adapt by reducing the impact of drought through the use of ecological buffering mechanisms, such as increased soil organic matter and water-holding capacity. This is one of two major themes in adaptation. Increasing the volume of operations and intensifying production methods are two ways that many farmers in conventional agricultural systems try to lower agricultural hazards. Imprecise market signals are another factor that needs to be considered in agriculture in developing countries due to governmental changes and infrastructure constraints (such as roads, educational systems, and communication networks). Although climate change could lead to more crop varieties, it could also make farmers more anxious and helpless in the face of plant and animal diseases due to a lack of information and delayed availability to necessary supplies. A massive, interdisciplinary effort is required to conduct a thorough review of these critical components (Garrett et al., 2011).

Long-term agricultural sustainability and dependability are crucial for maintaining both global and human health. Rapid implementation of strategies that strengthen agriculture's ability to withstand environmental stresses and climate change is crucial in light of the challenges this phenomenon poses. This is especially true of strategies that offer multiple benefits at once, such as integrated soil fertility management and diversification. Socioeconomic barriers, however, slow the spread of environmentally friendly technologies and behaviors. A better understanding of these challenges in context and the development of novel approaches to overcoming them are of the utmost importance. To make agriculture more reliable, sustainable, and climate resilient, people should eat healthier diets and try new foods. This would reduce the amount of land, water, and chemicals needed per person compared to now. It may also be helpful to put more money toward finding, adapting, and reducing the cost of new agricultural technologies (Yang et al., 2024). Biotechnological advancements are opening the door to the possibility of cultivating crops with enhanced resilience to both pests and the effects of climate change.

Genetically modified crops that are engineered to withstand drought, heat, and pest attacks offer a potential solution to reduce the need for chemical interventions (Kikuchi, 2020). The increased use of pesticides leads to harmful effects on ecosystems, particularly affecting non-target organisms such as pollinators, birds, and aquatic species. In addition, pesticide runoff into rivers and lakes threatens water quality and poses risks to human health (Aktar et al., 2009; Iyiola et al., 2023). This becomes more pronounced in regions already experiencing climate-related stress, such as droughts or floods, where the interaction between environmental strain and chemical contamination can be particularly damaging.

An unprecedented level of anthropogenic disturbances to the Earth's ecosystem has resulted from the rise in production activities caused by the growing global population and the societal demands that follow. The increasingly strained interaction between humans and environment has become a major roadblock to the regular advancement of human production and subsistence. In an endeavor to embrace a cleaner world and achieve sustainable growth of human civilization, studying the process of human–nature interaction has remained a key focus of research. However, most studies only look at environmental issues in isolation, which means we're learning a lot of different things and could end up with a completely disorganized environmental consciousness. Dong and Huang (2023) take a comprehensive and integrated view of Earth to investigate positive initiatives aimed at mitigating the feedback consequences of human–nature interactions. By integrating biological, cultural, and mechanical control measures, Integrated Pest Management (IPM) offers a potential alternative to chemical pesticides.

IPM uses ecosystem-based strategies, such as promoting natural pest predators, crop rotation, and habitat management, to maintain pest populations at manageable levels without excessive pesticide use (Ehler, 2006). In a changing climate, IPM offers flexibility by relying less on chemical inputs that may become ineffective.

From an agronomic standpoint, current methods for protecting crops and diversifying the vegetation are closely related (Sawyer et al., 2023). At least 6000 years have passed since field-level crop rotations were first employed to reduce pests and illnesses. In the eighteenth and nineteenth centuries, these practices completely changed agricultural production. Farms in industrialized nations have grown increasingly specialized with the introduction of synthetic chemical pesticides in the late twentieth century, and farmers there have simplified crop rotations, often toward very costly cash crops. This evolution has simplified and drastically altered the composition of landscapes. As a result of cash crop acceptance and the associated intensive production methods, technological breakthroughs have occurred, and agricultural systems today exhibit inflexible, stuck-in behaviors (Vialatte et al., 2021).

Humans have dispersed evidence from our planet's outer environments the hydrosphere, pedosphere, biosphere, and lithosphere in a spectacular fashion. But the disproportionate exchange of materials and energy that humans have with the natural world is steadily weakening nature's defenses. Global warming has already increased the planet's temperature by about 1.1 °C, and if it rises above 1.5 °C, global ecosystems will suffer significant and irreparable damage. The concentrations of carbon dioxide have surpassed levels not seen in the past two million years, while those of methane and nitrous oxide have reached their highest heights in at least 800,000 years. The degradation of more than 5% of Earth's land surface due to human-induced climate change has hastened desertification. Rising sea levels have occurred throughout the past century (Dong & Huang, 2023). Agroecological practices, such as organic farming, crop diversification, and conservation agriculture, are also key strategies for addressing the pesticide-climate feedback loop. These practices strengthen agricultural systems' ability to withstand the effects of climate change while simultaneously decreasing the need for pesticides. Pest control, biodiversity enhancement, and resource conservation are the tenets of agroecology (Altieri et al., 2017; Ogidi et al., 2024b).

This chapter provides a comprehensive overview of agrochemicals used in modern farming. The use of agrochemicals, including fertilizers, herbicides, insecticides, and fungicides, is crucial for farmers in order to enhance agricultural productivity, manage pests, and raise crop yields. While they have been useful in resolving global food security challenges since the Green Revolution, their improper and overuse can lead to soil degradation, environmental contamination, and injury to animals that aren't meant to be affected. Because of climate change, using agrochemicals is become more complicated. To ensure sustainability in the long run, it is necessary to implement integrated pest management, precision farming, and the use of environmentally friendly alternatives. Farmers and legislators need to prioritize the responsible use of agrochemicals while taking potential health and environmental effects into account (Ogidi et al., 2023; Ogidi & Akpan, 2023). We may work toward a more sustainable future in agriculture by weighing the advantages and disadvantages, preserving the environment and guaranteeing food supply (Shafiq et al., 2024).

It is important to improve the feedback loops that link ecosystems, food supply chain actors, and consumers. Trends that have been identified as significant obstacles to tighter feedback loops include intensification, specialization, homogenization, distance, and concentration. These patterns can interfere with communication throughout the food chain and make it more difficult to perceive or hear warning signals from sick ecosystems. Examples of tightened feedback loops have been recorded, and chances for improved feedback management at local to global levels are being investigated (Sundkvist et al., 2005). Potential areas for development include improved coordination among the participants in the food chain and awareness of ecological feedback, for example, by depending more on regional resources.

However, in cases when the distance between the resource and the resource user is too great, institutional channels must be used to channel input, such as through policy measures. Research should focus on understanding how pest populations and pesticide needs will evolve under different climate scenarios. Predictive models that integrate climate data, pest biology, and agricultural practices will be essential for planning future pesticide regulations and developing sustainable farming systems (Ziska et al., 2019).

2 The Role of Pesticides in Climate Change

Nematicides, rodenticides, fungicides, herbicides, insecticides, and molluscicides are all part of the pesticide class (Bernardes et al., 2015). Efficient pest control using a variety of pesticides is essential to combat pests and enhance crop yield, as insect infestations result in the loss of more than 45% of the world's food production every year. The critical need to enhance food production and fight insect-borne diseases spurred the invention of pesticides during World War II (1939–1945). Additionally, beginning in the 1940s, there was an even more significant increase in food production due to the increasing usage of synthetic crop protection agents (Bernardes et al., 2015).

The use of pesticides and other agricultural practices is responsible for around 30% of the world's greenhouse gas emissions, according to a study by the Intergovernmental Panel on Climate Change. One of the greatest problems that mankind is facing right now is climate change. In addition to increasing emissions of greenhouse gases, scientific evidence suggests that pesticides make our agricultural systems more vulnerable to the effects of climate change. As the growing season becomes extended and the climate becomes warmer, it is likely that weeds and insect pests will multiply which potentially resulting in increased pesticide usage. This increased pesticide usage is in turn responsible for emitting harmful substances that worsen climate change.

Pesticides influence climate change from start to finish, including their creation, packaging, shipping, use, and the harm they cause to the environment when disposed of. A large number of oil and gas businesses are actively involved in developing pesticide components, and it is important to mention that 99% of synthetic chemicals, including pesticides, are generated from fossil fuels (Drugmand et al., 2022). US Environmental Protection Agency (2022) research indicated that fumigant pesticides increased soil nitrous oxide generation seven to eight times more than that of nitrogen fertilizers, demonstrating that pesticides, such as nitrogen fertilizers, can lead to greenhouse gas emissions after application. In addition to harming human and environmental health, increasing pesticide use will cause weed and insect pests to develop resistance to these chemicals. Communities that are already struggling due to climate change issues, such as intense heat and smoke from wildfires, may feel the effects of increased usage of synthetic pesticides even more acutely (Ferguson et al., 2019).

Pesticides also affect the soil health. Pesticides entering the soil have the potential to change the diversity and biomass of soil microbes. Changes in soil microorganism behavior caused by pesticides result in disruption of soil ecosystem and decline in soil fertility (Handa et al., 1999). Moreover, worldwide pesticide production and utilization increased year by year. Table 1 shows a continuous increase in the use of pesticides during the period of 2018–2022.

According to Bernardes et al. (2015), hardly 1% of all pesticides are actually efficient in controlling insect pests that attack specific plants. However, large amounts of pesticides that did not reach their intended targets end up in non-target plants and other environmental media. Consequently, pesticide poisoning has caused environmental pollution and significant health impacts (Ogidi & Akpan, 2024).

Pesticides after their utilization for target plants are disposed of in the soil; they generally have the capacity to migrate within the soil through water by the process of transport or degradation (Singh, 2012). So the degradation of pesticides in the environment generates many new substances or chemicals. Depending on their behavior, pesticides can also travel by processes such as leaching, adsorption, spray drift, volatilization, and runoff from target locations to adjacent places devoid of vegetation (Robinson et al., 1999; Ogidi & Richard-Nwachukwu, 2024).

Pesticides are initially dispersed throughout the environment by the air, water, soil, animals, and plants; nevertheless, in the end, they are changed into their derivatives or more complicated structures. The amount of pesticides also plays a significant role, as their degradation sends these chemicals migrating into the soil, air, and ground water, where their mobility causes recirculation (Kerle et al., 2007). The many chemical kinds point to the variations in how they behave in the environment. Organochlorine chemicals, such as DDT, exhibit minimal acute toxicity but a notable capacity to accumulate in tissues and continue to cause harm over an extended period of time. Although their sale is prohibited in the majority of countries, their nature means that their remnants linger in the environment for a very long time. Organophosphate insecticides have a significant acute toxicity in mammals despite having a low persistence.

Table 1 Pesticide use during 2018–2022 in world

Year	Area	Pesticide (Tons)
2022	World	3,697,552.92
2021	World	3,568,335.92
2020	World	3,482,933.42
2019	World	3,441,026.9
2018	World	3,389,657.38

Source: Food and Agriculture Organization of the United Nations. FAOSTAT Online Database. https://www.fao.org/faostat/en/#data/RP. Accessed on September 19, 2024

Greenhouse gas emissions can be exacerbated by the release of volatile organic compounds (VOCs) caused by pesticide application. To create ground-level ozone, sulfur dioxide (SO_2), nitrogen oxides (NOx), and ultraviolet light react (Martin, 2013). Two major areas of contemporary scientific literature focus on the greenhouse gas emissions from insecticides. Greenhouse gas emissions from pesticide interactions with the environment after application are the subject of both short-term and long-term studies, but some studies focus on emissions from pesticide production, distribution, and field application. Agricultural production, which includes the manufacture of related inputs such as pesticides, is responsible for 31% of the world's greenhouse gas (GHG) emissions, which constitute more than one-third of all GHG emissions (Tubiello et al., 2022).

Experts have speculated that the high levels of N_2O emissions associated with some pesticide treatments may be due to their impact on soil bacteria. Soil microbe activity and soil health are both adversely affected by pesticide use, which in turn increases greenhouse gas emissions.

3 Climate Change and Pest Dynamics

Climate change is fundamentally altering the dynamics of pests and their interactions with crops, with significant implications for agricultural productivity and pesticide use. Warmer temperatures, changing precipitation patterns, and disruptions to natural pest control mechanisms are all contributing to increased pest pressures (Zhao et al., 2017). As a result, farmers may rely more heavily on chemical pesticides, leading to environmental and health risks. To mitigate these challenges, it is essential to adopt sustainable pest management strategies that incorporate the latest technologies and promote ecosystem resilience.

According to Anuradha and Singh (2017) and Kumar and Kumar (2019), this is the best way for agriculture to adjust to the changing climate while reducing the harmful effects of pesticides on people and the planet. The relationship between pests and crops is changing as a result of climate change's drastic effects on ecosystems around the world. Changes in precipitation patterns, increased frequency of extreme weather events, and overall global warming are all having profound effects on the dynamic between plants, pests, and their environments. The use of pesticides, methods of pest management, and agricultural output are all directly affected by this. Here we delve into the different ways that climate change affects pest dynamics, how it affects crop protection, and how these changing interactions lead to chemicals being used more frequently as pesticides (Leoci & Ruberti, 2021).

4 Classification of Pesticides and Their Pests

Pesticides can be classified according to their target organism, chemical composition (e.g., organic, biopesticide, insecticide, fungicide, insect spray, pediculicide, and rodenticide) (Gilden et al., 2010), or physical state (e.g., vaporous fumigant) (Amdur et al., 1997). Microbial and biochemical pesticides are both included in biopesticides. A new category of insecticides known as "botanicals" has recently emerged, those based on plants. A fourth category that includes scilliroside and strychnine is formed by the rotenoids, nicotinoids, and pyrethroids (Kamrin, 1997). Pesticides are categorized in Table 2.

A common way to identify pesticides is by the type of pest they kill. On the other hand, pesticides can be either biodegradable, meaning they break down into harmless compounds when broken down by microbes and other living things, or persistent, such as DDT, which accumulated in the food chain and killed birds at the top of the food web because it took months or years to break down.

5 Temperature Changes and Pest Development

One of the most significant ways climate changes affect pests is through temperature increases. Insects are ectothermic organisms, meaning their development, reproduction, and activity rates are directly influenced by the surrounding temperature. As global temperatures rise, many insect species, particularly pests, are experiencing accelerated life cycles, allowing them to reproduce more frequently in a growing season. Because of this, food yields may be negatively impacted by rising population densities (Cai et al., 2020). Warmer temperatures also allow many pests to expand their geographical range into previously cooler regions. For example, the migration of certain pests such as the mountain pine beetle in North America has been facilitated by warmer winters, allowing them to invade regions at higher

Table 2 Targeted pest-based classification of pesticides (Aktar et al., 2009; Kumar & Kumar, 2019)

Pests	Type of pesticides
Algae	Algicide
Bacteria	Bactericide
Birds	Avicide
Fishes	Piscicide
Fungi	Fungicide
Insects	Insecticide
Mites	Miticide
Nematodes	Nematicide
Rodents	Rodenticide
Snail & Slugs	Molluscicide
Weeds	Herbicide or weedicide

latitudes and altitudes. Similarly, tropical and subtropical pests, such as the fall armyworm and aphids, are expected to expand into temperate regions as these areas become warmer (Zhao et al., 2017). Increased temperature also extends the growing season for many crops, giving pests more opportunities to feed and reproduce. Longer growing seasons may result in a higher number of pest generations, thereby exacerbating pest pressure on crops. In turn, farmers may need to apply more pesticides or adopt new pest management strategies to cope with the increased infestation rates (Shang et al., 2018).

6 Shifts in Precipitation Patterns

Climate change not only alters temperatures but also influences precipitation patterns. In some regions, rainfall has become more erratic, with more intense and unpredictable precipitation events interspersed with periods of drought. These changes affect pest dynamics in multiple ways (Cai et al., 2020). Excessive rainfall can create favorable breeding conditions for pests such as mosquitoes, which breed in stagnant water. Increased humidity can also favor the growth of fungi, which may lead to the proliferation of fungal pathogens that weaken plants and make them more susceptible to insect damage. Conversely, prolonged drought conditions can reduce plant vigor, leaving crops more vulnerable to pests (Song et al., 2019). Drought also has indirect effects on pest populations. For example, in water stressed crops, plants may produce fewer natural defenses, such as phenolic compounds, making them easier targets for herbivorous insects. Additionally, drought may cause a shift in pest behavior, where pests that typically feed on moist areas of a plant may start attacking drier parts, intensifying the damage to the crop. Changes in rainfall patterns also influence pest distribution. Pests that thrive in humid environments may find new niches as previously arid regions experience increased rainfall. At the same time, pests that prefer dry conditions may expand into areas that are becoming more arid due to changing precipitation patterns (Wang et al., 2022).

7 Impacts on Natural Pest Predators

Climate change not only affects pest species but also disrupts the natural predators and parasitoids that help control pest populations. For instance, birds, amphibians, and beneficial insects, such as ladybugs and predatory beetles, play a crucial role in maintaining pest populations at manageable levels. However, climate-induced habitat loss and fragmentation, changes in prey availability, and shifting migration patterns can reduce the efficacy of these natural control agents (Relyea & Hoverman, 2008). Many beneficial insects have complex life cycles that are tightly linked to specific environmental conditions. The timing of pest emergence and the actions of natural enemies could be thrown off if these variables are altered by climate change.

In a world where warmer temperatures spur faster pest development without matching acceleration in the development of their predators, the former could lead to uncontrolled population growth (Helfrich et al., 2009). In addition, biological control services might be reduced when natural enemies are disproportionately affected by severe weather events, such as heat waves, floods, and storms. As a result, farmers may increasingly rely on pesticides to manage pests in the absence of these natural regulators (Relyea & Hoverman, 2008).

8 Increased Pesticide Use

As pest dynamics evolve in response to climate change, there is a growing concern that farmers may turn to more frequent and higher doses of chemical pesticides. The expansion of pest ranges, higher reproduction rates, and weakened natural control mechanisms collectively lead to higher pest pressures, necessitating more intensive crop protection measures. This increased reliance on pesticides presents several challenges (Kumar & Kumar, 2019). Initially, the use of chemicals may lead to the emergence of pesticide-resistant insect populations. Insects and other pests that are exposed to sublethal doses of pesticides over time can develop resistance, rendering these chemical treatments ineffective. This phenomenon, known as the pesticide treadmill, forces farmers to use stronger chemicals or higher doses, which can harm non-target organisms, including beneficial insects, and degrade ecosystem health (Bolognesi & Merlo, 2011). Secondly, the increasing use of pesticides is linked to environmental and health concerns. Pesticide runoff can contaminate water sources, harm aquatic ecosystems, and reduce biodiversity (Ogidi et al., 2024a; Ogidi & Akpan, 2022; Okosodo & Ogidi, 2023). In addition, pesticides can accumulate in the food chain, affecting the health of farmworkers and consumers who are exposed to these chemicals (Relyea & Hoverman, 2008) (Fig. 1).

9 Adaptation and Mitigation Strategies

Climate change is affecting pest dynamics and pesticide use in ways that traditional integrated pest management (IPM) approaches can't keep up with. Cultural practices, biological control, and the selective use of chemical pesticides are all part of integrated pest management (IPM), which advocates for a variety of sustainable approaches to pest control. To adjust integrated pest management (IPM) to the changing climate, we need to learn more about the changing dynamics of pests and create new tools and strategies to deal with them. One way to lessen reliance on chemical pesticides is to create crop types with increased resistance to pests and improved tolerance to environmental stresses such as drought (Chattopadhyay et al., 2017). Another strategy to reduce pest populations without resorting to chemical pesticides is to increase the use of predatory insects and parasitoids as biological

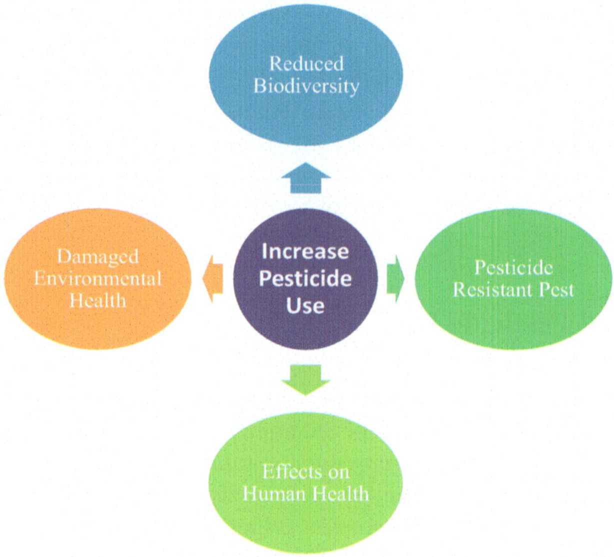

Fig. 1 Increased of pesticide

control agents. To lessen the blow that climate change will deal to pest dynamics, precision agriculture tools, such as predictive modeling and remote sensing, can be invaluable. Farmers can optimize the timing and application of pesticides by utilizing data-driven systems to monitor insect populations and environmental factors (Pathak et al., 2022). This helps to reduce the overall quantity of chemicals utilized. Lastly, agroforestry, intercropping, and the preservation of natural habitats are practices that can enhance landscape-level biodiversity and, by extension, agroecosystem resilience. Because they provide a variety of ecological niches that host natural pest predators, biodiverse systems are more resilient to pest outbreaks (Kumar & Kumar, 2019).

10 Sustainable Alternatives and Solutions to Pesticides in the Context of Climate Change

Sustainable alternatives to chemical pesticides are essential in addressing the environmental and agricultural challenges posed by climate change. Integrated pest management, agroecology, biopesticides, organic farming, precision agriculture, and genetic engineering offer viable solutions to reduce pesticide use, enhance biodiversity, and promote resilient ecosystems. As climate change intensifies, adopting these approaches will be crucial for sustaining agricultural productivity while minimizing the sector's contribution to environmental degradation. Through a combination of technological innovation, ecological stewardship, and policy support, a future with reduced pesticide dependence and increased climate resilience is achievable (Kumar & Kumar, 2019). As climate change exacerbates environmental stresses, the agricultural sector faces unprecedented challenges, including pest proliferation, reduced crop yields, and degraded ecosystems. Pesticides, while crucial for maintaining agricultural productivity, contribute to environmental degradation and can exacerbate climate change through greenhouse gas emissions during production and application. Sustainable alternatives to conventional pesticides have emerged as essential components in the battle against the dual crises of climate change and biodiversity loss. This section discusses various ecofriendly approaches and technologies that minimize pesticide use, enhance crop resilience, and promote a balance between agricultural productivity and environmental sustainability (Anuradha & Singh, 2017; Pathak et al., 2022).

11 Pest Management Through Integration (IPM)

By integrating chemical, mechanical, cultural, and biological approaches to pest management, Integrated Pest Management (IPM) aims to reduce the need for pesticides. Rather than relying solely on chemical pesticides, IPM promotes a comprehensive approach that prioritizes ecosystem health and pest prevention (Anuradha & Singh, 2017).

Components of IPM

Biological Control: Managing pest populations by the use of naturally occurring parasites, diseases, or predators. Ladybirds and nematodes are two examples of beneficial insects that prey on pests that live in soil.

Cultural Practices: These involve modifications in farming techniques to reduce pest establishment, such as crop rotation, intercropping, and selecting pest-resistant crop varieties.

Mechanical and Physical Control: Tools such as traps, barriers, and handpicking pests can reduce reliance on chemical interventions.

Chemical Control: When necessary, IPM allows for the judicious use of ecofriendly pesticides, often as a last resort, ensuring minimal environmental disruption.

Impact on Climate Change

By reducing the need for synthetic chemical inputs, IPM helps lower greenhouse gas emissions and preserves soil health, which in turn improves carbon sequestration. Furthermore, promoting biodiversity in agricultural systems enhances ecosystem resilience to climate-related stresses such as drought and extreme weather events (Anuradha & Singh, 2017).

Agroecology

Agroecology promotes sustainable farming by integrating ecological principles into agricultural practices. It emphasizes the use of renewable resources, biodiversity conservation, and minimizing external inputs, such as synthetic fertilizers and pesticides. By fostering a self-sustaining system, agroecology reduces the environmental footprint of farming (Anuradha & Singh, 2017; Choudhury & Saha, 2020).

Agroecological Practices

Diversification of Crops: Polyculture, or the cultivation of multiple crops, enhances soil fertility, suppresses pests, and mitigates the risk of crop failure due to climate variability.

Soil Health Management: Practices, such as no-till farming, cover cropping, and organic fertilization, boost soil organic matter, improving water retention and nutrient cycling while reducing the need for synthetic pesticides.

Agroforestry: Integrating trees and shrubs into farming landscapes not only provides habitat for beneficial insects and birds but also helps in carbon sequestration and reduces soil erosion.

Climate Change Mitigation

Agroecology enhances ecosystem resilience to climate change by fostering biodiversity and improving water and soil management. By reducing pesticide dependence, it lessens the release of harmful chemicals into the environment, further lowering carbon emissions (Wang et al., 2022).

12 Biopesticides

A less harmful substitute for traditional synthetic pesticides is biopesticides, which are made from natural sources such as plants, bacteria, fungus, and minerals. They target specific pests and pathogens while being less harmful to nontarget organisms, including humans, animals, and beneficial insects such as pollinators (Choudhury & Saha, 2020).

Types of Biopesticides

Microbial Pesticides: These utilize microorganisms such as *Bacillus thuringiensis* (Bt), which produces toxins that target specific insect larvae.

Botanical Pesticides: Extracts from plants such as neem, pyrethrum, and garlic possess natural pest-repelling or insecticidal properties.

Biochemical Pesticides: Compounds such as pheromones can disrupt pest mating cycles, reducing reproduction rates without harmful environmental effects.

Environmental and Climate Benefits

Biopesticides degrade more quickly in the environment compared to synthetic chemicals, reducing soil and water contamination. Their targeted action ensures minimal disruption to beneficial species, supporting biodiversity, which is crucial for ecosystem stability amid climate change. Moreover, biopesticides often require less energy to produce, contributing to lower carbon footprints (Leoci & Ruberti, 2021; Wang et al., 2022).

13 Organic Farming

Organic producers utilize natural methods to manage pests and weeds, and they never employ synthetic fertilizers or pesticides. By focusing on soil health, biodiversity, and sustainable practices, organic farming helps mitigate the negative environmental impacts associated with industrial agriculture (Shrestha, 2019).

Pest Control in Organic Farming

Companion Planting: Some plants actually keep pests away. For example, basil can protect plants from aphids while marigolds are thought to be nematode repellents.

Natural Predators: Organic farming often incorporates habitats for predators such as birds, spiders, and predatory insects to naturally control pest populations.

Crop Rotation and Diversification: Reducing the possibility of pest outbreaks and disrupting their life cycles, these techniques help keep chemical intervention to a minimum.

Climate Change Mitigation

Organic farming contributes to the mitigation of the effects of climate change by reducing dependence on energy-intensive synthetic inputs and enhancing soil carbon sequestration. Additionally, the promotion of biodiversity helps enhance ecosystem resilience to changing climatic conditions (Zhao et al., 2017).

Precision Agriculture

Precision agriculture employs state-of-the-art technology, including artificial intelligence-powered analytics, drones, GPS, and sensors, to improve agricultural practices. This approach allows farmers to apply inputs, such as water, fertilizers, and pesticides, in a highly targeted manner, reducing wastage and environmental harm (Shrestha, 2019).

14 Technological Innovations

Drone Surveillance: Drones equipped with multispectral sensors can monitor crop health and identify insect concentrations to ensure that pesticides are applied precisely where they are required.

Soil and Crop Sensors: These sensors enable you to make more informed decisions regarding the application of pesticides by monitoring soil moisture, nutrient levels, and plant health in real time.

AI and Machine Learning: Predictive models can forecast pest outbreaks based on weather patterns and crop conditions, allowing for proactive pest management with minimal chemical use.

Climate and Environmental Benefits

Precision agriculture can reduce greenhouse gas emissions by minimizing the over application of pesticides and fertilizers. Agricultural systems can more effectively withstand the effects of climate change if they optimize their resource utilization, as per Cooper and Dobson (2007) and Choudhury and Saha (2020). This contributes to the conservation of water, the preservation of soil health, and the reduction of pollution.

Genetic Engineering and Crop Breeding

Advances in genetic engineering and selective breeding have led to the development of pest-resistant crop varieties. These crops require fewer chemical inputs, reducing the reliance on pesticides and minimizing their environmental impact (Shrestha, 2019).

Techniques and Approaches

Genetically Modified Organisms (GMOs): Crops, such as Bt corn and cotton, have been engineered to express pest-resistant traits, reducing the need for external chemical control.

CRISPR Technology: This precise gene-editing tool allows the development of crops with enhanced resistance to pests and diseases, further minimizing pesticide dependence.

Traditional Breeding: Selective breeding programs focus on developing varieties that are naturally resistant to pests, improving crop resilience without genetic modification.

Climate Change Considerations

Crops bred for pest resistance reduce the need for synthetic pesticides, lowering emissions associated with their production and application. Additionally, pest-resistant crops are more resilient to climate-related stresses, such as increased pest pressure due to rising temperatures (Cooper & Dobson, 2007).

The Evolving Pest Threat in a Warming World

One of the primary concerns of climate change is its impact on pest populations. Warmer temperatures are expanding the geographical range of pests, allowing them to thrive in regions that were previously inhospitable. Additionally, pests are experiencing faster reproductive cycles and increased survival rates, posing an amplified threat to global agriculture. With such unpredictable pest behaviors, the reliance on chemical pesticides is expected to grow. However, this can exacerbate environmental degradation, reduce biodiversity, and contribute to pesticide resistance, further complicating pest management efforts (Yadav et al., 2019).

Biological control, crop rotation, and resistant crop types should be the cornerstones of future pest management methods that prioritize integrated pest management (IPM) systems. Advances in remote sensing and predictive modeling can also help farmers anticipate pest outbreaks, allowing them to minimize pesticide applications and thus reduce their environmental footprint. By embracing these adaptive measures, agriculture can maintain productivity while addressing the sustainability concerns tied to pesticide overuse in a warming climate (Shrestha, 2019).

Sustainable Alternatives to Pesticides

In response to the growing risks associated with pesticide use in a changing climate, there is an urgent need to explore and scale up sustainable alternatives. New biopesticides provide an environmentally benign alternative to traditional pesticides by drawing on substances found in nature, such as certain bacteria, fungi, and plant extracts. Degrading at a faster rate than synthetic chemicals, these biopesticides lessen the long-term contamination of ecosystems and are generally safer for non-target creatures and the environment (Wang et al., 2022). Genetically modified organisms (GMOs) with insect resistance could be another viable option, cutting down on chemical treatments. More resilient crops to pest attacks and environmental stresses may soon be in the works, thanks to developments in gene editing and genetic engineering tools such as CRISPR. However, it is crucial to approach the implementation of these technologies with caution, carefully considering both the potential ecological implications and the level of popular acceptance. Eliminating chemical pesticides is not the way to go; rather, a broad range of techniques that prioritize sustainability should be included into pesticide use going forward. Integrative pest management (IPM), precision agriculture, genetically modified organisms (GMOs), and biopesticides will help farmers use pesticides more wisely, which will reduce their negative effects on the environment and guarantee food security in a world where the weather is becoming more unpredictable (Dad et al., 2022).

15 Climate-Resilient Crop Varieties

Farmers may significantly contribute to the future by developing crop varieties resilient to climate change and pest infestations. The creation of drought- and heat-resistant crops is a continuous effort that integrates conventional breeding methods with advanced biotechnology approaches. These resilient crops will enable farmers to maintain yields despite climate-induced challenges, resulting in reduced total pesticide usage (Wang et al., 2022). Wheat, maize, and rice are fundamental crops, and researchers are now developing varieties that are resistant to heat and drought. The necessity for chemical pesticides will diminish because to crop resilience to specific diseases and pests. Future studies should prioritize enhancing nutritional value and yield potential, alongside these qualities (Leoci & Ruberti, 2021). The rate of acceptance of these varieties in various places and their accessibility to farmers will dictate the feasibility of climate-resilient crops. Given that small-scale farmers are disproportionately impacted by climate change, it is essential for public and commercial groups to collaborate in offering them robust crop alternatives (Shrestha, 2019).

16 Technological Innovations in Precision Agriculture

Technological innovations are reshaping the future of agriculture by enabling more precise and efficient use of resources. Precision agriculture, which employs tools such as drones, GPS mapping, and sensors, allows farmers to monitor soil conditions, water usage, and pest populations in real time. By using these tools, farmers can apply pesticides more accurately, minimizing waste and reducing the environmental impact of chemical inputs (Shrestha, 2019). For example, sensor-based irrigation systems can optimize water usage in regions affected by drought, while drones equipped with imaging technology can detect pest infestations early, allowing for targeted pesticide applications. The overuse of chemical inputs is reduced and natural resources are conserved through these technologies, which promote sustainability and increase farm output (Dad et al., 2022). Machine learning (ML) and artificial intelligence (AI) will also be more influential in farming decisions in the future. Planting times, pesticide applications, and irrigation schedules may be fine-tuned with the help of AI-powered systems that sift through mountains of data on weather trends, soil quality, and crop yields. Leoci and Ruberti (2021) state that farmers can adjust to climate unpredictability with the help of this data-driven method, which will also reduce their use of pesticides and other chemical inputs.

17 Regenerative Agriculture and Soil Health

The future of farming in a warming world depends on healthy soil. Important for climate resilience, degraded soils can't hold as much water, can't sustain plant life, and can't store as much carbon. Soil erosion and degradation are problems that have their roots in conventional farming methods, such as using too many pesticides and synthetic fertilizers. One promising approach to creating more robust agricultural systems is regenerative agriculture, which aims to turn this tendency around (Yadav et al., 2019). Methods, such as cover cropping, agroforestry, and not-till farming, are at the heart of regenerative agriculture's mission to improve soil quality. Soil organic matter enhancement, water retention, and microbial community diversity are all improved by these approaches. Soils that are in good health can endure natural disasters, such as floods and droughts, with less help from artificial fertilizers and pesticides (Dad et al., 2022). A radical change in farming philosophy is necessary for the implementation of regenerative agriculture. The promotion of techniques that prioritize long-term soil health over short-term returns should be a shared goal among farmers, lawmakers, and agribusinesses. Accelerating the shift to regenerative agricultural practices can be achieved through incentives including carbon credits, subsidies, and education programs.

18 Policy and Global Collaboration

Resilient policy frameworks and international collaboration are essential to reform agriculture in addressing climate change challenges. Policymakers should prioritize sustainable farming techniques, reduced greenhouse gas emissions from agriculture, and increased cultivation of crops resilient to adverse weather conditions. To address the global nature of climate change, international bodies and stakeholders must synchronize their activities and enable the transnational interchange of knowledge and resources. To motivate farmers to adopt practices that reduce their carbon footprint, carbon pricing mechanisms such as cap-and-trade systems or carbon taxes may be used. To foster innovation in agriculture, it is imperative to establish legislation that allocate funding for research on sustainable farming practices and technologies. Smallholder farmers in impoverished nations face distinct challenges due to their limited capacity to adapt to climate change, necessitating global collaboration for resolution. International aid programs, knowledge transfer, and capacity development initiatives are vital to help farmers enhance their climate resilience and adopt sustainable pest management practices (Yadav et al., 2019; Leoci & Ruberti, 2021).

19 Future Outlook: Reshaping Agriculture for a Changing Climate

The future of agriculture in a changing climate depends on a holistic approach that integrates technological innovations, sustainable farming practices, and strong policy support. As climate change intensifies, pest pressures will continue to evolve, necessitating a shift away from conventional pesticide-reliant systems toward more resilient and adaptive agricultural models. By embracing biopesticides, climate-resilient crops, precision agriculture, and regenerative farming, we can build a food system that not only mitigates climate impacts but also enhances long-term sustainability. Policymakers, researchers, and farmers must work together to ensure that these future outlooks become a reality, securing a stable food supply in the face of climate uncertainty (Dad et al., 2022). As climate change continues to alter ecosystems and agricultural landscapes, it is becoming increasingly clear that conventional agricultural practices, including pesticide use, must be adapted. Extreme weather events, changing precipitation patterns, and increasing temperatures are already wreaking havoc on crop yields, insect dynamics, and soil fertility. The intersection of sustainability and the requirement for global food security is agriculture, which is both a cause and an effect of climate change. A new agricultural paradigm based on technological advancements, environmentally friendly methods, and climate change adaptation policies is required to lessen the negative effects of global warming and to ensure the continued viability of food security systems (Yadav et al., 2019; Wang et al., 2022).

20 Conclusion

Pesticide use and climate change are deeply intertwined, with each amplifying the negative impacts of the other. To address the challenges posed by this feedback loop, we must pursue integrated solutions that emphasize sustainable agriculture, adaptive pest management strategies, and robust policy reforms. By altering agricultural practices and pest management, we can mitigate the adverse impacts of pesticide application and climate change on human health and the environment, thereby enhancing the resilience of global food systems in the long term. A feedback loop exists between pesticide use and climate change, which is intricate and concerning as it accelerates and exacerbates the detrimental impacts of each. The utilization of chemical pesticides in agriculture is increasing as ecosystems and pest populations are affected by climate change. Consequently, excessive pesticide application intensifies climate change by contributing to environmental degradation, greenhouse gas emissions, and ecological imbalances. This complex dynamic underscores the need for an integrated approach to agriculture that recognizes the interconnection between these challenges. The transition to more sustainable agricultural systems will also require global cooperation. Climate change is a global phenomenon, and its impacts on agriculture transcend national borders. International collaboration on agricultural research, policy development, and climate adaptation will be essential to ensure food security, environmental protection, and climate resilience for future generations. Ultimately, breaking the pesticide-climate feedback loop is not just about reducing chemical use, it is about transforming the very foundations of our agricultural systems to make them more sustainable, resilient, and adaptive to a changing world. By embracing innovation, sustainability, and collaboration, we can protect the planet's ecosystems, safeguard human health, and ensure that agriculture remains a cornerstone of human prosperity in a warming world.

References

Aktar, W., Sengupta, D., & Chowdhury, A. (2009). Impact of pesticides use in agriculture: Their benefits and hazards. *Interdisciplinary Toxicology, 2*(1), 1–12. https://doi.org/10.2478/v10102-009-0001-7

Altieri, M. A., Nicholls, C. I., & Montalba, R. (2017). Technological approaches to sustainable agriculture at a crossroads: An agroecological perspective. *Sustainability, 9*(3), 349.

Amdur, M. O., Doull, J., & Klaassen, C. D. (1997). Educational and informational strategies to reduce pesticide risks. *Preventive Medicine, 26*(2), 191–200. https://doi.org/10.1006/pmed.1996.0122

Anuradha, & Singh, J. (2017). Impact of agrochemicals on climate change: A study. *Journal of Emerging Technologies and Innovative Research (JETIR), 4*(11), 1394–1403.

Bernardes, M. F. F., Pazin, M., Pereira, L. C., & Dorta, D. J. (2015). Impact of pesticides on environmental and human health. In *Toxicology studies—Cells, drugs and environment* (pp. 195–233). IntechOpen.

Bolognesi, C., & Merlo, F. D. (2011). Pesticides: Human health effects. *Encyclopedia of Environmental Health, 16*, 438–453. https://doi.org/10.1016/B978-0-444-52272-6.00592-4
Cai, C., Li, G., Di, L., Ding, Y., Fu, L., Guo, X., Struik, P. C., Pan, G., Li, H., Chen, W., Luo, W., & Yin, X. (2020). The acclimation of leaf photosynthesis of wheat and rice to seasonal temperature changes in T-FACE environments. *Global Change Biology, 26*(2), 539–556.
Chattopadhyay, P., Banerjee, G., & Mukherjee, S. (2017). Recent trends of modern bacterial insecticides for pest control practice in integrated crop management system. *3 Biotech, 7*(1), 60. https://doi.org/10.1007/s13205-017-0717-6
Choudhury, P. P., & Saha, S. (2020). Dynamics of pesticides under changing climatic scenario. *Environmental Monitoring and Assessment, 192*(Suppl 1), 814. https://doi.org/10.1007/s10661-020-08719-y
Cooper, J., & Dobson, H. (2007). The benefits of pesticides to mankind and the environment. *Crop Protection, 26*, 1337–1348.
Dad, K., Zhao, F., Hassan, R., Javed, K., Nawaz, H., Saleem, M. U., Fatima, T., & Nawaz, M. (2022). Pesticides uses, impacts on environment and their possible remediation strategies. A review. *Pakistan Journal of Agricultural Research, 35*(2), 274–284. https://doi.org/10.17582/journal.pjar/2022/35.2.274.284
Dong, L., & Huang, Z. (2023). Some evidence and new insights for feedback loops of human-nature interactions from a holistic earth perspective. *Journal of Cleaner Production, 432*, 139667.
Drugmand, D., Feit, S., Fuhr, L., & Muffett, C. (2022). *Fossils, fertilizers, and false solutions: How laundering fossil fuels in agrochemicals puts the climate and the planet at risk*. The Center for International Law. https://www.ciel.org/wp-content/uploads/2022/10/Fossils-Fertilizers-andFalse-Solutions.pdf
Ehler, L. E. (2006). Integrated pest management (IPM): Definition, historical development, and implementation, and the other IPM. *Pest Management Science, 62*(9), 787–789.
Ferguson, R., Dahl, K., & DeLonge, M. (2019). *Farmworkers at risk: The growing dangers of pesticides and heat*. Union of Concerned Scientists. https://www.ucsusa.org/resources/farmworkers-at-risk
Garrett, K. A., Forbes, G. A., Savary, S., Skelsey, P., Sparks, A. H., Valdivia, C., & Yuen, J. (2011). Complexity in climate-change impacts: An analytical framework for effects mediated by plant disease. *Plant Pathology, 60*(1), 15–30.
Gilden, R. C., Huffling, K., & Sattler, B. (2010). Pesticides and health risks. *Journal of Obstetric, Gynecologic and Neonatal Nursing, 39*(1), 103–110. https://doi.org/10.1111/j.1552-6909.2009.01092.x
Handa, S. K., Agnihotri, N. P., & Kulshreshtha, G. (1999). Effect of pesticide on soil fertility. In *Pesticide residues; significance, management and analysis* (pp. 184–198). Research Periodicals and Book Publishing House.
Helfrich, L. A., Weigmann, D. L., Hipkins, P. A., & Stinson, E. R. (2009). *Pesticides and aquatic animals: A guide to reducing impacts on aquatic systems*. Virginia Polytechnic Institute and State University. Available from https://pubs.ext.vt.edu/420/420-013/420-013.html
Iyiola, A. O., Kolawole, A. S., Ajayi, F. O., Ogidi, O. I., & Ogwu, M. C. (2023). Sustainable water use and management for agricultural transformation in Africa. In *Water resources management for rural development: Challenges and mitigation* (Vol. 1, pp. 287–300). Elsevier. https://doi.org/10.1016/B978-0-443-18778-0.00001-5
Kamrin, M. A. (1997). *Pesticide profiles: Toxicity, environmental impact, and fate* (1st ed.). CRC Press. ISBN 978-1566701907. OCLC 35262311.
Kerle, E. A., Jenkins, J. J., & Vogue, P. A. (2007). *Understanding pesticide persistence and mobility for groundwater and surface water protection*. Oregon State University, Extension Service, EM8561-E.
Kikuchi, T. (2020). Genetically modified organisms and climate-resilient agriculture. *Biotechnology Advances, 40*, 107523.
Kumar, V., & Kumar, P. (2019). Pesticides in agriculture and environment: Impacts on human health. In V. Kumar, R. Kumar, J. Singh, & P. Kumar (Eds.), *Contaminants in agriculture and*

environment: Health risks and remediation (Vol. 1, pp. 76–95). Agro Environ Media. https://doi.org/10.26832/AESA-2019-CAE-0160-07

Leoci, R., & Ruberti, M. (2021). Pesticides: An overview of the current health problems of their use. *Journal of Geoscience and Environment Protection, 9*, 1–20.

Martin, T. (2013). *Volatile organic compound (VOC) emissions from pesticides*. University of California, Agriculture and Natural Resources. https://ucanr.edu/blogs/blogcore/postdetail.cfm?postnum=11273

Ogidi, O. I. (2023). Sustainable utilization of important medicinal plants in Africa. In *Sustainable utilization and conservation of Africa's biological resources and environment, sustainable development and biodiversity*. Springer Nature Singapore Pte Ltd. https://doi.org/10.1007/978-981-19-6974-4_12

Ogidi, O. I., & Akpan, U. M. (2022). Chapter 16: Aquatic biodiversity loss: Impacts of pollution and anthropogenic activities and strategies for conservation. In *Biodiversity in Africa: Potentials, threats and conservation*. Springer Nature Singapore Pte Ltd. https://doi.org/10.1007/978-981-19-3326-4_16

Ogidi, O. I., & Akpan, U. M. (2023). Impacts of chemical use in agricultural practices: Perspectives of soil microorganism and vegetation. In *One health implications of agrochemicals and their sustainable alternative*. Springer Nature Singapore Pte Ltd.

Ogidi, O. I., & Akpan, U. M. (2024). Microbial and phyto-remediation of crude oil contaminated soil. In *Eco-restoration of the environment*. CRC Press, Routledge Taylor and Francis Group.

Ogidi, O. I., & Emaikwu, N. G. (2024). Plant phenolic compound isolation and its bioinformatics approaches of molecular mechanisms in antimicrobial activities and resistance. In *Computational approaches in biomaterials, bioinformatics and biomedical engineering applications*. CRC Press, Taylor and Francis Group.

Ogidi, O. I., & Izah, S. C. (2024). Emerging technologies in the global south food industry: Prospects and challenges. In *Food safety and quality in global south*. Springer Nature Singapore Pte Ltd.

Ogidi, O. I., & Richard-Nwachukwu, N. (2024). Microbe assisted remediation of pesticide residues from soil and water. In *Eco-restoration of the environment*. CRC Press, Routledge Taylor and Francis Group.

Ogidi, O. I., Izah, S. C., & Akpan, U. M. (2023). Food quality and agrochemical use: Integrated monitoring, assessment, and management policies. In *One health implications of agrochemicals and their sustainable alternative*. Springer Nature Singapore Pte Ltd.

Ogidi, O. I., Angaye, T. C. N., & Carbom, H. E. (2024a). Microorganisms in the monitoring of terrestrial and aquatic ecosystems: Applications, challenges and future prospects. In *Biomonitoring of pollutants in the global south*. Springer Nature Singapore Pte Ltd.

Ogidi, O. I., Izah, S. C., & Tobia, P. S. (2024b). Biosurfactant production from cassava processing wastes. In *Sustainable cassava: Strategies from production through waste management*. Elsevier.

Okosodo, E. F., & Ogidi, O. I. (2023). Chapter 3: Biodiversity conservation strategies and sustainability. In *Sustainable utilization and conservation of Africa's biological resources and environment, sustainable development and biodiversity*. Springer Nature Singapore Pte Ltd.

Pathak, V. M., Verma, V. K., Rawat, B. S., Kaur, B., Babu, N., Sharma, A., Dewali, S., Yadav, M., Kumari, R., Singh, S., Mohapatra, A., Pandey, V., Rana, N., & Cunill, J. M. (2022). Current status of pesticide effects on environment, human health and it's eco-friendly management as bioremediation: A comprehensive review. *Frontiers in Microbiology, 13*, 962619.

Relyea, R. A., & Hoverman, J. T. (2008). Interactive effects of predators and a pesticide on aquatic communities. *Oikos, 117*(11), 1647–1658. https://doi.org/10.1111/j.1600-0706.2008.16933.x

Robinson, D. E., Mansingh, A., & Dasgupta, T. P. (1999). Fate and transport of ethoprophos in the Jamaican environment. *Science of the Total Environment, 238*, 373–378.

Sawyer, W. E., Jacob, D. E., Ogidi, O. I., Izah, S. C., & Popoola, A. M. (2023). Threats and conservation strategies of common edible vegetables that possess pharmacological potentials in

Nigeria. In *Herbal medicine phytochemistry* (Reference Series in Phytochemistry). Springer Nature Singapore Pte Ltd.
Shafiq, S., Haq, M. Z. U., Shahbaz, A., Shafique, S., Riaz, M., Hanif, M. T., et al. (2024). Agrochemicals and climate change. In *Water-soil-plant-animal nexus in the era of climate change* (pp. 49–77). IGI Global.
Shang, S., Shang, Z. T., Yu, J., Zhang, X. H., Wu, J., & Jiang, D. (2018). Impacts of climate change on Fusarium head blight in winter wheat. *Fresenius Environmental Bulletin, 27*(6), 3906–3913.
Sharma, A., Kumar, V., Shahzad, B., Tanveer, M., Sidhu, G. P. S., Handa, N., et al. (2019). Worldwide pesticide usage and its impacts on ecosystem. *Current Science, 117*(4), 569–578.
Shrestha, S. (2019). Effects of climate change in agricultural insect pest. *Acta Scientific Agriculture, 3*(12), 74–80.
Singh, D. K. (2012). Pesticides and environment. *Pesticide Chemistry and Toxicology, 1*, 114–122.
Song, Y., Linderholm, H. W., Wang, C., Tian, J., Huo, Z., Gao, P., Song, Y., & Guo, A. (2019). The influence of excess precipitation on winter wheat under climate change in China from 1961 to 2017. *Science of the Total Environment, 690*, 189–196.
Sundkvist, Å., Milestad, R., & Jansson, A. (2005). On the importance of tightening feedback loops for sustainable development of food systems. *Food Policy, 30*(2), 224–239.
Tubiello, F. N., Karl, K., Flammini, A., Gütschow, J., Conchedda, G., Pan, X., & Torero, M. (2022). Pre-and post-production processes increasingly dominate greenhouse gas emissions from agri-food systems. *Earth System Science Data, 14*(4), 1795–1809.
U.S. Environmental Protection Agency. (2022). *Overview of greenhouse gases*. https://www.epa.gov/ghgemissions/overview-greenhouse-gases-nitrous-oxide
Vialatte, A., Tibi, A., Alignier, A., Angeon, V., Bedoussac, L., Bohan, D. A., et al. (2021). Promoting crop pest control by plant diversification in agricultural landscapes: A conceptual framework for analysing feedback loops between agro-ecological and socio-economic effects. In *Advances in ecological research* (Vol. 65, pp. 133–165). Academic Press.
Wang, B.-X., Anouschka, R. H., & Chun-Sen, M. A. (2022). Impacts of climate change on crop production, pests and pathogens of wheat and rice. *Frontiers of Agricultural Science and Engineering, 9*(1), 4–18. https://doi.org/10.15302/J-FASE-2021432
Yadav, T., Baloda, A. S., Jakhar, B. L., & Yadav, A. (2019). Impact of climate changes on insect pest and integrated pest management: A review. *Journal of Entomology and Zoology Studies, 7*(4), 1150–1156.
Yang, Y., Tilman, D., Jin, Z., Smith, P., Barrett, C. B., Zhu, Y. G., & Zhuang, M. (2024). Climate change exacerbates the environmental impacts of agriculture. *Science, 385*(6713), eadn3747.
Zhao, C., Liu, B., Piao, S., Wang, X., Lobell, D. B., Huang, Y., Huang, M., Yao, Y., Bassu, S., Ciais, P., Durand, J. L., Elliott, J., Ewert, F., Janssens, I. A., Li, T., Lin, E., Liu, Q., Martre, P., Müller, C., Peng, S., Peñuelas, J., Ruane, A. C., Wallach, D., Wang, T., Wu, D., Liu, Z., Zhu, Y., Zhu, Z., & Asseng, S. (2017). Temperature increase reduces global yields of major crops in four independent estimates. *Proceedings of the National Academy of Sciences of the United States of America, 114*(35), 9326–9331.
Ziska, L. H., Blumenthal, D. M., Runion, G. B., Hunt, E. R., & Diaz-Soltero, H. (2019). Climate change and invasive plants in the United States: Anticipating the challenges. *Pest Management Science, 75*(9), 2209–2221.

Pesticide Contributions to Greenhouse Gas Emissions

Gabriel Gbenga Babaniyi, Ulelu Jessica Akor, and Abdulhameed Adewale Odeseye

1 Introduction

The energy-intensive nature of pesticide production, such as glyphosate, is a significant concern due to its environmental impact. Glyphosate production, as highlighted in a study on life cycle assessments (Maggi et al., 2020; Yang et al., 2024), requires substantial energy inputs, with the energy used globally in 2014 equivalent to fueling about 6.25 million cars for a year. Additionally, the use of pesticides contributes to greenhouse gas emissions, with some pesticides producing over 40 kg of CO_2e per kilogram, as discussed in the context of energy use and management alternatives in pest control (Helsel & Pimentel, 2007). Despite the energy efficiency improvements in pest control measures, the environmental implications of pesticide production remain relevant today, emphasizing the ongoing need for sustainable and cleaner production technologies to mitigate these impacts. Pesticides impact climate change throughout their lifecycle, including manufacturing, transportation, and disposal, and also lead to the production of harmful greenhouse gases.

Pesticides that are sprayed on to fields and used to fumigate soil can give off chemicals called volatile organic compounds, which can react with other chemicals and form a pollutant called ground-level ozone. Pesticide use accounts for about 6% of total ground-level ozone levels. On average, according to Maggi et al. (2020) and Solomon (2020), herbicide production creates 18.22–26.63 kg of CO_2e per kg,

G. G. Babaniyi (✉)
Department of Agricultural Development and Management (ADEM), Agricultural and Rural Management Training Institute (ARMTI), Ilorin, Nigeria

U. J. Akor
Department of Animal Production, University of Ilorin, Ilorin, Nigeria

A. A. Odeseye
Department of Rural Development and Gender Issues (RUDEG), Agricultural and Rural Management Training Institute (ARMTI), Ilorin, Nigeria

B. R. Babaniyi, E. E. Babaniyi (eds.), *The Interplay of Pesticides and Climate Change*, https://doi.org/10.1007/978-3-031-81669-7_8

insecticide production creates 14.79–18.91 kg of CO_2e per kg, and fungicide production creates 11.94–29.19 kg of CO_2e per kg. These numbers result in a significant contribution to the total greenhouse gas emissions that have played a role in a changing climate. A pesticide consists of both active and inert ingredients. Pesticides consist of active ingredients designed to eliminate pests and inert ingredients that aid in application and enhance product efficacy (Beane Freeman, 2022). Inert ingredients, including microplastics derived from fossil fuels, contribute to controlled release and prolonged effectiveness of the product (Cox & Zeiss, 2022; Ghosson et al., 2023). The synthesis of active ingredients from hydrocarbon feedstocks, combined with various elements, leads to the creation of chemical intermediates and subsequent greenhouse gas emissions (Straw et al., 2022). Additionally, inert ingredients added to pesticides can include substances recognized as active ingredients in other products, some of which possess biological activity and potential toxicity, highlighting the complexity and potential risks associated with pesticide formulations (Ghosson et al., 2023). Understanding the composition and interactions of these components is crucial for assessing the environmental and health impacts of pesticide use.

1.1 Overview of Greenhouse Gas (GHG) Emissions

Greenhouse gases (also known as GHGs) are gases in the earth's atmosphere that trap heat. During the day, the sun shines through the atmosphere, warming the earth's surface. At night the earth's surface cools, releasing heat back into the air. But some of the heat is trapped by the greenhouse gases in the atmosphere. Greenhouse gas (GHG) emissions stem from various sources such as fossil fuel combustion, agriculture, and industrial processes, contributing significantly to climate change (Vetter et al., 2023; Filonchyk et al., 2024; Bhurat et al., 2024). Transportation accounts for 3–4% of global GHG emissions, with mitigation technologies available to reduce these emissions. GHG emissions from the food system make up about one-third of global anthropogenic emissions, varying between developed and developing countries, emphasizing the need for region-specific mitigation strategies (Vetter et al., 2023). A citizen survey revealed that daily life sectors, such as household fuel use, electricity consumption, water use, solid waste discharge, and transportation, are key contributors to GHG emissions, with the transportation sector emitting the most GHGs per capita, followed by electricity and fuel consumption (Filonchyk et al., 2024). Efforts to reduce GHG emissions are crucial to combat climate change and its adverse effects on ecosystems and human health (Filonchyk et al., 2024) (Fig. 1).

Greenhouse gas emissions include carbon dioxide, methane, and nitrous oxide from all sources, including land-use change. They are measured in tons of carbon dioxide-equivalents over a 100-year timescale. Meanwhile, land-use change emissions can be negative. In addition, methane is a greenhouse gas that is significantly more potent than CO_2 when it comes to warming the Earth. Over a period of

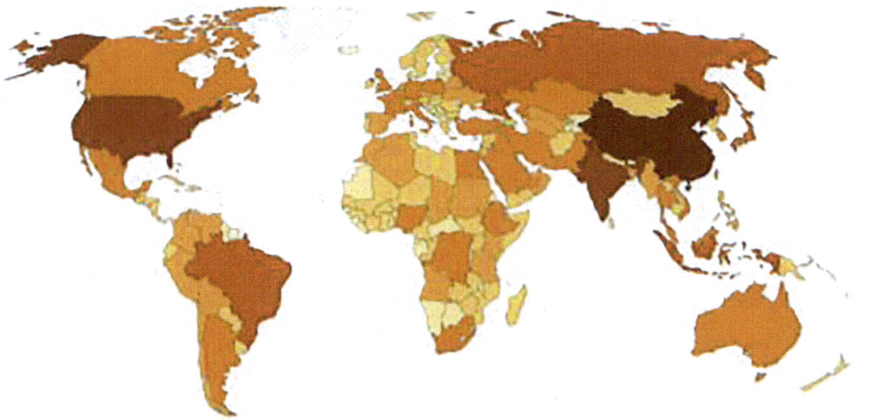

Fig. 1 Greenhouse gas emissions, 2022

100 years, and without considering climate feedback, one ton of methane could cause 28 times more warming than the same amount of CO_2 (Etminan et al., 2016; Ritchie et al., 2020). Even though methane makes up a smaller portion of greenhouse gas emissions by mass, it has been responsible for about a quarter of the radiative forcing since 1750 (Etminan et al., 2016). Methane is classified as a 'short-lived' greenhouse gas, which means it doesn't stay in the atmosphere for long; it's removed within decades, unlike CO_2, which can linger for hundreds or even thousands of years. On average, methane lasts around 12 years in the atmosphere (Balcombe et al., 2018). This means that if we reduce methane emissions, the levels of methane in the atmosphere will drop relatively quickly, leading to a decrease in its warming effects. Therefore, addressing methane emissions could be a swift and effective strategy for mitigating some climate change impacts within the coming decades (Allen et al., 2018).

1.2 The Role of Agriculture in GHG Emissions

Agriculture is a significant contributor to greenhouse gas (GHG) emissions, playing a multifaceted role in the global carbon cycle. One of the primary sources is enteric fermentation, a digestive process in ruminants like cattle, which produces methane (CH_4), a potent GHG with a global warming potential (GWP) approximately 28–36 times that of carbon dioxide (CO_2) over a 100-year period (Kumar & Bordoloi, 2024). Manure management also contributes to CH_4 and nitrous oxide (N_2O) emissions, the latter arising from the microbial denitrification and nitrification processes in soils. Soil management practices, particularly the application of synthetic nitrogen fertilizers, significantly amplify N_2O emissions due to increased nitrification and denitrification rates (Kabange et al., 2023). Additionally, land-use changes, such as deforestation for agricultural expansion, lead to substantial CO_2 emissions, as carbon stored in biomass and soil organic matter is released into the atmosphere (Arsyan Adlan et al., 2024). Agriculture also influences GHG emissions indirectly through the production and transportation of agrochemicals and the energy-intensive processes involved in mechanized farming. Furthermore, rice paddies, due to anaerobic decomposition of organic matter, are notable sources of CH_4 (Squalli & Adamkiewicz, 2023). Hence, the agricultural sector is intricately linked to GHG emissions through various biogeochemical pathways, necessitating comprehensive mitigation strategies to address climate change.

1.2.1 Methanogenesis in Enteric Fermentation

Enteric fermentation in ruminant animals leads to the production of methane (CH_4) via anaerobic digestion by methanogenic archaea in the digestive system. This process is a significant source of CH_4, a potent GHG with a global warming potential (GWP) far exceeding that of CO_2 over a 100-year period. Methane emissions from livestock, particularly cattle, represent a major component of agriculture's contribution to GHGs, accounting for approximately 44% of the sector's emissions (Palangi & Lackner, 2022; Li et al., 2024). However, methanogenesis in enteric fermentation is a biogenic process wherein methanogenic archaea within the rumen of ruminants, such as cattle and sheep, catalyze the reduction in carbon dioxide and hydrogen to methane (CH_4) under anaerobic conditions (Choudhury et al., 2022). This intricate microbial fermentation primarily involves the hydrolysis of polysaccharides, followed by microbial fermentation into volatile fatty acids (VFAs), carbon dioxide (CO_2), and hydrogen (H_2). The accumulation of H_2 in the rumen creates a reductive environment, favoring methanogenesis (Kader Esen et al., 2022). Methane, a potent greenhouse gas with a global warming potential vastly surpassing carbon dioxide, is subsequently eructated by the animal. This process accounts for a substantial fraction of global agricultural CH_4 emissions, contributing significantly to anthropogenic climate forcing (Pierre et al., 2022). The high methanogenic efficiency of these archaea underscores the complexity of mitigating emissions in ruminant livestock systems, where alterations to diet, microbiome composition, and fermentation pathways are explored for mitigation.

1.2.2 Nitrification and Denitrification in Soil Management

Nitrification and denitrification are critical microbial processes in soil nitrogen cycling, profoundly impacting greenhouse gas emissions. Nitrification, driven by chemoautotrophic bacteria like *Nitrosomonas* and *Nitrobacter*, oxidizes ammonia (NH_3) to nitrite (NO_2) and then to nitrate (NO_3), releasing nitrous oxide (N_2O) as a by-product (Wu et al., 2024). This nitrate, under anaerobic conditions, undergoes denitrification, where facultative anaerobes reduce it back to nitrogen gas (N_2)

through intermediate gaseous forms, including N_2O, a potent GHG with a global warming potential 298 times that of CO_2 (Hui et al., 2024; Ma et al., 2024). These processes are influenced by soil pH, moisture, and the carbon-to-nitrogen ratio, leading to complex spatiotemporal emission patterns. Intensive use of nitrogen fertilizers exacerbates N_2O emissions, especially under suboptimal conditions for plant uptake, creating a feedback loop of increased nitrogen inputs and elevated emissions (Ma et al., 2024). Thus, managing these biogeochemical processes is crucial for mitigating the agricultural sector's contribution to climate change. The application of synthetic fertilizers and manure to soils leads to nitrification and denitrification processes. Nitrification converts ammonia into nitrites and nitrates, while denitrification reduces nitrates to gaseous forms of nitrogen, including nitrous oxide (N_2O) (Wang et al., 2024). Therefore, N_2O, produced during denitrification, is a potent GHG with a GWP approximately 298 times that of CO_2 over 100 years. Agricultural soils, especially those with high nitrogen inputs, are a primary source of global N_2O emissions.

1.2.3 Deforestation and Land-Use Change

Deforestation and land-use change constitute profound perturbations in the terrestrial carbon cycle, precipitating significant greenhouse gas emissions. When forests are cleared for agriculture, the biomass carbon stored in trees and vegetation is oxidized, releasing vast quantities of carbon dioxide (CO_2) into the atmosphere (Wang et al., 2024). Moreover, the disruption of soil organic carbon (SOC) pools during land conversion further exacerbates CO_2 emissions, as soil carbon, once protected, becomes vulnerable to microbial decomposition. The conversion of carbon-rich ecosystems like peatlands and tropical forests amplifies these emissions, with long-lasting climatic repercussions (Damm et al., 2024). Land-use change not only depletes carbon sinks but also diminishes biodiversity, alters hydrological cycles, and disrupts ecosystem services. These transformations, driven by agricultural expansion, underscore the complex relationship between human activities and the global climate system, necessitating urgent mitigation through sustainable land management practices (Babaniyi et al., 2024). The conversion of forests to agricultural land leads to the release of carbon stored in biomass and soil organic carbon pools. This carbon is released primarily as CO_2, contributing to atmospheric GHG concentrations. Land-use change, particularly deforestation for agriculture, is a significant source of CO_2 emissions, contributing to about 12–20% of global anthropogenic GHG emissions.

1.2.4 Anaerobic Decomposition in Rice Paddies

Anaerobic decomposition in rice paddies is a pivotal process in methane (CH_4) emissions, driven by the waterlogged, anoxic conditions essential for rice cultivation. In this submerged environment, organic matter derived from plant residues and root exudates undergoes microbial degradation, leading to the accumulation of volatile fatty acids and hydrogen. Methanogenic archaea, thriving in these anaerobic niches, utilize these substrates to produce CH_4 via methanogenesis (Pereira-Mora et al., 2024). The resultant CH_4 diffuses through the soil and is emitted into the atmosphere via rice plant aerenchyma or directly through the water surface. This biogenic CH_4 contributes significantly to global greenhouse gas inventories, particularly in regions with intensive rice cultivation. The complexity of these biogeochemical processes highlights the challenge of mitigating emissions in rice systems while balancing agricultural productivity and climatic impacts (Luo et al., 2022; He et al., 2024). Waterlogged conditions in rice paddies create anaerobic environments conducive to methanogenesis, where methanogenic archaea decompose organic material to produce CH_4. Rice cultivation is responsible for a substantial portion of agricultural CH_4 emissions, contributing significantly to global GHG inventories, particularly in regions with intensive rice farming practices.

1.2.5 Carbon Sequestration Potential and Soil Organic Carbon Dynamics

Agricultural practices, such as conservation tillage, cover cropping, and agroforestry, can enhance carbon sequestration in soils by increasing the soil organic carbon (SOC) content. Conversely, practices like intensive tillage can deplete SOC, releasing CO_2. While agriculture has the potential to act as a carbon sink through enhanced sequestration practices, it also risks becoming a carbon source if unsustainable practices dominate (Pandey et al., 2024). Carbon sequestration in soil involves the dynamic interplay between carbon inputs and outputs, driven by organic matter decomposition and stabilization processes. Soil organic carbon (SOC) dynamics hinge on factors like climate, land use, and management practices (Sharma et al., 2024). Conservation agriculture, including no-till farming, cover cropping, and agroforestry, enhances SOC by promoting carbon accrual in stable organic fractions. Conversely, intensive tillage and monoculture cropping disrupt soil structure, accelerating organic matter oxidation and CO_2 release. The potential for carbon sequestration is vast, particularly in degraded soils, where restoring organic matter can offset the significant greenhouse gas emissions (Quijano et al., 2024). However, this sequestration is not permanent; carbon stored in soils can be rapidly lost under unfavorable conditions, necessitating continuous management. The delicate balance between sequestration and loss underscores the importance of sustainable practices to maintain soil health and mitigate climate change impacts.

1.2.6 Synthetic Fertilizer Production and Energy Use

The production of synthetic fertilizers, particularly nitrogen-based fertilizers, is an energy-intensive process that typically relies on fossil fuels, leading to indirect CO_2 emissions. The life cycle of synthetic fertilizers, from production to application, contributes significantly to the carbon footprint of agriculture, with substantial emissions arising from the Haber–Bosch process used in ammonia synthesis (Zhang et al., 2024a). Synthetic fertilizer production, epitomized by the Haber–Bosch process, is an energy-intensive endeavor, consuming vast amounts of fossil fuels for ammonia synthesis. This process, coupled with nitric acid production and feedstock extraction, drives significant CO_2 and N_2O emissions, exacerbating climate change. The embedded energy in transportation and application further amplifies its carbon footprint (Daramola & Hatzell, 2023). Despite advances in energy efficiency, the global reliance on synthetic fertilizers underpins a substantial and persistent contribution to anthropogenic greenhouse gas emissions.

Haber–Bosch Process The Haber–Bosch process, central to ammonia (NH_3) synthesis, is an energy-intensive chemical reaction combining nitrogen (N_2) from the air with hydrogen (H_2), typically derived from natural gas. This exothermic reaction occurs under high pressure and temperature, necessitating significant fossil fuel consumption, leading to substantial CO_2 emissions (Zhang et al., 2024b). The process is responsible for approximately 1.8% of global CO_2 emissions, reflecting its massive energy demand.

Feedstock Extraction The production of synthetic fertilizers requires the extraction of natural gas or other hydrocarbons as feedstock for hydrogen generation (Li et al., 2022). The extraction and refining processes are energy-intensive, often involving methane emissions, fugitive leaks, and flaring, further amplifying the carbon footprint associated with fertilizer production.

Nitric Acid Production Nitric acid (HNO_3), a precursor for nitrate-based fertilizers, is produced via the Ostwald process, which oxidizes ammonia using a platinum catalyst. This process releases nitrogen oxides (NO_x), including nitrous oxide (N_2O), a potent greenhouse gas with a global warming potential significantly higher than CO_2, contributing to the sector's overall GHG emissions (Xu et al., 2021).

Fertilizer Distribution and Application The transportation and application of synthetic fertilizers involve considerable fossil fuel use, particularly in mechanized agriculture (Meena et al., 2021). The energy consumed in transportation logistics and the operation of application machinery contributes indirectly to CO_2 emissions, embedding a hidden carbon cost in the fertilizer lifecycle.

Energy Use Efficiency The energy use efficiency in synthetic fertilizer production varies depending on technology and practices. Older, less efficient production facilities consume more energy and emit more CO_2 per unit of fertilizer produced (Xu et al., 2021). Modernizing plants and optimizing energy use are critical for reducing the carbon intensity of fertilizer production, yet such advancements are capital-intensive and slow to implement on a global scale.

1.2.7 Manure Management and Storage

The storage and handling of livestock manure under anaerobic conditions, such as in lagoons or pits, lead to CH_4 and N_2O emissions (Bleizgys et al., 2024; Im et al., 2024). Manure management practices are responsible for a significant share of CH_4 and N_2O emissions within the agricultural sector, particularly in intensive livestock operations. Manure management and storage are important in regulating greenhouse gas emissions from livestock operations, where anaerobic conditions in manure storage systems like lagoons, pits, and slurry tanks foster methanogenesis (Qu et al., 2023; Genedy et al., 2023). Methanogenic archaea metabolize organic carbon to produce methane (CH_4), a potent GHG. Additionally, nitrification and denitrification processes in manure lead to nitrous oxide (N_2O) emissions, further compounding the environmental impact. The volatility of emissions is influenced by temperature, moisture content, and storage duration (Li et al., 2022; Ólafsdóttir et al., 2023). Inefficient management exacerbates emissions, transforming nutrient-rich manure from a potential resource into a significant environmental liability. Strategies such as anaerobic digestion, composting, and cover systems aim to mitigate these emissions, balancing the dual need for nutrient recycling and GHG reduction in sustainable agricultural systems.

1.2.8 Agricultural Mechanization and Fossil Fuel Combustion

The use of machinery in agriculture, including tractors, harvesters, and irrigation systems, involves the combustion of fossil fuels, leading to CO_2 emissions. Although less significant than other sources, the direct CO_2 emissions from the combustion of fossil fuels in agricultural machinery contribute to the sector's overall GHG emissions (Silveira et al., 2022). Agricultural mechanization, integral to modern farming, relies heavily on the combustion of fossil fuels, contributing directly to CO_2 emissions. Tractors, harvesters, and irrigation systems, powered by diesel or gasoline, drive intensive energy consumption, embedding a carbon footprint within agricultural productivity (Zhang et al., 2024b, c). This mechanization enables large-scale operations but perpetuates the sector's dependence on non-renewable energy sources. Beyond direct emissions, the lifecycle of machinery from manufacturing to operation and maintenance exacerbates the environmental impact. The efficiency of these machines varies, with older models often being less fuel-efficient, amplifying emissions (Rakhra et al., 2022). While advancements in precision agriculture and alternative fuels offer mitigation potential, the entrenched reliance on fossil fuel-based mechanization underscores a critical challenge in decarbonizing the agricultural sector.

1.3 *Importance of Pesticides in Modern Agriculture*

Pesticides are pivotal in modern agriculture, safeguarding crops from pests, diseases, and weeds, thereby ensuring higher yields and quality. They facilitate large-scale farming by reducing labor-intensive pest control and enabling crop diversification, essential for food security. Pesticides also minimize post-harvest losses, preserving produce in storage (Kaur et al., 2024). While their environmental impact is debated, responsible use supports soil conservation by reducing the need for disruptive practices like tillage, contributing to sustainable agricultural productivity in a growing global population (Diaz et al., 2024). The importance of pesticides in modern agriculture is listed and discussed as follows.

Pest Control and Crop Protection Pesticides are crucial for managing a wide array of agricultural pests, including insects, weeds, fungi, and rodents. By reducing pest populations, pesticides protect crops from significant yield losses, ensuring a stable food supply (Rajak et al., 2023). Without pesticides, crop production would face severe threats from pest infestations, leading to reduced harvests and increased food insecurity. This protection is essential for meeting the global demand for food in the face of growing populations and changing climatic conditions.

Enhanced Crop Yield and Quality Pesticides directly contribute to increased crop yields by minimizing the damage caused by pests. By controlling diseases and weeds, pesticides help maintain the health and vigor of crops, resulting in higher productivity per unit area. This increased yield is critical not only for feeding a growing global population but also for improving the economic viability of farming (Rajak et al., 2023; Diaz et al., 2024). Additionally, by protecting crops from damage, pesticides ensure higher-quality produce, which is essential for both marketability and consumer health.

Economic Efficiency The use of pesticides enhances economic efficiency in agriculture by reducing the need for labor-intensive pest control methods, such as manual weeding or mechanical pest removal. This reduction in labor costs, coupled with higher yields, improves the profitability of farming operations. Pesticides also help prevent the economic losses that would result from widespread pest infestations (Khan et al., 2023; Rajak et al., 2023; Diaz et al., 2024). For farmers, the ability to protect crops with pesticides is essential for maintaining economic stability and competitiveness in the global market.

Food Security and Supply Stability By protecting crops from pests and diseases, pesticides play a critical role in ensuring food security. Stable crop production translates into a reliable food supply, which is essential for preventing food shortages and price volatility (Rajak et al., 2023). Pesticides are particularly important in regions where pest pressure is high and alternative pest control methods are either unavailable or ineffective (Diaz et al., 2024). In such contexts, pesticides are indispensable for maintaining consistent food production and avoiding the catastrophic losses that could lead to famine.

Reduction in Post-harvest Losses Pesticides are used not only in the field but also in storage facilities to protect harvested crops from pests. Post-harvest losses, often caused by insects, rodents, and fungi, can be substantial, especially in developing countries (Khan et al., 2023). Pesticides help preserve the quality and quantity of stored produce, ensuring that food remains available throughout the year and reducing the economic losses associated with spoilage (Diaz et al., 2024). Effective post-harvest pest management is crucial for maintaining food supply chains and preventing waste.

Facilitation of Large-Scale Farming The use of pesticides enables large-scale farming operations by making it feasible to manage extensive areas of crops with minimal labor. Without pesticides, the labor required to manually control pests in large fields would be prohibitive, making large-scale agriculture less viable. Pesticides thus support the efficiency and scalability of modern farming, allowing for the cultivation of vast monocultures that are essential for meeting global food demands (Rajak et al., 2023; Diaz et al., 2024). This scalability is the key to the economic success of modern agriculture.

Support for Crop Diversification Pesticides allow the cultivation of a wide variety of crops by controlling pests that are specific to different species. This capacity for pest management supports crop diversification, which is crucial for dietary diversity, soil health, and ecological balance. By enabling farmers to grow multiple types of crops in the same region, pesticides contribute to the resilience of agricultural systems against environmental stresses (Rajak et al., 2023; Diaz et al., 2024). Crop diversification, supported by effective pest control, is a cornerstone of sustainable agriculture.

Environmental and Soil Conservation While often criticized for their environmental impact, pesticides can contribute to soil and environmental conservation when used responsibly. For instance, herbicides reduce the need for tillage, which helps prevent soil erosion and degradation. This conservation tillage practice preserves soil structure, organic matter, and moisture, contributing to long-term soil health. Additionally, targeted pesticide applications can reduce the overall environmental footprint of farming by minimizing the need for more disruptive pest control methods (Rajak et al., 2023; Diaz et al., 2024). Proper pesticide use, integrated with other sustainable practices, can thus play a role in environmental stewardship.

2 Types of Pesticides and Their Composition

Pesticides are a broad category of chemicals used to control or eliminate pests, including insects, weeds, fungi, and other organisms, which can harm crops, livestock, and public health. These substances are essential in modern agriculture, but their composition and environmental impact vary significantly depending on the type. Below is a detailed discussion of different types of pesticides and their compositions.

2.1 Classification of Pesticides

Pesticides, a diverse arsenal of chemical agents, are meticulously classified based on target organisms, chemical structure, and mode of action. Broadly, they are divided into insecticides, targeting insects with neurotoxins like organophosphates and neonicotinoids; herbicides, which selectively or non-selectively annihilate weeds via mechanisms like the inhibition of photosynthesis (e.g., atrazine) or amino acid synthesis (e.g., glyphosate); and fungicides, combating fungal pathogens by disrupting critical processes such as ergosterol biosynthesis (e.g., azoles). Rodenticides like anticoagulants (e.g., warfarin) lethally inhibit blood clotting in rodents, while nematicides eliminate parasitic nematodes, often through highly toxic fumigants (Shi et al., 2022; Gaikwad & Mairal, 2022). Further sub-classifications, such as contact vs. systemic and selective vs. non-selective categories, reflect the specificity and systemic movement within target organisms, revealing the complicated, often perilous chemistry between efficacy and environmental impact.

2.1.1 Insecticides

Insecticides are pesticides specifically designed to kill or repel insects. Insecticides are complex agents of biochemical warfare, target the nervous systems or physiological processes of insects, often wielding neurotoxic effects. Organophosphates, notorious for acetylcholinesterase inhibition, wreak havoc on synaptic transmission, while pyrethroids, synthetic mimics of natural pyrethrins, induce hyperexcitation, leading to paralysis (Yang et al., 2015; Shi et al., 2022). Neonicotinoids, systemic and potent, bind irreversibly to nicotinic receptors, decimating pest populations but raising ecological alarms due to their impact on pollinators. Carbamates, sharing organophosphates' mode, yet distinct in toxicity profiles, contribute to this arsenal, which, though essential in pest control, poses significant environmental and non-target species risks, demanding judicious, regulated use (Wei et al., 2017; Xu et al., 2019). They can be classified into several categories based on their chemical structure and mode of action:

Organophosphates Derived from phosphoric acid, these insecticides, such as malathion and chlorpyrifos, function by inhibiting acetylcholinesterase, an enzyme critical for nerve function in insects. This leads to the accumulation of acetylcholine, causing paralysis and death. Organophosphates are known for their broad-spectrum activity but pose significant risks to non-target organisms and human health (Han et al., 2023).

Pyrethroids Synthetic analogs of natural pyrethrins derived from chrysanthemum flowers, pyrethroids like permethrin and cypermethrin act on the nervous system of insects, leading to hyperexcitation, paralysis, and death (Wang et al., 2014). They are widely used owing to their effectiveness and relatively low toxicity to mammals.

Neonicotinoids Structurally related to nicotine, neonicotinoids such as imidacloprid and thiamethoxam bind to nicotinic acetylcholine receptors in insects, causing overstimulation and eventual death. They are systemic insecticides, meaning they can be absorbed by plants and provide long-lasting protection (Wei et al., 2017; Xu et al., 2019). However, their environmental persistence and potential impact on pollinators like bees have raised concerns.

Carbamates Similar to organophosphates, carbamates such as carbaryl and methomyl inhibit acetylcholinesterase but are generally less toxic to mammals. They are used against a wide range of insect pests but can also affect non-target organisms (Wang et al., 2014; Wei et al., 2017; Xu et al., 2019).

2.1.2 Herbicides

Herbicides, the botanical executioners, selectively obliterate plant species, sculpting agricultural landscapes with chemical precision. Glyphosate, the ubiquitous non-selective agent, disrupts the shikimate pathway, starving plants of essential amino acids, yet stirs controversy due to its environmental persistence (Pei et al., 2021). Atrazine, a triazine compound, stifles photosynthetic machinery, while dicamba, a synthetic auxin, induces uncontrollable cellular growth, often straying into unintended crops, sparking legal and ecological disputes (Tran, 2018). The venerable 2,4-D, an enduring synthetic auxin, mirrors natural plant hormones, driving fatal growth distortions. These herbicides, each with distinct mechanisms, underpin modern agriculture's battle against weeds but provoke debates over drift, resistance, and long-term ecological impacts, necessitating careful, context-specific application and ongoing innovation in weed management strategies (Wang et al., 2014). However, herbicides are designed to control or eliminate unwanted vegetation, primarily weeds, and are categorized based on their mode of action and chemical composition:

Glyphosate A non-selective, systemic herbicide, glyphosate inhibits the enzyme 5-enolpyruvylshikimate-3-phosphate synthase (EPSPS), which is essential for the synthesis of certain amino acids in plants (Wei et al., 2017). It is widely used in agriculture, particularly with genetically modified crops designed to tolerate its application.

Atrazine A triazine herbicide, atrazine interferes with photosynthesis by inhibiting the photosystem II enzyme, leading to energy deprivation in plants. It is used primarily in corn and sorghum crops but has been controversial due to its persistence in soil and water and potential effects on aquatic life (Xu et al., 2019).

Dicamba A synthetic auxin herbicide, dicamba mimics natural plant hormones, causing uncontrolled growth and eventually plant death. It is effective against broadleaf weeds but has faced criticism for its tendency to drift, potentially damaging nearby crops (Han et al., 2023).

2,4-D Another synthetic auxin, 2,4-D (2,4-dichlorophenoxyacetic acid) was one of the first herbicides developed and is still widely used today. It disrupts cell division and growth in broadleaf plants, leading to their death. Its use has been controversial due to its persistence and potential health risks (Han et al., 2023).

2.1.3 Fungicides

Fungicides are used to prevent or treat fungal infections in plants. It is the silent sentinels of agriculture, wage war against the insidious fungi that threaten crops, using a variety of biochemical strategies. Azoles, a prominent class, inhibit lanosterol 14α-demethylase, crippling ergosterol synthesis, essential for fungal cell membranes. Strobilurins, derived from wood-decaying fungi, disrupt mitochondrial respiration by blocking the cytochrome bc1 complex, leading to cellular energy collapse (Tran, 2018; Han et al., 2023). Chloronitriles, like chlorothalonil, target multiple enzymatic pathways by reacting with thiol groups, acting as a broad-spectrum shield. Meanwhile, dithiocarbamates, through multi-site activity, thwart crucial fungal enzymes, preventing the synthesis of vital cellular components. Despite their efficacy, fungicides raise concerns about resistance development, environmental persistence, and non-target effects, prompting the continuous evolution of integrated management approaches to sustain their utility while minimizing ecological footprints (Wang et al., 2014). They are classified based on their chemical composition and whether they act systemically or on contact:

Azoles This class, including tebuconazole and propiconazole, inhibits the enzyme lanosterol 14α-demethylase, which is involved in ergosterol biosynthesis, a critical component of fungal cell membranes (Lyr, 2003; Wei et al., 2017). By disrupting cell membrane formation, these fungicides effectively prevent fungal growth.

Dithiocarbamates Fungicides such as mancozeb and thiram fall into this category. They work by inhibiting multiple enzymes involved in the energy production and synthesis of vital cellular components in fungi (Lyr, 2003; Wei et al., 2017). Dithiocarbamates are generally broad-spectrum but can be toxic to aquatic life.

Strobilurins Derived from a naturally occurring compound in wood-rotting fungi, strobilurins like azoxystrobin inhibit mitochondrial respiration in fungi by blocking the cytochrome bc1 complex. They are widely used due to their broad-spectrum activity and low toxicity to mammals (Lyr, 2003; Wei et al., 2017).

Chloronitriles Chlorothalonil, a representative of this class, inhibits various enzymes by reacting with thiol groups in proteins, disrupting vital processes in fungal cells. Chloronitriles are used as protectant fungicides, preventing fungal infections on the surface of plants (Lyr, 2003; Wei et al., 2017).

2.1.4 Rodenticides

Rodenticides are chemicals designed to kill rodents. Rodenticides, lethal agents of pest control, strategically target rodents, employing biochemical precision to disrupt vital physiological functions. Anticoagulants like warfarin and brodifacoum inhibit vitamin K epoxide reductase, crippling blood clotting processes and causing fatal internal hemorrhages over days (Semino, 2001). Bromethalin, a non-anticoagulant, disrupts mitochondrial oxidative phosphorylation, leading to ATP depletion, paralysis, and swift death. Zinc phosphide, an acute rodenticide, reacts with stomach acid to release phosphine gas, a potent cellular toxin (Pelfrène, 2010; McGee et al., 2020). While effective, these compounds pose significant risks of secondary poisoning to non-target species, including predators and humans, sparking ethical and ecological debates. The balance between efficacy and safety in rodenticide application necessitates careful management, with growing interest in alternative, less hazardous control methods as part of integrated pest management (IPM) strategies (Rached et al., 2020). They are often anticoagulants, which interfere with blood clotting:

Coumarins Rodenticides like warfarin and brodifacoum inhibit the enzyme vitamin K epoxide reductase, which is essential for synthesizing clotting factors in the blood. Rodents that ingest these poisons bleed internally and die (Pelfrène, 2010; McGee et al., 2020). However, the risk of secondary poisoning in non-target species is a significant concern.

Bromethalin Unlike anticoagulants, bromethalin disrupts the mitochondrial function in rodent cells, leading to a decrease in ATP production and eventual paralysis and death. It is highly toxic but acts faster than anticoagulants (Rached et al., 2020).

Zinc Phosphide This rodenticide reacts with stomach acid to release phosphine gas, which is highly toxic to cells, causing organ failure and death in rodents. Its acute toxicity makes it effective, but it also poses significant risks to non-target animals and humans (Rached et al., 2020).

2.1.5 Nematicides

Nematicides are specialized agents of agrarian defense, target the elusive nematodes microscopic worms that wreak havoc on plant roots. Fumigants like methyl bromide penetrate soil with toxic vapor, indiscriminately annihilating nematodes by disrupting cellular processes, though their environmental toll, including ozone depletion, has led to stringent regulation (Chen et al., 2020). Organophosphate nematicides, such as fenamiphos, inhibit acetylcholinesterase, paralyzing these pests but raising concerns about human and ecological toxicity. Carbamates like carbofuran act similarly, though with slightly reduced mammalian toxicity. The high potency and specificity of nematicides are counterbalanced by their environmental persistence and non-target effects, driving a shift toward integrated pest management (IPM) techniques that emphasize biological controls and reduced chemical reliance, striving for a delicate balance between efficacy and sustainability. Nematicides are used to control nematodes, parasitic worms that attack plant roots (Oka, 2020; Desaeger et al., 2020). They are often highly toxic and used in controlled environments:

Fumigants Nematicides like methyl bromide are volatile chemicals that diffuse through the soil, killing nematodes by disrupting their cellular processes. Due to their toxicity and potential to deplete the ozone layer, many fumigants are heavily regulated or banned (Desaeger et al., 2020).

Organophosphates Some nematicides are derived from organophosphates, which inhibit acetylcholinesterase in nematodes, leading to paralysis and death. Examples include fenamiphos and terbufos (Oka, 2020).

Carbamate Nematicides Carbofuran is an example of a carbamate nematicide, which acts similarly to its insecticidal counterparts by inhibiting acetylcholinesterase. These chemicals are effective but have raised environmental and health concerns (Chen et al., 2020).

Pesticides, with their diverse compositions and mechanisms of action, are indispensable tools in modern agriculture and public health. However, their environmental impact and potential risks to human health necessitate careful management, regulation, and ongoing research into safer, more sustainable alternatives. The complexity of their interactions with ecosystems underscores the importance of integrated pest management strategies and the continuous development of eco-friendly technologies.

Table 1 Chemical composition and production processes

Pesticide type	Chemical composition	Production process
Organophosphate insecticides	Contains phosphoric acid derivatives, e.g., Malathion and Chlorpyrifos	Synthesis involves esterification of phosphorus oxychloride with alcohols, followed by chlorination and purification.
Neonicotinoid insecticides	Structurally related to nicotine, e.g., Imidacloprid and Thiamethoxam	Derived from the reaction of chloronicotinic acid with nitromethylene or nitroguanidine, involving multi-step synthesis and purification.
Triazine herbicides	Composed of triazine rings, e.g., Atrazine and Simazine	Produced through a nucleophilic substitution reaction of cyanuric chloride with various amines, followed by crystallization and purification.
Synthetic pyrethroid insecticides	Analogous to natural pyrethrins, e.g., Permethrin and Cypermethrin	Synthesized by esterification of chrysanthemic acid with alcohol derivatives, followed by isomer separation and stabilization.
Dithiocarbamate fungicides	Contains sulfur compounds, e.g., Mancozeb and Thiram	Formed by the reaction of carbon disulfide with secondary amines, followed by oxidation, precipitation, and drying processes.
Anticoagulant rodenticides	Contains coumarin derivatives, e.g., Warfarin and Brodifacoum	Synthesized through the condensation of 4-hydroxycoumarin with benzylidene derivatives, followed by cyclization and recrystallization.

2.2 Chemical Composition and Production Processes (Table 1)

Organophosphate Insecticides These compounds, derived from phosphoric acid, function by inhibiting acetylcholinesterase. Their production involves esterification and chlorination processes, yielding potent neurotoxic agents with broad-spectrum insecticidal activity.

Neonicotinoid Insecticides Resembling nicotine in structure, neonicotinoids disrupt insect nervous systems by binding to nicotinic acetylcholine receptors. Their complex synthesis includes creating chloronicotinic acid intermediates, followed by further chemical transformations.

Triazine Herbicides These herbicides disrupt photosynthesis in plants by inhibiting photosystem II. They are produced through nucleophilic substitution reactions involving cyanuric chloride and various amines, forming stable triazine rings.

Synthetic Pyrethroid Insecticides Modeled after natural pyrethrins, these insecticides are synthesized through esterification, creating stable, highly effective insecticidal compounds.

Dithiocarbamate Fungicides These sulfur-containing fungicides are created by reacting carbon disulfide with secondary amines. The resultant compounds interfere with fungal enzyme activity, providing broad-spectrum protection.

Anticoagulant Rodenticides These are coumarin derivatives that inhibit blood clotting, causing fatal hemorrhages in rodents. Their synthesis involves complex condensation and cyclization reactions to produce stable, toxic compounds.

2.3 Environmental Persistence of Different Pesticides

The environmental persistence of pesticides, a critical factor in their ecological impact, varies dramatically depending on their chemical structure, mode of action, and environmental conditions.

Organophosphates Organophosphate insecticides, like Malathion and Chlorpyrifos, are characterized by their relatively rapid degradation in the environment. They hydrolyze in the presence of water and are broken down by microbial activity, leading to a shorter half-life compared to other classes (Ali et al., 2021). However, their initial toxicity is high, posing acute risks to non-target species, particularly aquatic organisms and pollinators, during the period of active application.

Neonicotinoids Neonicotinoids, such as Imidacloprid and Thiamethoxam, are systemic and persist in soil and water, where they can remain active for extended periods, often months to years. Their persistence allows prolonged protection against pests but also raises concerns about their accumulation in non-target organisms, particularly bees, leading to widespread ecological disruptions and contributing to the ongoing pollinator decline (Khan & Pathak, 2020).

Triazine Herbicides Herbicides like Atrazine, a member of the triazine family, are notorious for their environmental persistence. Atrazine is relatively stable and can persist in soil and water for extended periods, often leading to groundwater contamination. Its mobility through soil can result in widespread contamination, affecting both aquatic and terrestrial ecosystems, and has been linked to endocrine disruption in wildlife (Kumar et al., 2021).

Synthetic Pyrethroids Synthetic pyrethroids, including Permethrin and Cypermethrin, are more persistent than their natural counterparts. They tend to bind strongly to soil particles, reducing their mobility but increasing their persistence in the environment (Khan & Pathak, 2020). While they break down more slowly, their toxicity to aquatic life, particularly fish and invertebrates, remains a significant concern.

Dithiocarbamate Fungicides Dithiocarbamates, such as Mancozeb, exhibit moderate environmental persistence. They degrade relatively quickly under normal conditions but can persist in cooler, anaerobic environments. Their breakdown products, however, can be toxic, raising concerns about long-term soil health and the potential for bioaccumulation in food chains (Khan & Pathak, 2020; Kumar et al., 2021).

Anticoagulant Rodenticides Anticoagulant rodenticides like Brodifacoum are highly persistent in the environment due to their strong resistance to breakdown. They can remain active for months in both terrestrial and aquatic systems, leading to significant risks of secondary poisoning for predators that consume poisoned rodents (Khan & Pathak, 2020; Kumar et al., 2021). This persistence results in prolonged ecological impacts, including the potential for biomagnification in food webs.

The environmental persistence of pesticides is a double-edged sword, while it ensures prolonged efficacy in pest control, it also poses significant risks to non-target species, ecosystems, and human health. Understanding and mitigating these risks require a nuanced approach that balances pest management needs with environmental stewardship, emphasizing the development and adoption of less persistent, more targeted pest control methods.

3 Contributions of Pesticides to GHG Emissions

Pesticides, while crucial in modern agriculture, contribute to greenhouse gas (GHG) emissions in different and often underappreciated ways. The synthesis, application, and degradation of these chemicals generate emissions that ripple through ecosystems, exacerbating climate change.

Energy-Intensive Production Processes The manufacturing of pesticides, particularly synthetic varieties like organophosphates and neonicotinoids, is energy-intensive. The processes involve complex chemical reactions, high-temperature conditions, and the use of fossil fuel-derived inputs, leading to significant CO_2 emissions (Ali et al., 2021). Factories and plants involved in these processes release not only CO_2 but also other potent greenhouse gases, including methane (CH_4) and nitrous oxide (N_2O), into the atmosphere (Kumar et al., 2021).

Application-Related Emissions The application of pesticides, often requiring mechanical equipment like sprayers, further contributes to GHG emissions. The combustion of fossil fuels in these machines releases CO_2, while the volatilization of certain pesticides themselves can release volatile organic compounds (VOCs). VOCs, when interacting with nitrogen oxides (NOx) in the presence of sunlight, contribute to the formation of ground-level ozone, a potent GHG (Khan & Pathak, 2020; Kumar et al., 2021).

Soil Microbial Disturbances Pesticides disrupt soil microbiota, leading to altered microbial processes that influence GHG emissions. For instance, the application of certain herbicides and insecticides can reduce microbial diversity, particularly those involved in carbon sequestration (Manzetti & van der Spoel, 2015). This disturbance can lead to increased CO_2 release as organic matter decomposes without adequate microbial regulation, and, in some cases, can increase the production of N_2O, a GHG nearly 300 times more potent than CO_2 (Wisnujati, 2023).

Indirect Land-Use Changes The use of pesticides enables monocultures and large-scale agricultural practices that often require deforestation and land conversion. These changes release significant amounts of carbon stored in vegetation and soil. The carbon loss from deforestation, coupled with the reduced carbon sequestration potential of pesticide-dependent agricultural lands, results in substantial GHG emissions (Lima-Junior et al., 2024).

Degradation and Residual Impact The degradation of pesticides in the environment can also contribute to GHG emissions. For example, the breakdown of organochlorine compounds can lead to the release of chlorine, which, although not a GHG, can catalyze reactions that deplete ozone and indirectly affect climate systems (Hu & Liu, 2024). Furthermore, the long-term persistence of some pesticides means that their environmental impact, including GHG emissions, continues well beyond their initial application.

However, pesticides contribute to GHG emissions through their production, application, and post-application impacts on soil and ecosystems. These emissions, though often overshadowed by other agricultural practices, are significant and underscore the need for sustainable pest management strategies that minimize climate impacts while maintaining agricultural productivity. The integration of biopesticides, precision agriculture, and reduced chemical dependence are the key avenues for reducing these contributions to global GHG emissions.

3.1 Direct Contributions of Pesticides to GHG Emissions

Pesticides directly contribute to greenhouse gas (GHG) emissions through various pathways, including production, application, and degradation. The synthesis of pesticides, particularly synthetic types like organophosphates and neonicotinoids, involves energy-intensive chemical processes. These processes often require significant amounts of fossil fuels, leading to direct CO_2 emissions from the burning of these fuels (Cech et al., 2022). Moreover, the transportation and distribution of pesticides from manufacturing plants to agricultural sites also contribute to GHG emissions, with CO_2 being the primary gas released during these activities. During application, certain pesticides, particularly volatile organic compounds (VOCs), can vaporize and enter the atmosphere (Cornish et al., 2024). When VOCs interact with nitrogen oxides (NOx) under sunlight, they form ground-level ozone, a potent GHG and contributor to climate change. This process, although indirect, is a direct result of pesticide application.

Additionally, the degradation of certain pesticides in the environment can lead to the release of other GHGs. For example, the breakdown of nitrogen-based pesticides can produce nitrous oxide (N_2O), a GHG that is approximately 300 times more potent than CO_2 (Jezierska-Tys et al., 2021). The persistence of pesticides in the environment exacerbates these emissions, as ongoing degradation continues to release GHGs long after the initial application. In essence, while pesticides are designed to protect crops, their direct contributions to GHG emissions through production, application, and degradation pose significant challenges to climate change mitigation efforts, necessitating a re-evaluation of their role in sustainable agriculture (Shi et al., 2020).

3.1.1 Carbon Footprint of Pesticide Production

The carbon footprint of pesticide production is substantial, driven by the energy-intensive processes involved in synthesizing these chemicals. The production of pesticides, particularly synthetic varieties, requires high temperatures, pressure, and the use of fossil fuel-derived feedstocks, all of which contribute to significant CO_2 emissions (Shi et al., 2020; Jezierska-Tys et al., 2021). The carbon footprint is further compounded by the energy required to operate manufacturing facilities, where fossil fuels are often the primary energy source. For example, organophosphate insecticides are synthesized through complex chemical reactions that require chlorination and esterification, processes that consume large amounts of energy (Cech et al., 2022). This energy is typically sourced from coal, natural gas, or oil, all of which are major contributors to GHG emissions. The carbon footprint is not limited to CO_2 alone; methane (CH_4) and nitrous oxide (N_2O) are also emitted during these processes, particularly from the combustion of natural gas and other hydrocarbons. Transportation and distribution further increase the carbon footprint of pesticides. Once produced, pesticides must be transported to farms and agricultural centers, often over long distances (Cornish et al., 2024).

The logistics of this transportation rely heavily on diesel-powered vehicles, which emit significant amounts of CO_2 and other GHGs. Moreover, the packaging of pesticides, typically in plastic containers, adds to the carbon footprint. The production of plastic, itself an energy-intensive process, further exacerbates the overall GHG emissions associated with pesticide production (Fuentes-Ponce et al., 2022). Thus, the carbon footprint of pesticide production is a critical factor in the broader environmental impact of these chemicals. Reducing this footprint requires innovative approaches, such as improving production efficiency, utilizing renewable energy sources, and developing less energy-intensive pesticide formulations.

3.1.2 Emissions from Pesticide Application and Degradation

Emissions from pesticide application and degradation represent a significant, yet often overlooked, source of greenhouse gases (GHGs) in agriculture. During application, the use of machinery such as tractors and sprayers, which are typically powered by fossil fuels, directly emits CO_2 into the atmosphere. The intensity of these emissions is correlated with the frequency and scale of pesticide application, with large monoculture farms contributing more substantially to GHG outputs (Aneesha et al., 2020). The volatilization of pesticides during and after application is another critical concern. Volatile organic compounds (VOCs) present in certain pesticides, such as organophosphates, can evaporate into the atmosphere, where they contribute to the formation of ground-level ozone (Aneesha et al., 2020; Saxena et al., 2021). This ozone not only acts as a potent GHG but also exacerbates air pollution, with serious implications for human and environmental health.

Post-application, the degradation of pesticides in soil and water further contributes to GHG emissions. Microbial degradation of nitrogen-based pesticides, for instance, can produce nitrous oxide (N_2O), a greenhouse gas with a global warming potential significantly higher than CO_2 (Saxena et al., 2021). The persistence of pesticides in the environment means that these emissions can continue for extended periods, creating a long-lasting impact on the atmosphere. Moreover, the degradation process can lead to the formation of toxic by-products that contribute to environmental contamination (Srinivasa Rao et al., 2022). These by-products can persist in ecosystems, leading to secondary emissions through processes such as leaching and runoff, which transport degraded pesticides into water bodies, further amplifying their GHG footprint.

3.1.3 Role of Pesticides in Soil Carbon Emissions

Pesticides play a complex role in soil carbon emissions, influencing both the sequestration and release of carbon within agricultural ecosystems. The application of pesticides, particularly herbicides and insecticides, can disrupt soil microbial communities, which are integral to the carbon cycle (Yadav et al., 2021). Microorganisms in the soil, such as bacteria and fungi, decompose organic matter, converting it into stable forms of carbon that can be sequestered in the soil. However, pesticides can inhibit these microbial processes, reducing the rate of organic matter decomposition and potentially leading to an increase in soil carbon emissions (Mahanta et al., 2023). For instance, the use of broad-spectrum herbicides like glyphosate can reduce the diversity and activity of soil microbes, leading to a decline in soil organic matter. This decline not only reduces the soil's capacity to sequester carbon but also increases the likelihood of carbon being released as CO_2, contributing to atmospheric GHG levels (Yadav et al., 2021). Additionally, the application of pesticides can lead to soil compaction and reduced porosity, further inhibiting microbial activity and exacerbating carbon loss.

Pesticides can also indirectly contribute to soil carbon emissions by promoting agricultural practices that are less sustainable (Satpathy et al., 2022). For example, the reliance on chemical pesticides can discourage the adoption of crop rotation and cover cropping, practices that enhance soil health and carbon sequestration. Without these practices, soils are more likely to degrade over time, leading to increased carbon emissions and reduced soil fertility (Mishra et al., 2024). In summary, while pesticides are intended to protect crops, their impact on soil health and carbon emissions is profound. By disrupting microbial processes and promoting unsustainable agricultural practices, pesticides contribute to the release of soil carbon, undermining efforts to mitigate climate change through carbon sequestration.

3.2 Indirect Contributions of Pesticides to GHG Emissions

Pesticides indirectly contribute to greenhouse gas (GHG) emissions through their influence on agricultural practices, ecosystem health, and soil dynamics. The widespread use of pesticides supports monoculture farming systems, which often require intensive land use changes, such as deforestation and soil tillage. These changes result in the release of significant amounts of stored carbon into the atmosphere as CO_2 (Mishra et al., 2024). Additionally, the reliance on pesticides reduces the adoption of sustainable practices like crop rotation, cover cropping, and organic farming, which are crucial for maintaining soil carbon stocks and reducing GHG emissions (Fuentes-Ponce et al., 2022; Mishra et al., 2024). Moreover, pesticide-driven reduction in biodiversity affects the resilience of ecosystems, leading to imbalances that can exacerbate GHG emissions. For instance, the decline of beneficial insects and pollinators due to pesticide use disrupts natural pest control mechanisms, increasing the need for additional chemical inputs (Yadav et al., 2021). This cycle of increased pesticide use further intensifies GHG emissions through energy consumption in the production, transportation, and application processes.

Pesticides also indirectly affect GHG emissions by altering soil microbial communities. These changes can lead to reduced soil health and a decline in the soil's ability to sequester carbon. The disruption of microbial processes, such as nitrogen fixation and organic matter decomposition, contributes to the release of CO_2 and nitrous oxide (N_2O), a potent GHG (Kamala & Devanand, 2021; Skendžić et al., 2021). Thus, while pesticides are not directly emitting GHGs at the point of application, their cascading effects on agricultural systems and ecosystems contribute significantly to the overall carbon footprint of farming practices, highlighting the need for more integrated pest management strategies that minimize these indirect contributions.

3.2.1 Impact on Soil Microbial Activity and Organic Matter Decomposition

Pesticides profoundly impact soil microbial activity, which in turn influences organic matter decomposition a critical process in carbon cycling and soil health. Soil microbes, including bacteria, fungi, and archaea, play an essential role in breaking down organic material, transforming it into humus and stabilizing soil carbon (Sarkar et al., 2020; Verma et al., 2021). However, the application of pesticides, especially broad-spectrum fungicides and insecticides, disrupts this microbial community, reducing its diversity and functional capacity. The reduction in microbial activity leads to slower decomposition rates of organic matter (Roy et al., 2020; Singh, 2022). When microbial populations are suppressed, the breakdown of plant residues and other organic inputs into soil organic matter is hindered, resulting in an accumulation of undecomposed material. This accumulation can cause an imbalance in soil nutrient cycles, reducing the availability of nutrients necessary for plant growth and further compromising soil fertility.

Moreover, the altered decomposition process can shift the balance of carbon storage in soils. Instead of converting organic matter into stable carbon forms that can be sequestered in the soil for long periods, the disrupted microbial processes may result in the release of carbon as CO_2, contributing to increased GHG emissions (Roy et al., 2020). Additionally, the presence of pesticide residues can create a hostile environment for beneficial microbes, leading to the proliferation of more resilient, yet often less efficient, microbial species that may further disrupt soil health and carbon dynamics. The long-term impact of pesticides on soil microbial activity can lead to soil degradation, reduced carbon sequestration potential, and heightened vulnerability to climate change (Singh, 2022). Therefore, understanding and mitigating these effects through sustainable soil management practices is crucial for maintaining soil health and reducing the overall environmental footprint of agricultural systems.

3.2.2 Influence on Nitrogen Cycling and Nitrous Oxide Emissions

Pesticides have a significant influence on nitrogen cycling, a critical process in soil ecosystems, which directly affects the emissions of nitrous oxide (N_2O), a potent greenhouse gas. Nitrogen cycling involves the transformation of nitrogen in various forms, including ammonium, nitrate, and organic nitrogen, through processes like nitrification and denitrification (Chatterjee et al., 2021). These processes are largely driven by soil microorganisms, which are sensitive to the presence of pesticides. The application of pesticides can disrupt the microbial communities responsible for nitrogen cycling (Jasrotia et al., 2023). For instance, nitrifying bacteria, which convert ammonium to nitrate, can be inhibited by certain herbicides and insecticides. This inhibition leads to an accumulation of ammonium in the soil, which can alter plant nutrient uptake and soil pH. On the other hand, denitrifying bacteria, which convert nitrate to nitrogen gas (N_2) or nitrous oxide (N_2O), may also be affected, potentially increasing the production of N_2O due to incomplete denitrification processes.

The disruption of nitrogen cycling by pesticides not only affects soil fertility but also enhances N_2O emissions. Nitrous oxide is nearly 300 times more effective at trapping heat in the atmosphere compared to CO_2, making it a significant contributor to global warming (Singh, 2022). The increased N_2O emissions from soils impacted by pesticides represent a substantial indirect contribution of these chemicals to GHG emissions. Furthermore, the altered nitrogen dynamics can lead to nutrient imbalances in soils, necessitating additional fertilizer inputs, which themselves are sources of GHGs (Roy et al., 2020). This creates a feedback loop where pesticide use indirectly promotes further GHG emissions through both enhanced N_2O production and the increased need for synthetic nitrogen fertilizers.

3.2.3 Effects on Plant Growth and Carbon Sequestration

Pesticides, while primarily designed to control pests, have complex effects on plant growth and the broader carbon sequestration potential of agricultural ecosystems. The impact on plant growth is twofold: while pesticides can reduce pest-related damage, thereby enhancing crop yields, they can also impair plant physiological processes through phytotoxic effects, particularly when misapplied or overused (Saxena et al., 2021). Phytotoxicity, a condition where plants suffer from pesticide-induced stress, can lead to reduced photosynthetic efficiency, stunted growth, and lower biomass production. Since photosynthesis is the primary process through which plants sequester atmospheric carbon, any reduction in photosynthetic activity directly diminishes the plant's ability to capture and store carbon (Alemu, 2020). This reduction in carbon sequestration not only affects the immediate carbon balance of the ecosystem but also has long-term implications for soil carbon storage, as less organic matter is returned to the soil.

Moreover, pesticides can alter the composition of plant communities by selectively eliminating certain species while allowing others to dominate. This shift in plant diversity can reduce the resilience of ecosystems to climate change and other stressors, further impacting the carbon sequestration potential. For example, monocultures heavily reliant on pesticides may sequester less carbon than diverse, pesticide-free systems due to differences in root biomass, soil health, and microbial activity (Aggarwal et al., 2022). Additionally, pesticides can impact the symbiotic relationships between plants and soil microorganisms, such as mycorrhizal fungi, which play a crucial role in enhancing nutrient uptake and promoting plant growth. Disruption of these relationships can lead to decreased plant vigor and reduced carbon inputs into the soil, further diminishing the overall carbon sequestration potential of agricultural lands (Assad et al., 2021). Therefore, while pesticides are essential for protecting crops, their complex interactions with plant growth and soil carbon dynamics can undermine the carbon sequestration benefits of agricultural ecosystems, necessitating a more balanced approach to pest management.

4 Pesticide Use and Land-Use Change

Pesticide use and land-use change both reshape agricultural landscapes but through distinct mechanisms. Pesticides intensify farming by targeting pests, boosting yields, and maintaining monocultures, yet they risk biodiversity loss and soil health degradation. Conversely, land-use change transforms ecosystems forests to farmland driving biodiversity loss on a broader scale, increasing carbon emissions, and destabilizing climates (Lima-Junior et al., 2024; Hu & Liu, 2024). While pesticides offer short-term productivity gains, over-reliance breeds resistance and ecological harm. Land-use change, though expanding agricultural land, often leads to long-term environmental and economic costs due to degraded ecosystems (Manzetti & van der Spoel, 2015; Wisnujati, 2023). Both strategies underscore the tension between agricultural expansion and environmental sustainability.

4.1 Effect of Pesticides on Land-Use Changes

Pesticides drive land-use changes by facilitating monoculture farming, which often necessitates deforestation and habitat conversion. This transformation reduces carbon sinks, exacerbating greenhouse gas (GHG) emissions as stored carbon is released. Additionally, pesticide-reliant agricultural practices degrade soil health, prompting farmers to clear more land to maintain productivity, perpetuating a destructive cycle. The resulting ecosystem imbalance accelerates biodiversity loss and undermines long-term sustainability, complicating efforts to mitigate climate change.

4.1.1 Impact on Biodiversity

The impact of various human activities on biodiversity is profound and complex. Agricultural practices, such as land use and pesticide application, significantly reduce species and genetic diversity, leading to habitat fragmentation and extinction of soil organisms (Xin et al., 1999; Lima-Junior et al., 2024). Similarly, sludge deposition introduces persistent toxins into ecosystems, contributing to long-term biodiversity loss through nutrient pollution and the bioaccumulation of harmful substances in food webs (McLaughlin & Mineau, 1995; Hilary & Nnamdi, 2023; Hu & Liu, 2024). The introduction of genetically modified organisms (GMOs) raises concerns regarding their effects on biological diversity, with debates focusing on ethical and environmental assessments (Manzetti & van der Spoel, 2015; Wisnujati, 2023). Additionally, aquaculture often negatively impacts biodiversity through habitat loss and the introduction of non-native species, highlighting the need for sustainable practices in food production (Beveridge et al., 1994; Manzetti & van der Spoel, 2015; Skendžić et al., 2021; Hilary & Nnamdi, 2023). Collectively, these studies underscore the urgent need for integrated approaches to mitigate biodiversity loss across various sectors, emphasizing the importance of sustainable practices and environmental stewardship.

Pesticide Use Pesticides, particularly broad-spectrum ones, can negatively impact non-target species, leading to reduced biodiversity within agricultural ecosystems. This includes beneficial insects, pollinators, and soil microorganisms, which play crucial roles in ecosystem health.

Land-Use Change Land-use change, such as deforestation for agriculture, results in habitat loss, fragmentation, and the displacement of native species. It often leads to significant declines in biodiversity at a broader landscape level, affecting entire ecosystems rather than specific species.

4.1.2 Contribution to Climate Change

The relationship between pesticide use and climate change is complex and multifaceted. Climate change significantly impacts pesticide efficacy and environmental dynamics, as rising temperatures and altered precipitation patterns can enhance pest populations and their resistance to chemical controls (Skendžić et al., 2021; Hilary & Nnamdi, 2023). For instance, increased temperatures can heighten the toxicity of certain pesticides, leading to greater environmental risks and potential pesticide losses (Han et al., 2024). Conversely, the application of pesticides contributes to greenhouse gas emissions, exacerbating climate change (Ukhurebor et al., 2020). Moreover, the interaction between climate change and pest dynamics necessitates adaptive management strategies, such as integrated pest management (IPM), which emphasizes environmentally sensitive approaches to pest control (Masood et al., 2022). This dual influence highlights the need for sustainable agricultural practices that mitigate both pesticide-related environmental impacts and the adverse effects of climate change on pest populations (Skendžić et al., 2021; Hilary & Nnamdi, 2023). Overall, addressing these interconnected challenges is crucial for enhancing agricultural resilience and sustainability.

Pesticide Use The production and application of pesticides involve fossil fuel consumption, contributing indirectly to greenhouse gas (GHG) emissions. However, the overall impact on climate change is relatively minor compared to other agricultural practices.

Land-Use Change Land-use change, especially deforestation, is a major contributor to climate change, releasing significant amounts of stored carbon dioxide (CO_2) into the atmosphere. It is responsible for a substantial portion of global GHG emissions, exacerbating climate change.

4.1.3 Agricultural Productivity

Pesticides play a critical role in enhancing agricultural productivity by protecting crops from pests, diseases, and weeds, leading to higher yields and more stable food supplies. They enable large-scale monoculture farming by reducing the need for manual pest control, increasing efficiency, and lowering labor costs. However, over-reliance on pesticides can lead to resistance in pest populations, diminishing their effectiveness over time (Tahir et al., 2024; Zhang et al. 2024d). Additionally, the environmental impact of pesticides, including harm to non-target species and soil degradation, can undermine long-term productivity, highlighting the need for integrated pest management to sustain agricultural outputs while minimizing ecological damage.

Pesticide Use Pesticides directly enhance agricultural productivity by protecting crops from pests and diseases, leading to higher yields and more stable food supplies. They are essential in maintaining large-scale monocultures.

Land-Use Change Land-use change can increase agricultural productivity by converting natural ecosystems into farmland, expanding the area available for crop production. However, this often comes at the cost of long-term soil fertility and environmental degradation.

4.1.4 Soil Health

Pesticides significantly influence soil health, often disrupting the delicate balance of soil ecosystems. While they target pests, pesticides can also harm beneficial soil microorganisms, crucial for nutrient cycling and organic matter decomposition (Zhang et al., 2024d). Repeated pesticide use can reduce microbial diversity, leading to soil degradation, compaction, and diminished fertility over time. This impacts the soil's ability to support healthy crop growth, potentially requiring increased chemical inputs to maintain productivity (Skendžić et al., 2021; Hilary & Nnamdi, 2023; Tahir et al., 2024). Furthermore, pesticide residues can persist in the soil, affecting future crops and contaminating groundwater. Sustainable soil health requires careful management of pesticide use, emphasizing integrated pest management and organic practices.

Pesticide Use Pesticides can have detrimental effects on soil health by disrupting the microbial communities that are vital for nutrient cycling and soil structure. Repeated use can lead to soil degradation and reduced fertility. Pesticide use can negatively impact soil health by disrupting beneficial microorganisms essential for nutrient cycling and organic matter decomposition. Persistent pesticide residues may degrade soil structure, reduce fertility, and harm future crops (Skendžić et al., 2021). Over time, this can lead to soil compaction and erosion, requiring increased chemical inputs, ultimately undermining the long-term sustainability and productivity of agricultural soils.

Land-Use Change Land-use change, such as converting forests to farmland, often leads to soil erosion, loss of organic matter, and reduced soil fertility. These changes can result in long-term declines in agricultural productivity and ecosystem services.

4.1.5 Sustainability

Sustainability in relation to pesticide use is a complex and paradoxical issue. While pesticides enhance agricultural productivity and food security by controlling pests, their overuse poses significant risks to environmental and human health (García-Casillas et al., 2024). The indiscriminate application can lead to pesticide resistance, necessitating even higher doses, creating a vicious cycle that undermines long-term sustainability. Pesticides also disrupt ecosystems, harming non-target species, degrading soil health, and contaminating water bodies (Tahir et al., 2024). The challenge lies in balancing the immediate benefits of pesticides with their long-term ecological costs. Sustainable agriculture must integrate reduced pesticide use, emphasizing integrated pest management (IPM), organic farming, and biological control methods (Priyanka et al., 2024). This approach seeks to maintain productivity while preserving ecosystem health, ensuring the viability of farming systems for future generations.

Pesticide Use The sustainability of pesticide use is contentious; while they are effective in the short term, over-reliance can lead to pesticide resistance, environmental contamination, and non-target species harm, challenging long-term agricultural sustainability.

Land-Use Change Land-use change often leads to unsustainable agricultural practices, such as deforestation and overgrazing, which degrade land and reduce its capacity to support agriculture and ecosystem services in the long term.

4.1.6 Economic Implications

The economic implications of pesticide use are multifaceted, reflecting both short-term gains and long-term costs. Initially, pesticides enhance agricultural efficiency by reducing pest-induced crop losses, lowering labor costs, and boosting yields, which contributes to higher farm profitability (Hilary & Nnamdi, 2023; Tahir et al., 2024). However, these benefits come with hidden economic burdens. Over time, pests develop resistance, necessitating higher doses or newer, costlier pesticides, driving up production expenses (Diaz et al., 2024). Additionally, environmental degradation from pesticide runoff can lead to costly ecosystem restoration and lost ecosystem services. Human health impacts, including chronic illnesses linked to pesticide exposure, further strain public health systems and reduce worker productivity (Rajak et al., 2023; Diaz et al., 2024). Thus, while pesticides bolster immediate agricultural returns, their cumulative economic costs can erode these gains, challenging the sustainability of their use in the long run.

Pesticide Use Pesticides reduce the labor and costs associated with pest control, enhancing economic efficiency in farming. However, the cost of pesticide resistance, environmental damage, and human health impacts can undermine these economic benefits.

Land-Use Change While land-use change can initially boost agricultural output and economic gain, it often leads to long-term economic costs due to environmental degradation, loss of ecosystem services, and the potential need for expensive restoration efforts.

4.2 Pesticide-Driven Changes in Agricultural Practices

Pesticide-driven changes in agricultural practices have profoundly reshaped both pesticide use and land-use change, intertwining them in complex ways. The widespread adoption of pesticides has facilitated the expansion of monocultures, where single crop species dominate large tracts of land. This shift has intensified the reliance on chemical inputs to manage pests, leading to a cycle of increased pesticide application as pest resistance evolves (Mansfield et al., 2024). As pesticides became central to modern agriculture, they enabled the conversion of previously unarable or marginal lands into productive farmland, contributing to significant land-use changes, including deforestation and habitat destruction (Nagesh et al., 2023). However, this transformation comes at a cost. The heavy reliance on pesticides has led to soil degradation, reduced biodiversity, and contamination of water resources, pushing farmers to continually seek new land for cultivation as existing soils become less fertile. This land-use change exacerbates environmental degradation, creating a feedback loop where more land is cleared to maintain agricultural productivity, further necessitating pesticide use to control pests in newly disturbed ecosystems.

The global rise in pesticide use, particularly in herbicides, has transformed agricultural systems, with a reported increase of approximately 50% over the past three decades, reaching 2.7 million tons by 2020 (Nagesh et al., 2023). This intensification is largely attributed to the pressures of agricultural productivity and pest management, despite the emergence of resistance and environmental concerns (Mansfield et al., 2024; Dekker et al., 2023). In response, innovative farmer-led initiatives in regions like Central America have sought to address the negative impacts of pesticides through collaborative research and demonstration plots, promoting sustainable practices (Wisnujati, 2023). Furthermore, the development of Pesticide Agricultural Shared Socio-Economic Pathways (Pest-Agri-SSPs) highlights the importance of integrating socioeconomic and climate factors to inform future agricultural policies and practices, suggesting that sustainable approaches can lead to reduced pesticide reliance (Dekker et al., 2023). Simultaneously, the ecological consequences of pesticide-driven agricultural practices have sparked shifts toward more sustainable approaches, such as integrated pest management (IPM) and organic farming (Sims, 2021). These practices aim to reduce pesticide dependence, preserve biodiversity, and mitigate the adverse effects of land-use change, highlighting the urgent need for a balanced approach in agriculture.

4.3 Deforestation and Land Conversion for Pesticide-Intensive Crops

Deforestation and land conversion for pesticide-intensive crops are emblematic of the profound and often destructive interplay between pesticide use and land-use change. As global demand for agricultural products particularly those requiring heavy pesticide inputs, such as soybeans, palm oil, and cotton escalates, vast tracts of forests and natural ecosystems are cleared to make way for these high-yield monocultures (Dinesh et al., 2024). This deforestation is not merely a shift in land use; it is a radical transformation that disrupts entire ecosystems, releasing significant amounts of stored carbon dioxide (CO_2) and driving biodiversity loss (Estifanos et al., 2024). The conversion of forests into pesticide-intensive agricultural land accelerates the degradation of soil health, as the natural fertility provided by diverse plant and microbial communities is replaced by a reliance on chemical inputs. Pesticides, while controlling pests in these newly converted lands, simultaneously harm non-target species, including beneficial insects, birds, and soil microorganisms, further destabilizing the ecological balance (Qu et al., 2024). The loss of forested areas, which serve as critical carbon sinks, exacerbates climate change, creating a vicious cycle where the environmental costs of land conversion are magnified by the subsequent need for increased pesticide use.

Moreover, the large-scale application of pesticides in these deforested areas often leads to contamination of nearby water bodies through runoff, affecting aquatic ecosystems and local communities (Sims, 2021; Dinesh et al., 2024). The social implications are equally severe, as indigenous populations and local farmers are frequently displaced, losing their livelihoods and access to essential resources. The economic incentives driving this deforestation are short-term, focusing on immediate agricultural productivity, yet the long-term consequences include the irreversible loss of ecosystems, increased greenhouse gas emissions, and a decline in global biodiversity (Hosseini et al., 2024). In this context, the relationship between pesticide use and land-use change highlights a critical challenge: balancing agricultural demands with environmental stewardship. Sustainable practices, such as agroforestry, organic farming, and integrated pest management, offer potential pathways to mitigate these impacts, but they require significant shifts in policy, economic incentives, and global consumption patterns to be effective.

4.4 GHG Emissions from Land-Use Changes

Greenhouse gas (GHG) emissions from land-use changes are a significant driver of climate change, intricately linked to pesticide use and the expansion of agriculture. As forests, wetlands, and grasslands are converted into agricultural land, vast amounts of carbon stored in vegetation and soil are released into the atmosphere, primarily as carbon dioxide (CO_2) (Pandey et al., 2024; Sharma et al., 2024). This process is exacerbated when land is cleared for pesticide-intensive crops like soybeans, palm oil, and cotton, which thrive under monoculture systems heavily reliant on chemical inputs. Pesticide use, though not a direct source of GHGs, plays a crucial role in enabling and sustaining these land-use changes (Luo et al., 2022; Quijano et al., 2024). By controlling pests and diseases, pesticides allow the large-scale cultivation of monocultures on previously forested or marginal lands, driving further deforestation and land conversion. This expansion of agricultural frontiers results in the loss of carbon sinks natural ecosystems that absorb more CO_2 than they release, thereby intensifying GHG emissions (Luo et al., 2022; Quijano et al., 2024; He et al., 2024). Moreover, the degradation of soil health through repeated pesticide use can lead to increased emissions of nitrous oxide (N_2O), a potent GHG, from agricultural soils. As pesticides disrupt the balance of soil microbial communities, they can enhance the conditions for N_2O production during the nitrification and denitrification processes (Luo et al., 2022). The combined effect of land-use change and soil degradation thus contributes to a significant rise in GHG emissions.

The feedback loop created by land-use changes and pesticide use is particularly concerning. As more land is converted and degraded, the need for chemical inputs increases to maintain productivity, further perpetuating the cycle of emissions and environmental harm (Pereira-Mora et al., 2024; Wu et al., 2024). This cycle not only accelerates climate change but also undermines global efforts to reduce GHG emissions and protect ecosystems. Addressing these challenges requires a holistic approach that integrates sustainable land management practices, such as agroforestry and organic farming, with policies that limit deforestation and promote carbon sequestration (Hui et al., 2024; Ma et al., 2024; Ma et al., 2024). Such strategies are essential to breaking the link between pesticide use, land-use change, and GHG emissions, ensuring a more sustainable and climate-resilient future.

5 Comparative Analysis of Pesticides Versus Alternative Pest Management

Pesticides, the cornerstone of modern agriculture, offer immediate and effective pest control, ensuring high crop yields and economic returns. Their broad-spectrum action can swiftly eliminate a wide range of pests, making them indispensable in large-scale monoculture systems. However, this reliance on chemical pesticides comes with significant ecological and health costs (McCann, 2022; Priyanka et al., 2024). Over time, pests develop resistance, necessitating the use of higher doses or newer, more potent chemicals, which escalates production costs and environmental contamination. The non-selective nature of many pesticides also harms beneficial organisms, including pollinators and natural pest predators, leading to imbalanced ecosystems and further pest outbreaks. In contrast, alternative pest management strategies, such as Integrated Pest Management (IPM), organic farming, and biological control, emphasize sustainability and ecological balance (Bastakoti et al., 2024). IPM combines multiple tactics cultural, mechanical, biological, and chemical in a coordinated approach that minimizes reliance on synthetic pesticides (Daraban et al., 2023). For example, crop rotation and intercropping disrupt pest life cycles, while natural predators and parasitoids are introduced to control pest populations biologically. Organic farming eliminates synthetic pesticides altogether, relying instead on natural substances and techniques like composting and polyculture to maintain soil health and biodiversity (Ayilara et al., 2023). While these methods are generally more labor-intensive and require greater knowledge of local ecosystems, they reduce the risks of resistance development, environmental degradation, and health hazards.

The comparative analysis reveals a clear trade-off: pesticides provide immediate, scalable solutions but with long-term ecological and economic downsides, while alternative methods offer sustainable, environmentally friendly pest control with potentially lower yields and higher labor demands (Goulet, 2022). The decision between these approaches hinges on the balance between short-term productivity and long-term sustainability. For a resilient agricultural system, the integration of alternative pest management with judicious pesticide use reducing dependency on chemicals while enhancing ecological resilience presents a compelling pathway forward, mitigating the environmental impacts while maintaining agricultural productivity (McCann, 2022). The comparative analysis of pesticides and alternative pest management strategies reveals significant differences in their environmental impact, safety, and efficacy. While traditional chemical pesticides have been widely used, their adverse effects on health and ecosystems have prompted a shift toward organic and biopesticide alternatives.

Environmental Impact Chemical pesticides contribute to long-term environmental pollution, affecting soil health and aquatic life due to runoff and biomagnification (Daraban et al., 2023; Ayilara et al., 2023). In contrast, biopesticides, derived from natural sources, are generally safer for ecosystems and degrade more quickly, minimizing their environmental footprint (Daraban et al., 2023; Ayilara et al., 2023).

Health and Safety The use of synthetic pesticides is linked to various health issues, including cancer and neurological disorders (Ayilara et al., 2023). Biopesticides present a lower risk to human health, as they are less toxic and can be washed off easily (Daraban et al., 2023).

Efficacy and Adoption Integrated Pest Management (IPM) combines various strategies, including biopesticides, to optimize pest control while minimizing chemical use (Bastakoti et al., 2024). Despite the benefits of alternatives, the global market for chemical pesticides continues to grow, indicating a complex transition in agricultural practices (Goulet, 2022). While the shift toward biopesticides and IPM is promising, challenges such as pest resistance and the entrenched use of chemical pesticides remain significant hurdles in achieving sustainable pest management.

5.1 GHG Emissions from Organic Versus Conventional Farming Practices

The comparison of greenhouse gas (GHG) emissions from organic and conventional farming practices reveals significant differences influenced by agricultural methods and management strategies. Organic farming generally demonstrates lower GHG emissions due to its reliance on ecological processes and reduced chemical inputs.

GHG Emissions in Conventional Farming Conventional farming practices often lead to higher GHG emissions, particularly from soil. For instance, studies show average N_2O emissions ranging from 4.1 to 6.6 g ha^{-1} day^{-1} and CO_2 emissions between 73.8 and 122.3 kg ha^{-1} day^{-1} in conventionally managed soils (Valujeva et al., 2023). The application of mineral fertilizers significantly contributes to increased N_2O emissions, with peaks observed during specific years due to agronomic practices (Squalli & Adamkiewicz, 2023).

Benefits of Organic Farming Organic farming is associated with lower GHG emissions, as it emphasizes sustainable practices and reduced reliance on synthetic fertilizers. This approach not only mitigates emissions but also enhances soil health and biodiversity (Balode et al., 2024). The economic viability of organic farming, despite potentially lower yields, is supported by higher product prices and lower operating costs (Balode et al., 2024).

While organic farming presents a more sustainable alternative with lower GHG emissions, challenges such as lower yields and higher production costs may deter some farmers from transitioning. Balancing these factors is crucial for broader adoption of sustainable practices.

5.1.1 Differences Between GHG Emissions from Organic and Conventional Farming Practices

Greenhouse gas (GHG) emissions from organic and conventional farming practices differ significantly due to their contrasting approaches to land management, fertilizer use, and energy consumption. Organic farming generally exhibits lower GHG emissions per unit of area because it avoids synthetic fertilizers and pesticides, which are major sources of nitrous oxide (N_2O) and carbon dioxide (CO_2) emissions

Table 2 Differences between GHG emissions from organic and conventional farming practices

Aspect	Organic farming	Conventional farming
Fertilizer Use	Relies on organic compost, manure, and nitrogen-fixing crops	Utilizes synthetic fertilizers, leading to high N_2O emissions
Pesticide Use	Avoids synthetic pesticides, using biological controls instead	Heavy use of synthetic pesticides, contributing to soil and water contamination
Soil Carbon Sequestration	Enhances soil organic carbon through crop rotation and cover crops	Often depletes soil organic matter due to monocultures
Energy Consumption	Lower energy inputs, with reduced fossil fuel reliance	High energy consumption, particularly for fertilizer production and machinery
GHG Emissions	Generally lower but varies depending on manure management and crop types	Higher emissions, especially N_2O from synthetic fertilizers
Yield per Area	Typically lower, potentially requiring more land for the same output	Higher yields per area, but with greater environmental impact

in conventional farming (Fuentes-Ponce et al., 2022; Balode et al., 2024). Organic practices, such as crop rotation, cover cropping, and the use of compost, enhance soil carbon sequestration, effectively trapping CO_2 in the soil and mitigating climate change. However, the lower yield of organic farming can lead to a larger land footprint, potentially offsetting some of the carbon benefits (Valujeva et al., 2023). In contrast, conventional farming often relies on energy-intensive synthetic fertilizers, leading to higher N_2O emissions a potent GHG with a global warming potential far greater than CO_2. The heavy use of machinery and fossil fuels in conventional farming also contributes to higher CO_2 emissions. While conventional practices may produce higher yields on less land, they do so at the cost of increased GHG emissions, soil degradation, and reduced biodiversity (Squalli & Adamkiewicz, 2023).

Ultimately, the choice between organic and conventional farming presents a trade-off between yield efficiency and environmental impact. Organic farming offers a more sustainable approach with lower GHG emissions, but it requires more land, whereas conventional farming maximizes productivity with a higher carbon footprint, challenging global efforts to mitigate climate change (Table 2).

5.1.2 Similarities Between GHG Emissions from Organic and Conventional Farming Practices

Despite their differences, organic and conventional farming practices share several similarities in their contributions to greenhouse gas (GHG) emissions. Both systems can lead to carbon dioxide (CO_2) emissions through soil management practices. Tillage, common in both farming methods, disturbs soil structure, leading to the release of stored CO_2 into the atmosphere. Similarly, both organic and conventional livestock production emit significant amounts of methane (CH_4) through enteric

Table 3 Similarities between GHG emissions from organic and conventional farming practices

Aspect	Organic farming	Conventional farming
Soil Management	Both can result in soil disturbance that leads to CO_2 release	Both can result in soil disturbance that leads to CO_2 release
Livestock Production	Emissions from enteric fermentation and manure management	Emissions from enteric fermentation and manure management
Land Use	Both require land conversion, leading to potential carbon loss	Both require land conversion, leading to potential carbon loss
Irrigation Practices	Both may rely on energy-intensive irrigation systems	Both may rely on energy-intensive irrigation systems
Transport and Distribution	GHG emissions from the transportation of inputs and products	GHG emissions from the transportation of inputs and products
On-farm Energy Use	Some level of fossil fuel use for machinery and infrastructure	Some level of fossil fuel use for machinery and infrastructure

fermentation and manure management (Squalli & Adamkiewicz, 2023; Rodrigo García et al., 2023). These emissions contribute to the overall GHG footprint, irrespective of the farming approach. Land use is another common factor. Whether in organic or conventional systems, converting natural ecosystems to agricultural land often results in the loss of carbon sinks, releasing CO_2 stored in vegetation and soils (Squalli & Adamkiewicz, 2023). Additionally, both farming systems may rely on energy-intensive irrigation practices, which contribute to indirect GHG emissions due to the energy required for water extraction and distribution.

Transport and distribution also play a role in GHG emissions for both organic and conventional farming. The movement of inputs, such as seeds, fertilizers (organic or synthetic), and products to market, involves fossil fuel use, adding to the carbon footprint (Balode et al., 2024). Therefore, on-farm energy use, including the operation of machinery and the maintenance of infrastructure, generates CO_2 emissions in both farming methods (Valujeva et al., 2023). Despite their distinct approaches, organic and conventional farming practices share these fundamental processes that contribute to their GHG emissions, highlighting the need for innovative strategies to mitigate their environmental impact across both systems (Table 3).

5.2 *Potential of Integrated Pest Management (IPM) in Reducing Emissions*

Integrated Pest Management (IPM) represents a multifaceted approach to pest control that offers significant potential in reducing greenhouse gas (GHG) emissions within agricultural systems. Unlike conventional pest control, which often relies heavily on chemical pesticides, IPM integrates biological, cultural, mechanical, and, when necessary, chemical methods to manage pest populations at economically viable levels (Fuentes-Ponce et al., 2022; Squalli & Adamkiewicz, 2023). This holistic strategy not only enhances agricultural sustainability but also plays a crucial role in mitigating climate change. One of the primary ways IPM reduces emissions is through its emphasis on minimizing chemical pesticide use. The production, transportation, and application of synthetic pesticides are energy-intensive processes that contribute to carbon dioxide (CO_2) emissions (Rodrigo García et al., 2023). By reducing the reliance on these chemicals, IPM directly lowers the carbon footprint associated with agricultural practices. Furthermore, the use of biological controls such as natural predators, parasitoids, and pathogens within IPM can eliminate or significantly reduce the need for synthetic pesticides, further decreasing GHG emissions (Fuentes-Ponce et al., 2022).

Another critical aspect of IPM is its focus on improving soil health, which indirectly contributes to emission reductions. Healthy soils are better at sequestering carbon, trapping CO_2 that would otherwise contribute to global warming. Practices such as crop rotation, cover cropping, and reduced tillage, which are integral to IPM, enhance soil structure and organic matter content (Fuentes-Ponce et al., 2022; Balode et al., 2024). These practices not only reduce the need for chemical inputs but also increase the soil's capacity to store carbon, thereby mitigating GHG emissions. IPM also promotes the use of precision agriculture techniques, which optimize input use efficiency, including fertilizers and water (Valujeva et al., 2023; Rajak et al., 2023). By applying inputs more precisely and only when necessary, IPM reduces the excessive use of nitrogen-based fertilizers, which are significant sources of nitrous oxide (N_2O), a potent greenhouse gas with a global warming potential far exceeding that of CO_2 (Ritchie et al., 2020; Squalli & Adamkiewicz, 2023). Additionally, IPM's reliance on field monitoring and pest forecasting reduces unnecessary interventions, further lowering emissions associated with excessive input use. More so, the adoption of IPM can lead to a reduction in land-use change, another significant source of GHG emissions. Conventional farming practices often drive the conversion of forests and other natural ecosystems into agricultural land, releasing large amounts of stored carbon. IPM, by enhancing the resilience and productivity of existing farmland, reduces the pressure to clear additional land for agriculture, thereby preserving carbon sinks and preventing additional CO_2 emissions (Ritchie et al., 2020; Rajak et al., 2023). Therefore, Integrated Pest Management holds immense potential in reducing GHG emissions through its emphasis on reducing chemical inputs, enhancing soil health, optimizing resource use, and preventing land-use change. As the agricultural sector grapples with the dual challenges of feeding a growing population and mitigating climate change, IPM offers a sustainable pathway that aligns productivity with environmental stewardship.

5.3 Role of Biopesticides and Low-GHG Emission Strategies

Targeted Pest Control Biopesticides, derived from natural organisms such as bacteria, fungi, and plants, offer highly specific pest control, minimizing non-target effects and reducing the need for broad-spectrum chemical pesticides. By targeting only the harmful pests, biopesticides preserve beneficial insects and soil microorganisms, fostering a balanced ecosystem (Pandey et al., 2024). This specificity reduces the need for repeated chemical applications, which are often associated with higher GHG emissions due to production and transportation processes (Rajak et al., 2023).

Reduced Chemical Dependency The adoption of biopesticides lowers the agricultural sector's dependence on synthetic chemical pesticides, which are energy-intensive to produce and apply. The reduced need for synthetic inputs translates into lower overall GHG emissions, as biopesticides often require less energy to produce and are applied in smaller quantities. This shift also decreases the carbon footprint associated with the manufacture, transport, and application of conventional pesticides.

Enhanced Soil Carbon Sequestration Biopesticides contribute to soil health by minimizing chemical contamination, which can disrupt soil microbial communities. Healthier soils are more effective at sequestering carbon, a critical component in reducing atmospheric CO_2 levels. The use of biopesticides, alongside other low-GHG emission strategies like crop rotation and cover cropping, enhances soil organic matter, thereby boosting the soil's capacity to store carbon (Daramola & Hatzell, 2023; Pandey et al., 2024; Quijano et al., 2024).

Support for Integrated Pest Management (IPM) Biopesticides are integral to IPM strategies, which prioritize sustainable pest control methods that reduce environmental impact. By incorporating biopesticides into IPM, farmers can achieve effective pest management with minimal GHG emissions (Rajak et al., 2023; Pandey et al., 2024). IPM's holistic approach further supports low-emission practices by reducing over-reliance on any single control method, thus maintaining ecological balance and minimizing the need for energy-intensive interventions.

Lower Energy Use in Application The application of biopesticides generally requires less energy than synthetic pesticides, which often need heavy machinery for large-scale application. Biopesticides can be applied using more sustainable methods, such as targeted spraying or incorporation into organic farming practices, thereby reducing fossil fuel use and associated CO_2 emissions (Fuentes-Ponce et al., 2022). This energy efficiency makes biopesticides a key component of low-GHG emission strategies in agriculture.

Promotion of Sustainable Agriculture The use of biopesticides aligns with broader sustainable agricultural practices that focus on reducing the environmental impact of food production. As part of a low-GHG emission strategy, biopesticides encourage the adoption of organic and regenerative farming practices, which aim to restore ecosystem functions and enhance biodiversity (Balode et al., 2024). These practices not only reduce emissions but also build resilience against climate change, creating a more sustainable agricultural system.

6 Case Studies and Regional Analysis: Pesticide Use and GHG Emissions in Different Regions (Developed Versus Developing Countries)

6.1 Developed Countries: United States and Europe

In developed countries such as the United States and Europe, intensive agricultural practices and high reliance on synthetic pesticides have significantly contributed to greenhouse gas (GHG) emissions. The United States, for instance, leads the world in pesticide use, especially in crops like maize, soybeans, and cotton. These crops are grown in monoculture systems, heavily dependent on chemical inputs for pest control (Dekker et al., 2023; Huang & Li, 2024). The production, transportation, and application of these pesticides are energy-intensive processes, leading to substantial CO_2 emissions. Additionally, the overuse of nitrogen-based fertilizers, often in conjunction with pesticide application, contributes to the release of nitrous oxide (N_2O), a potent GHG. In Europe, the Common Agricultural Policy (CAP) has encouraged pesticide use through subsidies, leading to significant environmental impacts (Cech et al., 2022; Udias et al., 2023). However, there has been a growing shift toward sustainable farming practices and integrated pest management (IPM), which aim to reduce chemical dependency and lower GHG emissions. European countries are increasingly adopting biopesticides and precision agriculture techniques to minimize environmental damage and mitigate climate change.

6.2 Developing Countries: Nigeria and India

In contrast, developing countries like Nigeria and India face different challenges. While pesticide use is generally lower than in developed regions, the lack of regulatory frameworks and access to advanced pest management technologies often lead to inefficient and harmful pesticide application practices. In Nigeria, for example, farmers rely on older, more toxic pesticides, leading to severe environmental contamination and health risks (Ukhurebor et al., 2020; Daramola & Hatzell, 2023; Ayilara et al., 2023). The use of these chemicals, coupled with poor soil management practices, contributes to GHG emissions through soil degradation and deforestation as farmers' clear new land to maintain productivity. Additionally, in India, the Green Revolution significantly increased pesticide use, particularly in rice and wheat production. This shift led to an increase in GHG emissions due to the extensive use of chemical inputs and the energy required for irrigation and pesticide application (Ecobichon, 2003; Sharma et al., 2021; Tıraş, 2024). However, both countries are beginning to explore alternative pest management strategies, such as IPM and organic farming, to reduce their environmental impact and address the growing concerns of climate change.

6.3 *Specific Case Study: High-Pesticide-Use Crops in Nigeria—Soybeans and Maize*

Nigeria, a key agricultural player in sub-Saharan Africa, faces unique challenges with high-pesticide-use crops like soybeans and maize. These crops are vital for the country's food security and export economy, but their cultivation is heavily dependent on chemical pesticides, which have profound environmental and health implications.

Soybeans in the northern regions of Nigeria, where soybeans are predominantly grown, farmers apply large quantities of pesticides to combat pests like aphids and pod borers. The overuse of synthetic pesticides has led to soil degradation and the contamination of water bodies, affecting both ecosystems and human health (Masood et al., 2022). The reliance on chemical inputs also contributes to GHG emissions, primarily through the energy-intensive processes involved in the production and application of these pesticides (Amusat et al., 2023). Moreover, as farmers seek to expand soybean cultivation to meet rising demand, they often clear forests and grasslands, releasing stored carbon and exacerbating climate change.

Maize maize, another staple crop, is widely cultivated across Nigeria. The crop's vulnerability to pests like the fall armyworm has led to increased pesticide use, particularly in the southern and central regions. Smallholder farmers, who constitute the majority of maize producers, often lack access to proper training and equipment, leading to inefficient pesticide application and higher emissions (Adesina et al., 2024; Mojisola et al., 2024). Additionally, the expansion of maize farming into previously forested areas has resulted in significant deforestation, further contributing to GHG emissions. In both cases, the environmental cost of high pesticide use is compounded by the socioeconomic challenges faced by Nigerian farmers, including limited access to sustainable farming practices and markets (Bwala & Amaza, 2024; Oyeranti, 2024). To address these issues, there is a growing push toward adopting Integrated Pest Management (IPM) and agroforestry techniques that reduce chemical dependency, improve soil health, and lower GHG emissions. These approaches not only offer a more sustainable path for Nigerian agriculture but also align with global efforts to combat climate change.

7 Policy and Regulatory Framework

Policy and regulatory frameworks are essential in shaping agricultural practices, particularly when it has to do with the use of pesticide and its environmental impacts. These frameworks establish guidelines, restrictions, and incentives to manage pesticide application, ensuring that agricultural productivity does not come at the expense of ecological balance (Howard et al., 2023; Rashid & Uddin, 2024). In developed countries, stringent regulations often mandate the use of safer, less toxic pesticides, promote Integrated Pest Management (IPM), and set limits on permissible residue levels, thereby reducing greenhouse gas (GHG) emissions associated with chemical-intensive farming. Conversely, in developing nations, where regulatory oversight may be weaker, the lack of enforcement leads to indiscriminate pesticide use, exacerbating soil degradation, water contamination, and GHG emissions (Pham & Laurence, 2024; Akinbowale et al., 2024). To address these disparities, international agreements and national policies must converge, fostering global cooperation to harmonize standards, incentivize sustainable practices, and mitigate the environmental footprint of agriculture, particularly in regions most vulnerable to climate change.

7.1 Global and Regional Policies on Pesticide Use and GHG Emissions

Global and regional policies on pesticide use and greenhouse gas (GHG) emissions are increasingly intertwined, reflecting the urgent need to balance agricultural productivity with environmental sustainability. At the global level, initiatives such as the Stockholm Convention on Persistent Organic Pollutants (POPs) aim to eliminate or restrict the use of hazardous pesticides that not only harm ecosystems but also contribute to GHG emissions through their production and degradation (Howard et al., 2023; Akinbowale et al., 2024). Similarly, the United Nations' Sustainable Development Goals (SDGs), particularly SDG 13 (Climate Action) and SDG 15 (Life on Land), advocate for reducing the environmental impact of pesticides and promoting sustainable land management practices that sequester carbon and mitigate climate change. Regionally, policies vary significantly depending on economic development and agricultural practices (Bwala & Amaza, 2024). In the European Union, the Common Agricultural Policy (CAP) incentivizes the reduction in synthetic pesticide use through subsidies for organic farming and Integrated Pest Management (IPM), aligning with the EU's ambitious climate goals under the European Green Deal. These policies not only aim to cut GHG emissions but also to preserve biodiversity and enhance soil health.

In contrast, developing regions such as sub-Saharan Africa and Southeast Asia often struggle with weak regulatory frameworks, leading to the unchecked use of harmful pesticides. These regions face unique challenges: while they contribute less to global GHG emissions, the environmental degradation resulting from pesticide misuse, such as deforestation and soil erosion, exacerbates climate vulnerability (Rashid & Uddin, 2024). International organizations and regional bodies are increasingly working to harmonize pesticide regulations across borders, promote safer alternatives, and integrate climate resilience into agricultural policies (Bwala & Amaza, 2024; Rashid & Uddin, 2024). Moreover, trade policies also play a critical role in shaping pesticide use and emissions. Countries with stringent environmental standards often impose restrictions on imports of agricultural products with high pesticide residues, indirectly influencing farming practices in exporting nations (Adesina et al., 2024). Such policies can drive the adoption of lower-emission agricultural methods globally, albeit sometimes creating tensions between economic growth and environmental responsibility (Howard et al., 2023; Mojisola et al., 2024). Therefore, the convergence of global and regional policies on pesticide use and GHG emissions is crucial for achieving sustainable agriculture. However, disparities in regulatory capacity and enforcement across regions present challenges, necessitating greater international cooperation, knowledge sharing, and investment in green technologies to align agricultural practices with the goals of climate mitigation and environmental protection.

7.2 Role of International Agreements (e.g., Paris Agreement) in Regulating Pesticide Emissions

International agreements, particularly the Paris Agreement, play a critical role in shaping global efforts to regulate pesticide emissions, interlinking climate action with sustainable agricultural practices. The Paris Agreement, which aims to limit global warming to well below 2 °C, necessitates comprehensive strategies to reduce greenhouse gas (GHG) emissions across all sectors, including agriculture. Pesticide use, while primarily an agricultural concern, contributes indirectly to GHG emissions through energy-intensive production processes, land-use changes, and the degradation of ecosystems that act as carbon sinks. Under the Paris Agreement, countries submit Nationally Determined Contributions (NDCs), outlining their commitments to reduce GHG emissions. These NDCs increasingly recognize the need to address emissions from agriculture, including those related to pesticide use. For instance, some countries have committed to reducing reliance on synthetic pesticides by promoting organic farming, agroecological practices, and Integrated Pest Management (IPM), all of which are aligned with low-emission strategies. By integrating pesticide regulation into broader climate policies, the Paris Agreement fosters a holistic approach to reducing the agricultural sector's carbon footprint. Additionally, international frameworks like the Stockholm Convention and the Montreal Protocol, while not directly under the Paris Agreement, complement its goals by restricting or phasing out the use of particularly harmful pesticides that contribute to GHG emissions and climate change. These agreements enforce global standards that reduce the environmental and health impacts of pesticide use, thereby indirectly supporting the objectives of the Paris Agreement.

The role of international agreements extends to encouraging technology transfer and capacity building in developing countries, where regulatory frameworks may be weaker. By providing financial and technical assistance, these agreements help nations adopt sustainable practices, reduce pesticide emissions, and enhance resilience to climate impacts. Furthermore, they promote international cooperation and harmonization of standards, ensuring that efforts to curb pesticide emissions are effective and equitable across different regions. However, the challenge lies in the enforcement and monitoring of these commitments, as disparities in economic development, political will, and regulatory capacity can hinder progress. Therefore, the success of international agreements in regulating pesticide emissions depends on sustained global collaboration, robust governance, and the integration of climate goals with agricultural policies at both national and international levels.

7.3 *Incentives and Penalties for Sustainable Pesticide Practices*

Incentives and penalties are crucial mechanisms for promoting sustainable pesticide practices, guiding agricultural sectors toward environmentally responsible behaviors while discouraging harmful practices. These strategies are multifaceted, encompassing financial rewards, regulatory frameworks, and market-based approaches designed to foster sustainability in pesticide use.

7.3.1 Subsidies and Financial Assistance

Governments and international organizations offer subsidies to farmers who adopt sustainable practices, such as Integrated Pest Management (IPM), organic farming, or the use of biopesticides. These financial incentives lower the cost barrier for adopting eco-friendly methods, making it economically viable for farmers to reduce their reliance on synthetic pesticides (Mojisola et al., 2024; Rashid & Uddin, 2024). For instance, the European Union's Common Agricultural Policy (CAP) provides subsidies to farmers implementing green practices, directly linking financial support to environmental outcomes.

Tax Breaks and Credits Tax incentives are another powerful tool. Governments may offer tax breaks or credits to farms that demonstrate reduced pesticide use or shift to low-emission, sustainable alternatives. These incentives reduce operational costs for farmers who prioritize sustainability, thus encouraging wider adoption of best practices (Bwala & Amaza, 2024; Pham & Laurence, 2024). In some regions, tax incentives are tied to certification schemes, such as organic or eco-labels, that require adherence to strict pesticide usage guidelines.

Access to Premium Markets Sustainable farming practices, including reduced pesticide use, often grant access to premium markets where consumers are willing to pay more for products certified as organic or environmentally friendly. This market-driven incentive encourages farmers to adopt practices that align with consumer demand for sustainably produced food (Bwala & Amaza, 2024). Certification labels, such as USDA Organic or Fair Trade, provide market differentiation, giving sustainable producers a competitive edge and higher profit margins.

Research and Development Grants Governments and private sectors invest in research grants and development programs aimed at discovering and implementing sustainable pesticide alternatives (Akinbowale et al., 2024). These grants incentivize innovation in biopesticides, precision agriculture, and other technologies that minimize environmental impact (Masood et al., 2022; Oyeranti, 2024). Farmers who participate in pilot projects or early adoption of these innovations may receive financial support, further incentivizing the transition to sustainable practices.

7.3.2 Penalties for Unsustainable Pesticide Practices

Regulatory Fines and Sanctions To deter unsustainable pesticide practices, regulatory bodies impose fines on farmers or companies that violate pesticide regulations, such as exceeding permissible residue limits or using banned substances. These penalties are designed to create a financial deterrent against non-compliance, ensuring that the cost of violating environmental standards outweighs any potential economic benefits (Masood et al., 2022; Oyeranti, 2024; Akinbowale et al., 2024). For example, the Environmental Protection Agency (EPA) in the United States imposes hefty fines on violations of the Federal Insecticide, Fungicide, and Rodenticide Act (FIFRA).

Market Access Restrictions Non-compliance with sustainable pesticide practices can lead to restricted access to both domestic and international markets. Countries with strict pesticide residue regulations, like the European Union or Japan, may ban the import of products that fail to meet their safety standards (Adesina et al., 2024). This restriction serves as a significant penalty for exporters, particularly those in developing countries, where access to these markets is crucial for economic stability (Howard et al., 2023; Amusat et al., 2023). The threat of losing market access compels producers to adhere to sustainable practices.

Legal Liability and Compensation Claims Farmers and companies may face legal action and liability claims if their pesticide practices cause environmental harm or public health issues. These legal repercussions can include compensation for damages, which can be financially crippling. Lawsuits related to pesticide contamination of water sources, harm to non-target species, or public health impacts have become more common, particularly in countries with strong environmental protection laws (Howard et al., 2023; Adesina et al., 2024). The risk of legal action serves as a powerful deterrent against unsustainable practices.

Reputational Damage and Consumer Backlash Beyond legal and financial penalties, companies engaged in unsustainable pesticide practices risk significant reputational damage, leading to consumer boycotts and loss of brand loyalty. In today's highly connected world, news of environmental violations spreads quickly, and consumers increasingly favor brands that prioritize sustainability (Amusat et al., 2023; Adesina et al., 2024). The long-term financial impact of reputational damage can far exceed the cost of compliance with sustainable practices.

7.4 Integrated Approach

The most effective strategy for promoting sustainable pesticide practices often involves a combination of incentives and penalties, creating a balanced approach that encourages compliance while punishing non-compliance (Pham & Laurence, 2024). By aligning economic incentives with environmental goals, policymakers can drive meaningful change in agricultural practices, reducing the environmental impact of pesticide use and contributing to global efforts to mitigate climate change (Oyeranti, 2024; Bwala & Amaza, 2024). For instance, the European Green Deal integrates both incentives, such as subsidies for organic farming, and penalties, like stricter enforcement of pesticide regulations, creating a comprehensive framework that supports the transition to sustainable agriculture. Similarly, initiatives like the Sustainable Rice Platform (SRP) employ a mix of market incentives and regulatory guidelines to encourage sustainable pesticide use among rice farmers globally (Masood et al., 2022; Akinbowale et al., 2024). Therefore, the interplay of incentives and penalties forms a robust framework for promoting sustainable pesticide practices. By reducing financial and operational barriers to sustainable practices, while imposing significant consequences for non-compliance, these strategies create a powerful impetus for change, fostering a more sustainable and environmentally responsible agricultural sector.

8 Future Outlook and Mitigation Strategies

The future outlook for mitigating the contribution of pesticides to greenhouse gas (GHG) emissions hinges on a confluence of innovative practices, policy shifts, and technological advancements. As global agriculture grapples with the dual pressures of feeding a growing population and combating climate change, the role of pesticides in contributing to GHG emissions is increasingly scrutinized. Pesticides contribute indirectly to emissions through energy-intensive production processes, application, and the disruption of ecosystems that naturally sequester carbon.

Future Outlook

The future will likely see a paradigm shift toward reducing reliance on synthetic pesticides through the adoption of sustainable practices such as Integrated Pest Management (IPM) and organic farming. These approaches not only minimize pesticide use but also promote soil health and biodiversity, key factors in carbon sequestration (Allen et al., 2018). The development and adoption of biopesticides, derived from natural organisms, offer a promising alternative with a lower carbon footprint. Precision agriculture, utilizing drones and AI to apply pesticides only where needed, can drastically reduce the volume of chemicals used, thereby cutting associated emissions (Amusat et al., 2023).

Mitigation Strategies

Policy and Regulatory Frameworks: Governments worldwide are expected to tighten regulations on pesticide use, imposing stricter limits on chemical emissions and promoting alternatives. Incentivizing sustainable practices through subsidies and financial support will play a critical role in this transition.

Technological Innovation: The future of pesticide mitigation lies in cutting-edge technologies. Genetic engineering could lead to pest-resistant crops, reducing the need for chemical interventions. Additionally, advancements in precision agriculture tools will enable more targeted pesticide application, significantly reducing waste and emissions.

Carbon Offsetting and Sequestration: Implementing carbon offsetting strategies within agriculture, such as agroforestry and cover cropping, can help neutralize the emissions associated with pesticide use. These practices enhance the carbon-sequestering capacity of farmlands, creating a balance between emissions and absorption.

Global Collaboration and Education: International collaboration on research, technology transfer, and education will be crucial in spreading sustainable practices worldwide. Developing countries, where pesticide regulations may be less stringent, will benefit from global support in adopting low-emission strategies.

The path forward demands a different approach, blending regulatory measures, technological innovation, and sustainable practices to reduce the carbon footprint of pesticides. By embracing these strategies, the agricultural sector can make significant strides in mitigating GHG emissions, ensuring a sustainable future for both the environment and food production systems. The urgency of addressing pesticide-related emissions is clear, and with concerted global effort, a transition to greener, more resilient agricultural practices is within reach.

8.1 Advances in Pesticide Technology to Reduce

These advances in pesticide technology signify a crucial shift toward a more sustainable agricultural paradigm. By reducing the need for conventional pesticides and optimizing their use, these innovations contribute significantly to the reduction in GHG emissions in agriculture. However, the widespread adoption of these technologies requires supportive policies, investment in research, and global collaboration to ensure that all regions benefit from these advancements. The future of agriculture lies in these cutting-edge technologies, which promise not only to safeguard crop yields but also to protect the planet from the escalating threats of climate change.

8.1.1 GHG Emissions

Advances in pesticide technology are important in addressing the dual challenge of maintaining crop productivity while reducing greenhouse gas (GHG) emissions (Balcombe et al., 2018). These innovations are driving a transformation in agricultural practices, steering them toward sustainability and climate resilience with a high degree of complexity and adaptability.

8.1.2 Precision Agriculture and Smart Pesticides

One of the most significant advances is the development of precision agriculture technologies. These systems, which include GPS-guided equipment, drones, and AI-powered sensors, allow the precise application of pesticides, targeting only the areas where pests are present (Beane Freeman, 2022). This precision reduces the overall volume of pesticides used, thereby cutting down on the energy-intensive production and application processes that contribute to GHG emissions (Balode et al., 2024). Smart pesticides formulations that activate only in the presence of specific pests or environmental conditions are another leap forward, further minimizing unnecessary chemical use and environmental impact.

8.1.3 Biopesticides and Nano-technology

Biopesticides, derived from natural sources such as bacteria, fungi, and plant extracts, are emerging as a low-emission alternative to synthetic chemicals. These products are biodegradable, reducing the long-term environmental footprint and the GHG emissions associated with pesticide degradation. Additionally, nanotechnology is revolutionizing pesticide delivery (Bhurat et al., 2024). Nanoparticles can enhance the efficacy of active ingredients, allowing for lower dosages and more targeted action (Bleizgys et al., 2024). This reduces the frequency of applications and the overall chemical load on the environment, indirectly decreasing the GHG emissions tied to their lifecycle.

8.1.4 Gene Editing and Pest-Resistant Crops

Advances in gene editing techniques, such as CRISPR-Cas9, have opened the door to developing crops that are inherently resistant to pests, thus reducing the need for chemical interventions. These genetically modified crops can thrive with minimal pesticide use, leading to a significant reduction in GHG emissions from pesticide production, transport, and application (Beane Freeman, 2022; Bhurat et al., 2024).

8.1.5 Integrated Pest Management (IPM) and AI Integration

The integration of Artificial Intelligence (AI) with Integrated Pest Management (IPM) systems is another frontier in reducing pesticide-related emissions. AI can predict pest outbreaks with high accuracy, optimizing the timing and quantity of pesticide applications. By combining this with IPM strategies that prioritize biological controls and habitat management, the reliance on synthetic pesticides is drastically reduced, leading to lower GHG emissions (Beane Freeman, 2022; Bhurat et al., 2024).

8.2 Potential of Precision Agriculture in Minimizing Pesticide Use

Precision agriculture represents a ground-breaking shift in farming practices, offering immense potential to minimize pesticide use while enhancing crop productivity and environmental sustainability. This advanced farming technique leverages cutting-edge technologies, such as GPS-guided equipment, remote sensing, drones, and data analytics, to manage agricultural inputs with unprecedented accuracy and efficiency (Cox & Zeiss, 2022).

Hyper-Targeted Application At the core of precision agriculture is the principle of hyper-targeted application. Unlike traditional methods, which often involve blanket spraying of entire fields, precision agriculture enables farmers to apply pesticides only where and when they are needed. Through the use of GPS and geospatial mapping, farmers can identify specific areas within a field that are affected by pests, allowing for precise pesticide application that spares unaffected areas. This not only reduces the total volume of chemicals used but also minimizes the environmental impact, including the potential for pesticide runoff and non-target species exposure (Choudhury et al., 2022).

Data-Driven Decision-Making The data-driven nature of precision agriculture is another critical factor in reducing pesticide use. Advanced sensors and drones can monitor crop health in real-time, detecting early signs of pest infestations at a granular level. By analyzing this data, farmers can make informed decisions about when and how much pesticide to apply, often opting for lower doses or alternative treatments. This reactive approach contrasts with the preventative blanket applications typical in conventional farming, leading to a significant reduction in pesticide use (Etminan et al., 2016; Daraban et al., 2023).

Integration with AI and Machine Learning The integration of Artificial Intelligence (AI) and machine learning further amplifies the effectiveness of precision agriculture in minimizing pesticide use. AI algorithms can analyze vast datasets, including weather patterns, soil conditions, and pest behavior, to predict pest outbreaks with high accuracy. This predictive capability allows farmers to anticipate and address pest issues before they become severe, often reducing or even eliminating the need for chemical interventions (Etminan et al., 2016; Daraban et al., 2023).

Environmental and Economic Benefits The environmental benefits of reducing pesticide use through precision agriculture are substantial. Lower pesticide application reduces the risk of contamination of water bodies, soil degradation, and harm to beneficial insects and wildlife. Additionally, by using less pesticide, farmers can reduce their input costs, improving the economic viability of their operations (Ayilara et al., 2023; Babaniyi et al., 2024). The precision application also extends the lifespan of pesticides by reducing the development of resistant pest populations, ensuring that these tools remain effective for longer periods.

Challenges and Future Prospects Despite its potential, the widespread adoption of precision agriculture faces challenges, including high initial costs, the need for technical expertise, and the digital divide in rural areas. However, as technology becomes more accessible and affordable, and as farmers become more familiar with these tools, the adoption of precision agriculture is expected to accelerate.

However, precision agriculture holds the promise of revolutionizing pesticide use in farming. By enabling hyper-targeted, data-driven, and predictive approaches to pest management, it offers a pathway to significantly reduce the environmental and economic costs associated with pesticide use. As this technology continues to evolve, it will play a crucial role in advancing sustainable agricultural practices, ensuring food security, and mitigating the environmental impacts of farming in the face of global challenges such as climate change and population growth.

8.3 Global Strategies for Reducing the Environmental Impact of Pesticides

Global strategies for reducing the environmental impact of pesticides are multifaceted, involving regulatory frameworks, technological advancements, and sustainable agricultural practices. These strategies aim to balance the need for pest control with the imperative to protect ecosystems, human health, and the broader environment. The complexity and urgency of this challenge require a highly coordinated approach, integrating scientific innovation, policy, and on-the-ground practices.

Regulatory Frameworks and International Agreements Global regulatory frameworks play an essential role in mitigating the environmental impact of pesticides. International agreements, such as the Stockholm Convention on Persistent Organic Pollutants (POPs) and the Rotterdam Convention, aim to regulate and phase out the most hazardous pesticides (Howard et al., 2023; Huang & Li, 2024). These agreements enforce stringent controls on pesticide production, trade, and use, particularly for chemicals that persist in the environment and accumulate in the food chain. Moreover, the Paris Agreement indirectly influences pesticide use by promoting sustainable agriculture as a key component of climate change mitigation (Hui et al., 2024). Countries are encouraged to adopt practices that reduce pesticide reliance, thereby minimizing greenhouse gas (GHG) emissions and environmental contamination.

Adoption of Integrated Pest Management (IPM) Integrated Pest Management (IPM) is a cornerstone strategy for reducing pesticide use while maintaining crop productivity. IPM emphasizes the use of a diverse set of tools, including biological control agents, crop rotation, and resistant crop varieties, to manage pest populations (Hosseini et al., 2024). By prioritizing non-chemical methods and applying pesticides only as a last resort, IPM reduces the overall chemical load on the environment. This approach also mitigates the risk of pests developing resistance to pesticides, which often leads to the increased use of chemicals (Huang & Li, 2024). Governments and international organizations promote IPM through education, subsidies, and technical support, helping farmers transition to more sustainable practices.

Technological Innovations and Precision Agriculture Technological innovations are driving significant reductions in the environmental impact of pesticides. Precision agriculture, powered by GPS, drones, and artificial intelligence, allows the targeted application of pesticides, minimizing the volume of chemicals used (Kabange et al., 2023; Im et al., 2024). These technologies enable farmers to monitor pest populations and crop health in real-time, applying pesticides only where necessary. Additionally, biopesticides derived from natural organisms, such as bacteria, fungi, or plant extracts, are emerging as eco-friendly alternatives to synthetic chemicals (Kader Esen et al., 2022). These biologically based products are less persistent in the environment and pose fewer risks to non-target species.

Transition to Organic and Low-Input Farming The transition to organic and low-input farming systems is another critical strategy for reducing the environmental impact of pesticides. Organic farming prohibits the use of synthetic pesticides, relying instead on natural pest control methods, crop diversity, and soil health to manage pests (Allen et al., 2018; Akinbowale et al., 2024). While organic farming represents a small percentage of global agriculture, its principles are increasingly influencing mainstream farming practices. Low-input farming, which seeks to minimize the use of synthetic inputs, including pesticides, through better resource management, is also gaining traction. Both approaches contribute to a reduction in chemical runoff, soil degradation, and biodiversity loss (Bwala & Amaza, 2024).

Education and Capacity Building Education and capacity building are essential for implementing sustainable pesticide practices globally. Farmers, particularly in developing countries, often lack access to information and training on safe pesticide use and alternatives. International organizations, NGOs, and governments are working to bridge this gap by providing education on IPM, the risks associated with pesticide use, and the benefits of sustainable practices. Extension services, farmer field schools, and online platforms are increasingly used to disseminate knowledge and best practices, empowering farmers to make informed decisions (Allen et al., 2018; Akinbowale et al., 2024; Bwala & Amaza, 2024).

Financial Incentives and Market-Based Approaches Financial incentives and market-based approaches are powerful tools for encouraging the adoption of sustainable pesticide practices. Governments and international bodies offer subsidies, tax breaks, and grants to farmers who adopt IPM, organic farming, or other environmentally friendly practices. Certification schemes, such as organic or fair-trade labels, also create market incentives by allowing farmers to sell their products at a premium (Balode et al., 2024; Bastakoti et al., 2024). These economic tools help offset the initial costs associated with transitioning to sustainable practices, making them more accessible to a broader range of farmers.

International Cooperation and Research Collaboration International cooperation and research collaboration are vital for developing and disseminating new technologies and practices that reduce the environmental impact of pesticides. Collaborative research initiatives, often funded by governments and international organizations, focus on developing pest-resistant crops, biopesticides, and precision agriculture technologies. By sharing knowledge, technologies, and resources across borders, these collaborations accelerate the global adoption of sustainable practices.

Monitoring and Enforcement Monitoring and enforcement mechanisms are crucial for ensuring compliance with pesticide regulations and the effectiveness of reduction strategies. National governments, supported by international organizations, must establish robust monitoring systems to track pesticide use, environmental contamination, and health impacts. Enforcement of regulations, including penalties for non-compliance, ensures that both producers and users adhere to standards that protect the environment and public health.

Nevertheless, global strategies for reducing the environmental impact of pesticides are diverse and interconnected, requiring collaboration between governments, international organizations, the private sector, and farmers. By integrating regulatory frameworks, technological advancements, sustainable practices, and financial incentives, these strategies aim to create a more sustainable agricultural system. The challenge is immense, but with continued innovation and cooperation, it is possible to significantly reduce the environmental footprint of pesticides, safeguarding both ecosystems and human health for future generations.

9 Conclusion

The implications of pesticide use for agricultural sustainability and climate change are profound and multi-layered. Pesticides, while essential for modern agriculture, contribute directly and indirectly to GHG emissions through energy-intensive production processes, application, and soil degradation. Their impact on soil microbial activity, nitrogen cycling, and carbon sequestration disrupts the delicate balance necessary for maintaining healthy ecosystems, leading to increased emissions of CO_2 and N_2O. Moreover, pesticide-driven land-use changes, such as deforestation for monoculture crops, further exacerbate climate change by reducing carbon sinks and increasing atmospheric CO_2 levels. These environmental costs highlight the urgent need for sustainable pest management practices that prioritize ecological balance and long-term productivity over short-term gains. To address these challenges, future research should focus on developing and implementing alternative pest management strategies that minimize reliance on synthetic chemicals. Integrated Pest Management (IPM) and organic farming offer promising approaches that combine ecological principles with practical pest control methods, reducing the environmental impact of agriculture. Additionally, policymakers must prioritize the regulation of pesticide use, incentivizing sustainable practices, and investing in research to explore the environmental consequences of pesticides fully. Global cooperation is essential to ensure that these strategies are adopted worldwide, particularly in regions most vulnerable to climate change. By rethinking pesticide use and embracing sustainable agricultural practices, we can mitigate the adverse effects on the environment, protect biodiversity, and contribute to global efforts to combat climate change.

References

Adesina, G. O., Adelasoye, K. A., Akinjide, B. I., Abiola, S. O., & Adeniji, A. A. (2024). Assessment of pesticides residue in selected arable farm lands in Ogbomoso South local government area of Oyo state, Nigeria. *Agricultural Science & Technology (1313–8820), 16*(2), 75–85.

Aggarwal, P., Roy, J., Pathak, H., Kumar, S. N., Venkateswarlu, B., Ghosh, A., & Ghosh, D. (2022). Managing climatic risks in agriculture. *Indian Agriculture Towards, 2030*, 83–108.

Akinbowale, O. E., Klingelhöfer, H. E., Zerihun, M. F., & Mashigo, P. (2024). Development of a policy and regulatory framework for mitigating cyberfraud in the South African banking industry. *Heliyon, 10*(1), e23491.

Alemu, M. (2020). Trend of biotechnology applications in pest management: A review. *International Journal of Applied Sciences and Biotechnology, 8*(2), 108–131.

Ali, S., Ullah, M. I., Sajjad, A., Shakeel, Q., & Hussain, A. (2021). Environmental and health effects of pesticide residues. In *Sustainable agriculture reviews 48: Pesticide occurrence, analysis and remediation vol. 2 analysis* (pp. 311–336). Springer.

Allen, M. R., Shine, K. P., Fuglestvedt, J. S., Millar, R. J., Cain, M., Frame, D. J., & Macey, A. H. (2018). A solution to the misrepresentations of CO2-equivalent emissions of short-lived climate pollutants under ambitious mitigation. *npj Climate and Atmospheric Science, 1*(1), 16.

Amusat, A. I., Okewole, S. A., Rafiu, R. A., & Amusat, M. A. (2023). Overview of pesticide usage, misuse and its impact on environmental degradation in south-western states and some part of northern states in Nigeria. *Journal of Pesticide Science and Pest Control, 2*(1), 2833–0943.

Aneesha, V., Dhalin, D., Subhagan, S. R., Rani, O. P., & Khatawkar, D. S. (2020). Carbon footprint of electrostatic sprayer in comparison with air compression sprayer and mistblower. *Current Journal of Applied Science and Technology, 39*(29), 20–29.

Arsyan Adlan, C., Wicke, B., Hanssen, S. V., & Hendriks, C. (2024). Spatially-explicit greenhouse gas footprints of agricultural commodities from around the world. In *EGU general assembly conference abstracts* (p. 13479).

Assad, R., Bashir, I., Rafiq, I., Sofi, I. A., Mir, S. H., Reshi, Z. A., & Rashid, I. (2021). Global scenario of remediation techniques to combat pesticide pollution. In *Agricultural waste* (pp. 69–97). Apple Academic Press.

Ayilara, M. S., Adeleke, B. S., Akinola, S. A., Fayose, C. A., Adeyemi, U. T., Gbadegesin, L. A., et al. (2023). Biopesticides as a promising alternative to synthetic pesticides: A case for microbial pesticides, phytopesticides, and nanobiopesticides. *Frontiers in Microbiology, 14*, 1040901.

Babaniyi, G. G., Akor, U. J., & Adebomi, J. I. (2024). Legislation and programs supporting environmental protection. In S. A. Aransiola, B. R. Babaniyi, A. B. Aransiola, & N. R. Maddela (Eds.), *Prospects for soil regeneration and its impact on environmental protection* (pp. 3–14). Springer.

Balcombe, P., Speirs, J. F., Brandon, N. P., & Hawkes, A. D. (2018). Methane emissions: Choosing the right climate metric and time horizon. *Environmental Science: Processes & Impacts, 20*(10), 1323–1339.

Balode, L., Pakere, I., & Blumberga, D. (2024). Organic or non-organic agriculture: Comparison of organic and conventional farming sustainability. In *CONECT. International scientific conference of environmental and climate technologies* (p. 91). https://doi.org/10.7250/CONECT.2024.065

Bastakoti, B., Tiwari, S., Subedi, A. P., Giri, D., & Karki, A. (2024). Comparative analysis of knowledge and management practices of insect pests of maize among IPM adopters and non-adopters in Sindhupalchok, Nepal. *Archives of Agriculture and Environmental Science, 9*(1), 168–174.

Beane Freeman, L. E. (2022). Invited perspective: Pesticide adjuvants and inert ingredients—A missing piece of the puzzle. *Environmental Health Perspectives, 130*(8), 081301.

Beveridge, M. C., Ross, L. G., & Kelly, L. A. (1994). Aquaculture and biodiversity. *Ambio: A Journal of the Human Environment, 23*, 497–502.

Bhurat, S., Jaiswal, M., Ranjit, P. S., Kunwer, R., Gugolothu, S. K., & Bhurat, K. (2024). Greenhouse gas (GHG) emissions and its mitigation technology in transportation sector. In *Renewable energy innovations: Biofuels, solar, and other technologies* (pp. 159–180). John Wiley & Sons, Inc./Scrivener Publishing LLC.

Bleizgys, R., Povilaitis, A., Pekarskas, J., & Naujokienė, V. (2024). Effectiveness of the sustainable manure pile model for ammonia emission and soil. *Agronomy, 14*(7), 1475.

Bwala, M. A., & Amaza, P. (2024). Assessing the awareness and perception of the role of climate change factors in aggravating conflict among farmers and pastoralists in North Central Nigeria. Research Square. https://doi.org/10.21203/rs.3.rs-4355900/v1

Cech, R., Leisch, F., & Zaller, J. G. (2022). Pesticide use and associated greenhouse gas emissions in sugar beet, apples, and viticulture in Austria from 2000 to 2019. *Agriculture, 12*(6), 879.

Chatterjee, S., Gangopadhyay, C., Bandyopadhyay, P., Bhowmick, M. K., Roy, S. K., Majumder, A., et al. (2021). Input-based assessment on integrated pest management for transplanted rice (Oryza sativa) in India. *Crop Protection, 141*, 105444.

Chen, J., Li, Q. X., & Song, B. (2020). Chemical nematicides: Recent research progress and outlook. *Journal of Agricultural and Food Chemistry, 68*(44), 12175–12188.

Choudhury, P. K., Jena, R., Tomar, S. K., & Puniya, A. K. (2022). Reducing enteric methanogenesis through alternate hydrogen sinks in the rumen. *Methane, 1*(4), 320–341.

Cornish, C. M., Johnson, O. F., Bansal, S., Meier, J. A., Harris, T. D., & Sweetman, J. N. (2024). Common use herbicides increase wetland greenhouse gas emissions. *Science of the Total Environment, 933*, 172881.

Cox, C., & Zeiss, M. (2022). Health, pesticide adjuvants, and inert ingredients: California case study illustrates need for data access. *Environmental Health Perspectives, 130*(8), 085001.

Damm, Y., Cisneros, E., & Börner, J. (2024). Beyond deforestation reductions: Public disclosure, land-use change and commodity sourcing. *World Development, 175*, 106481.

Daraban, G. M., Hlihor, R. M., & Suteu, D. (2023). Pesticides vs. biopesticides: From pest management to toxicity and impacts on the environment and human health. *Toxics, 11*(12), 983.

Daramola, D. A., & Hatzell, M. C. (2023). Energy demand of nitrogen and phosphorus based fertilizers and approaches to circularity. *ACS Energy Letters, 8*(3), 1493–1501.

Dekker, S., Nagesh, P., Edelenbosch, O., de Boer, H., Mitter, H., & van Vuuren, D. P. (2023, May). Pesticide use under the influence of socio-economic and climate change: Pest-Agri-SSPs. In *EGU general assembly conference abstracts* (pp. EGU-2692).

Desaeger, J., Wram, C., & Zasada, I. (2020). New reduced-risk agricultural nematicides-rationale and review. *Journal of Nematology, 52*(1), 1–16.

Diaz, Y. F., Azmoun, S., & Lucchini, R. G. (2024). Fluoride and manganese. In *Textbook of children's environmental health* (p. 451). Oxford University Press.

Dinesh, G. K., Ravi, R., Venkatramanan, V., Cherukumalli, S., & Dubey, R. (2024). The potential of conservation agriculture and their ecosystem services for land degradation neutrality. *Authorea Preprints*.

Ecobichon, D. J. (2003). The pesticide dilemma in developing countries. *Journal of Science Technology and Humanities, 1*(1), 1–9.

Estifanos, T. K., Fisher, B., Galford, G. L., & Ricketts, T. H. (2024). Impacts of deforestation on childhood malaria depend on wealth and vector biology. *GeoHealth, 8*(3), e2022GH000764.

Etminan, M., Myhre, G., Highwood, E. J., & Shine, K. P. (2016). Radiative forcing of carbon dioxide, methane, and nitrous oxide: A significant revision of the methane radiative forcing. *Geophysical Research Letters, 43*(24), 12–614.

Filonchyk, M., Peterson, M. P., Zhang, L., Hurynovich, V., & He, Y. (2024). Greenhouse gases emissions and global climate change: Examining the influence of CO2, CH4, and N2O. *Science of the Total Environment, 953*, 173359.

Fuentes-Ponce, M. H., Gutiérrez-Díaz, J., Flores-Macías, A., González-Ortega, E., Mendoza, A. P., Sánchez, L. M. R., et al. (2022). Direct and indirect greenhouse gas emissions under conventional, organic, and conservation agriculture. *Agriculture, Ecosystems & Environment, 340*, 108148.

Gaikwad, T. S., & Mairal, P. B. (2022). The review on pesticides. *IJFMR-International Journal for Multidisciplinary Research, 4*, 1–7.

García-Casillas, A. C., Prado-Rebolledo, O. F., Martínez-González, S., & Téllez, I. G. (2024). Impact of pesticides on environmental, health, and agriculture sustainability Impacto de los pesticidas en la sustentabilidad ambiental, sanitaria y agrícola. *ECORFAN*, 71–93.

Genedy, R. A., Ogejo, J. A., Chung, M., & Shortridge, J. (2023). Integrating machine learning into process-based modeling to predict the ammonia loss from stored slurry dairy manure. In *2023 ASABE annual international meeting* (p. 1). American Society of Agricultural and Biological Engineers.

Ghosson, H., Raviglione, D., Salvia, M. V., & Bertrand, C. (2023). Characteristic response of formulation ingredients revealed by ultra high performance liquid chromatography-electrospray ionization-high resolution mass spectrometry-based untargeted screening of pesticides in soil. *Journal of Mass Spectrometry, 58*(7), e4962.

Goulet, F. (2022). The role of alternative technologies in the enactment of (dis) continuities. In *Technologies in decline* (pp. 167–184). Routledge.

Han, X., Guo, C., Xu, C., Shi, L., Liu, B., Zhang, Z., et al. (2023). Water-soluble metallo-supramolecular nanoreactors for mediating visible-light-promoted cross-dehydrogenative coupling reactions. *ACS Nano, 17*(4), 3723–3736.

Han, H., Liu, Y., Yang, S., Xue, H., Zhang, D., & Xie, Y. (2024). Temperature influences the toxicity, behavioral, and physiology response of chlorfenapyr, fipronil, imidacloprid, and indoxacarb to Reticulitermes flaviceps. *Journal of Asia-Pacific Entomology, 27*(2), 102230.

He, Z., Shen, J., Zhu, Y., Gao, J., Zhang, D., & Pan, X. (2024). Active anaerobic methane oxidation in the groundwater table fluctuation zone of rice paddies. *Water Research, 258*, 121802.

Helsel, Z. R., & Pimentel, D. (2007). Energy in pesticide production and use. In *Encyclopedia of pest management* (Vol. 2, pp. 157–160).

Hilary, E. A., & Nnamdi, O. S. (2023). *Ameliorating the impact of climate variability on pesticide dynamics and efficacy*. https://doi.org/10.32388/IK3NC2

Hosseini, M., Bahrami, H., Khormali, F., Khavazi, K., Mokhtassi-Bidgoli, A., & Rezaei, M. (2024). Effect of deforestation on soil organic carbon fate and pool, a case study in Mazandaran, Iran. *Catena, 237*, 107804.

Howard, K., Howard, K., Milne, C., Urch, A., Wilson, A., & Baker, D. (2023). Pore space as a resource: A discussion of the policy and regulatory framework for carbon capture, utilization, and storage. *Alberta Law Review, 61*, 183.

Hu, Y., & Liu, Y. (2024). Impact of fertilizer and pesticide reductions on land use in China based on crop-land integrated model. *Land Use Policy, 141*, 107155.

Huang, Y., & Li, Z. (2024). Assessing pesticides in the atmosphere: A global study on pollution, human health effects, monitoring network and regulatory performance. *Environment International, 187*, 108653.

Hui, D., Ray, A., Kasrija, L., & Christian, J. (2024). Impacts of climate change and agricultural practices on nitrogen processes, genes, and soil nitrous oxide emissions: A quantitative review of meta-analyses. *Agriculture, 14*(2), 240.

Im, S., Lee, M. K., Prakash, O., Shi, X., Chung, E. G., Na, E. H., & Kim, D. H. (2024). Keep manure fresh, get more! *Resources, Conservation and Recycling, 206*, 107629.

Jasrotia, P., Kumari, P., Malik, K., Kashyap, P. L., Kumar, S., Bhardwaj, A. K., & Singh, G. P. (2023). Conservation agriculture based crop management practices impact diversity and population dynamics of the insect-pests and their natural enemies in agroecosystems. *Frontiers in Sustainable Food Systems, 7*, 1173048.

Jezierska-Tys, S., Joniec, J., Bednarz, J., & Kwiatkowska, E. (2021). Microbiological nitrogen transformations in soil treated with pesticides and their impact on soil greenhouse gas emissions. *Agriculture, 11*(8), 787.

Kabange, N. R., Kwon, Y., Lee, S. M., Kang, J. W., Cha, J. K., Park, H., et al. (2023). Mitigating greenhouse gas emissions in agriculture: A review. *Sustainability, 15*(22), 15889.

Kader Esen, V., Palangi, V., & Esen, S. (2022). Genetic improvement and nutrigenomic management of ruminants to achieve enteric methane mitigation: A review. *Methane, 1*(4), 342–354.

Kamala, I. M., & Devanand, I. I. (2021). Impact of climate change on insects and their sustainable management. In *Sustainable intensification for agroecosystem services and management* (pp. 779–815). Springer.

Kaur, R., Choudhary, D., Bali, S., Bandral, S. S., Singh, V., Ahmad, M. A., et al. (2024). Pesticides: An alarming detrimental to health and environment. *Science of the Total Environment, 915*, 170113.

Khan, S. H., & Pathak, B. (2020). Zinc oxide based photocatalytic degradation of persistent pesticides: A comprehensive review. *Environmental Nanotechnology, Monitoring & Management, 13*, 100290.

Khan, B. A., Nadeem, M. A., Nawaz, H., Amin, M. M., Abbasi, G. H., Nadeem, M., et al. (2023). Pesticides: Impacts on agriculture productivity, environment, and management strategies. In *Emerging contaminants and plants: Interactions, adaptations and remediation technologies* (pp. 109–134). Springer International Publishing.

Kumar, R., & Bordoloi, N. (2024). Agriculture's contribution to the emission of greenhouse gas nitrous oxide (N 2 O) and its feasible mitigation strategies. In *Climate smart greenhouses-Innovations and impacts*. IntechOpen.

Kumar, M., Yadav, A. N., Saxena, R., Paul, D., & Tomar, R. S. (2021). Biodiversity of pesticides degrading microbial communities and their environmental impact. *Biocatalysis and Agricultural Biotechnology, 31*, 101883.

Li, J., Liu, Y., Liu, J., Cui, X., Hou, T., & Cheng, D. (2022). A novel synthetic slow release fertilizer with low energy production for efficient nutrient management. *Science of the Total Environment, 831*, 154844.

Li, Q., Ma, Z., Huo, J., Zhang, X., Wang, R., Zhang, S., et al. (2024). Distinct microbial hydrogen and reductant disposal pathways explain interbreed variations in ruminant methane yield. *The ISME Journal, 18*(1), wrad016.

Lima-Junior, D. P., Lima, L. B., Carnicer, C., Zanella, R., Prestes, O. D., Floriano, L., & Júnior, P. D. M. (2024). Exploring the relationship between land-use and pesticides in freshwater ecosystem: A case study of the Araguaia River Basin, Brazil. *Environmental Advances, 15*, 100497.
Luo, X., Liu, Y., Muhmood, A., Zhang, Q., Wang, J., Ruan, R., et al. (2022). Effect of time and temperature of pretreatment and anaerobic co-digestion of rice straw and swine wastewater by domesticated paddy soil microbes. *Journal of Environmental Management, 323*, 116218.
Lyr, H. (2003). Fungicides, aromatic hydrocarbons. In *Encyclopedia of agrochemicals*. https://doi.org/10.1002/047126363X.agr092
Ma, F., Li, Y., Liu, Y., Dong, L., Han, X., Zhang, X., et al. (2024). Elevated atmospheric CO2 promotes the contribution of autotrophic nitrification to N2O emissions in a typical summer maize field. *European Journal of Agronomy, 153*, 127041.
Maggi, F., La Cecilia, D., Tang, F. H., & McBratney, A. (2020). The global environmental hazard of glyphosate use. *Science of the Total Environment, 717*, 137167.
Mahanta, D. K., Samal, I., Komal, J., Bhoi, T. K., Majhi, P. K., & Ahmad, M. A. (2023). Understanding anthropogenic climate change, its consequences on insect pests, and techniques in forecasting and monitoring pest dynamics: A contemporary scenario. In *Climate change and insect biodiversity* (pp. 44–67). CRC Press.
Mansfield, B., Werner, M., Berndt, C., Shattuck, A., Galt, R., Williams, B., et al. (2024). A new critical social science research agenda on pesticides. *Agriculture and Human Values, 41*(2), 395–412.
Manzetti, S., & van der Spoel, D. (2015). Impact of sludge deposition on biodiversity. *Ecotoxicology, 24*, 1799–1814.
Masood, N., Akram, R., Fatima, M., Mubeen, M., Hussain, S., Shakeel, M., et al. (2022). Insect pest management under climate change. In *Building climate resilience in agriculture: Theory, practice and future perspective* (pp. 225–237). Springer.
McCann, L. (2022). *Three essays on the economics of pesticide use* [Doctoral dissertation]. University of Missouri, Columbia.
McGee, C. F., McGilloway, D. A., & Buckle, A. P. (2020). Anticoagulant rodenticides and resistance development in rodent pest species–A comprehensive review. *Journal of Stored Products Research, 88*, 101688.
McLaughlin, A., & Mineau, P. (1995). The impact of agricultural practices on biodiversity. *Agriculture, Ecosystems & Environment, 55*(3), 201–212.
Meena, B. P., Biswas, A. K., Singh, M., Das, H., Chaudhary, R. S., Singh, A. B., et al. (2021). Energy budgeting and carbon footprint in long-term integrated nutrient management modules in a cereal-legume (Zea mays–Cicer arietinum) cropping system. *Journal of Cleaner Production, 314*, 127900.
Mishra, R., Tripathi, P., Kumar, P., Rajpoot, P. K., Verma, S., & Aman, A. S. (2024). Innovations and future trends in storage pest management. *Journal of Experimental Agriculture International, 46*(5), 155–165.
Mojisola, A., Toyin, B., & Kemi, S. I. (2024). Assessment of health impact on agricultural workers and rural inhabitants exposed to pesticide in selected towns in southern region of Ekiti State, Nigeria. *International Journal of Research and Scientific Innovation, 11*(4), 224–236.
Nagesh, P., Edelenbosch, O. Y., Dekker, S. C., de Boer, H. J., Mitter, H., & van Vuuren, D. P. (2023). Extending shared socio-economic pathways for pesticide use in Europe: Pest-Agri-SSPs. *Journal of Environmental Management, 342*, 118078.
Oka, Y. (2020). From old-generation to next-generation nematicides. *Agronomy, 10*(9), 1387.
Ólafsdóttir, S. S., Jensen, C. D., Lymperatou, A., Henriksen, U. B., & Gavala, H. N. (2023). Effects of different treatments of manure on mitigating methane emissions during storage and preserving the methane potential for anaerobic digestion. *Journal of Environmental Management, 325*, 116456.
Oyeranti, O. A. (2024). Effect of carbon footprint on agricultural productivity in Nigeria: An empirical analysis. *International Journal of Social Science and Economic Research, 9*(5), 1518–1535.

Palangi, V., & Lackner, M. (2022). Management of enteric methane emissions in ruminants using feed additives: A review. *Animals, 12*(24), 3452.

Pandey, P., Swaroopa, V. J., Sahu, P. L., & Parmar, B. (2024). A comprehensive review on carbon sequestration potential and addition of organic carbon to soil. *International Journal of Environment and Climate Change, 14*(7), 724–733.

Pei, M. S., Liu, H. N., Wei, T. L., Yu, Y. H., & Guo, D. L. (2021). Detection and characterization of genome-wide genetic variation associated with the early-ripening phenotype of grape mutants. *Scientia Horticulturae, 285*, 110195.

Pelfrène, A. F. (2010). Rodenticides. In *Hayes' handbook of pesticide toxicology* (pp. 2153–2217). Academic Press.

Pereira-Mora, L., Guerrero, L. D., Erijman, L., & Fernández-Scavino, A. (2024). Tartrate fermentation with H2 production by a new member of Sporomusaceae enriched from rice paddy soil. *Applied and Environmental Microbiology, 90*(4), e02351–e02323.

Pham, H., & Laurence, R. (2024). Regulatory framework and policy challenges for connected and automated vehicles adoption in Vietnam. In *International conference series on geotechnics, civil engineering and structures* (pp. 1522–1533). Springer Nature Singapore.

Pierre, F. D., Friuli, M., Rollé, L., Fortina, R., Tassone, S., & Dinuccio, E. (2022). Development and evaluation of an in vitro procedure to assess enteric methane production. In *2022 IEEE workshop on metrology for agriculture and forestry (MetroAgriFor)* (pp. 33–37). IEEE.

Priyanka, K., Yadav, B. S., Kote, A., Yadav, R., & Shejwal, P. (2024). Comparative analysis of organic and chemical pesticides: Impacts on crop health and environmental sustainability. *Uttar Pradesh Journal of Zoology, 45*(11), 93–96.

Qu, Q., Zhang, K., & Groot, J. C. (2023). A model to identify entry points to curb emissions from complex manure management chains. *Journal of Cleaner Production, 415*, 137787.

Qu, X., Li, X., Bardgett, R. D., Kuzyakov, Y., Revillini, D., Sonne, C., et al. (2024). Deforestation impacts soil biodiversity and ecosystem services worldwide. *Proceedings of the National Academy of Sciences, 121*(13), e2318475121.

Quijano, C. C., Filipiak, M., Nieder, R., & Stichnothe, H. (2024). Unlocking the carbon sequestration potential of agricultural soils through deep tillage—A dynamic life cycle analysis perspective.

Rached, A., Moriceau, M. A., Serfaty, X., Lefebvre, S., & Lattard, V. (2020). Biomarkers potency to monitor non-target fauna poisoning by anticoagulant rodenticides. *Frontiers in Veterinary Science, 7*, 616276.

Rajak, P., Roy, S., Ganguly, A., Mandi, M., Dutta, A., Das, K., et al. (2023). Agricultural pesticides–friends or foes to biosphere? *Journal of Hazardous Materials Advances, 10*, 100264.

Rakhra, M., Deb, P., Dahiya, O., Chandel, S. S., Bhutta, B., Badotra, S., et al. (2022). An analytical study of the types of implements used by farmers in mechanized agriculture. In *In 2022 international mobile and embedded technology conference (MECON)* (pp. 683–687). IEEE.

Rashid, M. A., & Uddin, M. S. (2024). Legal, regulatory, and policy framework hypothesis testing in a proposed e-procurement implementation assessment model framework. *GSC Advanced Research and Reviews, 19*(2), 219–225.

Ritchie, H., Roser, M., & Rosado, P. (2020). CO_2 and greenhouse gas emissions. In *Our world in data*. https://ourworldindata.org/greenhouse-gas-emissions

Rodrigo García, E., Murillo Peña, R., Pérez Álvarez, E. P., Garde Cerdán, T., & Martínez Vidaurre, J. M. (2023). The influence of climatic conditions and agronomic practices on greenhouse gas emissions in a conventional vineyard (DOCa. Rioja, Spain). *Agronomy, 13*(9), 2199.

Roy, S., Barooah, A. K., Ahmed, K. Z., Baruah, R. D., Prasad, A. K., & Mukhopadhyay, A. (2020). Impact of climate change on tea pest status in northeast India and effective plans for mitigation. *Acta Ecologica Sinica, 40*(6), 432–442.

Sarkar, D., Kar, S. K., Chattopadhyay, A., Rakshit, A., Tripathi, V. K., Dubey, P. K., & Abhilash, P. C. (2020). Low input sustainable agriculture: A viable climate-smart option for boosting food production in a warming world. *Ecological Indicators, 115*, 106412.

Satpathy, S., Gotyal, B. S., & Babu, V. R. (2022). Impact of climate change on insect pest dynamics and its mitigation. In *Conservation agriculture and climate change* (pp. 107–118). CRC Press.
Saxena, P., Singh, A. K., & Gupta, R. (2021). Adverse environment and pest management for sustainable plant production. In *Plant performance under environmental stress: Hormones, biostimulants and sustainable plant growth management* (pp. 535–557). Springer.
Semino, G. (2001). Rodenticides. In *Patty's toxicology* (pp. 1–39). https://doi.org/10.1002/0471125474.tox139
Sharma, G. D., Shah, M. I., Shahzad, U., Jain, M., & Chopra, R. (2021). Exploring the nexus between agriculture and greenhouse gas emissions in BIMSTEC region: The role of renewable energy and human capital as moderators. *Journal of Environmental Management, 297*, 113316.
Sharma, R., Levi, M. R., Ricker, M. C., Thompson, A., King, E. G., & Robertson, K. (2024). Scaling of soil organic carbon in space and time in the Southern Coastal Plain, USA. *Science of the Total Environment, 933*, 173060.
Shi, L., Guo, Y., Ning, J., Lou, S., & Hou, F. (2020). Herbicide applications increase greenhouse gas emissions of alfalfa pasture in the inland arid region of northwest China. *PeerJ, 8*, e9231.
Shi, W., Yang, J., Chen, D., Yin, C., Zhang, H., Xu, X., et al. (2022). The rice blast fungus SR protein 1 regulates alternative splicing with unique mechanisms. *PLoS Pathogens, 18*(12), e1011036.
Silveira, F. D., Machado, F. M., Farias, M. S. D., & Schlosser, J. F. (2022). Fuel consumption by agricultural machinery: A review of pollutant emission control technologies. *Ciência Rural, 53*(5), e20220029.
Sims, L. (2021). Farmer-led change: Addressing environmental and health problems caused by widespread pesticide use in Costa Rica, Nicaragua and Honduras. In *Peer research in health and social development* (pp. 205–218). Routledge.
Singh, T. V. K. (2022). 09. Climate change and its impact on agricultural pests. In *Sustainable agricultural innovations for resilient agri-food systems* (p. 26). The Indian Ecological Society.
Skendžić, S., Zovko, M., Živković, I. P., Lešić, V., & Lemić, D. (2021). The impact of climate change on agricultural insect pests. *Insects, 12*(5), 440.
Solomon, K. R. (2020). Estimated exposure to glyphosate in humans via environmental, occupational, and dietary pathways: An updated review of the scientific literature. *Pest Management Science, 76*(9), 2878–2885.
Squalli, J., & Adamkiewicz, G. (2023). The spatial distribution of agricultural emissions in the United States: The role of organic farming in mitigating climate change. *Journal of Cleaner Production, 414*, 137678.
Srinivasa Rao, M., Mani, M., Prasad, Y. G., Prabhakar, M., Sridhar, V., Vennila, S., & Singh, V. K. (2022). Climate change and pest management strategies in horticultural and agricultural ecosystems. In *Trends in horticultural entomology* (pp. 81–122). Springer.
Straw, E. A., Thompson, L. J., Leadbeater, E., & Brown, M. J. (2022). 'Inert' ingredients are understudied, potentially dangerous to bees and deserve more research attention. *Proceedings of the Royal Society B, 289*(1970), 20212353.
Tahir, R., Afzal, F., Jamil, H., Razzaq, M., & Khan, M. S. (2024). Physiological impacts of pesticidal contamination: Challenge to sustainable agriculture and biodegradation methods. *Pakistan Journal of Agriculture, Agricultural Engineering and Veterinary Sciences, 40*(1), 24–37.
Tıraş, H. H. (2024). Tarım İlaçlarının Sera Gazı Emisyonlarına Etkisi: BRICS Ülkeleri Örneği. *Türk Tarım ve Doğa Bilimleri Dergisi, 11*(1), 130–139.
Tran, T. T. M. (2018). *Impact of trade openness on the environment: An assessment of CO_2 emissions in Vietnam: A thesis submitted in partial fulfilment of the requirements for the degree of Doctor of Philosophy at Lincoln University* [Doctoral dissertation]. Lincoln University.
Udias, A., Galimberti, F., Dorati, C., & Pistocchi, A. (2023). Emissions of pesticides in the European Union: A new regional-level dataset. *Scientific Data, 10*(1), 869.
Ukhurebor, K. E., Aigbe, U. O., Olayinka, A. S., Nwankwo, W., & Emegha, J. O. (2020). Climatic change and pesticides usage: A brief review of their implicative relationship. *AU eJournal of Interdisciplinary Research, 5*(1). ISSN: 2408-1906.

Valujeva, K., Grinfelde, I., & Pilecka-Ulcugaceva, J. (2023). GHG emission variability from soil in conventional farming system. *International Multidisciplinary Scientific GeoConference: SGEM, 23*(4.2), 235–241.

Verma, K., Sharma, P., Kumar, D., Vishwakarma, S. P., & Meena, N. K. (2021). Strategies sustainable management of agroforestry in climate change mitigation and adaptation. *International Journal of Current Microbiology and Applied Sciences, 10*(01), 2439–2449.

Vetter, S., Nayak, D., McBey, D., Dondini, M., Kuhnert, M., & Oyesiku-Blakemore, J. (2023). Environmental issues: Greenhouse gas emissions. In *Sustainable food science: A comprehensive approach* (pp. 216–248). Elsevier.

Wang, Z. X., Hu, X. B., Xu, Z. C., Cai, L. M., Wang, J. N., Zeng, D., & Hong, H. J. (2014). Cadmium in agricultural soils, vegetables and rice and potential health risk in vicinity of Dabaoshan Mine in Shaoguan, China. *Journal of Central South University, 21*, 2004–2010.

Wang, Q., Wei, J., Wan, B., An, Q., Gao, J., & Zhuang, G. (2024). The regulation effect of preventing soil nitrogen loss using microbial quorum sensing inhibitors. *Environmental Research, 246*, 118136.

Wei, S., Bian, Y., Zhao, Q., Chen, S., Mao, J., Song, C., et al. (2017). Salinity-induced palmella formation mechanism in halotolerant algae Dunaliella salina revealed by quantitative proteomics and phosphoproteomics. *Frontiers in Plant Science, 8*, 810.

Wisnujati, N. S. (2023). Unveiling the epoch: Exploring pesticide utilization and trade trends globally and regionally from 1990 to 2020. *SCIENTIA: Journal of Multi Disciplinary Science, 2*(2), 108–121.

Wu, D., Song, Y., Yu, Y., & Dörsch, P. (2024). *Nitrite stimulates HONO and NOx but not N2O emissions in Chinese agricultural soils during nitrification* (No. EGU24-3699). Copernicus Meetings.

Xin, C. H. E. N., Jian-Jun, T. A. N. G., & Zhao-Qian, W. A. N. G. (1999). The impacts of agricultural activities on biodiversity. *Biodiversity Science, 7*(3), 234.

Xu, S., Zhou, S., Xing, L., Shi, P., Shi, W., Zhou, Q., et al. (2019). Fate of organic micropollutants and their biological effects in a drinking water source treated by a field-scale constructed wetland. *Science of the Total Environment, 682*, 756–764.

Xu, B., Zhong, R., Liu, D., & Liu, Y. (2021). Investigating the impact of energy consumption and nitrogen fertilizer on NO x emissions in China based on the environmental Kuznets curve. *Environment, Development and Sustainability, 23*(12), 17590–17605.

Yadav, P. K., Sharma, S., & Sharma, A. (2021). Management trends of rice insect pests in South Asia: A review. *Reviews in Food and Agriculture (RFNA), 2*, 46–53.

Yang, G., Chen, R., Yin, C., Gao, J., Zhang, L., Wu, Y., et al. (2015). Available online at www.sciencedirect.com

Yang, K., Jing, D., Wang, X., Zhou, S., Zhang, B., Qin, L., et al. (2024). Life cycle assessment perspective on waste resource utilization and sustainable development: A case of glyphosate production. *Journal of Environmental Management, 350*, 119584.

Zhang, Y., Wu, L., Jebari, A., & Collins, A. L. (2024a). Impacts of reduced synthetic fertiliser use under current and future climates: Exploration using integrated agroecosystem modelling in the upper River Taw observatory, UK. *Journal of Environmental Management, 351*, 119732.

Zhang, L., Yang, Y., Li, Z., Li, F. M., Huang, J., & Zhang, F. (2024b). Identifying synergistic solutions for the food-energy-water nexus via plastic film mulching cultivation. *Science of the Total Environment, 920*, 171046.

Zhang, J., Wang, F., & Ding, X. (2024c). Can agricultural mechanization promote carbon reduction in countries along the Belt and Road?. *Journal of Environmental Planning and Management*, 1–23. https://doi.org/10.1080/09640568.2024.2311821

Zhang, B., Lv, F., & Yang, J. (2024d). Pesticides toxicity, removal and detoxification in plants: A review. *Agronomy, 14*(6), 1260.

Ecotoxicological Impacts of Pesticide

Ojo Emmanuel Ige, Folorunso Patrick Aliu, Adesuyi Emmanuel Omole, and Oluwaseun Omotayo Alabi

1 Introduction

Pesticides are chemical substances used to prevent, destroy, or control pests that threaten agricultural productivity. They play a crucial role in modern agriculture by protecting crops from insects, weeds, fungi, and other harmful organisms. The use of pesticides has significantly contributed to the increase in agricultural yields and food production, which is essential to feed the growing global population.

The development and widespread adoption of pesticides have revolutionized agricultural practices since the mid-twentieth century. Herbicides, insecticides, and fungicides are among the most commonly used types of pesticides, each targeting specific pests that can cause significant crop losses. For instance, herbicides control unwanted vegetation, insecticides manage insect pests, and fungicides protect plants from fungal infections (Policy Department for External Relations, Directorate General for External Policies of the Union, 2021). Despite their benefits, the extensive use of pesticides has raised concerns about their environmental and health impacts. Non-target organisms, including pollinators like bees, beneficial insects such as ladybugs, and various wildlife species, can be adversely affected by pesticide exposure. The disruption of these organisms can lead to ecological imbalances and the decline of ecosystem services that are vital for sustainable agriculture and biodiversity (Samanta et al., 2023).

O. E. Ige (✉) · O. O. Alabi
Department of Chemistry, School of Physical Sciences, Federal University of Technology, Akure, Ondo State, Nigeria

F. P. Aliu
Bioresource Development Centre, National Biotechnology Research and Development Agency, Abuja, Nigeria

A. E. Omole
Department of Marine Biology, University of Padova, Padua, Italy

B. R. Babaniyi, E. E. Babaniyi (eds.), *The Interplay of Pesticides and Climate Change*, https://doi.org/10.1007/978-3-031-81669-7_9

The persistence and bioaccumulation of certain pesticides in the environment can result in long-term ecological consequences. Pesticide residues are known to contaminate soil, water bodies, and air, impacting a broad spectrum of non-target organisms and contributing to reduced biodiversity (van der Sluijs et al., 2015). Studies have confirmed that such contamination exacerbates the decline in ecosystem health, with significant impacts on species diversity and ecosystem services (Tang et al., 2018). As climate change progresses, it is likely to worsen the effects of pesticides on non-target organisms and ecosystems. Changes in temperature, precipitation patterns, and extreme weather events can alter the behavior, distribution, and vulnerability of both pests and non-target species. Additionally, climate change may influence the degradation rates and distribution patterns of pesticides, potentially increasing their environmental persistence and toxicity (Goulson, 2014).

Pesticides play a crucial role in modern agriculture by safeguarding crops and enhancing productivity. However, the widespread use of these chemicals has raised significant environmental concerns, particularly regarding their impact on non-target organisms. This review seeks to comprehensively assess both the direct and indirect effects of pesticides on vital components of ecosystems, such as pollinators, beneficial insects, and wildlife. In addition, it will explore strategies for mitigating these impacts to preserve biodiversity and maintain essential ecosystem services.

The primary aim of this review is to thoroughly assess the ecotoxicological effects of pesticides on non-target organisms. Specifically, it seeks to understand both the direct and indirect impacts of pesticides on crucial groups such as pollinators, beneficial insects, and wildlife. By integrating current research findings, we aim to reveal the mechanisms through which pesticides affect these organisms and their broader ecological implications.

Direct effects of pesticides are critical to consider, as these chemicals can immediately harm non-target species. Such effects include acute toxicity, which may cause death, as well as sublethal impacts that disrupt reproduction and development. Additionally, pesticides can lead to behavioral changes that interfere with feeding and mating patterns (Samanta et al., 2023). Pollinators like bees are vital for crop pollination, and their decline due to pesticide exposure can significantly impact agricultural productivity and biodiversity (van der Sluijs et al., 2015). The indirect effects of pesticides also pose serious threats to ecosystems and ecological interactions. These effects can include disruptions in food web dynamics, habitat degradation, and a decrease in resource availability for non-target species. For instance, the reduction in insect populations caused by pesticide use can impact birds and other wildlife that depend on insects as a primary food source, leading to cascading ecological consequences (Goulson, 2014).

Furthermore, this chapter considers how climate change might exacerbate the effects of pesticides on non-target organisms. Changes in temperature, precipitation patterns, and the frequency of extreme weather events can influence the distribution, persistence, and toxicity of pesticides in the environment. Understanding these interactions is critical for developing adaptive management strategies to mitigate the adverse effects of pesticides in a changing climate (Bolan et al., 2024).

2 Pesticide Regulations

2.1 Review of Current Regulations and Suggested Improvements

Current regulations governing pesticide use vary widely across different regions and countries, aiming to balance the benefits of pest control with the need to protect human health and the environment. Regulatory frameworks such as the United States Environmental Protection Agency (EPA), the European Union's Registration, Evaluation, Authorization, and Restriction of Chemicals (REACH), and Nigeria's National Agency for Food and Drug Administration and Control (NAFDAC) set standards for the approval, usage, and monitoring of pesticides (EPA, 2021; European Commission, 2021).

These regulations generally involve rigorous testing of pesticides for their potential toxicological effects on non-target organisms before approval. This includes evaluating acute toxicity, long-term exposure effects, and environmental persistence. Despite these measures, many non-target species continue to suffer from pesticide exposure due to issues such as improper application, non-compliance with guidelines, and cumulative impacts from multiple chemicals (Ahmad et al., 2024).

One major issue with current regulations is the focus on individual pesticide effects rather than the cumulative and synergistic effects of multiple chemicals. Furthermore, existing regulations may not fully address the sublethal and chronic impacts on non-target organisms, which can be as detrimental as acute toxicity. There is also a need for enhanced monitoring and enforcement to ensure adherence to regulations and mitigate adverse effects (Goulson, 2014).

To better protect non-target organisms, several improvements can be suggested:

(i) *Enhanced testing protocols*: Regulatory bodies should adopt more comprehensive testing protocols that evaluate the long-term, sublethal, and synergistic effects of pesticide exposure on a wider range of non-target organisms.
(ii) *Integrated pest management (IPM) promotion*: It encourages the adoption of IPM practices that reduce reliance on chemical pesticides through the use of biological controls, crop rotation, and other sustainable agricultural practices.
(iii) *Stricter enforcement and monitoring*: It improves the enforcement of existing regulations and increase monitoring efforts to ensure compliance and to gather data on pesticide impacts in real-world conditions.
(iv) *Climate change considerations*: Climate change considerations should be integrated into pesticide regulation frameworks to account for how shifting environmental conditions may influence pesticide behavior and their resulting impacts. As temperature, precipitation patterns, and other climatic factors evolve, they can modify the efficacy, degradation, and toxicity of pesticides, necessitating adaptive regulatory measures (Motta et al., 2020).
(v) *Public awareness and education*: It increases public awareness and education about the proper use and risks of pesticides, promoting safer practices among farmers and gardeners.

3 Pollinators and Their Role in Ecosystems

Pollinators, which include a wide variety of insects, birds, and other animals, are essential to the functioning of both natural ecosystems and agricultural systems worldwide. Their primary role is to facilitate the reproduction of flowering plants by transferring pollen between flowers, enabling fertilization and the production of seeds and fruit. This process, known as pollination, supports the reproduction of over 75% of the world's flowering plants and contributes to approximately 35% of global food crops (Klein et al., 2007; IPBES, 2016). In natural ecosystems, pollinators promote genetic diversity and enhance the resilience of plant populations, which supports biodiversity and stabilizes ecosystems (Winfree et al., 2011). In agriculture, pollinators are critical for increasing both the quantity and quality of crops such as fruits, vegetables, and nuts (Gallai et al., 2009). Their economic value is significant, with pollinators contributing billions of dollars annually to global crop production (Breeze et al., 2014).

Despite their vital role, pollinators are increasingly threatened by anthropogenic factors such as habitat destruction, climate change, and widespread pesticide use (Goulson et al., 2015). Pesticides, while essential in modern agriculture to control pests and increase crop yields, can inadvertently harm non-target species, including pollinators (Pisa et al., 2015). Studies have confirmed that pesticide exposure, combined with habitat loss and climate change, poses a multifaceted threat to pollinator populations, with significant consequences for global food security (Dicks et al., 2019; Sánchez-Bayo & Wyckhuys, 2019).

The complex interdependence between pollinators, plant reproduction, and food production highlights the urgent need to address these threats. Effective conservation strategies must be implemented to preserve these invaluable ecosystem services and ensure the sustainability of both natural and agricultural landscapes.

4 Pesticide Use and Exposure Pathways in Agriculture

4.1 Types of Pesticides Used in Agriculture

Pesticides used in modern agriculture can be broadly classified into several categories, each aimed at controlling specific types of pests or pathogens:

Insecticides Insecticides are designed to control insect pests that cause damage to crops. Common classes include organophosphates, carbamates, pyrethroids, neonicotinoids, and organochlorines (Goulson, 2013). Neonicotinoids, in particular, have been the subject of significant scrutiny due to their systemic properties and potential negative impacts on pollinators (Simon-Delso et al., 2015). Recent research continues to highlight their role in pollinator decline, especially in combination with other stressors like habitat loss and climate change (Wood & Goulson, 2017; Tsvetkov et al., 2017).

Herbicides Herbicides are used to eliminate or control weeds that compete with crops for essential resources like nutrients, water, and light. Glyphosate, one of the most widely used herbicides globally, has faced increasing concern over its environmental and health impacts (Benbrook, 2016). Recent studies emphasize the need for careful management of herbicides to prevent unintended consequences on non-target organisms, including soil microbiota and pollinators (Kniss, 2018).

Fungicides Fungicides are chemicals commonly applied to crops to manage or prevent fungal infections that negatively affect both yield and quality. These compounds exhibit a broad spectrum of modes of action and vary in terms of their environmental persistence, which has important implications for both agricultural efficiency and ecological impact (Zubrod et al., 2019).

Rodenticides and Other Pesticides Although rodenticides are primarily used to control rodents that pose a threat to agricultural productivity, they can also affect non-target species, including wildlife and birds of prey, through secondary poisoning (Rattner et al., 2017). The broader implications of these chemicals on ecosystems are increasingly recognized, calling for more sustainable pest management practices (Elliott et al., 2020).

4.2 Exposure Pathways of Pollinators to Pesticides

Pollinators can encounter pesticides through several pathways, which can be direct or indirect.

Direct Contact Pollinators may experience direct exposure to pesticides through contact with sprays applied to crops or nearby vegetation. This can happen during pesticide application or when pollinators land on surfaces treated with these chemicals (Bortolotti et al., 2017).

Ingestion of Contaminated Nectar and Pollen As pollinators feed on nectar and pollen, they may ingest pesticide residues present in these floral resources. Systemic pesticides, such as neonicotinoids, can be absorbed by plants and subsequently transfer to floral tissues, where they are accessible to pollinators (Botías et al., 2015; Wood & Goulson, 2018). This route of exposure is particularly concerning due to the high likelihood of residue accumulation in these critical resources.

Water Contamination Pesticides can also enter water bodies through runoff or leaching, contaminating puddles and other water sources that pollinators use for drinking. This indirect exposure can be significant, especially in agricultural areas where pesticide runoff from treated fields can impact adjacent aquatic habitats. Such contamination not only affects pollinators directly but can also have broader ecological consequences, disrupting entire aquatic ecosystems (Goulson et al., 2021; Woodcock et al., 2019).

Residual Effects Many pesticides persist in the environment for extended periods, posing a risk to pollinators long after their initial application. Residues may remain on plant surfaces, in soil, or in plant exudates, continuing to affect pollinators and other non-target organisms (Goulson, 2013; Tsvetkov et al., 2017).

These exposure pathways is essential for evaluating the risks pesticides pose to pollinators and for developing effective strategies to mitigate these risks and protect these crucial organisms.

4.2.1 Impact on Bees

Pesticides, particularly insecticides like neonicotinoids and pyrethroids, have been linked to significant mortality in bee populations.

Acute Toxicity Studies Research has shown that exposure to sublethal doses of neonicotinoids can impair bee health and increase mortality rates. For instance, studies have found that even low concentrations of neonicotinoids can cause immediate mortality in bees, affecting both honeybees (*Apis mellifera*) and wild bee species (Potts et al., 2010; Rundlöf et al., 2015).

Field Studies Field observations and experimental studies have demonstrated higher mortality rates among bees in areas where systemic pesticides like neonicotinoids are used extensively. This includes not only direct exposure during foraging but also through contaminated pollen and nectar brought back to the hive (Sanchez-Bayo & Goka, 2014; Woodcock et al., 2017).

Reproductive Effects

Pesticides can also have profound effects on bee reproduction, impacting the queen fertility, colony growth, and overall reproductive success:

Queen Fertility Research indicates that exposure to neonicotinoids and other systemic pesticides can reduce queen bee fertility and reproductive output. This is critical because queens are essential for colony sustainability and growth (Whitehorn et al., 2012; Laycock et al., 2014).

Colony Development Pesticides can disrupt the development of bee larvae and affect the health and longevity of adult bees within the colony. Chronic exposure to sublethal doses of pesticides may weaken colony resilience to other stressors, such as pathogens and parasites (Henry et al., 2012; Tsvetkov et al., 2017).

Behavioral Changes

Pesticides can alter various aspects of bee behavior, which are crucial for colony survival and foraging efficiency.

Foraging Patterns Bees exposed to pesticides may exhibit altered foraging behaviors, including reduced pollen collection and altered preferences for floral resources. This can impact the nutritional intake of the colony and reduce overall productivity (Cutler & Scott-Dupree, 2007; Tsvetkov et al., 2017).

Navigation Abilities Neonicotinoids, known for their neurotoxic effects, can impair bee navigation abilities, affecting their ability to find and return to the hive. This can lead to increased energy expenditure and decreased foraging efficiency (Fischer et al., 2014; Stanley et al., 2015).

Communication Within Colonies Pesticides can disrupt communication systems within bee colonies, such as pheromone signaling, and waggle dances used to communicate food sources. This disruption can impair colony cohesion and coordination, further compromising their ability to respond to environmental challenges (Guez et al., 2001; Wu-Smart & Spivak, 2016).

Understanding these impacts on bee mortality, reproduction, and behavior underscores the complex interactions between pesticides and bee health. Mitigation strategies should focus on reducing pesticide exposure in agricultural landscapes while promoting pollinator-friendly practices.

4.2.2 Direct Effects of Pesticides on Pollinators

Pesticides, particularly neonicotinoids and other systemic insecticides, have profound direct effects on pollinators such as bees, butterflies, and other insects crucial for pollination. These impacts manifest in increased mortality rates, reproductive challenges, and significant behavioral alterations, ultimately affecting pollinator populations and the vital ecosystem services they offer.

Mortality

Neonicotinoids are highly toxic to bees and other pollinators. Numerous studies have shown that exposure to these chemicals leads to increased mortality. For instance, Tsvetkov et al. (2017) found that chronic exposure to neonicotinoids near corn crops significantly reduced honeybee colony health and survival. Similarly, Rundlöf et al. (2015) reported that wild bee populations experienced increased mortality and reduced reproductive success when exposed to neonicotinoid-coated seeds.

Reproductive Effects

The reproductive health of pollinators is adversely affected by pesticide exposure. Neonicotinoids and other pesticides can impair queen production and reduce colony growth. Whitehorn et al. (2012) demonstrated that exposure to neonicotinoids led to a significant reduction in bumblebee colony growth and queen production. Wu-Smart and Spivak (2016) also showed that dietary exposure to neonicotinoids resulted in sublethal effects on honeybee queen fecundity, impacting overall colony development and productivity.

Behavioral Changes
Behavioral disruptions due to pesticide exposure are another critical concern. These changes can affect foraging efficiency, navigation, and social interactions among pollinators. Fischer et al. (2014) found that neonicotinoids interfere with specific components of honeybee navigation, making it challenging for them to return to their hives. Tsvetkov et al. (2017) discovered that neonicotinoid exposure disrupted bumblebee nest behavior, social networks, and thermoregulation, which are vital for colony health and productivity.

Moreover, Henry et al. (2012) reported that a common pesticide decreased foraging success and survival in honeybees, highlighting how even sublethal doses can lead to significant behavioral impairments. These behavioral changes reduce the effectiveness of pollination services provided by these insects, as evidenced by Stanley et al. (2015), who found that neonicotinoid exposure impaired crop pollination services provided by bumblebees.

The direct effects of pesticides on pollinators, including increased mortality, reproductive challenges, and behavioral changes, underscore the critical need for sustainable agricultural practices. The continued use of harmful pesticides poses a significant threat to pollinator populations and the essential ecosystem services they provide, such as pollination. The findings from various studies highlight the urgency for policymakers and stakeholders to implement practices that mitigate these risks to ensure the sustainability of pollinator populations and the broader ecosystem.

4.2.3 Effects on Predators and Parasitoids—Beneficial Insect

Beneficial insects, including predators and parasitoids, play a crucial role in controlling pest populations in agricultural and natural ecosystems. However, the widespread use of pesticides, particularly systemic insecticides like neonicotinoids, has profound negative impacts on these beneficial species.

Predators
Predatory insects such as ladybugs, lacewings, and ground beetles are essential for maintaining the balance of pest populations. Pesticides can reduce their populations directly through acute toxicity and indirectly by diminishing their prey base. Desneux et al. (2007) documented that sublethal exposure to pesticides can impair the hunting and reproductive behaviors of predatory insects, reducing their effectiveness in pest control. Moreover, the reduction in prey due to pesticide applications can lead to food scarcity for predators, further diminishing their populations.

Pesticide applications can lead to significant declines in natural predator populations, which may result in pest resurgence due to the absence of natural control mechanisms. This phenomenon, as highlighted by Zhang et al. (2020), underscores the importance of maintaining healthy populations of predatory insects for sustainable pest management.

Parasitoids

Parasitoids, such as certain wasps and flies, are another group of beneficial insects adversely affected by pesticides. These insects lay their eggs in or on other insects, with the developing larvae consuming the host. Pesticides can directly kill adult parasitoids or disrupt their ability to locate and parasitize hosts.

Research by Desneux et al. (2007) revealed that sublethal doses of pesticides could affect the olfactory and locomotors functions of parasitoids, hindering their ability to find hosts. Additionally, the study by Beketov et al. (2013) found that pesticide exposure reduces the regional biodiversity of stream invertebrates, including many parasitoid species. The loss of parasitoid biodiversity can lead to an imbalance in ecosystem functions and increased pest outbreaks.

4.2.4 Wildlife: Impacts on Birds, Mammals, Amphibians, and Other Species

Pesticides exert a wide range of effects on wildlife, including birds, mammals, amphibians, reptiles, and invertebrates. The impacts range from acute toxicity, which can lead to immediate mortality, to sublethal effects that impair key biological functions such as reproduction, growth, and behavior. These consequences not only affect individual species but also destabilize entire ecosystems.

Birds

Birds are particularly susceptible to pesticide exposure through various pathways, including ingestion of pesticide-treated seeds, consumption of contaminated prey, and drinking water containing pesticide residues. A study by Hallmann et al. (2014) identified a strong link between neonicotinoid concentrations and declining populations of insectivorous birds, noting that the reduction in their insect prey is a major factor. Additionally, Eng et al. (2019) demonstrated that neonicotinoid exposure not only causes direct toxicity in birds but also reduces reproductive success by impairing egg production and chick survival.

Mammals

Mammals, ranging from small rodents to large herbivores, are affected by pesticide contamination in their diet and water sources. Acute exposure can lead to death, while sublethal exposure may result in a variety of negative health outcomes such as impaired growth, fertility, and immune function. Kaur et al. (2024) and Shore et al. (2017) emphasized that even low-level exposure to pesticides can disrupt mammalian reproductive and behavioral systems, leading to population declines. Moreover, mammalian predators consuming contaminated prey may face bioaccumulation, further amplifying the toxic effects through the food chain.

Amphibians

Amphibians are especially vulnerable to pesticide exposure due to their permeable skin and aquatic habitats, which make them more prone to absorbing toxic chemicals from their environment. Acute toxicity can result in large-scale amphibian mortality, while sublethal effects such as developmental deformities, immune suppression, and endocrine disruption are also prevalent. Beketov et al. (2013) documented significant amphibian population declines in pesticide-polluted areas, with far-reaching consequences for ecosystem balance as these species play crucial roles as both predators and prey.

Other Wildlife

Other wildlife, including reptiles, invertebrates, and small vertebrates, also face significant risks from pesticide exposure. Acute toxicity leads to direct mortality, while sublethal impacts may result in impaired growth, reproduction, and disrupted behavioral patterns. The broad-spectrum nature of many pesticides, particularly organophosphates and neonicotinoids, inadvertently affects non-target species, reducing biodiversity and disrupting ecosystem services such as pollination and natural pest control (Sánchez-Bayo & Wyckhuys, 2019).

In summary, pesticides pose profound risks to wildlife, both through acute and sublethal effects. The resulting population declines of birds, mammals, amphibians, and other species disrupt ecosystem services and biodiversity. These outcomes highlight the urgent need for integrated pest management practices that minimize harm to non-target organisms while ensuring ecological sustainability.

5 Indirect Effects of Pesticides

5.1 Food Web Disruption

Pesticides, while effective at controlling target pest populations, can have significant unintended consequences on non-target organisms, leading to substantial disruptions in food webs. A food web represents the complex network of feeding relationships within an ecosystem, where organisms are interconnected through predator-prey dynamics and other ecological interactions. When pesticide exposure reduces populations of certain organisms, these disturbances can cascade through the food web, affecting multiple trophic levels and threatening the overall stability of ecosystems.

One of the primary ways pesticides disrupt food webs is by reducing populations of herbivorous insects, which are essential primary consumers. These insects play a critical role in transferring energy from plants to higher trophic levels, including insectivorous birds, mammals, and other predators. A decline in herbivorous insect populations due to pesticide exposure results in a reduced food supply for predators, negatively affecting their reproductive success, survival rates, and overall population numbers (Benton et al., 2002; Sánchez-Bayo & Wyckhuys, 2019).

Neonicotinoid insecticides, widely used in agriculture, are a key example of this disruption, as they cause direct and indirect harm to bee populations. Bees, in addition to being vital pollinators, are an important food source for many insectivorous species. As bee populations decline, both plant pollination and the availability of prey for predators are impacted, leading to broader ecological consequences (Goulson, 2013). Other pollinators and beneficial insects, such as butterflies and beetles, are also affected, further exacerbating ecological disruptions (Wood & Goulson, 2017).

Pesticides can also build up within primary consumers and be passed up the food chain through increasing concentrations. Predators that feed on contaminated prey may end up with higher levels of pesticides in their tissues, leading to toxic effects. Birds of prey and carnivorous mammals that consume pesticide-contaminated insects or small vertebrates may experience reproductive failures, neurological impairments, and higher mortality rates (Hallmann et al., 2014).

Aquatic ecosystems are particularly vulnerable to pesticide-induced food web disruptions. Agricultural runoff carries pesticides into water bodies, affecting aquatic invertebrates, which are critical food sources for fish and amphibians. The reduction in aquatic invertebrate populations can lead to declines in fish populations, which in turn affects piscivorous birds and mammals. For example, studies have shown that insecticides such as pyrethroids significantly reduce populations of aquatic insects, leading to decreased growth and survival rates in fish species that depend on these insects for food (Beketov et al., 2013; Morrissey et al., 2020).

Moreover, the indirect effects of pesticides on food webs extend beyond direct trophic interactions. Sublethal effects, such as impaired foraging behavior, reduced fecundity, and weakened immune responses, can diminish the fitness of non-target organisms and disrupt their ecological roles. These physiological and behavioral impairments can lead to shifts in species composition and abundance, further destabilizing food webs (Desneux et al., 2007; Pestana et al., 2019).

The indirect effects of pesticides on food webs are profound and far-reaching. By eliminating or reducing populations of primary consumers and other non-target organisms, pesticides disrupt the delicate balance of ecological interactions within ecosystems. This disruption can have cascading effects across multiple trophic levels, ultimately threatening biodiversity and ecosystem stability. Understanding and mitigating these indirect effects are crucial for developing sustainable agricultural practices and safeguarding ecosystem health.

5.2 Habitat Alteration

Pesticide use can significantly alter habitats, creating indirect effects that ripple through ecosystems and impact non-target species. When pesticides are applied, they often disperse beyond the target area, contaminating surrounding environments through air, water, and soil. This contamination results in habitat changes that indirectly affect a wide range of non-target species.

One significant way pesticides alter habitats is by reducing plant diversity and abundance. Herbicides, in particular, drastically decrease the variety of plant species in an area. This reduction in plant diversity affects the entire food web, as plants form the base of most terrestrial ecosystems. As plant diversity decreases, species dependent on a variety of plants for food and habitat, such as insects, birds, and small mammals, also decline, triggering a cascade of effects that reduce predator populations higher up the food chain (Desneux et al., 2007). Recent studies have shown that herbicide drift onto non-target plants can have lasting impacts on biodiversity in agricultural landscapes (Buhler et al., 2020).

Aquatic ecosystems are equally vulnerable. Pesticides that run off into water bodies harm aquatic plants and algae, essential for maintaining the stability of these environments. The loss of these primary producers decreases oxygen levels in the water, which can harm fish and other aquatic organisms. Moreover, the decline in plant and algal populations disrupts aquatic food webs, reducing species diversity and abundance (Beketov et al., 2013). For instance, pyrethroid insecticides have been shown to cause significant damage to aquatic plants and invertebrates, with repercussions for entire aquatic ecosystems (Morrissey et al., 2020).

Pesticides also alter soil composition and health, impacting essential soil organisms such as earthworms, bacteria, and fungi, which are critical for nutrient cycling and soil structure. Disruption of these organisms leads to decreased soil fertility and compromised structure, which in turn affects plant growth and ecosystem health (Silva et al., 2021). Degraded soils hinder plant establishment and growth, further diminishing habitat quality for numerous species. Additionally, soil-dwelling organisms like nematodes and mycorrhizal fungi, essential for maintaining soil health, are highly sensitive to pesticide exposure, potentially leading to long-term soil degradation.

The alteration of habitats through pesticide use also exacerbates the effects of other environmental stressors. For example, climate change is already placing significant stress on many ecosystems. When coupled with habitat degradation caused by pesticides, the resilience of ecosystems is further compromised. Species that might have adapted to changing temperatures or precipitation patterns may find it more difficult to survive when their habitats are simultaneously degraded by pesticide use (Goulson et al., 2015). This compounded impact of pesticide use and climate change is being increasingly recognized in recent studies as a key factor driving biodiversity loss (Sánchez-Bayo & Wyckhuys, 2019).

The indirect effects of pesticides on non-target species through habitat alteration are profound and multifaceted. Pesticides reduce plant diversity, disrupt aquatic ecosystems, and degrade soil health, causing cascading effects throughout food webs. These changes are compounded by other environmental stressors, such as climate change, highlighting the need to consider broader ecological impacts. To protect biodiversity and maintain healthy ecosystems, implementing sustainable agricultural practices and developing strategies to mitigate the adverse effects of pesticides are crucial.

Pesticides have far-reaching implications not only for targeted pests but also for broader ecosystem services, which are vital for maintaining ecological balance and supporting agriculture. Essential ecosystem services such as pollination, pest control, and soil health are crucial for the productivity and sustainability of agricultural systems. However, pesticide use disrupts these services, diminishing the resilience of ecosystems and their ability to support human agriculture (Garibaldi et al., 2018).

5.3 *Pollination*

Pollinators are crucial to the reproduction of numerous plants, particularly crops. These species play a pivotal role in maintaining biodiversity and ensuring food security. However, the widespread use of pesticides, especially neonicotinoids, has been shown to negatively impact pollinator populations. Exposure to neonicotinoids impairs key behaviors in bees, such as foraging efficiency, navigation, and overall colony health (Goulson et al., 2015). Honeybees and wild bee populations are particularly susceptible to these effects. The decline in pollinator numbers due to pesticide exposure has a direct impact on pollination services, leading to reduced crop yields and a loss of biodiversity. Klein et al. (2007) emphasized the critical role of pollinators in the productivity of 87 leading global food crops, underscoring their economic and ecological importance. The reduction in pollination services due to pesticide-induced population declines could, therefore, have far-reaching consequences for food security and ecosystem health (Woodcock et al., 2017).

Recent studies have further illustrated how systemic pesticides not only affect bees but also other important pollinators like butterflies and hoverflies, contributing to their population declines. These trends have raised significant concerns about the long-term sustainability of agricultural systems dependent on pollinators (Siviter & Muth, 2020).

5.4 *Pest Control*

Natural pest control is a critical ecosystem service provided by predatory insects, birds, and other wildlife that helps regulate pest populations, reducing the need for chemical interventions. However, many pesticides are non-selective, killing both harmful pests and beneficial predators. The decline of beneficial insect populations, such as ladybugs, lacewings, and predatory beetles, due to pesticide exposure, has caused an imbalance in pest populations (Desneux et al., 2007). This reduction in natural predators results in more frequent pest outbreaks, leading farmers to become increasingly dependent on chemical pesticides, creating a self-perpetuating cycle where pesticide overuse exacerbates pest issues rather than solving them.

Over the past decade, the widespread application of pesticides to target a variety of pests has led to significant challenges. Target pests have adapted in various ways, such as migrating to new environments or evolving through genetic mutations and changes in population growth rates (Gill & Garg, 2014). As a result, an increasing number of pesticide-resistant species have emerged. Resistant individuals, once rare in a population, are now proliferating as susceptible pests are eradicated by chemical treatments. This natural selection process allows resistant pests to dominate, rendering previously effective pesticides useless (Bass et al., 2015).

The issue of pesticide resistance continues to grow. Intensive pesticide use has led to resistance in numerous insect and mite species worldwide. Resistance has been documented across various insecticide classes, including cyclodienes, DDT, organophosphates, carbamates, pyrethroids, and fumigants. Agricultural pests, livestock parasites, urban pests, and disease vectors have all evolved resistance, making their control increasingly challenging (Sparks & Nauen, 2015). In the case of *Helicoverpa armigera*, one of the world's most destructive agricultural pests, widespread resistance has rendered many conventional insecticides ineffective, making integrated pest management strategies increasingly important (Dewar, 2020).

Several factors contribute to pesticide resistance, including genetic variation, biological interactions, and control methods. Advances in resistance detection, using biochemical and molecular techniques such as enzyme assays, PCR, and enzyme electrophoresis, have improved our ability to monitor and manage resistant pest populations (Gill & Garg, 2014). These developments are essential for preventing the further spread of resistance and ensuring the continued effectiveness of pest control measures.

5.5 Soil Health

Soil health is foundational to plant growth, water filtration, and carbon sequestration. Soil biodiversity, including bacteria, fungi, earthworms, and other invertebrates, plays a crucial role in nutrient cycling and soil structure formation. Earthworms, in particular, are essential for enhancing soil fertility through the breakdown of organic matter into humus. Their activities promote soil aeration, drainage, and nutrient availability. However, earthworm populations are highly sensitive to agricultural practices, particularly the indiscriminate use of pesticides, which make them valuable indicators of soil quality in agricultural ecosystems (Lavelle et al., 2019).

Pesticides, especially carbamate insecticides, have been shown to be highly toxic to earthworms, and organophosphates can also reduce their populations. Pesticide exposure negatively impacts earthworm growth, reproduction, and cocoon production, affecting the number of hatchlings and their development. Earthworms exposed to pesticides often show physical damage such as ruptured cuticles, swelling, and tissue softening. In more severe cases, pesticide exposure can cause neurotoxic and physiological damage, including cellular dysfunction and protein degradation, as observed in studies from France (Gill & Garg, 2014).

The decline in soil biodiversity caused by pesticide use impairs essential ecosystem functions, reducing soil fertility and degrading soil structure (Pimentel, 2005). Earthworms, for instance, are highly sensitive to pesticide exposure, and their decline compromises their role in enhancing soil aeration and nutrient availability. As a result, agricultural productivity suffers, with poorer soil quality directly affecting crop yields (Pelosi et al., 2021). This reduction in soil health underscores the importance of adopting sustainable soil management practices that minimize the adverse effects of pesticides on soil organisms.

Recent studies have highlighted the critical need for integrated pest management (IPM) approaches that reduce the dependency on chemical pesticides while preserving soil biodiversity (Aktar et al., 2020). These strategies promote sustainable agricultural practices that maintain soil health, ensuring that the vital ecological services provided by soil organisms are protected for long-term agricultural productivity.

Broader Impacts
The disruption of these ecosystem services has cascading effects on the environment and human well-being. Reduced pollination can lead to lower crop yields and decreased availability of fruits, vegetables, and nuts, impacting food security and nutrition. The decline in natural pest control can increase agricultural costs and chemical use, posing additional environmental and health risks. Furthermore, compromised soil health can lead to erosion, decreased water retention, and lower resilience to climate change, affecting long-term agricultural sustainability.

In conclusion, the broader impact of pesticides on ecosystem services is profound and multifaceted. Pollination, pest control, and soil health are all critical services that support agricultural productivity and ecological balance. Disruption of these services through pesticide use not only threatens biodiversity but also undermines the foundations of sustainable agriculture and food security. To mitigate these impacts, it is essential to adopt more sustainable pest management practices, such as integrated pest management (IPM), and to reduce reliance on harmful chemical pesticides. Promoting organic farming practices and enhancing habitat diversity can also help support and restore these vital ecosystem services.

6 Impact of Climate Change

6.1 Exacerbation of Effects

Climate change can significantly exacerbate the impacts of pesticides on non-target organisms. Rising temperatures can increase the volatility and degradation rates of pesticides, leading to more frequent applications and higher overall usage. This heightened exposure can amplify the toxicity effects on pollinators, beneficial insects, and wildlife. Additionally, altered precipitation patterns, such as more intense and irregular rainfall, can cause runoff that carries pesticides into new areas, contaminating habitats and water sources that were previously unaffected. The combination of increased pesticide uses and changing environmental conditions can intensify the negative impacts on biodiversity and ecosystem services (Singh et al., 2020).

6.2 Species Vulnerability

Certain species are more vulnerable to the combined stressors of pesticides and climate change. For instance, pollinators such as bees and butterflies, which are already declining due to pesticide exposure, may face additional stress from habitat changes and resource scarcity caused by climate change. Amphibians, which are highly sensitive to both pollutants and temperature fluctuations, may also experience heightened vulnerability. These species, already on the brink due to pesticide exposure, could see accelerated declines as climate change further disrupts their habitats and life cycles.

6.3 Geographical Variations

The impact of climate change and pesticide use will vary geographically. Regions with hotter climates may experience faster degradation of pesticides, leading to more frequent applications and higher risks of exposure for non-target species. Conversely, areas with increased rainfall may suffer from greater runoff and water contamination, affecting aquatic ecosystems. Additionally, regions with significant agricultural activities may face compounded effects, as both climate change and heavy pesticide use can synergistically harm local biodiversity. Understanding these geographical variations is crucial for developing targeted mitigation strategies that address the specific challenges faced by different regions.

7 Case Studies

Successful Mitigation Case studies of successful mitigation strategies offer valuable lessons in reducing pesticide impacts while maintaining agricultural productivity. Integrated pest management (IPM) strategies have proven particularly effective. In Europe, for example, IPM practices have been widely adopted, combining biological control methods, crop rotation, and resistant crop varieties to minimize reliance on chemical pesticides. A notable case is Denmark, where IPM implementation led to a significant reduction in pesticide use, resulting in lower environmental contamination and improved biodiversity (Lamichhane et al., 2017). Similarly, the adoption of buffer zones and vegetative strips around agricultural fields in the United States has successfully reduced pesticide runoff into water bodies, protecting aquatic ecosystems and water quality (Zhang et al., 2017).

In Nigeria, there is growing interest in IPM practices to address pesticide overuse. Recent initiatives emphasize biological control and cultural practices to manage pests while promoting environmental sustainability. For instance, the Nigerian government has been encouraging the adoption of IPM techniques, such as the use of natural predators and non-chemical alternatives, to improve agricultural practices and reduce pesticide dependence (Adedire et al., 2021). These efforts have shown promise in protecting local biodiversity and enhancing food security.

8 Lessons Learned

Insights from both successful and unsuccessful mitigation efforts provide valuable guidance for future strategies. Successful cases, such as Denmark's IPM program, highlight the importance of a holistic approach involving regulatory support, farmer education, and community engagement. Government backing and incentives were crucial in fostering the adoption of sustainable practices. In Nigeria, there have been efforts to implement similar integrated pest management strategies, though challenges remain due to limited farmer education and resources. Some regions have made progress by incorporating indigenous knowledge with modern pest control practices, but a lack of widespread regulatory enforcement and infrastructure still allows pesticide misuse and runoff into waterways, impacting both local biodiversity and human health.

Conversely, in regions where mitigation has failed, such as certain areas of Nigeria, non-compliance with buffer zone regulations—due to insufficient monitoring and a lack of awareness—has resulted in continued pesticide contamination of ecosystems. These lessons highlight the need for better regulatory frameworks, robust monitoring systems, continuous farmer education, and locally adapted strategies that address specific environmental and agricultural conditions. By learning from both global and local contexts, future actions can better address the complexities of pesticide impacts, particularly in the face of climate change.

9 Conclusion

This review has highlighted the profound ecotoxicological effects that pesticides exert on non-target organisms, with a particular emphasis on pollinators, beneficial insects, and wildlife. Pesticides, particularly neonicotinoids, have been shown to significantly impact pollinator populations, leading to biodiversity loss and disruptions in vital ecosystem services such as pollination and natural pest control. These effects are compounded by the pressures of climate change, which alters environmental conditions and heightens the vulnerability of species already at risk from pesticide exposure.

Although mitigation strategies like integrated pest management (IPM) and enhanced regulatory frameworks offer promise, their incomplete implementation and weak enforcement continue to pose significant barriers to progress. To address these challenges, research must increasingly focus on understanding the combined impacts of climate change and pesticide exposure, not only at the molecular level but also across entire ecosystems. Long-term, field-based studies are critical for assessing the cumulative and cascading effects of these stressors on biodiversity and ecological resilience.

Policymakers must take decisive steps to strengthen pesticide regulations, improve monitoring and enforcement practices, and promote sustainable agricultural techniques that minimize chemical dependence. A coordinated approach involving scientists, governments, farmers, and the public is crucial for developing and implementing solutions that protect both biodiversity and food security.

Protecting non-target organisms and maintaining biodiversity are essential for sustaining healthy ecosystems, which, in turn, underpin human well-being. Immediate, concerted action is required to mitigate the harmful effects of pesticides and to adapt to the accelerating impacts of climate change. By prioritizing cutting-edge scientific research, enacting robust policy reforms, and embracing sustainable agricultural practices, we can safeguard the delicate balance of ecosystems and protect the vast array of species that depend on them. It is a shared responsibility to build a more resilient and sustainable future, ensuring the well-being of ecosystems for generations to come.

References

Adedire, C. O., Fasoranti, J. O., & Olaifa, J. I. (2021). Adoption of integrated pest management practices in Nigeria: Challenges and prospects. *Journal of Sustainable Agriculture and Environmental Sciences, 12*(3), 245–258.

Ahmad, M. F., Ahmad, F. A., Alsayegh, A. A., Zeyaullah, M., AlShahrani, A. M., Muzammil, K., Saati, A. A., Wahab, S., Elbendary, E. Y., Kambal, N., Abdelrahman, M. H., & Hussain, S. (2024). Impact of pesticides on human health and the environment: Mechanisms of action and possible countermeasures. *Heliyon, 10*, 29128. https://doi.org/10.1016/j.heliyon.2024.e29128

Aktar, W., Sengupta, D., & Chowdhury, A. (2020). Impact of pesticides use in agriculture: Their benefits and hazards. *Interdisciplinary Toxicology, 2*(1), 1–12. https://doi.org/10.2478/v10102-009-0001-7

Bass, C., Puinean, A. M., Zimmer, C. T., Denholm, I., Field, L. M., Foster, S. P., et al. (2015). Insecticide resistance mechanisms and their management in Helicoverpa armigera populations. *Current Opinion in Insect Science, 4*, 71–77. https://doi.org/10.1016/j.cois.2015.03.010

Beketov, M. A., Kefford, B. J., Schäfer, R. B., & Liess, M. (2013). Pesticides reduce regional biodiversity of stream invertebrates. *Proceedings of the National Academy of Sciences, 110*(27), 11039–11043. https://doi.org/10.1073/pnas.1305618110

Benbrook, C. M. (2016). Trends in glyphosate herbicide use in the United States and globally. *Environmental Sciences Europe, 28*(1), 3. https://doi.org/10.1186/s12302-016-0070-0

Benton, T. G., Bryant, D. M., Cole, L., & Crick, H. Q. P. (2002). Linking agricultural practice to insect and bird populations: A historical study over three decades. *Journal of Applied Ecology, 39*(4), 673–687. https://doi.org/10.1046/j.1365-2664.2002.00740.x

Bolan, S., Padhye, L. P., Jasemizad, T., Govarthanan, M., Karmegam, N., Wijesekara, H., Amarasiri, D., Hou, D., Zhou, P., Biswal, B. K., Balasubramanian, R., Wang, H., Siddique, K. H. M., Rinklebe, J., Kirkham, M. B., & Bolan, N. (2024). Impacts of climate change on the fate of contaminants through extreme weather events. *Science of the Total Environment, 909*, 168388. https://doi.org/10.1016/j.scitotenv.2022.168388

Bortolotti, L., Bortolotti, A., & Barros, C. (2017). Acute and chronic effects of insecticides on honeybee survival and behavior. *Journal of Apicultural Research, 56*(4), 349–357. https://doi.org/10.1080/00218839.2017.1343807

Botías, C., David, A., Hill, E. M., & Goulson, D. (2015). Neonicotinoid residues in wildflowers, a potential route of chronic exposure for bees. *Environmental Science and Technology, 49*(21), 12731–12740. https://doi.org/10.1021/acs.est.5b03459

Breeze, T. D., Bailey, A. P., Balcombe, K. G., & Potts, S. G. (2014). Agricultural policies exacerbate honeybee pollination service supply-demand mismatches across Europe. *PLoS One, 9*(1), e82996. https://doi.org/10.1371/journal.pone.0082996

Buhler, D. D., Coulson, S. J., & Ferris, H. (2020). Herbicide drift and non-target plant diversity in agricultural ecosystems. *Agricultural and Environmental Letters, 5*(1), e20035. https://doi.org/10.1002/ael2.20035

Cutler, G. C., & Scott-Dupree, C. D. (2007). Exposure to clothianidin seed-treated canola has no long-term impact on honey bees. *Journal of Economic Entomology, 100*(3), 765–772. https://doi.org/10.1093/jee/100.3.765

Desneux, N., Decourtye, A., & Delpuech, J. M. (2007). The sublethal effects of pesticides on beneficial arthropods. *Annual Review of Entomology, 52*, 81–106. https://doi.org/10.1146/annurev.ento.52.110405.091440

Dewar, A. M. (2020). The adverse impact of the neonicotinoid seed treatment ban on crop protection in sugar beet in the United Kingdom. *Pest Management Science, 76*(3), 710–715. https://doi.org/10.1002/ps.5617

Dicks, L. V., Breeze, T. D., Ngo, H. T., Senapathi, D., An, J., Aizen, M. A., et al. (2019). A global-scale expert assessment of drivers and risks associated with pollinator decline. *Nature Ecology & Evolution, 4*(1), 23–33. https://doi.org/10.1038/s41559-019-0972-8

Elliott, J. E., Hindmarch, S., Albert, C. A., Emery, J., Mineau, P., & Maisonneuve, F. (2020). Exposure pathways of anticoagulant rodenticides to nontarget wildlife. *Environmental Monitoring and Assessment, 192*(4), 211. https://doi.org/10.1007/s10661-020-8187-x

Eng, M. L., Stutchbury, B. J., & Morrissey, C. A. (2019). A neonicotinoid insecticide reduces fueling and delays migration in songbirds. *Science, 365*(6458), 1177–1180. https://doi.org/10.1126/science.aaw9419

European Commission. (2021). *REACH*. Retrieved from https://www.ec.europa.eu

Fischer, J., Müller, T., Spatz, A. K., Greggers, U., Grünewald, B., & Menzel, R. (2014). Neonicotinoids interfere with specific components of navigation in honeybees. *PLoS One, 9*(3), e91364. https://doi.org/10.1371/journal.pone.0091364

Gallai, N., Salles, J. M., Settele, J., & Vaissière, B. E. (2009). Economic valuation of the vulnerability of world agriculture confronted with pollinator decline. *Ecological Economics, 68*(3), 810–821. https://doi.org/10.1016/j.ecolecon.2008.06.014

Garibaldi, L. A., Carvalheiro, L. G., Vaissière, B. E., & Gemmill-Herren, B. (2018). Pollination services for sustainable agriculture: The role of pesticides and pollinators. *Nature Sustainability, 1*(6), 262–271. https://doi.org/10.1038/s41893-018-0077-z

Gill, H. K., & Garg, H. (2014). Chapter 8: Pesticides: Environmental impacts and management strategies. In H. K. Gill & H. Garg (Eds.), *Pesticides: Environmental impacts and management strategies* (pp. 188–189). IntechOpen. https://doi.org/10.5772/57399

Goulson, D. (2013). An overview of the environmental risks posed by neonicotinoid insecticides. *Journal of Applied Ecology, 50*(4), 977–987. https://doi.org/10.1111/1365-2664.12111

Goulson, D. (2014). Pesticides linked to bird declines. *Nature, 511*(7509), 295–296. https://doi.org/10.1038/511295a

Goulson, D., Nicholls, E., Botías, C., & Rotheray, E. L. (2015). Bee declines driven by combined stress from parasites, pesticides, and lack of flowers. *Science, 347*(6229), 1255957. https://doi.org/10.1126/science.1255957

Goulson, D., Nicholls, E., Botías, C., & Rotheray, E. L. (2021). An overview of the environmental and human health impacts of neonicotinoids. *Environmental Science and Pollution Research, 28*(3), 3501–3526. https://doi.org/10.1007/s11356-020-11980-5

Guez, D., Suchail, S., Gauthier, M., Maleszka, R., & Belzunces, L. P. (2001). The effects of sublethal doses of synthetic pyrethroid insecticides on honeybee (Apis mellifera) behavior. *Laboratory Investigation, 81*(10), 1251–1257. https://doi.org/10.1038/labinvest.3780339

Hallmann, C. A., Foppen, R. P. B., van Turnhout, C. A. M., de Kroon, H., & Jongejans, E. (2014). Declines in insectivorous birds are associated with high neonicotinoid concentrations. *Nature, 511*(7509), 341–343. https://doi.org/10.1038/nature13531

Henry, M., Béguin, M., Requier, F., Rollin, O., Odoux, J. F., Aupinel, P., et al. (2012). A common pesticide decreases foraging success and survival in honey bees. *Science, 336*(6079), 348–350. https://doi.org/10.1126/science.1215039

Intergovernmental Science-Policy Platform on Biodiversity and Ecosystem Services (IPBES). (2016). *Summary for policymakers of the assessment report of the Intergovernmental Science-Policy Platform on Biodiversity and Ecosystem Services on pollinators, pollination and food production*. Secretariat of the Intergovernmental Science-Policy Platform on Biodiversity and Ecosystem Services, Bonn, Germany. Retrieved from https://www.ipbes.net

Kaur, R., Choudhary, D., Bali, S., Bandral, S. S., Singh, V., Ahmad, M. A., Rani, N., Singh, T. G., & Chandrasekaran, B. (2024). Pesticides: An alarming detrimental to health and environment. *Science of the Total Environment, 887*, 170113. https://doi.org/10.1016/j.scitotenv.2024.170113

Klein, A. M., Vaissière, B. E., Cane, J. H., Steffan-Dewenter, I., Cunningham, S. A., Kremen, C., & Tscharntke, T. (2007). Importance of pollinators in changing landscapes for world crops. *Proceedings of the Royal Society B: Biological Sciences, 274*(1608), 303–313. https://doi.org/10.1098/rspb.2006.3721

Kniss, A. R. (2018). Genetically engineered herbicide-resistant crops and herbicide-resistant weed evolution in the United States. *Weed Science, 66*(2), 260–273. https://doi.org/10.1017/wsc.2017.70

Lamichhane, J. R., Dachbrodt-Saaydeh, S., Kudsk, P., & Messean, A. (2017). Toward a reduced reliance on conventional pesticides in European agriculture. *Plant Disease, 101*(4), 632–644. https://doi.org/10.1094/PDIS-05-16-0749-FE

Lavelle, P., Decaëns, T., Aubert, M., Barot, S., Blouin, M., Bureau, F., et al. (2019). Soil invertebrates and ecosystem services. *European Journal of Soil Biology, 94*, 1–15. https://doi.org/10.1016/j.ejsobi.2019.103103

Laycock, I., Cotterell, K. C., O'Shea-Wheller, T. A., & Cresswell, J. E. (2014). A meta-analysis of impacts of neonicotinoid pesticides on honeybees. *Journal of Applied Ecology, 51*(4), 847–857. https://doi.org/10.1111/1365-2664.12229

Morrissey, C. A., Mineau, P., Devries, J. H., & Sanchez-Bayo, F. (2020). The risk of pesticides to non-target organisms: A review. *Environmental Pollution, 267*, 115629. https://doi.org/10.1016/j.envpol.2020.115629

Motta, R., Tesoro, G., & Marchetti, R. (2020). Climate change and pesticide regulation: Evolving risks and adaptive management strategies. *Environmental Science and Policy, 104*, 98–105. https://doi.org/10.1016/j.envsci.2019.11.012

Pelosi, C., Bertrand, M., Roger-Estrade, J., & Mougin, C. (2021). Earthworm responses to agricultural practices and pesticide applications: A review. *Agronomy for Sustainable Development, 41*(1), 1–22. https://doi.org/10.1007/s13593-021-00671-3

Pestana, J. L. T., Alexander, A. C., Culp, J. M., Baird, D. J., & Soares, A. M. V. M. (2019). Pesticide-induced alterations in aquatic food webs: Current research and future directions. *Ecological Applications, 29*(7), e01995. https://doi.org/10.1002/eap.1995

Pimentel, D. (2005). Environmental and economic costs of the application of pesticides primarily in the United States. Environment, *Development and Sustainability, 7*(2), 229–252. https://doi.org/10.1007/s10668-005-7314-2

Pisa, L. W., Amaral-Rogers, V., Belzunces, L. P., Bonmatin, J. M., Downs, C. A., Goulson, D., et al. (2015). Effects of neonicotinoids and fipronil on non-target invertebrates. *Environmental Science and Pollution Research, 22*(1), 68–102. https://doi.org/10.1007/s11356-014-3471-x

Policy Department for External Relations, Directorate General for External Policies of the Union. (2021). *The use of pesticides in developing countries and their impact on health and the right to food (PE 653.622 – January 2021)*. Study requested by the DEVE committee. Directorate-General for External Policies, European Union. EP/EXPO/DEVE/FWC/2019-01/LOT3/R/06 EN.

Potts, S. G., Biesmeijer, J. C., Kremen, C., Neumann, P., Schweiger, O., & Kunin, W. E. (2010). Global pollinator declines: Trends, impacts and drivers. *Trends in Ecology & Evolution, 25*(6), 345–353. https://doi.org/10.1016/j.tree.2010.01.007

Rattner, B. A., Lazarus, R. S., Elliott, J. E., Shore, R. F., & van den Brink, N. W. (2017). Adverse outcome pathway and risks of anticoagulant rodenticides to predatory wildlife. *Environmental Science & Technology, 51*(15), 8878–8891. https://doi.org/10.1021/acs.est.7b02280

Rundlöf, M., Andersson, G. K. S., Bommarco, R., Fries, I., Hederström, V., Herbertsson, L., et al. (2015). Seed coating with a neonicotinoid insecticide negatively affects wild bees. *Nature, 521*(7550), 77–80. https://doi.org/10.1038/nature14420

Samanta, S., Maji, A., Sutradhar, B., Banerjee, S., Shelar, V. B., Khaire, P. B., Yadav, S. V., & Bansode, G. D. (2023). Impact of pesticides on beneficial insects in various agroecosystems: A review. *International Journal of Environment and Climate Change, 13*(8), 1928–1936. https://doi.org/10.9734/IJECC/2023/v13i82149

Sanchez-Bayo, F., & Goka, K. (2014). Pesticide residues and bees – A risk assessment. *PLoS One, 9*(4), e94482. https://doi.org/10.1371/journal.pone.0094482

Sánchez-Bayo, F., & Wyckhuys, K. A. G. (2019). Worldwide decline of the entomofauna: A review of its drivers. *Biological Conservation, 232*, 8–27. https://doi.org/10.1016/j.biocon.2019.01.02

Shore, R. F., Pereira, M. G., Walker, L. A., & Thompson, D. R. (2017). Pesticides and birds: Integrating regulatory and conservation approaches. *Ecotoxicology, 26*, 1073–1080. https://doi.org/10.1007/s10646-017-1843-6

Silva, V., Mol, H. G. J., Zomer, P., Tienstra, M., Ritsema, C. J., & Geissen, V. (2021). Pesticide residues in European agricultural soils – A hidden reality unfolded. *Science of the Total Environment, 755*, 142–145. https://doi.org/10.1016/j.scitotenv.2020.142-145

Simon-Delso, N., Amaral-Rogers, V., Belzunces, L. P., Bonmatin, J. M., Chagnon, M., Downs, C., et al. (2015). Systemic insecticides (neonicotinoids and fipronil): Trends, uses, mode of action and metabolites. *Environmental Science and Pollution Research, 22*(1), 5–34. https://doi.org/10.1007/s11356-014-3470-y

Singh, A., Raza, S. K., & Singh, M. P. (2020). Climate change and its impact on the environment: Current scenario and future perspectives. *Environmental Monitoring and Assessment, 192*, 184. https://doi.org/10.1007/s10661-020-8155-x

Siviter, H., & Muth, F. (2020). Do novel insecticides pose a threat to beneficial insects? *Proceedings of the Royal Society B, 287*(1924), 20201265. https://doi.org/10.1098/rspb.2020.1265

Sparks, T. C., & Nauen, R. (2015). IRAC: Mode of action classification and insecticide resistance management. *Pesticide Biochemistry and Physiology, 121*, 122–128. https://doi.org/10.1016/j.pestbp.2014.11.014

Stanley, D. A., Garratt, M. P. D., Wickens, J. B., Wickens, V. J., Potts, S. G., & Raine, N. E. (2015). Neonicotinoid pesticide exposure impairs crop pollination services provided by bumblebees. *Nature, 528*(7583), 548–550. https://doi.org/10.1038/nature16167

Tang, F. H. M., Lenzen, M., McBratney, A., & Maggi, F. (2018). Risk of pesticide pollution at the global scale. *Nature Geoscience, 11*(4), 251–255. https://doi.org/10.1038/s41561-018-0071-2

Tsvetkov, N., Samson-Robert, O., Sood, K., Patel, H. S., Malena, D. A., Gajiwala, P. H., et al. (2017). Chronic exposure to neonicotinoids reduces honey bee health near corn crops. *Science, 356*(6345), 1395–1397. https://doi.org/10.1126/science.aam7472

United States Environmental Protection Agency (EPA). (2021). *Overview of risk assessment in the pesticide program*. Retrieved from https://www.epa.gov

van der Sluijs, J. P., Simon-Delso, N., Goulson, D., Maxim, L., Bonmatin, J. M., & Belzunces, L. P. (2015). Neonicotinoids, bee disorders and the sustainability of pollinator services. *Current Opinion in Environmental Sustainability, 7*, 28–30. https://doi.org/10.1016/j.cosust.2013.07.005

Whitehorn, P. R., O'Connor, S., Wäckers, F. L., & Goulson, D. (2012). Neonicotinoid pesticide reduces bumble bee colony growth and queen production. *Science, 336*(6079), 351–352. https://doi.org/10.1126/science.1215025

Winfree, R., Aguilar, R., Vázquez, D. P., LeBuhn, G., & Aizen, M. A. (2011). Native pollinators in anthropogenic habitats. *Annual Review of Ecology, Evolution, and Systematics, 42*, 1–22. https://doi.org/10.1146/annurev-ecolsys-102710-145042

Wood, T. J., & Goulson, D. (2017). The environmental risks of neonicotinoid pesticides: A review of the evidence post-2013. *Environmental Science and Pollution Research, 24*(21), 17285–17325. https://doi.org/10.1007/s11356-017-9240-x

Wood, T. J., & Goulson, D. (2018). The impacts of neonicotinoid pesticides on pollinators and their environment: A review. *Journal of Environmental Management, 221*, 89–102. https://doi.org/10.1016/j.jenvman.2018.01.026

Woodcock, B. A., Bullock, J. M., Shore, R. F., Heard, M. S., Pereira, M. G., Redhead, J., et al. (2017). Country-specific effects of neonicotinoid pesticides on honey bees and wild bees. *Science, 356*(6345), 1393–1395. https://doi.org/10.1126/science.aaa1190

Woodcock, B. A., Isaac, N. J. B., & Bullock, J. M. (2019). Impacts of neonicotinoid use on pollinators and pollination. *Nature Sustainability, 2*(1), 10–22. https://doi.org/10.1038/s41893-019-0273-8

Wu-Smart, J., & Spivak, M. (2016). Sub-lethal effects of dietary neonicotinoid insecticide exposure on honey bee queen fecundity and colony development. *Scientific Reports, 6*, 32108. https://doi.org/10.1038/srep32108

Zhang, X., Liu, X., & Zang, W. (2017). Mitigating pesticide contamination through buffer zones and vegetative strips: A case study in the U.S. *Journal of Environmental Management, 196*, 63–71. https://doi.org/10.1016/j.jenvman.2017.02.063

Zhang, W., Jiang, F., & Ou, J. (2020). Global pesticide use: Profile, trend, cost/benefit, and more. *Frontiers of Agricultural Science and Engineering, 7*(1), 7–20. https://doi.org/10.15302/J-FASE-2019307

Zubrod, J. P., Bundschuh, M., Arts, G., Brühl, C. A., Imfeld, G., Knäbel, A., et al. (2019). Fungicides: An overlooked pesticide class? *Environmental Science & Technology, 53*(7), 3347–3365. https://doi.org/10.1021/acs.est.8b04392

Technological Innovations Aimed at Reducing the Environmental Impact of Pesticides and Increasing the Resilience of Agriculture to Climate Change

Barbara Sawicka, Piotr Barbaś, Piotr Pszczółkowski, and Barbara Krochmal-Marczak

1 Introduction

Climate change is a current threat to the entire world. It is the occurrence of extreme events such as droughts, floods, and storms. These conditions cause additional stress, which in turn leads to food distribution (Singh et al., 2017; Hasegawa et al., 2022). Climate change causes yield losses in rice by 35%, wheat by 20%, sorghum by 50%, barley by 13%, and maize by 60% (Porter et al., 2014a, b). Each 1 °C increase in temperature by 2050 could cause a 6% decrease in global wheat production, while maize yield could decrease by up to 30% (Tomar et al., 2023). The changing climate also affects the expansion of pests and diseases through an increase in pest populations (Lehmann et al., 2020a) as well as the number of their generations (Taylor et al., 2018). The response of plants to increased temperature is shown in Fig. 1. Global climate change also has a decisive impact on the quality of crops,

B. Sawicka (✉)
Department of Plant Production Technology and Commodities Science, University of Life Sciences in Lublin, Lublin, Poland
e-mail: barbara.sawicka@up.lublin.pl

P. Barbaś
Department Agronomy of Potato, Plant Breeding and Acclimatization Institute—National Research Institute, Branch Jadwisin, Serock, Poland
e-mail: p.barbas@ihar.edu.pl

P. Pszczółkowski
Research Centre for Cultivar Testing in Słupia Wielka, Słupia Wielka, Poland
e-mail: p.pszczolkowski.inspektor@coboru.gov.pl

B. Krochmal-Marczak
Department of Plant Production and Food Safety, Carpathian State College in Krosno, Krosno, Poland
e-mail: barbara.marczak@pans.krosno.pl

B. R. Babaniyi, E. E. Babaniyi (eds.), *The Interplay of Pesticides and Climate Change*, https://doi.org/10.1007/978-3-031-81669-7_10

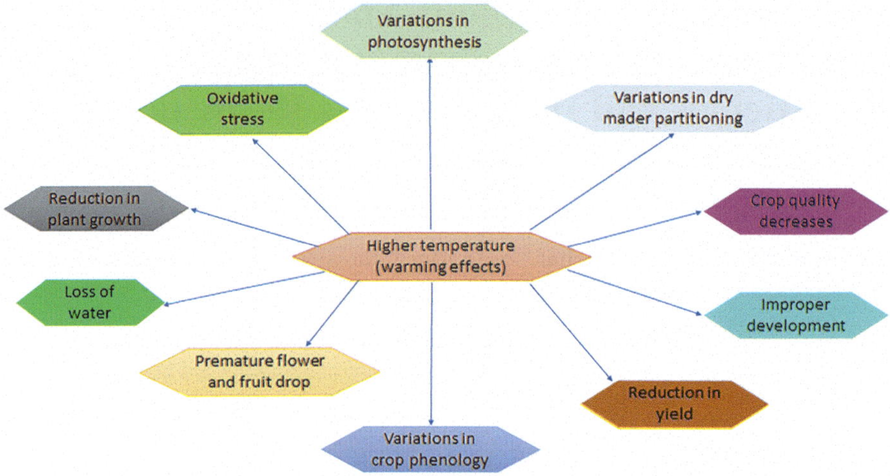

Fig. 1 Effects of higher temperatures on plant growth and development. (Source: Own study)

both in terms of nutritional value and taste of crops. Increased CO_2 concentration and temperature can cause a decrease in protein content and deterioration of crop quality (Wang et al., 2020; Feng et al., 2023a, b). The changing climate also affects the expansion of pests and diseases through an increase in pest populations (Lehmann et al., 2020a, b), as well as an increase in the number of their generations (Taylor et al., 2018). Extreme climate phenomena are observed all over the world, but these changes are more visible in the Southern Hemisphere than in the Northern Hemisphere (Stock, 2020). This situation requires a deeper analysis of potential changes in exposure to individual and simultaneous extreme phenomena at the population and country level.

Climate change is mainly caused by greenhouse gas (GHG) emissions such as CO_2, CH_4 (methane) and N_2O (nitrous oxide) (Chataut et al., 2023). Intensive industrial development and rapidly increasing population lead to drastic deforestation and demand for land, which in turn led to intensive GHG emissions (Stock, 2020). Reducing greenhouse gas emissions is one of the most important tasks for agriculture in the case of climate change. In order to manage this problem, solutions are available, such as creating new varieties of crops or developing smart agriculture. The second important issue in the era of climate change is reducing the use of pesticides. Food and Agriculture Organization of the United Nations (FAO), the term pesticide means any substance or mixture of substances intended for the prevention, destruction, or control of any pest, including vectors of human diseases and undesirable plant and animal species that cause damage during the production, processing, storage, transport, or commercialization of food and agricultural products (Sharma et al., 2020). Pesticides are divided into different categories, depending on the mode of transfer, including surface absorption, percolation, spray drift, volatilization, and surface runoff, which can protect against threats and make non-target plants available (Sharma et al., 2020). Pesticides effectively control pests, but their effects can be harmful to other organisms, such as beneficial insects, aquatic organisms, birds, or non-target units (Darban et al., 2023). Pesticides can also pollute air, water, and soil (Poudel et al., 2020).

An alternative to pesticides are biopesticides, which are of plant origin, considered metabolites, plant growth regulators, or animal pheromones (Mołoń & Durak, 2018a, b). Biopesticides, as a result of the use of which there is no risk of phytochemical changes, risk to human health and the environment, which do not cause greenhouse emissions (Lengai & Muthomi, 2018).

Therefore, the aim of the work is to analyze the impact of climate change on plants, taking into account yield losses, the spread of pests, and deterioration of crop quality. The work also aims to discuss the role resulting from the reduction in greenhouse gas emissions and alternative methods of plant protection, such as biopesticides, in the context of reducing climate risk. Special application will be used for the needs of new plant varieties resistant to stress and for intelligent development, available for the management of resources, defined and protecting the environment.

2 Innovation in Agriculture

Innovation is the process of research and development of new or improvement in existing tests, up to their implementation (Mirkowska, 2010). In agriculture, innovations include new approaches and changes in processes that can comply with new techniques and technologies (Wysocki & Ryś-Jurek, 2008). Agriculture is a sector in which innovations are key to further development. There are several types of progress in agriculture:

- Biological progress—It consists in the elimination of plant and animal organisms by increasing the number of species, varieties and breeds (Uciechowski & Sadowski, 2018).
- Technical progress—It is supplementation of new devices and technical devices (Michałek & Grotkiewicz, 2010).
- Organizational progress—It is progress in changes in work organization (Mirkowska, 2010; Feng et al., 2023a, b).
- Technological progress—It is a new production and manufacturing technologies (Bartkowiak & Bartkowiak, 2017a, b; Periakaruppan et al., 2023).
- Socio-economic progress—It aims to introduce changes in the agricultural system (Dash et al., 2022b).

The progress in agriculture is illustrated in Fig. 2. Progress in agriculture has promoted the development of new disciplines, including biotechnology (Wahab, 2009; Adlak et al., 2019) and genetic engineering (Chandra et al., 2020). The innovation in agriculture around the world is Agriculture 4.0, which is based on digital technologies such as artificial intelligence (AI), Big Data, Cloud Computing, robotics, Internet of Things (IoT), RS, and Blockchain (Sonka, 2016; Narmilan & Niroash, 2020a, b; Zahi et al., 2020; Ahmed et al., 2022). Artificial intelligence (AI)

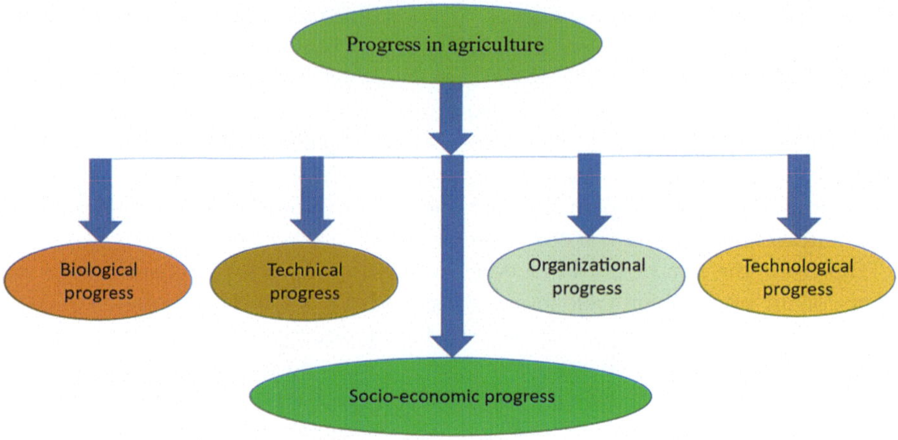

Fig. 2 Progress in agriculture. (Source: own study)

and machine learning (ML) technologies are modifying agribusiness and deciding on the selection of more informed decisions, increasing their operational efficiency. As reported by Benos et al. (2021a, b) and Vijayakumar et al. (2022), they can analyze huge amounts of data from satellite images, weather forecasts, and soil sensors, providing farmers with precise insight into crop status, soil conditions, and optimal planting time. Internet of Things—Internet of Things (IoT) is a connection between biocenosis and biotope, in which smart devices communicate and work synergistically, enabling optimization of agricultural operations.

IoT sensors deployed in fields allow for readings of critical parameters such as soil moisture, temperature, and crop status, transmitting data in real time to centralized systems or directly to farmers' smartphones (Rushchitskaya et al., 2024).

Such connectivity allows for immediate adjustment of irrigation, fertilization, and pest control strategies, significantly improving resource efficiency and crop yields. In recent years, the introduction of modern technologies, including precision farming, has also been an innovation in agriculture (Rusmayadi et al., 2023a, b). Remote sensing and integrated pest management (IPM) are used to reduce the adverse impact of pesticides on the environment (Awad Fahad, 2023a, b; Pachori et al., 2023a, b), while green chemistry and biopesticides can replace the use of natural pesticides (Ubuoh, 2016; Samada & Tambunan, 2020a, b; Peeriga et al., 2022a, b; Ayilara et al., 2023a, b), which is extremely important in the era of changing climate. The climate change and the development of technology in agriculture pose a huge challenge for modern society. Implementing innovations such as biopesticides and agricultural practices can help mitigate the effects of climate change and improve production efficiency.

3 Technological Innovations

Innovation is a process of long-term research and development of a new solution, improvement of existing solutions, up to its implementation and implementation into practice (Mirkowska, 2010; Samada & Tambunan, 2020a; Gupta et al., 2023). Innovation in the field of agriculture was defined by Wysocki and Ryś-Jurek (2008) as a new concept and change in processes, the effect of which may be a new technique and technology. Agriculture is a branch of the economy in which progress, that is, innovation, determines the dynamics of development. In the case of agriculture, we are talking about biological, organizational, and technical progress, as well as economic progress, which should appear as a result of the application of the previously mentioned (Mirkowska, 2010). Biological progress consists in improving plant and animal organisms useful for humans, as well as increasing the number of their species, varieties, and breeds (Uciechowski & Sadowski, 2018); technical progress introduces improved or new devices and technical means into practice (Michałek & Grotkiewicz, 2010); organizational progress is focused on changes in the organization of the farm and work in order to save production forces and increase management efficiency (Mirkowska, 2010); technological progress introduces new manufacturing technologies and production methods (Bartkowiak & Bartkowiak, 2017a, b); and socioeconomic progress is aimed at inducing beneficial changes in the agricultural system (Dash et al., 2022a, b).

4 Technological Innovations in Agriculture

Innovation in agriculture is becoming increasingly important due to global challenges like climate change, sustainable development, and food security. Wysocki and Ryś-Jurek (2008) defined it as introducing new concepts and processes that lead to advancements in techniques and technologies. Innovations today extend beyond traditional methods, incorporating digitalization, automation, and biotechnology (Gupta et al., 2023). Figure 3 presents technological innovations in agriculture.

Digitalization through technologies such as drones, sensors, and AI enables precision agriculture, improving resource efficiency by optimizing processes like irrigation and fertilization (Sonka, 2016; Ahmed et al., 2022). Biotechnology plays a key role in developing genetically modified organisms (GMOs) that resist pests, diseases, and extreme climates. This contributes to sustainable practices by reducing the need for chemical inputs (Kumar et al., 2021a; Azizoglu et al., 2023). Automation in agriculture, through robotic harvesting and autonomous systems, increases productivity and resource sustainability (Balasundram et al., 2023a, b).

Agriculture 4.0, a cutting-edge approach, combines AI, IoT, and big data to enable more informed decision-making and better crop management. IoT sensor networks continuously monitor soil and plant conditions, facilitating real-time adjustments to maximize yield and minimize environmental impact (Narmilan &

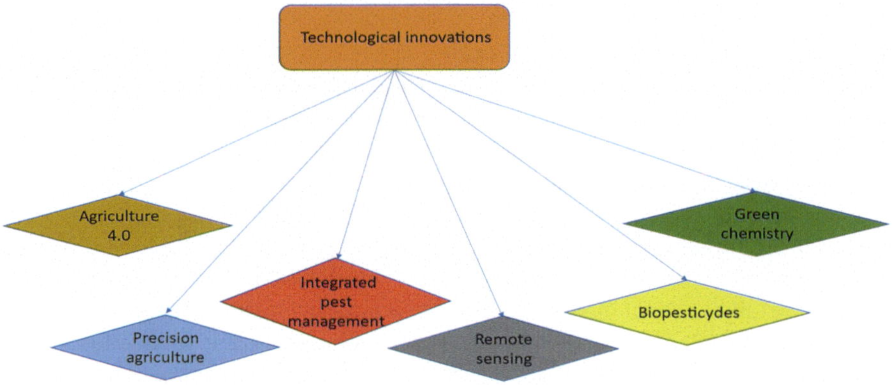

Fig. 3 Technological innovations. (Source: own study)

Niroash, 2020b; Rushchitskaya et al., 2024). Furthermore, precision farming techniques, along with biopesticides and green chemistry, minimize pesticide use and environmental harm (Pachori et al., 2023a, b).

Finally, modern breeding techniques like marker-assisted back crossing (MABC) and genomic selection (GS) accelerate the development of stress-resistant crop varieties, addressing the urgent need for crops that can withstand climate fluctuations (Kadirvel et al., 2015; Bhargava et al., 2023). These innovations collectively enhance sustainable farming and food security.

In summary, modern innovations in agriculture are multidimensional and include not only mechanical technologies but also digital, biotechnological, and automation solutions. The development of these technologies affects the increase in efficiency, effectiveness, and sustainable development in agriculture, which is crucial in the face of growing global challenges.

5 Precision Agriculture

Over the past 30 years, technological advances have significantly improved crop yields and quality. Farmers have gained access to advanced automation, control, and data analysis systems that have enabled them to optimize agricultural operations on an unprecedented scale. Key technologies include global positioning system (GPS) systems, satellite maps, ground-based platforms, and local measurement devices such as data loggers that allow for field monitoring and identification of potential management errors (Balasundram et al., 2023a, b).

Unmanned aerial vehicle (UAV) systems, robotics, the Internet of Things (IoT), and digital agriculture (DA) have played a key role in the development of precision agriculture, contributing to both increased food production and more sustainable agricultural practices (MacPherson et al., 2022a, b). UAVs equipped with advanced sensors enable real-time monitoring of key parameters such as soil moisture, plant growth, crop health, and weather conditions. These technologies are invaluable in predicting drought, optimizing pesticide and fertilizer application, and assessing soil nutrient needs (Olson & Anderson, 2021a, b).

The use of unmanned aerial vehicles (drones) with real-time sensors allows farmers to precisely spray crops, which allows for a significant reduction in the use of chemicals. In research conducted by Zanin et al. (2022a, b), a precision spraying system for soybeans and corn reduced the number of pesticides, including herbicides and fungicides, which led to a reduction in costs by up to two to three times, compared to traditional spraying methods. Thanks to the precision in the application of these chemicals, it is possible to minimize the negative impact on the environment while maintaining high crop yields.

Artificial intelligence (AI) and machine learning (ML) also play an important role in modern agriculture. These technologies enable the processing of huge amounts of data, which contributes to more precise crop management, yield forecasting, and optimization of irrigation and fertilization (Benos et al., 2021a, b). The latest digital technologies not only improve production efficiency but also contribute to reducing the consumption of resources such as water and energy, which is crucial in the context of climate change.

6 Remote Sensing

Remote sensing is an advanced tool for monitoring of Earth's resources that uses space technologies, including satellite images, to complement ground-based observations. This allows for greater precision and accuracy in environmental analyses (Sangeetha et al., 2024). Remote sensing uses different ranges of the electromagnetic spectrum, including visible, infrared, and microwave radiation, to analyze Earth's surface features. This science integrates various fields, such as optics, spectroscopy, computer science, photography, electronics, and telecommunications (Roy et al., 2017; Schäfer & Török, 2021). Effective remote sensing requires coordination of several steps, including the emission of electromagnetic radiation (EMR) delivered by the Sun or self-emission, transmission of this energy to the Earth's surface, interaction of EMR with the Earth's surface in the form of reflection or emission, and then transmission of energy to sensors and data processing by these sensors (Khargharia, 2021a, b). With this technology, it is possible to monitor plant growth, land use, land cover changes, detect weeds, diseases, pest infestations, and forecast crop yields and harvest dates (Shanmugapriya et al., 2019; Jurišić et al., 2021a, b). Remote sensing plays a key role in detecting plant diseases in their early stages, which is especially important for precision agriculture. Early detection helps to limit the spread of diseases and minimize production losses. Many traditional disease detection methods, such as visual assessment, are time-consuming, labor-intensive, and prone to human error (Too et al., 2019). Remote sensing allows to automate these processes and detect diseases even in asymptomatic phases, which is difficult to achieve with ground-based methods. In precision agriculture, remote sensing is key in concepts such as Digital Farming, Agriculture 5.0 and Smart Farming, which increase the accuracy of agricultural operations and support critical decision-making (Saiz-Rubio & Rovira-Más, 2020). The use of technologies such as American Landsat satellites, the European Sentinel-2 system, Indian IRS satellites, or the RapidEye constellation allows for monitoring of agriculture at a global level. An example is the Sentinel-2 platform, used, among others, for estimating chlorophyll content in potatoes (Clevers et al., 2017), radiation efficiency (Peng et al., 2021a, b), yield forecasting (Gómez et al., 2019), and crop mapping (Ashourloo et al., 2020a, b). Indian IRS satellites have enabled precise estimation of sugarcane yields (Pandey et al., 2019a, b), while the RapidEye constellation has been used to estimate leaf area and biomass index in maize and soybean crops (Kross et al., 2015). Recent studies also use advanced remote sensing techniques to analyze crop performance and monitor plant health, allowing for precise resource management in the context of changing climatic conditions.

6.1 Remote Sensing of Crop Yields

Remote sensing of crop yields is becoming a key tool in managing spatial variability in crop fields. Crop growth is often uneven due to biotic variables, such as the presence of pests, and abiotic variables, including soil conditions and water availability (Rivero et al., 2022a, b; Dwivedi & Sopory, 2023a, b). Monitoring spatial variability of crop yields using images from unmanned aerial vehicles (UAVs) and analyses of high-resolution multi-temporal data allow for precise tracking of crop health and yield prediction.

One of the key elements of this technology is the use of vegetation indices (e.g., normalized difference vegetation index (NDVI), soil-adjusted vegetation index (SAVI), enhanced vegetation index (EVI), crop water stress index (CWSI), which allow for the assessment of plant growth stages and their condition. Such indices allow for inferences about yield potential based on observed plant traits, such as biomass or chlorophyll content. An important research issue is to find strong correlations between these indicators and actual yields, which can facilitate precise resource management and timely interventions. Table 1 presents examples of studies that describe the use of UAV images to monitor crop yield and development:

These studies demonstrate that UAV-based remote sensing is an effective tool for monitoring plant development and predicting yields, which enables optimizing agricultural practices and better crop management.

Table 1 Estimating crop yield and development using UAV images

Crop	References
Wheat	Kanning et al. (2018), Tao et al. (2020), Feng et al. (2022), and Wang et al. (2022)
Potato	Liu et al. (2022, 2024) and Luo et al. (2022)
Corn	Han et al. (2019) and Zhang et al. (2021)
Rice	Xu et al. (2022)
Oats	Dhakal et al. (2023)
Barley	Wengert et al. (2021)

6.2 Remote Sensing of Irrigation and Water Stress

Remote sensing of irrigation and monitoring of water stress is becoming increasingly advanced, providing farmers with the tools to precisely analyze the water requirements of crops. Water is a key factor influencing plant growth and yield. Monitoring soil moisture has a direct impact on the effectiveness of fertilization and on preventing water stress, which occurs when the amount of available water does not meet the evapotranspiration needs of plants (Katimbo et al., 2022).

Remote sensing in irrigation is based on a variety of techniques. One of the most commonly used methods is the use of thermal imaging, which allows for the detection of changes in the surface temperature of plant leaves, that is a strong indicator of their water status. Increased leaf temperature is often a sign of water stress, and remote sensing technologies enable rapid response to its early symptoms (Ferrante & Mariani, 2018a, b). Other techniques, such as multispectral and hyperspectral imaging, allow for the analysis of soil moisture and plant health based on various vegetation indices (e.g., NDVI, CWSI), which are strongly correlated with plant water requirements (Barbedo, 2021). Modern approaches to irrigation monitoring include the increasing use of UAVs (unmanned aerial vehicles), which can monitor soil moisture, plant temperature, and other key parameters in real time using sensors. These technologies enable precise management of water resources and can lead to significant reductions in water use in crop fields, while improving plant productivity and health (Acharya et al., 2021a, b). Thanks to the integration with IoT (Internet of Things), data collected by UAVs can be analyzed in real time, which enables automatic and precise control of irrigation systems (Evans & Sadler, 2008). Remote sensing also allows monitoring problems related to excess water. Excessive irrigation can lead to soil erosion, nutrient leaching, and deterioration of plant health. Therefore, technologies based on thermal and multispectral analysis help to precisely adjust irrigation to the actual needs of crops, thus reducing the risk of resource loss (Katimbo et al., 2022).

6.3 Remote Sensing of Nutrients

Remote sensing of nutrients in crops has become a key tool in monitoring the nutritional status of crops and detecting nutrient deficiencies. These deficiencies can significantly affect plant growth, leading to a decrease in leaf area, reduced chlorophyll content, and discoloration, which are visible in the plant's reflectance signatures. These reflectance's can be used by spectral sensors, allowing for indirect assessment of the plant's nutritional status in real time (Croft & Chen, 2018; Sims & Gamon, 2002). This makes remote sensing an effective alternative to time-consuming laboratory analyses.

The most commonly monitored nutrient using remote sensing is nitrogen, which has a direct impact on chlorophyll content and, consequently, on the reflectance of light in the visible and near-infrared range. In nitrogen studies, sensors mounted on UAVs can accurately assess the nutritional status of crops in the field. Much less frequently, phosphorus and potassium deficiencies, which also play a key role in plant development, have been studied in the literature (Darvishzadeh et al., 2008).

Modern research on remote sensing of nutrient deficiencies in crops includes an increasingly wide range of applications of UAV technology, including in crops such as maize, wheat, potatoes, and rice. For example, Lu et al. (2021) used UAV images to monitor nitrogen levels in rice crops, which allowed for an accurate estimation of fertilization needs. In studies on wheat, Fu et al. (2022) assessed the level of nitrogen and phosphorus deficiencies, which allowed for precise adjustment of fertilizer applications. Table 2 presents a list of studies on monitoring nutrient deficiencies in various crops using UAV remote sensing:

The use of remote sensing to assess the nutritional status of plants allows for the rapid detection of deficiencies, which in turn allows farmers to take immediate action to optimize fertilization and increase yields.

Table 2 Monitoring of nutrient stress in crops using UAV images

Crop	References
Corn	Quemada et al. (2014), Cai et al. (2019), and Xu et al. (2021)
Wheat	Lu et al. (2019) and Fu et al. (2022)
Potato	Nigon et al. (2014), Hunt et al. (2018), Yang et al. (2021), and Zhou et al. (2022)
Apples	Yang et al. (2021) and Li et al. (2022)
Cotton	Yin et al. (2022)
Grapes	Peng et al. (2022)
Rice	Cuevas et al. (2021) and Lu et al. (2021)

6.4 Remote Sensing of Weeds

Remote sensing of weeds is an innovative tool in the fight against weeds, which pose one of the greatest threats to crops. Weeds compete for water, light, and nutrients, which can lead to an average decrease in yields of up to 34% (Wang et al., 2019). Traditional methods of weed control, such as manual or mechanical weeding, have their drawbacks—they are time-consuming, expensive, and in the case of mechanical methods can damage crops. In turn, commonly used herbicides, despite their high effectiveness, carry the risk of environmental pollution, including soil and water (Woyessa, 2022). In view of the increasing regulations restricting the use of pesticides in many countries, advanced technologies using UAVs (unmanned aerial vehicles) and spectral and color sensors are becoming an alternative.

The use of UAVs in weed identification is based on precise monitoring of fields using high-resolution images, which allows for the precise location of individual weed species. Based on this data, it is possible to implement site-specific application (SSA) methods, which allow for the application of herbicides only where weeds occur, thus limiting the unnecessary use of chemicals on bare fields or target crops (Zhang et al., 2021). This selective method promotes both environmental protection and reducing the costs associated with the purchase of plant protection products.

In weed identification, UAVs are specialized in various types of sensors, most often spectral, enabling the detection of weeds based on their light reflection signature in different wave ranges (Peteinatos et al., 2014; Louargant et al., 2018). In addition, color sensors are increasingly used, which allow for image analysis based on the color and shape of plants (Gai et al., 2020). These technologies allow for the creation of three-dimensional maps of the distribution of weeds in the field, which in turn allows for precise application of herbicides where they are really needed (Zisi et al., 2018; Dobbs et al., 2022). The weed mapping method and site-specific herbicide spraying allowed for 23–89% herbicide savings in a series of field experiments on cereals, corn, sugar beet, and pea, which is confirmed by research by many authors (Gerhards & Oebel, 2006; Christensen et al., 2009; Berge et al., 2012).

6.5 Remote Sensing of Crop Diseases

Remote sensing of crop diseases has gained importance in recent years, mainly due to the development of technologies such as unmanned aerial vehicles (UAV), the Internet of Things (IoT), and artificial intelligence (AI). Traditional methods of diagnosing plant diseases, such as visual assessment or isolation of pathogen cultures, are time-consuming and expensive, which leads to the search for more effective solutions. Modern remote sensing technologies enable precise monitoring of plant health in real time, which allows for early detection of diseases and faster taking of appropriate preventive measures (Singh et al., 2020; Shahi et al., 2023; Jafar et al., 2024; Pszczółkowski et al., 2024).

The use of UAVs in disease detection is particularly effective due to the possibility of using various sensors: digital, multispectral, hyperspectral, thermal, and fluorescent. These sensors enable monitoring of changes in plant reflectance patterns and temperature, which allows for the detection of signs of disease stress, often before visible symptoms appear (Chin et al., 2023). Drones equipped with digital, multispectral, hyperspectral, thermal, and fluorescent sensors enable more precise resolution of plant diseases and allow for disease recognition at earlier developmental stages than is possible with satellite systems (Chin et al., 2023). Examples of UAV applications in plant disease monitoring:

- Potato: UAVs with multispectral and thermal sensors have been used to monitor the occurrence of diseases such as potato blight (*Phytophthora infestans*), where early detection of pathogens allows for the optimization of chemical protection and reduction in yield losses (Gold et al., 2020; León-Rueda et al., 2024; Zarrouk et al., 2024).
- Maize: Detection of diseases such as root rot using drones with hyperspectral sensors has enabled early detection and spatial mapping of the disease, enabling precise intervention (Jia et al., 2023).
- Tomato: UAVs equipped with thermal and multispectral cameras have been used to detect diseases such as tomato mosaic virus, enabling monitoring of plant health at the level of individual leaves (Abdulridha et al., 2020).
- Wheat: The use of drones to monitor diseases such as wheat brown rust has enabled earlier control measures and limiting the spread of the disease on a larger scale (Abdulridha et al., 2023; Feng et al., 2024).
- Rapeseed: Detection of fungal diseases such as rapeseed black spot using UAVs was effective in precisely identifying areas that required fungicide applications (Jełowicki et al., 2020).

New research in this area of knowledge focuses on integrating UAVs with artificial intelligence (AI) and machine learning (ML) systems, which enables automated analysis of remote sensing data and faster identification of anomalies in plant health. The introduction of AI methods significantly increases the efficiency of these systems while reducing the amount of data that must be analyzed by humans (Shahi et al., 2023; Jafar et al., 2024).

6.6 Integrated Pest Management

Integrated pest management (IPM) refers to a multi-faceted plant protection system that was initiated in the 1950s. According to the definition of the Food and Agriculture Organization of the United Nations (FAO), IPM involves the use of all methods of plant protection against pests and then integrating available measures that prevent the development of pests (FAO, 2020). The aim of IPM is to maintain the use of pesticides and other intervention measures at an economically justified level, with minimal risk to human health and the environment (FAO, 2023).

The IPM function is divided into two integrated units: the vertical and horizontal function. Vertical integration is combated with a single dangerous species, which is particularly effective in monoculture crops. By way of horizontal integration associated with intelligent management of pest classes, such as phytopathogenic microorganisms, arthropods, phytopathogenic nematodes, and weeds, which is a more advanced solution (Coll & Wajnberg, 2017). These authors point out that the term integration, according to many authors, refers to all pests of a given crop (phytopathogenic microorganisms, arthropods, phytopathogenic nematodes/pests, weeds/parasitic plants). According to them, integration consists in controlling one or a small number of pest species, which is very often referred to as vertical integration, in contrast to horizontal integration, which involves the simultaneous management of several pest classes (Ehler, 2006; Angon et al., 2023).

The progress of technology in agriculture, including the development of biotechnology, is primarily based on the development of IPM. The introduction of pest-resistant plant varieties and the use of pests to control pests are benefits in this progress. Additional new monitoring technologies, such as remote sensing, artificial intelligence, and the Internet of Things (IoT), enable more advanced monitoring of pests and environmental conditions, which support crop management decisions (Angon et al., 2023).

The effectiveness of IPM implementation strongly depends on the comprehensive application of agro-ecosystem components, including crop species, pests, natural hazards, and conditions, and depends on understanding the interplay between different elements of the basic agro-ecosystems, such as crops, pests, natural enemies, and environmental conditions. It is increasingly necessary to use advanced mathematical models that accurately predict disease dynamics under different application and cropping conditions (Han et al., 2019; Too et al., 2019; Raja et al., 2022; Angon et al., 2023). These models can provide useful tools that can be incorporated into field decisions (Böckmann & Meyhöfer, 2015; Raja et al., 2022).

Recent studies also demonstrate that IPM programs must be available in ecological and protected conditions. Angon et al. (2023) point out that effective IPM implementation requires a holistic approach that excludes undesirable occurrences.

7 Green Chemistry

Green chemistry includes the application of release synthesis, the aim of which is to design a device that minimizes, including the use of chemicals and the use of safe residues, (bio)catalysts, and switches. The correct operation of electrical energy is crucial here, which is possible through the use of technologies such as microwave activation, mechanochemistry, ultrasound, or photocatalysis (Schäfer & Török, 2021; Zuin et al., 2021). Modern technologies enable access to a number of devices, through the design of devices that efficiently use raw materials, as well as the design of biodegradable end products, which directly supports sustainable development (Mulvihill et al., 2011; Zimmerman et al., 2020; Abdullah & Zahoor, 2023).

Contemporary research in green chemistry focuses on several key areas: first, catalysis and biocatalysis, starting the power supply in the engine, as a result of their operation and reducing energy. Biocatalysts such as enzymes are particularly used because they are activated under specific conditions, which is more environmentally friendly. Second, increasing importance is being attached to biomass, which is an alternative to fossil fuels (Vaz, 2018; Dhakal et al., 2023; Liu et al., 2022, 2024). One of the leading solutions for the development of green chemistry is also the research and development of new photocatalytic reactions. Photocatalysis, supported by sunlight, can be used for more ecological products in the production of chemicals, while reducing the demand for energy from conventional sources. Additional alternative methods of action, such as radio waves, mechanochemistry, or ultrasound, are applications for future green chemistry, because these methods are active at different frequencies and different temperatures (Pandey et al., 2019a, b; Schäfer & Török, 2021; Abdullah & Zahoor, 2023). Green chemistry is following technological innovations in line with the idea of development. By using the catalysis method, alternative switches and intelligent technology, additional control of heating devices in the environment can be enabled, while bringing economic benefits. Alternative goals of green chemistry are evolving, and research on this is becoming more and more effective and ecological, and is crucial for future environmental protection and application development (Samada & Tambunan, 2020a, b; Ngegba et al., 2022; Ayilara et al., 2023a, b; Chowdhury et al., 2024).

7.1 *Biopesticides*

Biopesticides are increasingly being used as an ecological alternative to the use of pesticides, which are used to address threats such as environmental hazards, environmental impacts, and water degradation. The protection of biopesticides is their non-toxicity and rapid biodegradability, which means they can be effectively integrated into pest management programs (IPM) (Kumar et al., 2021a, b). Biopesticides are also used to disinfect soil and additionally support soil protection in organic farming (Lengai & Muthomi, 2018).

Biopesticides can be divided into several groups: phytopesticides (of plant origin), microbial pesticides (of microbial origin), and nanobiopesticides (containing nanoparticles used with biological substances) (Yin et al., 2023). Phytopesticides include phytochemicals that do not cause greenhouse gas emissions, which is important for human health and the environment, compared to known synthetic pesticides. These compounds can be obtained from various parts of plants, such as leaves, rhizomes, roots, seeds, fruits, and flowers, which can be used both in fresh and dried form. Plant extracts and essential oils, often used in phytopesticides, are obtained by different extraction methods. The use of dried plant parts is recommended, which increases the risk of using ingredients and thus their effectiveness (Chouhan et al., 2017; Kumar et al., 2021b). The advantage of biopesticides is their lack of toxicity, which means that they can be effectively incorporated into integrated pest management (IPM). In addition, they decompose quickly and are used to disinfect agricultural soils (Lengai & Muthomi, 2018). Microbial pesticides include organisms such as bacteria, fungi, viruses, and nematodes that are infected with diseases that cause their death or spread. Examples are available based on the bacterium Bacillus thuringiensis (Egbuna et al., 2020; Bandopadhyay, 2020; Azizoglu et al., 2023; Shukla et al., 2023). Nanobiopesticides are characterized by high biocompatibility and biodegradability compared to synthetic pesticides (Shangguan et al., 2024). However, the disadvantage of biopesticides may be their low efficacy, which is closely related to the influence of numerous biotic and abiotic factors (Swapan et al., 2024). For this reason, biopesticides are often used as part of extended protection, rather than as a sole pest control tool.

7.2 *Microbiological Pesticides*

Microbiological pesticides are compounds used to eliminate pests whose production is related to the presence of microorganisms, which include, among others, bacteria, viruses, fungi, protozoa, and algae (Fig. 4) (Ruiu, 2018; Egbuna et al., 2020; Vero et al., 2023).

The most commonly used biopesticides include bacteria; fungi *Lecanicillium, Hirsutella, Metarhizium, Paecilomyces, Beauveria, Verticillium*; microalgae *Chlorella vulgaris, Anabaena laxa;* nematodes of the genera *Steinernema, Heterorhabd*, and protozoa *Nosema pyrausta* and viruses *Plutella xylostella* granulovirus (Ayilara et al., 2023a, b; Daraban et al., 2023; Chowdhury et al., 2024). However, the most commonly used bacteria are *Bacillus thuringiensis* (Bt), which produces Cry toxins that selectively act on insect larvae, and bacteria of the genera *Chromobacterium*, *Pseudomonas*, and *Yersinia* (Aioub et al., 2024). Among fungi, species from the genera *Beauveria bassiana*, *Metarhizium robertsii*, and

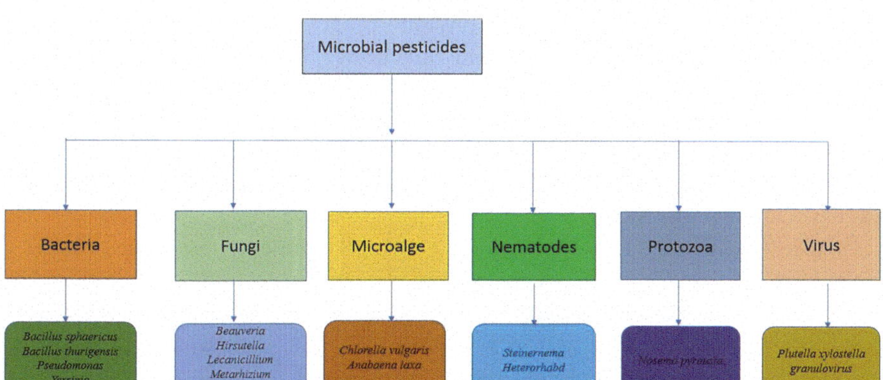

Fig. 4 Sources of microbial-based biopesticides. (Source: own study based on Ruiu (2018))

Lecanicillium play a special role, attacking insects and other pests. Biopesticides based on these organisms are widely used in agriculture, mainly against insects (BioMed Central). In addition, studies on microalgae, such as *Chlorella vulgaris* or *Anabaena laxa*, show that their bioactive compounds can act as natural insecticides, especially in relation to some mosquitoes (Aioub et al., 2024).

Nematodes from the genera *Steinernema* and *Heterorhabditis* are also effectively used in pest control, mainly in the soil, attacking their larvae. Viruses, such as *Plutella xylostella* granulovirus, effectively control specific insect species (Hamrouni et al., 2024).

An assessment of the global market sales of agricultural chemicals and biopesticides indicates the growing potential of the latter. In 2023, the market sales of pesticides and other agricultural chemicals amounted to over USD 98 billion and the expected sales in 2027 amounted to USD 139 billion with the resulting compound annual growth rate (CAGR) of 9.1% (Anonymous, 2023).

Biopesticides, especially those based on microorganisms, are a promising alternative with high biodegradability, specificity, suitability for incorporation into integrated pest management practices, low probability of resistance development and practically no risk to human health; however, their limitations include shorter shelf life and susceptibility to environmental conditions, which requires further research and innovation (Hamrouni et al., 2024).

7.3 *Trichoderma*

Fungi of the genus *Trichoderma* are widely used as biological plant protection agents due to their rapid growth, adaptation to various conditions, and antagonistic properties towards pathogens (Kosicka et al., 2014; Saravanakumar & Wang, 2020; Guzmán-Guzmán et al., 2023). They act through hyperparasitism, antibiosis, and the production of hydrolytic enzymes such as chitinases and glucanases, which destroy the cell walls of pathogens (Walid et al., 2022). *Trichoderma* also produces metabolites (gliotoxin, harzivin) and phytohormones that support plant growth, improving their resistance to stress and infections (Rakibuzzaman et al., 2021). "Harzianic acid", a well-documented metabolite produced by *Trichoderma harzianum*. Harzianic acid has antifungal properties and promotes plant growth by increasing resistance to stress and infections. Studies show an increase in potato yields from 11.3% to 23.8%, with a simultaneous decrease in infection by 1.6% (Rakibuzzaman et al., 2021).

Thanks to its ability to colonize roots and high competitiveness for resources, *Trichoderma* is an effective tool in integrated plant protection, especially in the control of diseases caused by Rhizoctonia solani, with an efficiency of up to 78% (Walid et al., 2022).

The latest studies also emphasize the effectiveness of *Trichoderma* in the control of diseases such as *Rhizoctonia* solani and pre-emergence seedling dieback, achieving biological effectiveness of over 64%. It is worth adding that the development of biocontrol technology and the formulation of *Trichoderma*-based products allows for further improvement in their effectiveness, especially in the context of changing climatic conditions and soil types (Rakibuzzaman et al., 2021; Gonçalves, 2021; Guzmán-Guzmán et al., 2023).

7.4 Bacillus thuringiensis *(Bt)*

Bacillus thuringiensis (Bt) belongs to spore-forming bacteria characterized by the ability to produce crystalline proteins (Cry). A characteristic feature of Cry proteins is toxicity to many insects, and therefore, Bt is used as a microbiological insecticide improving plant resistance through genetic modification (Sajid et al., 2018). When larvae consume the protein present in the intestines, digestive enzymes activate tyrosine, which is followed by paralysis and death of the larva (Azizoglu et al., 2023; Abdullah & Zahoor, 2023). The advantage of Cry proteins is their specificity of action for specific groups of insects and the lack of negative impact on non-target organisms, which include humans and animals and beneficial insects (Osman et al., 2015; Sánchez-Yáñez et al., 2022). The beneficial properties of Bt also include biofertilizing and plant growth stimulating (PGP) activity, biodegradation and bioregeneration of various pollutants, such as heavy metals, pesticides, and biosynthesis of various value-added biopolymers, including polyhydroxyalkanoate, which are new environmental and industrial features (Kumar et al., 2021a; Azizoglu et al., 2023). The beneficial effects of Bt PGP activity were confirmed in studies on tomato (Kouam et al., 2023), *Abelmoschus esculentus* (Bandopadhyay, 2020), corn (Amogou et al., 2019), pepper (Guenoun et al., 2019), and wheat (Delfm & Dijoo, 2021). In addition, some Bt strains exhibit increased resistance to drought and salinity stress by altering antioxidant signal parameters, dehalogenase production, and bacteriocin synthesis (Kahtani et al., 2021; Hossain & Islam, 2022). *Bacillus thuringiensis* can be used to control various important plant pests, including mosquito larvae. In addition to producing endotoxins that cause the destruction of larval intestinal tissues, microbial pesticides also compete with pathogens at the site of infection; this is possible due to the use of nutrients and better colonization than pathogens. Such action of microbial agents reduces post-harvest spoilage of fruits and vegetables (Yousuf et al., 2018; Elik et al., 2019; Azizoglu et al., 2023). Recent studies also highlight the benefits of Bt as a plant growth promoter (PGP) that supports yield and increases resistance to stress, including drought and salinity, through the production of antioxidants, dehalogenases, and bacteriocins. Bt strains can also degrade heavy metals and pesticides, making them a valuable tool in bioremediation and sustainable food production (Bandopadhyay, 2020). The use of Bt has shown positive results in crops such as tomatoes, okra (*Abelmoschus esculentus*), maize, peppers, and wheat (Delfm & Dijoo, 2021; Hossain & Islam, 2022). In addition to its insecticidal properties, Bt helps reduce post-harvest spoilage of fruits and vegetables by better competing with pathogens, both in terms of nutrients and colonization space. In this way, Bt acts not only as a biopesticide but also as a crop durability enhancer (Hossain & Islam, 2022; Azizoglu et al., 2023). Kouam et al. (2023) demonstrated that more resistant strains are being developed that can better cope with difficult climatic conditions, making them even more effective tools in sustainable agriculture.

7.5 Microalgae and Cyanobacteria

Due to the introduction of regulations restricting the use of mineral fertilizers and chemical products in agriculture, including pesticides, more sustainable and environmentally friendly solutions are being sought to improve agricultural productivity. Therefore, products of biological origin are increasingly used, including microalgae and cyanobacteria, which are a valuable source for the production and protection of crops due to the potential of biofertilization and biostimulation (Gonçalves, 2021). Microalgae and cyanobacteria are a source of biologically active compounds, including proteins, phenolic compounds, polysaccharides with antibiotic, antioxidant, antifungal, and plant growth-promoting effects (Asimakis et al., 2022). Microalgae and cyanobacteria participate in stimulating the immune system of plants thanks to the production of enzymes with a defense mechanism (Encarnação et al., 2015). In addition, living organisms, both prokaryotes (e.g., nitrogen-fixing cyanobacteria) and eukaryotes (e.g., microalgae and macroalgae, seaweeds), are used in fertilizing soils and stimulating plant growth (Viana, 2022).

7.6 Protozoa

Protozoa (protozoa) are a diverse group of unicellular organisms that are used in pest biocontrol due to their ability to weaken insects. The main mechanism of action of protozoa is that their spores are ingested by the host and then germinate in the insect's midgut. The released sporoplasma attacks the target cells of the insect and leads to infection. This causes a gradual weakening of the insect, affecting its feeding, development, fertility, and shortening its lifespan (Kesho, 2020).

As a result of protozoal infection, the metabolic processes and immune system of the insect are weakened, making it more susceptible to other pathogens and biocontrol agents. Additionally, protozoa such as *Nosema* can act for long term, which makes them a useful tool in integrated plant protection (Meshjel, 2024). The most frequently studied protozoa in biocontrol are *Nosema* and *Vavraia,* which are effective in controlling pests, especially insects from the orders *Lepidoptera and Diptera.* New studies also indicate the possibility of using these organisms in more diverse cropping systems (Kesho, 2020; Meshjel, 2024).

7.7 Nematode

Like other microbial biopesticides, entomopathogenic nematodes (EPNs) mainly include a group of soil nematodes characterized by low requirements needed for their life and survival (Kumar et al., 2021a; Angon et al., 2023). They have the ability to search for target insects in the soil, on the surface of plants, partially embedded insects in plant tissues and quickly kill target insects by releasing symbiotic bacteria (Shapiro-Ilan et al., 2002; Kozak-Ciężczyk, 2005; Van Lenteren, 2012; Egbuna et al., 2020). They are also completely safe for the environment, humans, and animals. The most common entomopathogenic nematodes (EPNs) include *Steinernema scapterisci, Symbiobacterium thermophilum, Steinernema carpocapsae, Steinernema feltiae, Heterorhabditis megidis*, and *Steinernema riobravis* (Egbuna et al., 2020; Angon et al., 2023). New research directions focus on improving the effectiveness of nematodes by combining them with other biocontrol methods and optimizing application in different cultivation conditions. Studies indicate the possibility of synergy between EPNs and entomopathogenic fungi, such as *Beauveria bassiana*, which increases the effectiveness of biocontrol of both soil and ground pests. In addition, new technologies for storing and applying nematodes are being developed, which extend their durability and improve the efficiency of application in difficult climatic conditions (Angon et al., 2023; Hamrouni et al., 2024). New studies also indicate the possibility of genetic modification of nematodes, which may improve their pest-seeking abilities and increase their resistance to unfavorable soil conditions, such as drought and salinity (Hamrouni et al., 2024).

7.8 Viral Biopesticides

The main viruses infecting insects (butterflies, wasps, dipteran, and orthopteran beetles) classified as biopesticides include baculoviruses (BV), nucleopolyhedrosis viruses (NPV), and granuloviruses (GV) (Moscardi et al., 2011). After ingestion by larvae, the protein coat dissolves in the midgut, and the virions penetrate the epithelial cells of the midgut, where they then infect the fat body, epidermis, trachea, muscles, gonads, hemocytes, nervous and endocrine systems (Bhati et al., 2016). After an incubation period of 5–7 days (sometimes 20 days), infected larvae stop feeding, become white (NPV) or very dark (GV), climb to the top of the canopy, the body contents liquefy, and they die within 3–8 days (Padavala et al., 2017).

Recent research on insect viruses as biopesticides has expanded our understanding of their role in sustainable control of insect populations (Irsad et al., 2023). Baculoviruses, including nucleopolyhedroviruses (NPV) and granuloviruses (GV), target pest insects such as butterflies and beetles, infecting their midgut cells upon ingestion, ultimately leading to death through widespread infection of key systems. Baculoviruses are highly host-specific, limiting the risk to non-target species, including beneficial insects and vertebrates. NPV and GV, both members of the *Baculoviridae* family, have circular double-stranded DNA genomes. NPVs typically produce larger occlusion bodies (OBs) with multiple virions, while GVs produce smaller OBs with a single virion. These OBs protect the virus particles in the environment until they are ingested by the host insect, which ultimately leads to infection. Recent studies show promising results in the control of *Helicoverpa armigera* and *Helicoverpa zea*, pests that have developed resistance to chemical pesticides. Field experiments have shown significant reductions in pest populations using NPV-based products. Further studies are optimizing virus application rates and formulations and exploring synergies with other biological control methods to increase efficacy and minimize environmental impact. These advances in knowledge position insect viruses as key tools in sustainable pest control systems (Kesho, 2020; Irsad et al., 2023).

7.9 Effective Microorganisms

EM preparations (effective microorganisms) concern the main, naturally occurring microorganisms, such as lactic acid bacteria (*Lactobacillus casei, Streptococcus lactis*), yeasts (*Saccharomyces albus, Candida utilis*), actinomycetes (*Streptomyces albus, S. griseus*), photosynthetic bacteria (*Rhodopseudomonas palustris, Rhodobacter* sp.), and mold fungi (*Aspergillus oryzae, Mucor hiemalis*) (Martyniuk & Księżniak, 2011; Paśmionka & Kotarba, 2015; Pszczółkowski & Sawicka, 2018; Egbuna et al., 2020). EM and herbal extracts can be effectively used as fungicides in plant production, reducing the use of plant protection products. However, in recent years, questions have arisen about the safety of using some biological preparations for human health and the environment. Therefore, it is necessary to unconditionally comply with the regulations concerning the impact on health and the environment and to register biological preparations in plant protection (Martyniuk, 2010; Martyniuk & Księżniak, 2011; Kołodziejczyk, 2014; Netala et al., 2016; Pszczółkowski & Sawicka, 2018). In the European Union, such regulations concerning viruses, microbiological agents and other solutions, while decisions concerning beneficial organisms, such as parasitic insects, predatory mites, and insecticidal nematodes, are left to the discretion of individual member states (Pacholczak et al., 2015; Pan et al., 2023). A special place in international work on biological plant protection agents is occupied by macroorganisms, which include, among others, parasitic and predatory insects, predatory mites, and insecticidal nematodes (Table 3).

Table 3 Biopreparations based on living organisms (macroorganisms) available on the European market for use in crops under cover

Name and characteristics a living organism	The scope of application
Adalia bipunctata Predatory ladybird (colon)	Aphids in crops vegetables and decorative plants under covers
Amblyseius andersoni Predatory mite	Spider mites, other mites herbivores in crops vegetables and decorative plants under covers
Amblyseius barkeri Predatory mite	Trips in crops vegetables and decorative plants under cover
Amblyseius californicus Predatory mite	Spider mites, other mites herbivorous and trips in crops vegetables and decorative plants after shields
Amblyseius cucumeris Predatory mite	Trips in crops vegetables and decorative plants under covers
Amblyseius swirskii Predatory mite	Whiteflies, spider mites and trips in crops vegetables and decorative plants under covers
Aphelinus abdominalis Predatory diphtheria	Aphids in crops vegetables and decorative plants under covers
Aphidius colemani Parasitic wasp (greenhouse aphid)	Aphids in crops vegetables and decorative plants under covers
Aphidius matricariae Parasitic wasp	Aphids in crops vegetables and decorative plants under covers
Aphidoletes aphidimyza, a predatory dipter feeding on aphids	Aphids in crops of vegetables and decorative plants under cover
Atheta coriaria Predatory ladybird	Ground and pupae trips in crops vegetables and decorative plants under covers
Chrysoperla carnea predatory networkers (golden eyed) common)	Trips, spider mites and whiteflies in up
Cryptolaemus montrouzieri Predatory ladybird	Mealybugs and June w plant crops decorative under covers
Dacnusa sibirica + *Diglyphus* ISAE Parasitic wasps	Miners in crops vegetables and decorative plants under covers
Delphastus pusillus Predatory ladybird (whitefly)	Whiteflies in the cultivation of vegetables and decorative plants under covers
Encarsia Formosa Parasitic wasp (greenhouse hygrometer)	Whiteflies in crops vegetables and decorative plants under covers
Encarsia formosa + *Eretmocerus fremitus* Parasitic wasps	Whiteflies in crops vegetables and decorative plants under covers
Epicurus Batiatus Predatory dipteran	Aphids in the cultivation of vegetables and ornamental plants under covers
Removers eremicus, E. *ervi* Parasitic wasps	Whiteflies in crops vegetables and decorative plants under covers

(continued)

Table 3 (continued)

Name and characteristics a living organism	The scope of application
Eretmocerus mundus Parasitic wasp	Whiteflies in crops vegetables and decorative plants under covers
Euseius gallicus Predatory mite	Whiteflies and trips in plant crops decorative under covers
Feltiella acarisuga Predatory diptera (minced pimples)	Spider mites in crops vegetables and decorative plants under covers
H. miles, H. aculeifer Predatory mites	Ground and pupae trips in crops vegetables and decorative plants under covers
Heterorhabditis bacteriophora Parasitic nematode	Beetle larvae (mainly gardeners, mycelium and cockchafer May) in crops vegetable
Leptomastix dactylopii Parasitic wasp (yellowish reddish)	Mealybugs and June w plant crops decorative under covers
Macrolophus melanotoma or *pygmeus*, the predatory bug (whitefly perennial)	Whiteflies, spider mites, trips in crops vegetables and decorative plants under covers
Neoseiulus degenerans Predatory mite (benefactor brown)	Trips in crops vegetables and decorative plants under covers
Neoseiulus fallacis Predatory mite	Spider mites, other mites herbivores in crops vegetables and decorative plants after shields
Nesidiocoris tenuis Predatory bugs	Whiteflies and *Tuta absoluta* in the cultivation of greenhouse tomato
O. laevigatus, O. majusculus, O. insidiosus, O. strigicollis Predatory bugs	Trips in vegetable and ornamental crops under covers
Phasmarhabditis hermaphrodita Parasitic nematode	Snails in crops vegetables and decorative plants under covers
Phytoseiulus persimilis Predatory mite (benefactor greenhouse)	Spider mites in crops vegetables and decorative plants under covers
Sphaerophoria rueppellii Predatory diptera	Aphids, whiteflies, thrips and spider mites in vegetable and plant crops decorative under covers
Steinernema carpocapsae Parasitic nematode	Caterpillars of butterflies in greenhouse cultivation
Steinernema feltiae Parasitic nematode	*Diptera* from the family ground worms in vegetable and plant crops decorative under covers and in mushroom farms
Stethorus punctillum Predatory ladybug	Spider mites in vegetable crops under covers

Source: Own based on Shapiro-Ilan et al. (2002), Kozak-Ciężczyk (2005), van Lenteren (2012), Egbuna et al. (2020), and Hamrouni et al. (2024)

8 Botanical Pesticides

The advantage of botanical pesticides is their low toxicity to mammals and non-target organisms, and their easy biodegradability (Toan et al., 2021). In addition, they contain a mixture of many substances whose effects can be attributed to different pest activities, and therefore it is difficult to develop pesticide resistance. Botanical pesticides are a mixture of chemicals obtained from different parts of plants, including leaves, stems, shoots, bark, flowers, and fruits (Kumar et al., 2021a). Botanical pesticides include plant extracts and essential oils, natural plant growth regulators, and secondary metabolites (Ngegba et al., 2022). Essential oils are a complex mixture of many volatile organic compounds, including, among others: alcohols, esters, aldehydes, ketones, viruses, amides, amines, phenols, and terpenes (Toan et al., 2021). Considering the type of plant and insect species and their morphological characteristics, essential oils and plant extracts can inhibit respiration, making it difficult for insects to identify host plants, as well as reduce the hatching of adults through ovicidal and larvicidal effects (Gupta et al., 2023). Botanical pesticides are gaining importance in research on alternative methods of plant protection, especially due to their low toxicity and biodegradability, which makes them safe for the environment. The latest research directions focus on the optimization of the use of plant extracts and essential oils and on the mechanisms of their action on pests. Particular attention is paid to the discovery of new bioactive compounds in plants and studies on the synergistic effects of various substances in mixtures, which can increase the effectiveness of biopesticides and reduce the development of resistance in pests (Toan et al., 2021; Kumar et al., 2021b; Gupta et al., 2023). Recent studies also indicate the growing importance of plants from the families *Lamiaceae, Rutaceae, Apiaceae,* and *Myrtaceae* (Ngegba et al., 2022). Terpenoid compounds, flavonoids and alkaloids, are intensively studied for their toxicity to crop pests such as aphids and beetles, but also in the context of their potential in the control of soil pathogens. New research on botanical nanopesticides, which combine natural plant extracts with nanomaterials, has shown their increased stability and prolonged action. The development of nanotechnology in the context of biopesticides also allows for a reduction in the application rate and minimization of the impact on non-target organisms. New directions also include research on resistance to environmental conditions and effectiveness in a wide temperature range, which may increase their use in organic farming (Ngegba et al., 2022; Gupta et al., 2023).

8.1 Nanopesticides

Synthetic pesticides are characterized by poor water solubility and low effectiveness after their application, which leads to their accumulation in the environment (Yadav & Devi, 2017). Therefore, the incorporation of nanotechnology into the production of nanopesticides characterized by controlled release kinetics and increased permeability, stability, and solubility is an important method that can be used in various agricultural systems, which is an important attribute supporting its mass application (Yin et al., 2023; Bhandari et al., 2023; El-Ramady et al., 2023). Nanotechnology is a guarantee of new agrochemicals enabling more efficient mechanisms of their delivery in order to improve crop productivity, and can also contribute to reducing the use of plant protection products. According to Usman et al. (2020), the use of nanotechnology can increase agricultural production and plant protection products in the following areas: (1) nanoformulation of agrochemicals for the use of plant protection products and fertilizers; (2) the use of nanosensors in crop protection to identify diseases and agrochemical residues; (3) nanodevices for plant genetic engineering; (4) plant disease diagnostics; (5) improving animal health in farming; (6) post-harvest crop management. An important advantage of nanopesticides is their excellent thermal stability and biodegradability, which is very important for their mass use (Periakaruppan et al., 2023; Shangguan et al., 2024). Nanoparticles (NPs) are characterized by new and improved properties, which include, among others, their size (1–100 nm), shape, and structure (Yin et al., 2023). Nanoparticles sprayed on leaves are taken up by plant cells by endocytosis through stomata or cuticles on the leaf surface (Hu et al., 2020). The uptake of nanoparticles by the foliar route, as reported by the authors, depends on the leaf morphology, pore size, and stomata density. In addition, the penetration of nanoparticles into the cuticle and leaf mesophyll tissue is also influenced by their hydrophobicity, size, and chemical coating. Two types of nanoparticles are known in the environment that can be synthesized: (a) organic; (b) inorganic. Organic nanoparticles include carbon nanoparticles, while inorganic nanoparticles include metallic nanoparticles such as (Au, Ag, Cu, Al), magnetic (Co, Fe, Ni), and semiconducting (ZnO, ZnS, CdS) (Jeevanandam et al., 2018; Sajid & Płotka-Wasylka, 2020). The use of inorganic nanoparticles as pesticide carriers increases their antimicrobial properties, which is why they can be used as bioactive substances (Xin et al., 2020). Silver nanoparticles have the strongest effect of this type. Scientific studies have confirmed the antibacterial, antifungal, and antiviral effects of silver nanoparticles against plant pathogens (Netala et al., 2016). Other nanomaterials also have antibacterial and antifungal properties, including metallic copper nanoparticles (Rai et al., 2018), titanium dioxide (Pan et al., 2023), and zinc oxide (Sardar et al., 2022). In addition, nanoparticle silicon oxide (El-Naggar et al., 2020), nanosilver (Dhaka et al., 2023), and nanoaluminum (Jasrotia et al., 2022) exhibit strong insecticidal activity. Nanopesticides, which are advanced forms of chemical plant protection products, are gaining popularity due to their improved physicochemical properties, such as greater stability, solubility, and controlled release kinetics. This makes them more effective and has a smaller environmental impact than traditional synthetic pesticides (Dhaka et al., 2023). New research focuses on the wide application of organic nanoparticles (e.g., carbon nanoparticles) and inorganic nanoparticles (e.g., metal nanoparticles such as silver, copper and zinc oxide), which exhibit strong antimicrobial, insecticidal, and antifungal properties. Silver nanoparticles, which are distinguished by their antibacterial and antiviral activity against plant pathogens, attract special attention in research. Other materials, such as copper nanoparticles, titanium oxide, zinc oxide and nanosilver, also exhibit strong biocidal properties (Ayilara et al., 2023a, b; Pan et al., 2023). New directions of research on nanopesticides include the development of nanosensors and nanodevices that can precisely identify plant diseases and optimize the dosage of nanopesticides. There are also studies on the possibility of integrating nanopesticides with other technologies, such as plant biotechnology or air and water quality control in agricultural systems (Bhandari et al., 2023; Dhaka et al., 2023).

9 Summary

This chapter discusses the wide application of various microbiological and botanical plant protection products, as well as modern nanopesticides. The benefits of using bacteria, fungi, insecticidal nematodes, botanical pesticides, and nanopesticides as an alternative to chemical plant protection products are presented. The mechanisms of their action are described, including the production of toxins and enzymes by *Bacillus thuringiensis* (Bt), the antagonistic properties of *Trichoderma fungi,* and the effectiveness of nematodes in pest control. Nanopesticides are distinguished by the use of nanotechnology, which improves their efficiency, stability, and biodegradability. Nanoparticles such as silver, copper, zinc oxide, and others have strong biocidal properties, which make them increasingly used in agriculture. This chapter draws attention to new directions of research that focus on improving the safety and efficiency of biopesticides and nanopesticides, while minimizing their impact on the environment.

References

Abdullah, S., & Zahoor, I. (2023). Biopesticides: A green substitute to chemical pesticide. *International Journal of Chemical and Biochemical Sciences, 24*(4), 141–156.

Abdulridha, J., Ampatzidis, Y., Kakarla, S. C., & Roberts, P. (2020). Detection of target spot and bacterial spot diseases in tomato using UAV-based and benchtop-based hyperspectral imaging techniques. *Precision Agriculture, 21*, 955–978.

Abdulridha, J., Min, A., Rouse, M. N., Kianian, S., Isler, V., & Yang, C. (2023). Evaluation of stem rust disease in wheat fields by drone hyperspectral imaging. *Sensors, 23*(8), 4154.

Acharya, B. S., Bhandari, M., Bandini, F., Pizarro, A., Perks, M., Joshi, D. R., Wang, S., Dogwiler, T., Ray, R. L., Kharel, G., & Sharma, S. (2021a). Unmanned aerial vehicles in hydrology and water management: Applications, challenges, and perspectives. *Water Resources Research, 57*(11), e2021WR029925.

Acharya, S., et al. (2021b). Innovative UAV applications for precision irrigation. *Sensors, 21*(12), 4077.

Adlak, T., Tiwari, S., Tripathi, M. K., Gupta, N., Sahu, V. K., Bhawar, P., & Kandalkar, V. S. (2019). Biotechnology: An advanced tool for crop improvement. *Current Journal of Applied Science and Technology, 33*(1), 1–11.

Ahmed, R. A., Hemdan, E. E. D., El-Shafai, W., Ahmed, Z. A., El-Rabaie, E. S. M., & Abd El-Samie, F. E. (2022). Climate-smart agriculture using intelligent techniques, blockchain and Internet of Things: Concepts, challenges, and opportunities. *Transactions on Emerging Telecommunications Technologies, 33*(11), e4607.

Aioub, A. A. A., Ghosh, S., AL-Farga, A., et al. (2024). Back to the sources: Biopesticides as promising alternatives to conventional agrochemicals. *European Journal of Plant Pathology, 169*, 697–713.

Amogou, O., Agbodjato, N. A., Dagbénonbakin, G., Noumavo, P. A., Sina, H., Sylvestre, A. A., Adoko, M. Y., Nounagnon, M., Kakaï, R. G., Adjanohoun, A., & Baba-Moussa, L. (2019). Improved maize growth in condition controlled by PGPR inoculation on Ferruginous soil in Central Benin. *Food and Nutrition Sciences, 10*(12), 1433.

Angon, P. B., Mondal, S., Jahan, I., Datto, M., Antu, U. B., Ayshi, F. J., & Islam, M. S. (2023). Integrated pest management (IPM) in agriculture and its role in maintaining ecological balance and biodiversity. *Advances in Agriculture, 2023*(1), 5546373. https://doi.org/10.1155/2023/5546373

Anonymous. (2023). Research_and_market. *Global pesticides and other agricultural chemicals market report 2023*. Available at: https://www.researchandmarkets.com/reports/5240332/pesticide-and-other-agricultural-chemicals-global#src-pos-2

Ashourloo, D., et al. (2020a). Mapping crop types using Sentinel-2 data. *Journal of Applied Remote Sensing, 14*(3), 034517.

Ashourloo, D., Shahrabi, H. S., Azadbakht, M., Rad, A. M., Aghighi, H., & Radiom, S. (2020b). A novel method for automatic potato mapping using time series of Sentinel-2 images. *Computers and Electronics in Agriculture, 175*, 105583. https://doi.org/10.1016/j.compag.2020.105583

Asimakis, E., Shehata, A. A., Eisenreich, W., Acheuk, F., Lasram, S., Basiouni, S., Emekci, M., Ntougias, S., Taner, G., May-Simera, H., Yilmaz, M., & Tsiamis, G. (2022). Algae and their metabolites as potential bio-pesticides. *Microorganisms, 10*(2), 307.

Awad Fahad, A. (2023a). Modern techniques in integrated pest management to achieve sustainable agricultural development. *International Journal of Family Studies, Food Science and Nutrition Health, 4*(1), 1–14.

Awad Fahad, M. (2023b). Integrated pest management: Current trends and future perspectives. *Pest Management Science, 79*(4), 134–147.

Ayilara, M. S., Adeleke, B. S., Akinola, S. A., Fayose, C. A., Adeyemi, U. T., Gbadegesin, L. A., Omole, R. K., Johnson, R. M., Uthman, Q. O., & Babalola, O. O. (2023a). Biopesticides as a promising alternative to synthetic pesticides: A case for microbial pesticides, phytopesticides, and nanobiopesticides. *Frontiers in Microbiology, 14*, 1040901.

Ayilara, M. S., et al. (2023b). Biopesticides for sustainable agriculture: Potential and future prospects. *Journal of Plant Pathology, 105*(2), 401–416.

Azizoglu, U., Salehi Jouzani, G., Sansinenea, E., & Sanchis-Borja, V. (2023). Biotechnological advances in *Bacillus thuringiensis* and its toxins: Recent updates. *Reviews in Environmental Science and Bio/Technology, 22*(2), 319–348.

Balasundram, S. K., Shamshiri, R. R., Sridhara, S., & Rizan, N. (2023a). The role of digital agriculture in mitigating climate change and ensuring food security: An overview. *Sustainability, 15*(6), 5325. https://doi.org/10.3390/su15065325

Balasundram, S. K., et al. (2023b). Automation and data analytics in modern agriculture. *Precision Agriculture Journal, 19*(2), 145–162.

Bandopadhyay, S. (2020). Application of plant growth promoting *Bacillus thuringiensis* as biofertilizer on *Abelmoschus esculentus* plants under field condition. *Journal of Pure and Applied Microbiology, 14*(2), 1287–1294.

Barbedo, J. G. A. (2021). Agricultural applications of UAV-based thermal imaging: A review. *Computers and Electronics in Agriculture, 183*, 105999.

Bartkowiak, A., & Bartkowiak, P. (2017a). Technical and technological progress in the context of sustainable development of agriculture in Poland. *Procedia Engineering, 182*, 66–75.

Bartkowiak, P., & Bartkowiak, J. (2017b). Zmiany klimatyczne a wydajność upraw rolnych w Polsce. *Wiadomości Rolnicze, 27*(5), 110–119. (in Polish).

Benos, L., Tagarakis, A. C., Dolias, G., Berruto, R., Kateris, D., & Bochtis, D. (2021a). Machine learning in agriculture: A comprehensive updated review. *Sensors, 21*, 3758. https://doi.org/10.3390/s21113758

Benos, L., et al. (2021b). Machine learning algorithms for yield prediction: A review. *Computers and Electronics in Agriculture, 191*, 106486.

Berge, T. W., Goldberg, S., Kaspersen, K., & Netland, J. (2012). Towards machine vision based site-specific weed management in cereals. *Computers & Electronics in Agriculture, 81*, 79–86.

Bhandari, G., Dhasmana, A., Chaudhary, P., Gupta, S., Gangola, S., Gupta, A., Rustagi, S., Shende, S. S., Rajput, V. D., Minkina, T., Malik, S., & Slama, P. (2023). A perspective review on green nanotechnology in agro-ecosystems: Opportunities for sustainable agricultural practices & environmental remediation. *Agriculture, 13*(3), 668.

Bhargava, K., Abhishek, E., Madhusudhan, B., Naveen, A., Akhil, V. S., Yadav, T. V., Mudhale, A., & Kumar, G. A. (2023). A review of rapid generation advancement (RGA) in crop improvement. *International Journal of Plant & Soil Science, 35*(7), 138–145.

Bhati, S. S., Baheti, B. L., & Jangid, K. C. (2016). Bio-pesticides and their role in agriculture. *Marumegh, 2*(2), 40–46.

Böckmann, E., & Meyhöfer, R. (2015). AEP – Eine automatische Entscheidungshilfe-Software für den integrierten Pflanzenschutz. *Gesunde Pflanzen, 67*(1), 1–10.

Cai, Y., Guan, K., Nafziger, E., Chowdhary, G., Peng, B., Jin, Z., Wang, S., & Wang, S. (2019). Detecting in-season crop nitrogen stress of corn for field trials using UAV-and CubeSat-based multispectral sensing. *IEEE Journal of Selected Topics in Applied Earth Observations and Remote Sensing, 12*(12), 5153–5166.

Chandra, A. K., Kumar, A., Bharati, A., Joshi, R., Agrawal, A., & Kumar, S. (2020). Microbial-assisted and genomic-assisted breeding: A two way approach for the improvement of nutritional quality traits in agricultural crops. *3 Biotech, 10*(1), 2.

Chataut, G., Bhatta, B., Joshi, D., Subedi, K., & Kafle, K. (2023). Greenhouse gases emission from agricultural soil: A review. *Journal of Agriculture and Food Research, 11*, 100533.

Chin, R., Catal, C., & Kassahun, A. (2023). Plant disease detection using drones in precision agriculture. *Precision Agriculture, 24*(5), 1663–1682.

Chouhan, S., Sharma, K., & Guleria, S. (2017). Antimicrobial activity of some essential oils—present status and future perspectives. *Medicines, 4*(3), 58.

Chowdhury, S. K., Banerjee, M., Basnett, D., & Mazumdar, T. (2024). Natural pesticides for pest control in agricultural crops: An alternative and eco-friendly method. *Plant Science Today, 11*(1), 433–450. https://doi.org/10.14719/pst.2547

Christensen, S., Søgaard, H. T., Kudsk, P., Nørremark, M., Lund, I., Nadimi, E. S., & Jørgensen, R. (2009). Site specific weed control technologies. *Weed Research, 49*, 233–241.

Clevers, J. G. P. W., Kooistra, L., & Van den Brande, M. M. M. (2017). Using Sentinel-2 data for retrieving LAI and leaf and canopy chlorophyll content of a potato crop. *Remote Sensing, 9*, 405. https://doi.org/10.3390/rs9050405

Coll, M., & Wajnberg, E. (2017). Environmental pest management: A call to shift from a pest-centric to a system-centric approach. In *Environmental pest management: Challenges for agronomists, ecologists, economists and policymakers* (pp. 1–17). Wiley.

Croft, H., & Chen, J. M. (2018). Leaf pigment content. *Comprehensive Remote Sensing, 3*, 117–142.

Cuevas, C. M., Saludes, R., Dorado, M., & Cruz, P. S. (2021). Assessment of nutrient status of lowland irrigated rice (*Oryza sativa* L.) using low altitude remote sensing. *Philippine Journal of Agricultural and Biosystems Engineering, 17*(1), 39–48.

Daraban, G. M., Hlihor, R. M., & Suteu, D. (2023). Pesticides vs. biopesticides: From pest management to toxicity and impacts on the environment and human health. *Toxics, 11*(12), 983.

Darban, S., et al. (2023). Climate-resilient crop varieties: Strategies for adaptation. *Journal of Crop Science, 13*(1), 77–94.

Darvishzadeh, R., Skidmore, A., Atzberger, C., & van Wieren, S. (2008). Estimation of vegetation LAI from hyperspectral reflectance data: Effects of soil type and plant architecture. *International Journal of Applied Earth Observation and Geoinformation, 10*(3), 358–373.

Dash, P. B., Naik, B., Nayak, J., & Vimal, S. (2022a). Socio-economic factor analysis for sustainable and smart precision agriculture: An ensemble learning approach. *Computer Communications, 182*, 72–87.

Dash, S. K., et al. (2022b). Predicting crop yields under climate variability using machine learning techniques. *Computers and Electronics in Agriculture, 175*(3), 105612.

Delfm, J., & Dijoo, Z. K. (2021). *Bacillus thuringiensisas* a biofertilizer and plant growth promoter. In G. H. Dar, R. A. Bhat, M. A. Mehmood, & K. R. Hakeem (Eds.), *Microbiota and biofertilizers* (Vol. 2). Springer. https://doi.org/10.1007/978-3-030-61010-4_12

Dhaka, A., Mali, S. C., Sharma, S., & Trivedi, R. (2023). A review on biological synthesis of silver nanoparticles and their potential applications. *Results in Chemistry*, 101108. https://doi.org/10.1016/j.rechem.2023.101108

Dhakal, R., Maimaitijiang, M., Chang, J., & Caffe, M. (2023). Utilizing spectral, structural and textural features for estimating oat above-ground biomass using UAV-based multispectral data and machine learning. *Sensors, 23*(24), 9708.

Dobbs, A. M., Ginn, D., Skovsen, S. K., Bagavathiannan, M. V., Mirsky, S. B., Reberg-Horton, C. S., & Leon, R. G. (2022). New directions in weed management and research using 3D imaging. *Weed Science, 70*, 641–647. https://doi.org/10.1017/wsc.2022.56

Dwivedi, P., & Sopory, S. K. (2023a). Unravelling the mechanism of stress responses in crop plants. *Journal of Plant Growth Regulation, 42*(10), 5961–5964.

Dwivedi, S., & Sopory, S. K. (2023b). Advances in crop stress management. *Plant Science Reviews, 42*(1), 33–49.

Egbuna, C., Sawicka, B., Tijjani, H., Kryeziu, T. L., Ifemeje, J. C., Skiba, D., & Lukong, C. B. (2020). Chapter 4 – Biopesticides, safety issues and market trends. In C. Egbuna & B. Sawicka (Eds.), *Natural remedies for pest, disease and weed control* (Vol. 2020, pp. 43–53). Academic Press. https://doi.org/10.1016/B978-0-12-819304-4.00004-X. ISBN 9780128193044.

Ehler, L. E. (2006). Integrated pest management (IPM) definition, historical development and implementation, and the other IPM. *Pest Management Science, 62*, 787–789.

Elik, A., Yanik, D. K., Istanbullu, Y., Guzelsoy, N. A., Yavuz, A., & Gogus, F. (2019). Strategies to reduce post-harvest losses for fruits and vegetables. *Strategies, 5*(3), 29–39.

El-Naggar, M. E., Abdelsalam, N. R., Fouda, M. M., Mackled, M. I., Al-Jaddadi, M. A., Ali, H. M., Siddiqui, M. H., & Kandil, E. E. (2020). Soil application of nano silica on maize yield and its insecticidal activity against some stored insects after the post-harvest. *Nanomaterials, 10*(4), 739.

El-Ramady, H., Abdalla, N., Sári, D., Ferroudj, A., Muthu, A., Prokisch, J., Fawzy, Z. F., Brevik, E. C., & Solberg, S. Ø. (2023). Nanofarming: Promising solutions for the future of the global agricultural industry. *Agronomy, 13*, 1600. https://doi.org/10.3390/agronomy13061600

Encarnação, T., Pais, A. A., Campos, M. G., & Burrows, H. D. (2015). Cyanobacteria and microalgae: A renewable source of bioactive compounds and other chemicals. *Science Progress, 98*(2), 145–168.

Evans, R. G., & Sadler, E. J. (2008). Methods and technologies to improve efficiency of water use. *Water Resources Research, 44*(7). https://doi.org/10.1029/2007WR006200

FAO. (2020). *NSP–Integrated Pest Management, FAO definition*. http://www.fao.org/agriculture/crops/thematic-sitemap/theme/pests/ipm/en/

FAO. (2023). *FAO Statistical Yearbook 2023 – World Food and Agriculture*. https://reliefweb.int/report/world/fao-statistical-yearbook-2023-world-food-and-agriculture

Feng, H., Tao, H., Fan, Y., Liu, Y., Li, Z., Yang, G., & Zhao, C. (2022). Comparison of winter wheat yield estimation based on near-surface hyperspectral and UAV hyperspectral remote sensing data. *Remote Sensing, 14*(17), 4158.

Feng, X., Tian, H., Cong, J., & Zhao, C. (2023a). A method review of the climate change impact on crop yield. *Frontiers in Forests and Global Change, 6*, 1198186.

Feng, L., et al. (2023b). Climate-smart agriculture: Policy innovations and field practices. *Agronomy Journal, 18*(4), 230–245.

Feng, G., Gu, Y., Wang, C., Zhou, Y., Huang, S., & Luo, B. (2024). Wheat fusarium head blight automatic non-destructive detection based on multi-scale imaging: A technical perspective. *Plants, 13*(13), 1722.

Ferrante, A., & Mariani, L. (2018a). Crop water stress index as a tool for irrigation scheduling. *Journal of Agricultural and Food Chemistry, 66*(12), 3068–3077.

Ferrante, A., & Mariani, L. (2018b). Agronomic management for enhancing plant tolerance to abiotic stresses: High and low values of temperature, light intensity, and relative humidity. *Horticulture, 4*(3), 21. https://doi.org/10.3390/horticulturae4030021

Fu, Z., Yu, S., Zhang, J., Xi, H., Gao, Y., Lu, R., Zheng, H., Zhu, Y., Cao, W., & Liu, X. (2022). Combining UAV multispectral imagery and ecological factors to estimate leaf nitrogen and grain protein content of wheat. *European Journal of Agronomy, 132*, 126405.

Gai, J., Tang, L., & Steward, B. L. (2020). Automated crop plant detection based on the fusion of color and depth images for robotic weed control. *Journal of Field Robotics, 37*(1), 35–52.

Gerhards, R., & Oebel, H. (2006). Practical experiences with a system for site-specific weed control in arable crops using real-time image analysis and GPS-controlled patch spraying. *Weed Research, 46*, 185–193.

Gold, K. M., Townsend, P. A., Chlus, A., Herrmann, I., Couture, J. J., Larson, E. R., & Gevens, A. J. (2020). Hyperspectral measurements enable pre-symptomatic detection and differentiation of contrasting physiological effects of late blight and early blight in potato. *Remote Sensing, 12*, 286.

Gómez, D., Salvador, P., Sanz, J., & Casanova, J. L. (2019). Potato yield prediction using machine learning techniques and Sentinel 2 data. *Remote Sensing, 11*, 1745. https://doi.org/10.3390/rs11151745

Gonçalves, A. L. (2021). The use of microalgae and cyanobacteria in the improvement of agricultural practices: A review on their biofertilising, biostimulating and biopesticide roles. *Applied Sciences, 11*, 871. https://doi.org/10.3390/app11020871

Guenoun, K., Chattaoui, M., Bouri, M., Rhouma, A., Naghmouchi, K., & Raies, A. (2019). Biological control of growth promoting rhizobacteria against verticillium wilt of pepper plant. *Biologia, 74*, 237–250.

Gupta, I., Singh, R., Muthusamy, S., Sharma, M., Grewal, K., Singh, H. P., & Batish, D. R. (2023). Plant essential oils as biopesticides: Applications, mechanisms, innovations, and constraints. *Plants, 12*, 2916. https://doi.org/10.3390/plants12162916

Guzmán-Guzmán, P., Kumar, A., de los Santos-Villalobos, S., Parra-Cota, F. I., Orozco-Mosqueda, M. D. C., Fadiji, A. E., Hyder, S., Babalola, O. O., & Santoyo, G. (2023). *Trichoderma* species: Our best fungal allies in the biocontrol of plant diseases—A review. *Plants, 12*, 432. https://doi.org/10.3390/plants12030432

Hamrouni, R., Regus, F., Da Silva, A.-M. F., Orsiere, T., Boudenne, J. L., Laffont-Schwob, I., Christen, P., & Dupuy, N. (2024). Current status and future trends in microorganism- and nematode-based biopesticides for the biocontrol of crop pathogens. *Critical Reviews in Biotechnology, 2024*(10), 1–20. https://doi.org/10.1080/07388551.2024.2370370

Han, L., Yang, G., Dai, H., Xu, B., Yang, H., Feng, H., Li, Z., & Yang, X. (2019). Modeling maize above-ground biomass based on machine learning approaches using UAV remote-sensing data. *Plant Methods, 15*, 1–19.

Hasegawa, T., et al. (2022). Extreme climate events increase the risk of global food insecurity and adaptation needs. *Nature Food, 3*(2), 587–595. (Springer Nature).

Hossain, M. T., & Islam, T. (2022). Amelioration of salinity stress by *Bacillus* species as promoters of plant growth in saline soil. In *Bacilli in agrobiotechnology: Plant stress tolerance, bioremediation, and bioprospecting* (pp. 199–208). Springer International Publishing.

Hu, P., An, J., Faulkner, M. M., Wu, H., Li, Z., Tian, X., & Giraldo, J. P. (2020). Nanoparticle charge and size control foliar delivery efficiency to plant cells and organelles. *ACS Nano, 14*(7), 7970–7986.

Hunt, E. R., Horneck, D. A., Spinelli, C. B., Turner, R. W., Bruce, A. E., Gadler, D. J., Brungardt, J. J., & Hamm, P. B. (2018). Monitoring nitrogen status of potatoes using small unmanned aerial vehicles. *Precision Agriculture, 19*, 314–333.

Irsad, Shahid, M., Haq, E., Mohamed, A., Rizvi, P. Q., & Kolanthasamy, E. (2023). Biopesticides on entomopathogens application: Insights into exploring their application in insect pest control. *Frontiers in Microbiology, 2023*, 14. https://doi.org/10.3389/fmicb.2023.1208237

Jafar, A., Bibi, N., Naqvi, R. A., Sadeghi-Niaraki, A., & Jeong, D. (2024). Revolutionizing agriculture with artificial intelligence: Plant disease detection methods, applications, and their limitations. *Frontiers in Plant Science, 15*, 1356260. https://doi.org/10.3389/fpls.2024.1356260

Jasrotia, P., Nagpal, M., Mishra, C. N., Sharma, A. K., Kumar, S., Kamble, U., Bhardwaj, A. K., Kashyap, P. L., Kumar, S., & Singh, G. P. (2022). Nanomaterials for postharvest management of insect pests: Current state and future perspectives. *Frontiers in Nanotechnology, 3*, 811056. https://doi.org/10.3389/fnano.2021.811056

Jeevanandam, J., Barhoum, A., Chan, Y. S., Dufresne, A., & Danquah, M. K. (2018). Review on nanoparticles and nanostructured materials: History, sources, toxicity and regulations. *Beilstein Journal of Nanotechnology, 9*(1), 1050–1074.

Jełowicki, Ł., Sosnowicz, K., Ostrowski, W., Osińska-Skotak, K., & Bakuła, K. (2020). Evaluation of rapeseed winter crop damage using UAV-based multispectral imagery. *Remote Sensing, 12*(16), 2618.

Jia, X., Yin, D., Bai, Y., Yu, X., Song, Y., Cheng, M., Liu, S., Bai, Y., Meng, L., Liu, Y., Nan, F., Nie, C., Shi, L., Dong, P., Guo, W., & Jin, X. (2023). Monitoring maize leaf spot disease using multi-source UAV imagery. *Drones, 7*(11), 650.

Jurišić, M., et al. (2021a). Remote sensing and precision agriculture: A comprehensive review. *Precision Agriculture, 27*(4), 1–20.

Jurišić, M., Radočaj, D., Šiljeg, A., Antonić, O., & Živić, T. (2021b). Current status and perspective of remote sensing application in crop management. *Journal of Central European Agriculture, 22*(1), 156–166.

Kadirvel, P., Senthilvel, S., Geethanjali, S., Sujatha, M., & Varaprasad, K. S. (2015). Genetic markers, trait mapping and marker-assisted selection in plant breeding. In *Plant biology and biotechnology: Volume II: Plant genomics and biotechnology* (pp. 65–88). Springer.

Kahtani, A. L., Hafez, M. Y., Attia, K., Al-Ateeq, T., Ali, M. A., Hasanuzzaman, M., & Abdelaal, K. (2021). Bacillus thuringiensis and silicon modulate antioxidant metabolism and improve the physiological traits to confer salt tolerance in lettuce. *Plants, 10*(5), 1025.

Kanning, M., Kühling, I., Trautz, D., & Jarmer, T. (2018). High-resolution UAV-based hyperspectral imagery for LAI and chlorophyll estimations from wheat for yield prediction. *Remote Sensing, 10*(12), 2000.

Katimbo, A., Rudnick, D. R., DeJonge, K. C., Lo, T. H., Qiao, X., Franz, T. E., Nakabuye, H. N., & Duan, J. (2022). Crop water stress index computation approaches and their sensitivity to soil water dynamics. *Agricultural Water Management, 266*, 107575.

Kesho, A. (2020). Microbial bio-pesticides and their use in integrated pest management. *Chemical and Biomolecular Engineering, 5*(1), 26–34. https://doi.org/10.11648/j.cbe.20200501.15

Khargharia, G. (2021a). Fundamentals of electromagnetic radiation in remote sensing. *Journal of Space Research, 32*(2), 112–125.

Khargharia, R. (2021b). An overview on remote sensing: Principle and applications. *Agri-India Today, 1*(5), 12–15.

Kołodziejczyk, M. (2014). Effect of nitrogen fertilization and microbial preparations on potato yielding. Plant, Soil and Environment 60(8): 379-386, doi: 10.17221/7565-PSE.

Kosicka, D., Wolna-Maruwka, A., & Trzeciak, M. (2014). Aspekty stosowania *Trichoderma* sp. w ochronie roślin i rozkładzie materii organicznej. *Kosmos, 63*(4), 635–642. (in Polish).

Kouam, I. D., Mabah, J., Germain Ntsoli, P., Tchamani, L., Yaouba, A., Katte, B., & Bitom, D. (2023). Growth promotion potential of *Bacillus* spp. isolates on two tomato (*Solanum lycopersicum* L.) varieties in the West region of Cameroon. *Open Agriculture, 8*(1), 20220154.

Kozak-Ciężczyk, M. (2005). Molecular diagnostics in parasitology. *Cosmos. Problems of Biological Sciences, 54*(1) (266), 49–60. (in Polish).

Kross, A., McNairn, H., Lapen, D., Sunohara, M., & Champagne, C. (2015). Assessment of RapidEye vegetation indices for estimation of leaf area index and biomass in corn and soybean crops. *International Journal of Applied Earth Observation and Geoinformation, 34*, 235–248.
Kumar, J., Ramlal, A., Mallick, D., & Mishra, V. (2021a). An overview of some biopesticides and their importance in plant protection for commercial acceptance. *Plants, 10*(6), 1185.
Kumar, P., Kamle, M., Borah, R., Mahato, D. K., & Sharma, B. (2021b). *Bacillus thuringiensis* as microbial biopesticide: Uses and application for sustainable agriculture. *Egyptian Journal of Biological Pest Control, 31*(1), 95.
Lehmann, P., Ammunét, T., Barton, M., Battisti, A., Eigenbrode, S. D., Jepsen, J. U., Kalinkat, G., Neuvonen, S., Niemelä, P., Terblanche, J. S., Økland, B., & Björkman, C. (2020a). Complex responses of global insect pests to climate warming. *Frontiers in Ecology and the Environment, 18*(3), 141–150.
Lehmann, J., et al. (2020b). Biochar in agriculture: Prospects and challenges under a changing climate. *Global Change Biology, 26*(2), 291–309.
Lengai, G. M., & Muthomi, J. W. (2018). Mycotoxin contamination in food and its impacts on health and food security. *African Journal of Food Science, 12*(1), 1–10.
León-Rueda, W. A., Gómez-Caro, S., Mendoza-Vargas, L. A., León-Sánchez, C. A., & Ramírez-Gil, J. G. (2024). Linking the laboratory and the field in potato early dying detection: From spectral signatures to vegetation indices obtained with multispectral cameras coupled to drones. *Agronomy, 14*, 1569. https://doi.org/10.3390/agronomy14071569
Li, W., Zhu, X., Yu, X., Li, M., Tang, X., Zhang, J., Xue, Y., Zhang, C., & Jiang, Y. (2022). Inversion of nitrogen concentration in apple canopy based on UAV hyperspectral images. *Sensors, 22*(9), 3503.
Liu, Y., Feng, H., Yue, J., Fan, Y., Jin, X., Zhao, Y., Song, X., Long, H., & Yang, G. (2022). Estimation of potato above-ground biomass using UAV-based hyperspectral images and machine-learning regression. *Remote Sensing, 14*(21), 5449.
Liu, Y., Feng, H., Fan, Y., Yue, J., Chen, R., Ma, Y., Bian, M., & Yang, G. (2024). Improving potato above ground biomass estimation combining hyperspectral data and harmonic decomposition techniques. *Computers and Electronics in Agriculture, 218*, 108699.
Louargant, M., Jones, G., Faroux, R., Paoli, J. N., Maillot, T., Gée, C., & Villette, S. (2018). Unsupervised classification algorithm for early weed detection in row-crops by combining spatial and spectral information. *Remote Sensing, 10*(5), 761.
Lu, N., Wang, W., Zhang, Q., Li, D., Yao, X., Tian, Y., Zhu, Y., Cao, W., Baret, F., Liu, S., & Cheng, T. (2019). Estimation of nitrogen nutrition status in winter wheat from unmanned aerial vehicle based multi-angular multispectral imagery. *Frontiers in Plant Science, 10*, 1601. https://doi.org/10.3389/fpls.2019.01601
Lu, J., Eitel, J. U., Engels, M., Zhu, J., Ma, Y., Liao, F., Zheng, H., Wang, X., Yao, X., Cheng, T., Zhu, Y., Cao, W., & Tian, Y. (2021). Improving Unmanned Aerial Vehicle (UAV) remote sensing of rice plant potassium accumulation by fusing spectral and textural information. *International Journal of Applied Earth Observation and Geoinformation, 104*, 102592.
Luo, S., Jiang, X., He, Y., Li, J., Jiao, W., Zhang, S., Xu, F., Han, Z., Sun, I., Yang, Y., Wang, X., Ma, X., & Lin, Z. (2022). Multi-dimensional variables and feature parameter selection for aboveground biomass estimation of potato based on UAV multispectral imagery. *Frontiers in Plant Science, 13*, 948249.
MacPherson, J., Voglhuber-Slavinsky, A., Olbrisch, M., Schöbel, P., Dönitz, E., Mouratiadou, I., & Helming, K. (2022a). Future agricultural systems and the role of digitalization for achieving sustainability goals. A review. *Agronomy for Sustainable Development, 42*(4), 70.
MacPherson, M., et al. (2022b). Emerging technologies in agriculture: UAVs and IoT. *Journal of Agricultural Engineering, 34*(3), 78–92.
Martyniuk, S. (2010). Production of microbial preparations: Symbiotic bacteria of legumes as an example. *Journal of Research and Applications in Agricultural Engineering, 55*(4), 20–23.
Martyniuk, S., & Księżniak, J. (2011). Evaluation of pseudo-microbial biopreparations used in crop production. *Polish Journal of Agronomy, 6*, 27–33.

Meshjel, M. H. (2024). Diversity of free-living protozoa. *Wasit Journal for Pure Science, 2790–5241*, 2790–5233. https://www.researchgate.net/publication/381907835_Diversity_of_Free-Living_Protozoa#fullTextFileContent. Accessed 24 Sept 2024.

Michałek, R., & Grotkiewicz, K. (2010). Miejsce i rola postępu naukowego w warunkach rolnictwa zrównoważonego. *Problemy Inżynierii Rolniczej. Nr, 1*(67), 5–12. (in Polish).

Mirkowska, Z. (2010). Innowacje i innowacyjna gospodarka a rolnictwo. *Zagadnienia Ekonomiki Rolnej, 4*, 122–134. (in Polish).

Mołoń, A., & Durak, R. (2018a). Biopesticides as plant resistant stimulators. *Polish Journal for Sustainable Development, 22*(10), 69–74. https://doi.org/10.15584/pjsd.2018.22.1.9. [in Polish].

Mołoń, M., & Durak, R. (2018b). Wpływ zmian klimatu na plonowanie roślin rolniczych w Polsce. *Acta Agrophysica, 25*(1), 29–40. (in Polish).

Moscardi, F., de Souza, M. L., de Castro, M. E. B., Lara Moscardi, M., & Szewczyk, B. (2011). Baculovirus pesticides: Present state and future perspectives. In I. Ahmad, F. Ahmad, & J. Pichtel (Eds.), *Microbes and microbial technology*. Springer. https://doi.org/10.1007/978-1-4419-7931-5_16

Mulvihill, M. J., Beach, E. S., Zimmerman, J. B., & Anastas, P. T. (2011). Green chemistry and green engineering: A framework for sustainable technology development. *Annual Review of Environment and Resources, 36*(1), 271–293.

Narmilan, A., & Niroash, K. (2020a). Digital farming: Smart agriculture with IoT. *Journal of Agricultural Technology, 8*(2), 215–232.

Narmilan, A., & Niroash, G. (2020b). Reduction techniques for consequences of climate change by Internet of Things (IoT) with an emphasis on the agricultural production: A review. *International Journal of Science, Technology, Engineering and Management – A VTU Publication, 2*(3), 6–13.

Netala, V. R., Kotakadi, V. S., Domdi, L., Gaddam, S. A., Bobbu, P., Venkata, S. K., Ghosh, S. B., & Tartte, V. (2016). Biogenic silver nanoparticles: Efficient and effective antifungal agents. *Applied Nanoscience, 6*, 475–484.

Ngegba, P. M., Cui, G., Khalid, M. Z., & Zhong, G. (2022). Use of botanical pesticides in agriculture as an alternative to synthetic pesticides. *Agriculture, 12*(5), 600.

Nigon, T. J., Mulla, D. J., Rosen, C. J., Cohen, Y., Alchanatis, V., & Rud, R. (2014). Evaluation of the nitrogen sufficiency index for use with high resolution, broadband aerial imagery in a commercial potato field. *Precision Agriculture, 15*, 202–226.

Olson, C., & Anderson, T. (2021a). Drones and sensors in modern farming: Real-time data for better yields. *Journal of Agricultural Technology, 25*(1), 31–47.

Olson, D., & Anderson, J. (2021b). Review on unmanned aerial vehicles, remote sensors, imagery processing, and their applications in agriculture. *Agronomy Journal, 113*(2), 971–992.

Osman, G. E. H., El-Ghareeb, D., Already, R., Assaeedi, A. S. A., Organji, S. R., Abulreesh, H. H., & Althubiani, A. S. (2015). Bioinsecticide *Bacillus thuringiensis* a comprehensive review. *Egyptian Journal of Biological Pest Control, 25*(1), 271–288.

Pacholczak, A., Petelewicz, P., Jagiełło-Kubiec, K., & Ilczuk, A. (2015). The effect of two biopreparations on rhizogenesis in stem cuttings of *Cotinus coggygria* Scop. *European Journal of Horticultural Science, 80*(4), 183–189. ISSN: 1611-4426, https://doi.org/10.17660/eJHS.2015/80.4.6. Accessed: 12.02.2019.

Pachori, S., Thakur, S., Barela, A., Tomar, A., Nagre, S. P., Anand, K. J., Patel, T., & Sharma, S. K. (2023a). Remote sensing for crop management: A comprehensive review. *Biological Forum – An International Journal, 15*(2), 382–387.

Pachori, P., et al. (2023b). Teledetection and pest management in agriculture. *Journal of Plant Protection, 12*(2), 102–119.

Padavala, S., Saxena, K., & Nandeha, N. (2017). Bio-pesticides: Future to insect pest management. *Indian Farming, 67*(8), 29–31.

Pan, X., Nie, D., Guo, X., Xu, S., Zhang, D., Cao, F., & Guan, X. (2023). Effective control of the tomato wilt pathogen using TiO2 nanoparticles as a green nanopesticide. *Environmental Science: Nano, 10*(5), 1441–1452.

Pandey, S., Patel, N. R., Danodia, A., & Singh, R. (2019a). Discrimination of sugarcane crop and cane yield estimation using Landsat and IRS resourcesat satellite data. *The International Archives of the Photogrammetry, Remote Sensing and Spatial Information Sciences, 42*, 229–233.

Pandey, S. N., et al. (2019b). Estimation of sugarcane yield using IRS satellites. *Journal of Indian Remote Sensing, 47*(2), 289–303.

Paśmionka, I., & Kotarba, K. (2015). Possible application of effective microorganisms in environmental protection. *Cosmos, Problems of Biological Sciences, 64*(1), 173–184. (in Polish).

Peeriga, R., Balagani, P. K., & Atmakuri, L. R. (2022a). Prominence of biopesticides to supersede chemical pesticides in sustainable agriculture. *International Journal of Pharmacognosy and Pharmaceutical Sciences, 5*(1), 81–85.

Peeriga, L. S., et al. (2022b). Plant-based biopesticides: A sustainable alternative to chemical pest control. *Journal of Crop Protection, 6*(1), 105–119.

Peng, J., Manevski, K., Kørup, K., Larsen, R., Zhou, Z., & Andersen, M. N. (2021a). Environmental constraints to net primary productivity at northern latitudes: A study across scales of radiation interception and biomass production of potato. *International Journal of Applied Earth Observation and Geoinformation, 94*, 102232. https://doi.org/10.1016/j.jag.2020.102232

Peng, J., et al. (2021b). Radiation use efficiency in crops using Sentinel-2 data. *Field Crops Research, 271*, 108229.

Peng, X., Chen, D., Zhou, Z., Zhang, Z., Xu, C., Zha, Q., Wang, F., & Hu, X. (2022). Prediction of the nitrogen, phosphorus and potassium contents in grape leaves at different growth stages based on UAV multispectral remote sensing. *Remote Sensing, 14*(11), 2659.

Periakaruppan, R., Romanovski, V., Thirumalaisamy, S. K., Palanimuthu, V., Sampath, M. P., Anilkumar, A., Sivaraj, D. K., Ahamed, N. A. N., Murugesan, S., Chandrasekar, D., & Selvaraj, K. S. V. (2023). Innovations in modern nanotechnology for the sustainable production of agriculture. *Chemical Engineering, 7*(4), 61.

Peteinatos, G. G., Weis, M., Andújar, D., Rueda Ayala, V., & Gerhards, R. (2014). Potential use of ground-based sensor technologies for weed detection. *Pest Management Science, 70*(2), 190–199.

Porter, J. R., Xie, L., Challinor, A. J., Cochrane, K., Howden, S. M., Iqbal, M. M., Lobell, D. B., & Travasso, M. I. (2014a). *Food security and food production systems*. Cambridge University Press.

Porter, J. R., et al. (2014b). Food security and food production systems. In *Climate change 2014: Impacts, adaptation, and vulnerability* (pp. 485–533). Cambridge University Press.

Poudel, S., Poudel, B., Acharya, B., & Poudel, P. (2020). Pesticide use and its impacts on human health and environment. *Environmental and Ecosystem Sciences, 4*(1), 47–51.

Pszczółkowski, P., & Sawicka, B. (2018). The effect of using fungicides, microbiological preparations and herbal extracts on shaping of potato yield. *Fragmenta of Agronomica, 35*(1), 81–93. (in Polish).

Pszczółkowski, P., Barbaś, P., Farhan, A. K., Krochmal-Marczak, B., & Sawicka, B. (2024). Toward enhanced seed potato yield: Exploring ultrasonication techniques in agricultural engineering. *Preprints 2024*, 2024040389. https://doi.org/10.20944/preprints202404.0389.v1

Quemada, M., Gabriel, J. L., & Zarco-Tejada, P. (2014). Airborne hyperspectral images and ground-level optical sensors as assessment tools for maize nitrogen fertilization. *Remote Sensing, 6*(4), 2940–2962.

Rai, M., Ingle, A. P., Pandit, R., Paralikar, P., Shende, S., Gupta, I., Biswas, J. K., & da Silva, S. S. (2018). Copper and copper nanoparticles: Role in management of insect-pests and pathogenic microbes. *Nanotechnology Reviews, 7*(4), 303–315.

Raja, S. P., Sawicka, B., Stamenkovic, Z., & Mariammal, G. (2022). Crop prediction based on characteristics of the agricultural environment using various feature selection techniques and classifiers. *IEEE Access, 10*, 23625–23641.

Rakibuzzaman, M., Akand, M. H., Siddika, M., & Uddin, A. J. (2021). Impact of *Trichoderma* application as bio-stimulator on disease suppression, growth and yield of potato. *Journal of Bioscience and Agriculture Research, 27*(01), 2252–2257.

Rivero, R. M., Mittler, R., Blumwald, E., & Zandalinas, S. I. (2022a). Developing climate-resilient crops: Improving plant tolerance to stress combination. *The Plant Journal, 109*(2), 373–389.

Rivero, R. M., et al. (2022b). Biotic and abiotic stress tolerance in crops. *Agricultural Research, 11*(2), 65–77.

Roy, P. S., Behera, M. D., & Srivastav, S. K. (2017). Satellite remote sensing: Sensors, applications and techniques. *Proceedings of the National Academy of Sciences, India Section A: Physical Sciences, 87*, 465–472.

Ruiu, L. (2018). Microbial biopesticides in agroecosystems. *Agronomy, 8*(11), 235. https://doi.org/10.3390/agronomy8110235

Rushchitskaya, O., Kulikova, E., Kot, E., & Kruzhkova, T. (2024). Sustainable practices and technological innovations transforming agribusiness dynamics. In *E3S web of conferences* (Vol. 542, p. 03003). EDP Sciences. https://doi.org/10.1051/e3sconf/202454203003

Rusmayadi, G., Mulyanti, D. R., & Alaydrus, A. Z. A. (2023a). Revolutionizing agrotechnology: Meeting global food demand through sustainable and precision farming innovations. *West Science Interdisciplinary Studies, 1*(08), 619–628.

Rusmayadi, R., et al. (2023b). Precision agriculture: Optimizing farming practices through technology. *Journal of Agricultural Science, 14*(1), 67–85.

Saiz-Rubio, V., & Rovira-Más, F. (2020). Smart farming: Advances in agricultural technologies. *Agricultural Systems, 177*, 102671.

Sajid, M., & Płotka-Wasylka, J. (2020). Nanoparticles: Synthesis, characteristics, and applications in analytical and other sciences. *Microchemical Journal, 154*, 104623. https://doi.org/10.1016/j.microc.2020.104623

Sajid, M., Geng, C., Li, M., Wang, Y., Liu, H., Zheng, J., Peng, D., & Sun, M. (2018). Whole-genome analysis of *Bacillus thuringiensis* revealing partial genes as a source of novel Cry toxins. *Applied and Environmental Microbiology, 84*(14), e00277–e00218.

Samada, L. H., & Tambunan, U. S. F. (2020a). Biopesticides as promising alternatives to chemical pesticides: A review of their current and future status. *Journal of Biological Sciences, 20*(2), 66–76. https://doi.org/10.3844/ojbsci.2020.66.76

Samada, L. H., & Tambunan, R. (2020b). Biopesticides: Trends and opportunities. *Agriculture, Ecosystems & Environment, 294*, 106858.

Sánchez-Yáñez, J. M., Rico, J. L., & Ulíbarri, G. (2022). *Bacillus thuringiensis* (Bt) is more than a special agent for biological control of pests. *Journal of Applied Biology & Biotechnology, 9*, 33–39.

Sangeetha, C., Moond, V., Rajesh, G. M., Damor, J. S., Pandey, S. K., Kumar, P., & Singh, B. (2024). Remote sensing and geographic information systems for precision agriculture: A review. *International Journal of Environment and Climate Change, 14*(2), 287–309.

Saravanakumar, K., & Wang, M. H. (2020). Isolation and molecular identification of *Trichoderma* species from wetland soil and their antagonistic activity against phytopathogens. *Physiological and Molecular Plant Pathology, 109*, 101458. https://doi.org/10.1016/j.pmpp.2020.101458

Sardar, M., Ahmed, W., Al Ayoubi, S., Nisa, S., Bibi, Y., Sabir, M., Khan, M. M., Ahmed, W., & Qayyum, A. (2022). Fungicidal synergistic effect of biogenically synthesized zinc oxide and copper oxide nanoparticles against Alternaria citri causing citrus black rot disease. *Saudi Journal of Biological Sciences, 29*(1), 88–95.

Schäfer, C., & Török, B. (2021). Application of nontraditional activation methods in green and sustainable chemistry: Microwaves, ultrasounds, electro-, photo-, and mechanochemistry, and high hydrostatic pressure. In *Nontraditional activation methods in green and sustainable applications* (pp. 1–26). Elsevier.

Shahi, T. B., Xu, C. Y., Neupane, A., & Guo, W. (2023). Recent advances in crop disease detection using UAV and deep learning techniques. *Remote Sensing, 15*, 2450. https://doi.org/10.3390/rs15092450

Shangguan, W., Chen, H., Zhao, P., Cao, C., Yu, M., Huang, Q., & Cao, L. (2024). Scenario-oriented nanopesticides: Shaping nanopesticides for future agriculture. *Advanced Agrochem, 3*, 265. https://doi.org/10.1016/j.aac.2024.07.002

Shanmugapriya, P., et al. (2019). Application of remote sensing in crop growth monitoring. *Journal of Agrotechnology, 22*(3), 58–67.

Shapiro-Ilan, D. I., Gaugler, R., Tedders, W. L., Brown, I., & Lewis, E. E. (2002). Optimization of inoculation for in vivo production of entomopathogenic nematodes. *Journal of Nematology, 34*(4), 343–350.

Sharma, A. K., Sharma, D., & Chopra, A. K. (2020). An overview of pesticides in the development of agriculture crops. *Journal of Applied and Natural Science, 12*(2), 101–109.

Shukla, P., Bankar, D. R., Kumar, A., Aman, A. S., Mishra, P. K., Raghuvanshi, H. R., & Gayithri, M. (2023). Advancements in the use of entomopathogenic microbes for pest and disease management-A review. *International Journal of Environment and Climate Change, 13*(10), 945–953.

Sims, D. A., & Gamon, J. A. (2002). Relationships between leaf pigment content and spectral reflectance across a wide range of species, leaf structures and developmental stages. *Remote Sensing of Environment, 81*(2–3), 337–354.

Singh, M., Poonia, M. K., & Kumhar, B. L. (2017). Climate change: Impact, adaptation and mitigation: A review. *Agricultural Reviews, 38*(1), 67–71.

Singh, V., Sharma, N., & Singh, S. (2020). A review of imaging techniques for plant disease detection. *Artificial Intelligence in Agriculture, 4*, 229–242.

Sonka, S. (2016). Big data: Fueling the next evolution of agricultural innovation. *Journal of Innovation Management, 4*(1), 114–136.

Stock, J. H. (2020). Climate change, climate policy, and economic growth. *NBER Macroeconomics Annual, 34*(1), 399–419.

Swapan, C., Mainak, B., Deewa, B., & Tanmoy, M. (2024). Natural pesticides for pest control in agricultural crops: An alternative and eco-friendly method. *Plant Science Today, 11*(1), 433–450.

Tao, H., Feng, H., Xu, L., Miao, M., Yang, G., Yang, X., & Fan, L. (2020). Estimation of the yield and plant height of winter wheat using UAV-based hyperspectral images. *Sensors, 20*(4), 1231.

Taylor, R. A. J., Herms, D. A., Cardina, J., & Moore, R. H. (2018). Climate change and pest management: Unanticipated consequences of trophic dislocation. *Agronomy, 8*(1), 7.

Toan, D. H., Van Hoang, D., Hoang, V. D., & Dai Lam, T. (2021). Application of botanical pesticides in organic agriculture production: Potential and challenges. *Vietnam Journal of Science and Technology, 59*(6), 679–701.

Tomar, B., Parihar, S. S., & Patle, T. (2023). Climate change's effects on agriculture. *Just Agriculture, 3*, 92–98.

Too, E. C., Yujian, L., Njuki, S., & Yingchun, L. (2019). A comparative study of fine-tuning deep learning models for plant disease identification. *Computers and Electronics in Agriculture, 161*, 272–279.

Ubuoh, E. A. (2016). Green chemistry: A panacea for environmental sustainability agriculture in global perspective. *Global Journal of Pure and Applied Chemistry Research, 4*(1), 21–29.

Uciechowski, P., & Sadowski, A. (2018). Implementing biological advancements in agriculture as an innovation. *Intercathedra, 3*(36), 315–321. https://doi.org/10.17306/J.INTERCATHEDRA.2018.00027

Usman, M., Farooq, M., Wakeel, A., Nawaz, A., Cheema, S. A., Ur Rehman, H., Ashraf, I., & Sanaullah, M. (2020). Nanotechnology in agriculture: Current status, challenges and future opportunities. *Science of the Total Environment, 721*, 137778.

Van Lenteren, J. C. (2012). The state of commercial augmentative biological control: Plenty of natural enemies, but a frustrating lack of uptake. *BioControl, 57*, 1–20. https://doi.org/10.1007/s10526-011-9395-1

Vaz, S. (2018). Biomass and the green chemistry principles. In S. Vaz (Ed.), *Biomass and green chemistry. Building a renewable pathway* (pp. 1–9). https://doi.org/10.1007/978-3-319-66736-2_1

Vero, S., Garmendia, G., Allori, E., Sanz, J. M., Gonda, M., Alconada, T., Cavello, I., Dib, J. R., Diaz, M. A., Nally, C., Pimenta, R. S., da Silva, J. F. M., Vargas, M., Zaccari, F., & Wisniewski, M. (2023). Microbial biopesticides: Diversity, scope, and mechanisms involved in plant disease control. *Diversity, 15*(3), 457.

Viana, C. (2022). *Study of the biopesticide and biostimulant activity of microalgae in horticultural species* (Doctoral dissertation). Faculdade de Ciências e Tecnologia -Universidade do Algarve.

Vijayakumar, S., Kumar, R. M., Choudhary, A. K., Deiveegan, M., Tuti, M. D., Sreedevi, B., & Sundaram, R. M. (2022). Artificial intelligence (AI) and its application in agriculture. *Chronicle of Bioresource Management, 6*(1), 25–31.

Wahab, S. (2009). Biotechnological approaches in the management of plant pests, diseases and weeds for sustainable agriculture. *Journal of Biopesticides, 2*(2), 115–134.

Walid, N., Al-Jaramany, L., Elbenay, A., & Al-Mhethawi, R. (2022). Biological control of tomato damping-off and potato black scurf by seed treatment with *Trichoderma harzianum*. *Jordan Journal of Biological Sciences, 15*(3), 373–380.

Wang, A., Zhang, W., & Wei, X. (2019). A review on weed detection using ground-based machine vision and image processing techniques. *Computers and Electronics in Agriculture, 158*, 226–240.

Wang, X., Zhao, C., Müller, C., Wang, C., Ciais, P., Janssens, I., Penuelas, J., Asseng, S., Li, T., Elliott, J., Huang, Y., Li, L., & Piao, S. (2020). Emergent constraint on crop yield response to warmer temperature from field experiments. *Nature Sustainability, 3*(11), 908–916.

Wang, D., Li, R., Zhu, B., Liu, T., Sun, C., & Guo, W. (2022). Estimation of wheat plant height and biomass by combining UAV imagery and elevation data. *Agriculture, 13*(1), 9.

Wengert, M., Piepho, H. P., Astor, T., Graß, R., Wijesingha, J., & Wachendorf, M. (2021). Assessing spatial variability of barley whole crop biomass yield and leaf area index in silvoarable agroforestry systems using UAV-borne remote sensing. *Remote Sensing, 13*, 275.

Woyessa, D. (2022). Weed control methods used in agriculture. *American Journal of Life Science and Innovation, 1*(1), 19–26.

Wysocki, F., & Ryś-Jurek, R. (2008). Upowszechnienie doradztwa rolniczego i postępu biologicznego w indywidualnych gospodarstwach rolnych. [W:] M. Adamowicz (red.), *Innowacje i innowacyjność w sektorze agrobiznesu* (s. 147–158). Wyd. SGGW. (in Polish).

Xin, X., Judy, J. D., Sumerlin, B. B., & He, Z. (2020). Nano-enabled agriculture: From nanoparticles to smart nanodelivery systems. *Environmental Chemistry, 17*(6), 413–425.

Xu, X., Fan, L., Li, Z., Meng, Y., Feng, H., Yang, H., & Xu, B. (2021). Estimating leaf nitrogen content in corn based on information fusion of multiple-sensor imagery from UAV. *Remote Sensing, 13*(3), 340.

Xu, T., Wang, F., Xie, L., Yao, X., Zheng, J., Li, J., & Chen, S. (2022). Integrating the textural and spectral information of UAV hyperspectral images for the improved estimation of rice aboveground biomass. *Remote Sensing, 14*(11), 2534.

Yadav, I. C., & Devi, N. L. (2017). Pesticides classification and its impact on human and environment. *Environmental Science and Engineering, 6*(7), 140–158.

Yang, F., Feng, H., Li, Z., Pan, J., & Xie, R. (2021). Comparison of hyperspectral remote sensing inversion methods for apple leaf nitrogen content. *Remote Sensing Technology and Application, 36*, 353–361.

Yin, C., Lv, X., Zhang, L., Ma, L., Wang, H., Zhang, L., & Zhang, Z. (2022). Hyperspectral UAV images at different altitudes for monitoring the leaf nitrogen content in cotton crops. *Remote Sensing, 14*(11), 2576.

Yin, J., Su, X., Yan, S., & Shen, J. (2023). Multifunctional nanoparticles and nanopesticides in agricultural application. *Nanomaterials, 13*(7), 1255.

Yousuf, O., Alam, I., Bhat, M. I., & Titikshya, S. (2018). Assessment of microbial spoilage and techniques to avert the deterioration in fruits and vegetables. *International Journal of Chemical Studies, 6*(3), 2230–2235.

Zahi, M., et al. (2020). Blockchain for sustainable agriculture: Challenges and opportunities. *Journal of Agribusiness and Rural Development, 7*(3), 45–59.

Zanin, A. R. A., Neves, D. C., Teodoro, L. P. R., da Silva Júnior, C. A., da Silva, S. P., Teodoro, P. E., & Baio, F. H. R. (2022a). Reduction of pesticide application via real-time precision spraying. *Scientific Reports, 12*(1), 5638.

Zanin, G., et al. (2022b). Real-time sensor-based spraying systems: Efficiency in reducing pesticide use in soybean and maize. *Precision Agriculture, 23*(1), 1–15.

Zarrouk, Y., Yandouzi, M., Grari, M., Bourhaleb, M., Rahmoune, M., & Hachami, K. (2024). Revolutionizing potato late blight surveillance: UAV-driven object detection innovations. *Journal of Theoretical and Applied Information Technology, 102*(7), 2934–2944.

Zhang, Y., Xia, C., Zhang, X., Cheng, X., Feng, G., Wang, Y., & Gao, Q. (2021). Estimating the maize biomass by crop height and narrowband vegetation indices derived from UAV-based hyperspectral images. *Ecological Indicators, 129*, 107985.

Zhou, J., Wang, B., Fan, J., Ma, Y., Wang, Y., & Zhang, Z. (2022). A systematic study of estimating potato N concentrations using UAV-based hyper-and multi-spectral imagery. *Agronomy, 12*(10), 2533.

Zimmerman, J. B., Anastas, P. T., Erythropel, H. C., & Leitner, W. (2020). Designing for a green chemistry future. *Science, 367*(6476), 397–400.

Zisi, T., Alexandridis, T. K., Kaplanis, S., Navrozidis, I., Tamouridou, A. A., Lagopodi, A., Moshou, D., & Polychronos, V. (2018). Incorporating surface elevation information in UAV multispectral images for mapping weed patches. *Journal of Imaging, 4*(11), 132. https://doi.org/10.3390/jimaging4110132

Zuin, V. G., Eilks, I., Elschami, M., & Kümmerer, K. (2021). Education in green chemistry and in sustainable chemistry: Perspectives towards sustainability. *Green Chemistry, 23*(4), 1594–1608.

Synergies Between Green Chemistry Principles and the Sustainable Development Goals (SDGs) in Developing Eco-Friendly Pesticides

Rodrigo Duarte-Casar, Juan Carlos Romero-Benavides, Natalia Bailon-Moscoso, Marell Navarro-Rojas, and Marlene Rojas-Le-Fort

1 Introduction

The global need for sustainable agriculture and—by association—pest management solutions has become a pressing concern in recent decades. As the world population continues to grow, expected to exceed 9.7 billion people by 2050 (Sadigov, 2022), the demand for food is increasing at an unprecedented rate. This creates immense pressure on agricultural systems to produce more food without further depleting natural resources or continue to cause environmental damage. Consequently, a shift towards sustainable agriculture practices that optimize resource use, enhance food security, and minimize agrochemical use and environmental harm has gained significant global attention (Damalas, 2021; van der Werf & Bianchi, 2022).

The traditional agricultural model, which heavily relies on chemical inputs such as synthetic fertilizers and pesticides, has proven unsustainable. The model has contributed to a range of environmental issues, including soil degradation, water pollution, loss of biodiversity, and the acceleration of climate change. Over-reliance on chemical pesticides has led to pesticide resistance in many pest species, rendering some treatments ineffective and driving farmers to use even more chemicals, creating a vicious cycle of dependency (Jørgensen et al., 2020).

In many parts of the world and due to several factors, farmers face dwindling arable land, water scarcity, and soil fertility loss. These challenges are exacerbated by climate change, which is altering precipitation patterns, increasing the frequency

R. Duarte-Casar (✉) · M. Rojas-Le-Fort
Pontificia Universidad Católica del Ecuador Sede Manabí, Portoviejo, Ecuador
e-mail: rduarte@pucesm.edu.ec

J. C. Romero-Benavides · N. Bailon-Moscoso
Universidad Técnica Particular de Loja, Loja, Ecuador

M. Navarro-Rojas
Universidad Politécnica Salesiana, Cuenca, Ecuador

B. R. Babaniyi, E. E. Babaniyi (eds.), *The Interplay of Pesticides and Climate Change*, https://doi.org/10.1007/978-3-031-81669-7_11

of extreme weather events, and impacting crop yields. Moreover, poor agricultural practices can deplete soil nutrients, leading to long-term degradation of land (Prăvălie et al., 2021). Thus, to meet future food demands, agriculture must become more efficient, resilient, and environmentally sound.

Sustainable agriculture is a system of farming that seeks to produce food while maintaining the ecological balance and ensuring the long-term viability of farming. It emphasizes resource efficiency, environmental stewardship, and social equity (Harwood, 1990; Khan et al., 2021). Key practices in sustainable agriculture include polycrops, crop rotation, agroforestry, reduced tillage, organic farming, and integrated pest management (IPM). These methods aim to reduce dependency on chemical inputs, improve soil health, and increase biodiversity, thereby making farms more resilient to pests, diseases, and changing climate conditions (Martinez et al., 2024).

One critical aspect of sustainable agriculture is improving soil health. Practices such as cover cropping, composting, and conservation tillage help to restore soil fertility, sequester carbon, and reduce erosion. Healthy soils are better able to retain water and nutrients, making them more productive and resilient to droughts and floods.

Pest management is a crucial component of agriculture that directly impacts crop productivity and food security. However, conventional pest management, which often involves the heavy use of synthetic pesticides, is unsustainable. Overuse of chemical pesticides can lead to environmental contamination, loss of beneficial species, and the emergence of pesticide-resistant pests.

Sustainable pest management solutions, particularly integrated pest management (IPM), offer a balanced approach to controlling pest populations. IPM combines biological, cultural, physical, and chemical tools to minimize pest damage in an environmentally responsible way. Biological controls, such as introducing natural predators or using pheromones to disrupt pest mating cycles, can reduce pest populations without harmful chemicals. Cultural practices, like rotating crops and using pest-resistant plant varieties, can also help mitigate pest issues. When chemical pesticides are used, IPM advocates for targeted applications and minimal quantities to reduce environmental impact (Angon et al., 2023).

Sustainable agriculture and pest management are no longer optional, but imperative if food security and environmental sustainability are to be attained. The shift towards more sustainable practices can not only mitigate environmental harm but also create a more resilient food system that can better withstand the pressures of climate change and growing global demand. Sustainable pest management solutions, particularly IPM and agroecological crop protection (ACP), play a critical role in this transformation by offering a holistic approach to reducing pest-related crop losses without compromising ecosystem health. The global adoption of these practices is essential for a future where agriculture can thrive in harmony with nature (Deguine et al., 2021).

2 The Environmental Impacts of Conventional Pesticides and the Need for Greener Alternatives

The widespread use of conventional pesticides in modern agriculture has contributed significantly to increased food production and improved food security, but it has also caused profound and often detrimental effects on the environment (Gomes et al., 2020). Conventional pesticides, which include insecticides, herbicides, and fungicides, are designed to target specific pests or weeds, but their impacts extend far beyond the intended organisms. The environmental consequences of conventional pesticide use jumped to the forefront with the proof of the effects of DDT (Fig. 1) on the environment in the first edition of *Silent Spring*, which has become a classic for environmentalists (Carson et al., 2002). These effects include soil degradation, water contamination, biodiversity loss, and disruption of ecosystems (AbuQamar et al., 2024). These harmful effects and their accumulation in time have led to an urgent call for greener, more sustainable alternatives in pest management with voices hoping for avoiding the need for chemical pesticides entirely (Rajendran & Singh, 2016). It is noteworthy that almost 60 years after the discovery of the effects of DDT, it was and still is in use (Arévalo-Jaramillo et al., 2019).

When evaluating the environmental impacts of pesticides, several studies identify organophosphorus compounds, neonicotinoids, and chlorinated insecticides as particularly harmful. Organophosphates such as diazinon and chlorpyrifos, widely used in agriculture, have been shown to significantly disrupt riverine ecosystems and aquatic life, posing risks in surface waters (Wee & Aris, 2017). Neonicotinoids, long known for their detrimental impact on pollinators, are being replaced by substances like sulfoxaflor—a sulfoximine—which still exhibit harmful effects, including the reduction in survival rates in bumblebees when combined with nutritional stress (Siviter et al., 2018).

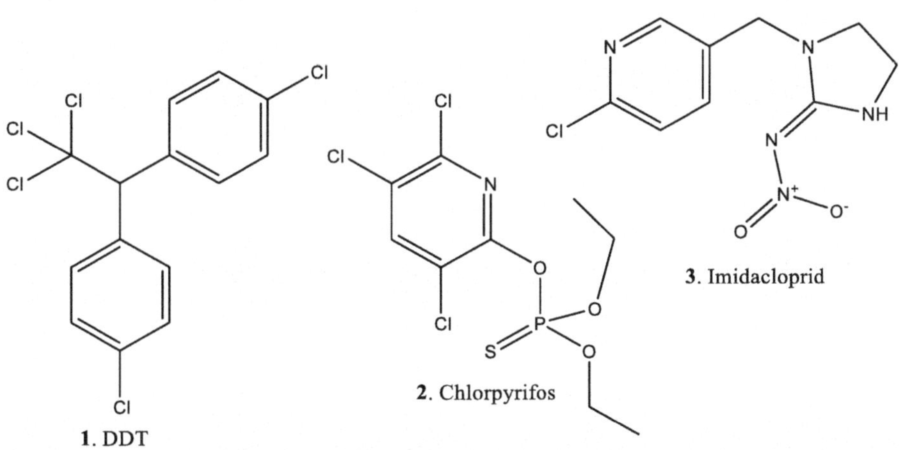

Fig. 1 Well-known environmentally harmful pesticides

In soil ecosystems, certain insecticides (i.e., esfenvalerate and cyclaniliprole) and fungicides (picoxystrobin and fenamidone) have been found to negatively impact non-target organisms like earthworms, leading to loss of crucial soil ecosystem services (Pistocchi et al., 2023; Urionabarrenetxea et al., 2022). There is consensus in that the most harmful pesticides, environmentally speaking, are those that severely affect aquatic systems, pollinators, and soil organisms.

Eliminating the harmful effects of pesticides and pesticide production in human health is mandatory. Potent examples of the very present dangers of the pesticide industry are the Seveso disaster (1976), Bhopal disaster (1984), Kasargod endosulfan poisoning, and too many others (Deepa et al., 2011; Pradyumna & Chelaton, 2018).

2.1 Soil Degradation and Ecosystem Disruption

Conventional pesticides can have a lasting effect on soil health, which is essential for sustainable agriculture. Several chemical pesticides, especially those that persist in the environment (known as persistent organic pollutants, or POPs), can accumulate in the soil over time. These substances disrupt the complex web of interactions among soil organisms, including beneficial microbes, fungi, and insects that are crucial for maintaining soil fertility and structure. For example, insecticides designed to kill crop pests can also eliminate non-target organisms such as pollinators and soil-dwelling insects that play a role in nutrient cycling and plant health. Over time, the decline of these beneficial organisms and the ecological services they provide can reduce soil fertility, diminish crop yields, and create a dependency on more chemical inputs to sustain production (Elizalde et al., 2020; Karimi et al., 2022).

Additionally, the widespread use of herbicides has contributed to the problem of herbicide-resistant weeds, sometimes called "superweeds." These resistant species evolve rapidly under the selective pressure of repeated chemical applications, rendering some pesticides ineffective and forcing farmers to use more chemicals or turn to stronger, more toxic alternatives. This, in turn, together with climate change, exacerbates the cycle of environmental harm (Ma et al., 2021).

2.2 Water Contamination and Aquatic Life

Another major concern surrounding conventional pesticides is their potential to contaminate water sources. Persistent pesticides such as atrazine, glyphosate, and endosulfan can enter rivers, lakes, and groundwater through agricultural runoff, where they can persist and accumulate over time. Once in aquatic systems, these chemicals pose a significant threat to both the environment and animal and human health (Panis et al., 2022).

In water bodies, pesticides can disrupt ecosystems by altering and harming aquatic life, including fish, amphibians, and beneficial microorganisms: classical examples are Atrazine—an endocrine disruptor in frogs, and endosulfan—highly toxic to fish. Herbicides can damage aquatic plants, disrupting the balance of aquatic ecosystems and leading to reduced oxygen levels in water, which in turn harms fish and other species. Additionally, pesticide-contaminated water can pose risks to human populations through contaminated drinking water or exposure via irrigation (Arrobo, 2023).

2.3 *Biodiversity Loss*

One of the most alarming effects of conventional pesticide use is its contribution to biodiversity loss. Pesticides often harm non-target plant and animal species, including important pollinators like bees, butterflies, and birds, which are critical for the reproduction of many plants. Studies have shown that neonicotinoid insecticides, in particular, have played a significant role in the global decline of pollinator populations (Mehdizadeh et al., 2021; Urionabarrenetxea et al., 2022).

Biodiversity loss is not only an ecological concern but also a threat to food security. Reduced biodiversity can lead to weaker ecosystems that are less resilient to changes, such as climate variability or the introduction of new pests such as the Fusarium wilt in banana (Westerhoven et al., 2022). A diverse ecosystem is better able to naturally regulate pest populations, reducing the need for chemical interventions. The loss of these natural systems places additional burdens on agricultural systems that then rely more heavily on synthetic pesticides, creating a vicious cycle of environmental degradation that affects not only the environment but also human populations (Zhou et al., 2020).

3 The Need for Greener Alternatives

To mitigate the environmental impacts of conventional pesticides, there is a growing need for greener, more sustainable pest management alternatives. Integrated pest management (IPM), organic farming practices, and the development of biopesticides offer promising pathways to reducing the reliance on harmful chemicals (Barzman et al., 2015; Ortega et al., 2023).

Biopesticides, derived from natural materials such as plants, bacteria, and minerals, offer a less harmful way to control pests while minimizing environmental damage. These alternatives target specific pests, reducing the collateral damage to beneficial organisms (Liu et al., 2021). Similarly, IPM promotes a combination of biological, cultural, and chemical tools to manage pests with minimal environmental impact. This approach emphasizes the use of natural predators, crop rotation, and habitat management to control pest populations, with reduced use of chemical pesticides only as a last resort. By adopting practices such as IPM, organic farming, and the use of biopesticides, agriculture can reduce its environmental footprint while still meeting the global demand for food. Greener alternatives are not only beneficial for the environment but are also essential for the long-term sustainability of agriculture (Angon et al., 2023).

This chapter shows how green chemistry practices intersect with the sustainable development goals (SDGs), providing a roadmap for the development of sustainable chemical technologies that not only meet industrial and commercial needs but also advance global efforts toward a healthier and more sustainable future.

4 Sustainable Development Goals

The sustainable development goals (SDGs) are a global framework established by the United Nations in 2015, designed to address critical environmental, social, and economic challenges by 2030 (Pizzi et al., 2020). The 17 SDGs shown in Fig. 2 aim to promote sustainable development, eradicate poverty, protect the planet, and ensure prosperity for all. These goals represent an integrated approach to sustainable growth, recognizing the interdependence of environmental stewardship, economic progress, and social well-being.

Even though all SDGs are intertwined, some are more central to the object of this chapter than others. The SDGs that more closely align with green chemistry in the production of sustainable pesticides are SDG 2 (Zero Hunger), SDG 3 (Good Health and Well-being), SDG 6 (Clean Water and Sanitation), SDG 12 (Responsible Consumption and Production), SDG 13 (Climate Action), SDG 14 (Life Below Water), and SDG 15 (Life on Land). Together, these goals emphasize the

Fig. 2 Sustainable development goals. (Source: United Nations)

importance of sustainable agriculture, human and environmental health, and the protection of natural ecosystems in pesticide production.

The goal of SDG 2: Zero Hunger is to end hunger, achieve food security, and promote sustainable agriculture. Sustainable pesticides and pest management help increase crop yields while minimizing damage to the environment and non-target species. By reducing reliance on harmful chemicals, sustainable pest management fosters food security (Batra et al., 2024).

SDG 3: Good Health and Well-being seeks to ensure healthy lives and promote well-being for all. Reducing the toxicity and harmful side effects of pesticides directly contributes to better health for agricultural workers, consumers, and surrounding communities. Safer, biodegradable pesticides help avoid the long-term health risks associated with toxic chemical exposure (Raimi et al., 2021).

SDG 6: Clean Water and Sanitation aims to ensure availability and sustainable management of water and sanitation for all. Sustainable pesticides reduce water pollution by preventing the contamination of water bodies from chemical runoff. By using biodegradable or less toxic chemicals, clean water supplies are protected (Zhang et al., 2022).

SDG 12: Responsible Consumption and Production seeks to ensure sustainable consumption and production patterns. Sustainable pesticides promote responsible production by minimizing hazardous waste and environmental contamination, avoiding overloading the productive areas with agrochemicals (Shaikh et al., 2024).

SDG 13: Climate Action. This is a call to take urgent action to combat climate change and its impacts. Pesticide effectivity and behavior are subject to variations due to climate change. Climate action together with sustainable pesticides will mitigate resistances, occupational hazards, and environmental effects (Bloomfield et al., 2006; Delcour et al., 2015; Gatto et al., 2016).

SDG 14: Life Below Water aims to conserve and sustainably use the oceans, seas, and marine resources. By minimizing chemical runoff and preventing the accumulation of toxic pesticides in marine ecosystems, sustainable pesticides protect aquatic life and help preserve the health of oceans and coastal areas (Raimi et al., 2021).

SDG 15: Life on Land seeks to protect, restore, and promote sustainable use of terrestrial ecosystems, combat desertification, and halt biodiversity loss. Sustainable pesticides protect terrestrial ecosystems by reducing harmful effects on soil quality, non-target species, and biodiversity. Biodegradable pesticides and natural pest control agents help maintain healthy ecosystems and prevent land degradation in an interlocking scenario (Måren, 2019).

SDG 9: Industry, Innovation, and Infrastructure is regarded as one of secondary relevance and at the same time a nexus for most other SDGs: an innovative industry will permit the attainment of the goals of successfully feeding an ever-growing population while reducing the environmental footprint. The development and adoption of sustainable pesticides involves innovative chemical processes and technologies. By investing in green chemistry, industries can advance sustainable agricultural and manufacturing systems.

SDGs provide references and applicability, aligning scientific innovation with broader sustainability objectives in the development of pest control strategies. Green chemistry—focused on reducing or eliminating hazardous substances in the design, manufacture, and application of chemical products—supports several key SDGs by advancing safer, more efficient, and environmentally responsible practices. Together, these goals emphasize the importance of sustainable agriculture, human and environmental health, and the protection of natural ecosystems in pesticide production (Fenibo et al., 2021, 2022).

5 Green Chemistry Principles

Green chemistry is the science of designing chemical processes and products that reduce or eliminate the use and generation of hazardous substances, making chemistry more environmentally friendly, sustainable, and safe, by preventing rather than cleaning up pollution. Green chemistry aims to mimic the way chemistry is done in nature, using renewable, biodegradable, non-persistent materials in a safer way (Zimmerman et al., 2020).

Green chemistry is guided by 12 interlocking key principles that promote sustainability and safety in chemical processes:

1. Prevention: It is better to prevent waste than to manage it after production.
2. Atom economy: It maximizes the incorporation of materials into the final product.
3. Less hazardous chemical syntheses: It designs processes that minimize toxicity to humans and the environment.
4. Designing safer chemicals: It creates chemicals with minimal toxicity while maintaining their intended function.
5. Safer solvents and auxiliaries: It avoids or minimizes the use of harmful auxiliary substances like solvents.
6. Energy efficiency: It minimizes energy use by conducting reactions at ambient temperatures and pressures.
7. Use of renewable feedstocks: It prefers renewable raw materials over depleting resources when possible.
8. Reduce derivatives: It avoids unnecessary chemical modifications that generate waste.
9. Catalysis: Catalysts are preferred over stoichiometric reagents due to their efficiency.
10. Design for degradation: It creates products that break down into harmless substances after use.
11. Real-time analysis: It develops methods for in-process monitoring to prevent hazardous by-products.

12. Safer chemistry for accident prevention: It chooses substances and processes that reduce the risk of accidents, such as spills, explosions, or fire.

In the production of sustainable pesticides, eight green chemistry principles (GCPs) stand out as particularly relevant. These aim to reduce environmental impact, enhance safety, and improve the overall sustainability of pesticide production. The principles of design for degradation, less hazardous chemical syntheses, safer solvents, atom economy, and use of renewable feedstocks are the most closely aligned with the goals of producing sustainable pesticides. These principles ensure that the development and use of pesticides are efficient, safe, and environmentally responsible, promoting a healthier relationship between agriculture and the environment.

Principle 10. Design for Degradation. Pesticides should be designed to break down into harmless, non-toxic substances after use. This minimizes the accumulation of harmful chemicals in the environment and prevents long-term ecological damage. This ensures that pesticides do not persist in soil, water, or the atmosphere, protecting non-target organisms and ecosystems (van Dijk et al., 2022). Current pesticides, not designed for degradation, must be targeted with bacterial strains and metallic nanoparticles among other methods (Burbano et al., 2008; Pratap et al., 2024).

Principle 3. Less Hazardous Chemical Syntheses. The synthesis of pesticides should involve processes that use and generate fewer toxic substances. This reduces risks to workers, consumers, and the environment. Safer production methods result in fewer hazardous by-products, making the entire lifecycle of the pesticide—from manufacturing to application—more sustainable (Tiwari & Brahmpurkar, 2021).

Principle 5. Safer Solvents and Auxiliaries. The use of solvents in the synthesis of pesticides is often necessary, but choosing safer, less toxic solvents can significantly reduce environmental and human health risks. Using water or bio-based solvents instead of petroleum-based or toxic solvents can make pesticide production greener and less harmful to the environment.

Principle 2. Atom Economy. It strives to maximize the incorporation of starting materials into the final product. For pesticides, this means designing more efficient reactions that generate minimal waste. Improved atom economy reduces the generation of harmful by-products and makes production processes more resource-efficient. Bioprocesses are a growing field of atom economy in chemical processes (Fenibo et al., 2021).

Principle 7. Use of Renewable Feedstocks. Sustainable pesticides should be derived from renewable sources, such as plant-based or bio-derived chemicals, rather than non-renewable, fossil-based materials. Utilizing renewable feedstocks ensures that pesticide production has a lower environmental impact and promotes the use of sustainable resources.

Principle 1. Prevention of Waste. Preventing waste throughout the pesticide development process—from raw material acquisition to manufacturing—reduces the environmental burden associated with waste disposal. Minimizing waste production conserves resources and reduces contamination of air, soil, and water during pesticide manufacturing.

Principle 6. Energy Efficiency. Sustainable pesticide production should minimize energy use, particularly in synthesis reactions that can be performed at lower temperatures and pressures. Reducing the energy footprint of pesticide production lowers greenhouse gas emissions and contributes to climate action.

Principle 12. Real-Time Analysis for Pollution Prevention. Implementing real-time monitoring and analysis during pesticide production allows for the immediate detection and reduction of pollutants and hazardous by-products. Pollution prevention strategies ensure that harmful substances are managed before they are released into the environment (Musarurwa & Tavengwa, 2020). The advanced analytical methods devised to enforce this principle are also used in detection for remediation (Kaur et al., 2021).

Synergies between SDGs and green chemistry principles are many, as both endeavor to conduct human production to cleaner, safer results and a reduced footprint.

6 Synergies and Examples

There is a "less is more" paradox in IPM, where the best pesticide is one that is not required. For example, the mating disruption of moths through synthetic hormones reduces the need for chemical pesticides as the reduced amount of larvae is easily controllable without chemicals, maximizing resources and minimizing chemical by-products and potential for pollution (Benelli et al., 2019). The greenest chemistry and the most aligned with SDGs would be one that is required in the least amounts.

The relationship between green chemistry and the SDGs is synergistic. By adhering to the principles of green chemistry, scientists and industries can directly contribute to achieving the SDGs, particularly in areas such as clean water (SDG 6), responsible production (SDG 12), climate action (SDG 13), and ecosystem protection (SDGs 14 and 15). Conversely, the SDGs offer a global mandate for the chemical sciences to innovate solutions that address urgent challenges like pollution, resource depletion, and climate change. The greatest synergies between GCPs and SDGs are summarized in Fig. 3.

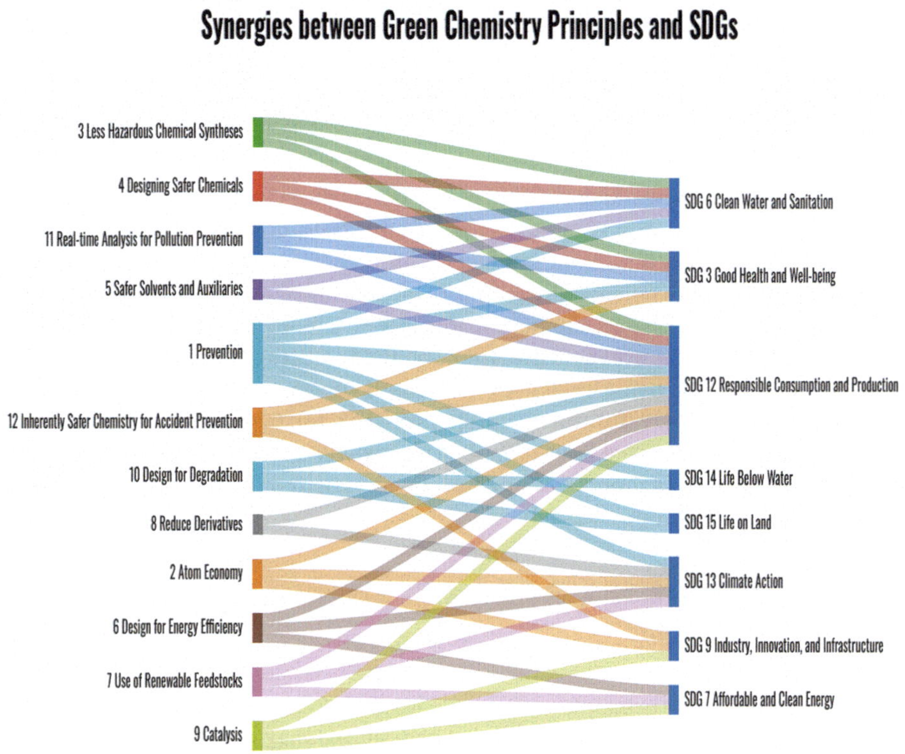

Fig. 3 Synergies between green chemistry principles and SDGs

6.1 Prevention

The "Prevention" principle of green chemistry is the widest and, as such, the one that presents the most synergies with SDGs. It emphasizes avoiding the creation of waste rather than treating or cleaning up waste after it has been created, synergizes with the following SDGs:

SDG 3. By reducing hazardous chemicals and minimizing waste, green chemistry prevents pollution and exposure to toxic substances, promoting healthier environments and reducing health risks.

SDG 6. Prevention of chemical waste ensures that water sources are not contaminated, helping to maintain clean water and sanitation systems.

SDG 12. This principle directly aligns with the goal of ensuring sustainable production patterns by minimizing waste, reducing chemical use, and encouraging efficient resource management.

SDG 13. Preventing waste and pollution reduces the need for energy-intensive waste treatment processes, contributing to lower greenhouse gas emissions and supporting climate mitigation efforts.

SDG 14. By preventing chemical pollution, green chemistry helps protect aquatic ecosystems from harmful substances that can disrupt marine life.

SDG 15. Prevention of waste contributes to the protection of terrestrial ecosystems by reducing soil and biodiversity degradation from chemical pollutants.

A pertinent example of the prevention principle synergizing with SDGs is the use of natural products as pesticides. There are no by-products to dispose of, the precursors are of biological origin, and the production is outsourced to biological organisms, so to speak. Natural plant products with bioactivity toward insects include several classes of molecules, for example: terpenes, flavonoids, alkaloids, polyphenols, cyanogenic glucosides, quinones, to name a few. Research in this field has great potential for environmentally friendly food security (Souto et al., 2021). Natural products are being mined in the search of new active ingredients. This search has proved more fruitful for pesticides than for herbicides (Sparks et al., 2023). The field needs more long-term data on toxicity and overall safety of natural product pesticides, so support and monitoring are required for it to reach its potential (Damalas & Koutroubas, 2020).

6.2 Atom Economy

The "Atom Economy" principle of green chemistry aims to maximize the efficiency of chemical reactions by ensuring that most or all the reactants end up in the final product, minimizing waste. The main synergies are:

SDG 12. Atom economy directly supports the goal of reducing waste and promoting resource efficiency. By optimizing chemical processes to maximize material use, this principle ensures that resources are used more sustainably, reducing the need for waste treatment and minimizing environmental impact from production.

SDG 13. Efficient chemical reactions reduce energy demands and resource consumption, which in turn lowers greenhouse gas emissions associated with industrial processes. By minimizing waste, less energy is needed for waste management and disposal, contributing to climate change mitigation.

SDG 9. Atom economy fosters innovation in chemical synthesis and industrial processes by encouraging the development of more efficient, less wasteful technologies. It promotes sustainable industrialization and drives the creation of eco-friendly materials and products, contributing to more resilient and sustainable infrastructures.

An example of atom economy is the development of semi-synthetic herbicides and pesticides by modification of natural products in the same way that drugs have been produced and improved by modification of natural products (Berestetskiy, 2023; Najmi et al., 2022).

6.3 Less Hazardous Chemical Syntheses

The principle of "Less Hazardous Chemical Syntheses" in green chemistry emphasizes designing chemical reactions that reduce or eliminate the use and generation of toxic or hazardous substances. This principle synergizes with several UN sustainable development goals (SDGs) as follows:

SDG 3: Good Health and Well-being. By reducing the use of hazardous chemicals, this principle directly supports public health. It minimizes human exposure to toxic substances during production, use, and disposal of chemicals, thus reducing the risk of diseases and health issues related to chemical exposure, improving overall well-being.

SDG 12: Responsible Consumption and Production. This principle encourages safer production methods that lead to less environmental harm and safer products. Reducing hazardous substances in production means less toxic waste, more sustainable consumption patterns, and reduced risks for ecosystems, aligning with SDG 12's aim for responsible and sustainable production.

SDG 6: Clean Water and Sanitation. Less hazardous chemical syntheses help prevent the contamination of water sources by reducing toxic chemical runoff and pollutants in wastewater. This contributes to achieving clean and safe water, which is essential for sanitation and public health.

Biomanufacturing of pesticides and insect deterrents aligns green chemistry and SDGs. The manufacture of terpenoid pesticides from *Nicotiana* species by modified *Escherichia coli* using agricultural waste products as substrate is a good example (Mischko et al., 2018).

6.4 Designing Safer Chemicals

SDG 3: Good Health and Well-being; this principle aims to minimize or eliminate toxicity in chemical products, reducing human health risks from exposure to harmful substances during product use and disposal. This promotes safer living and working environments.

SDG 12: Responsible Consumption and Production. Safer chemicals lead to safer products that can be manufactured, used, and disposed of with minimal environmental impact, supporting sustainable production and consumption patterns.

SDG 6: Clean Water and Sanitation. By reducing toxic chemical release into water systems, safer chemicals help prevent water pollution, contributing to the availability of clean and safe water for all.

In silico design for safety is a way of enacting this principle from the earliest stage or in the modification of existing agrochemicals for reduced toxicity toward non-target species contributing to public health, responsible production, and cleaner water (Devineni et al., 2023).

6.5 Safer Solvents and Auxiliaries

SDG 6: Clean Water and Sanitation. Reducing the use of harmful solvents and auxiliaries in chemical processes prevents the contamination of water resources, supporting the goal of maintaining clean and safe water.

SDG 12: Responsible Consumption and Production. This principle promotes using safer, more environmentally benign substances in chemical processes, thereby contributing to more sustainable industrial practices and reducing hazardous waste.

Again, the use of bioprocesses greatly reduces the need for solvents and their inherent problems and allows for the production of more environmentally friendly products in aqueous phase (Mischko et al., 2018).

6.6 Design for Energy Efficiency

SDG 7: Affordable and Clean Energy. This principle encourages minimizing energy consumption in chemical processes. Efficient processes reduce energy demand, contributing to the transition towards affordable, clean, and renewable energy.

SDG 12: Responsible Consumption and Production Energy. Efficient processes contribute to sustainable production by lowering the environmental footprint and reducing resource consumption.

SDG 13: Climate Action By optimizing energy use, especially from non-renewable sources. This principle reduces greenhouse gas emissions, contributing to efforts to combat climate change.

The "less is more" paradox applies in this case also, where IPM increases the productivity and energy efficiency of agricultural systems while reducing the need for chemicals (Rathore et al., 2022).

6.7 Use of Renewable Feedstocks

SDG 12: Responsible Consumption and Production Utilizing renewable resources reduces reliance on finite materials, fostering sustainable production practices and resource management.

SDG 13: Climate Action Shifting towards renewable feedstocks helps decrease the carbon footprint of chemical production, aiding in climate change mitigation efforts.

SDG 7: Affordable and Clean Energy Renewable feedstocks, often derived from biological sources, can provide a sustainable alternative to fossil fuels, supporting the development of clean and renewable energy sources.

The use of natural products, modified natural products, and chemical manufactured through bioprocesses mentioned above satisfy this principle and the associated SDGs.

6.8 Reduce Derivatives

SDG 12. Reducing unnecessary steps or chemical derivatives in processes leads to less waste and fewer by-products, supporting more efficient and sustainable production systems.
SDG 13. Fewer steps and less waste in chemical processes lead to reduced energy consumption and lower emissions, contributing to climate change mitigation.

Where applicable, the use of natural pesticides—either directly or minimally modified—satisfies this principle. Neem (*Azadirachta indica*) derivatives are being used and researched (Chatterjee et al., 2023), as well as chrysanthemum (Ravula & Yenugu, 2021), garlic (Hayat et al., 2022), and others.

6.9 Catalysis

SDG 12. Catalysts increase reaction efficiency, reducing the need for excess reactants and minimizing waste generation, which supports more sustainable production practices.
SDG 9. Catalytic processes drive innovation by enabling more efficient, scalable, and environmentally friendly industrial processes, contributing to sustainable industrialization.
SDG 7. Catalysis is key in many energy-efficient and renewable energy technologies, including the production of biofuels and the improvement in energy storage, contributing to the development of clean energy solutions.

Most efforts are directed to the use of catalysts not to produce, but to degrade residual pesticides (Burbano et al., 2008; Ejeta & Imae, 2021). Nevertheless, there are enzymes such as chitinase and lipases with pesticide activity (Kate et al., 2022).

6.10 Design for Degradation

SDG 14. Chemicals designed for degradation will break down into harmless substances in the environment, helping to prevent marine pollution and protect aquatic ecosystems.

SDG 15. Designing chemicals to degrade ensures they do not persist in the environment, reducing the risk of soil contamination and safeguarding terrestrial ecosystems.

SDG 12. Products that degrade safely after use contribute to reducing environmental pollution and promoting sustainable consumption by minimizing long-term waste.

Natural pesticides are naturally degradable (Souto et al., 2021) and are thus an example of this principle.

6.11 Real-Time Analysis for Pollution Prevention

SDG 6. Monitoring and preventing pollution in real-time help avoid the contamination of water sources, ensuring access to clean water.

SDG 3. By detecting and addressing hazardous substances early in chemical processes, this principle helps protect human health by preventing exposure to pollutants.

SDG 12. Real-time monitoring enables more efficient processes by reducing waste, ensuring safer production practices, and preventing pollution before it occurs.

There is advance in real-time analysis that includes biosensors, electrochemical sensors, wearables, AI, and other technologies that enable and improve real-time monitoring of pesticide presence and application (Tewari et al., 2020; Umapathi et al., 2022; Wahab et al., 2022; Zhao et al., 2020).

6.12 Inherently Safer Chemistry for Accident Prevention

SDG 3. Designing processes and chemicals to be less hazardous reduces the risk of accidents and exposure to harmful substances, protecting both workers and the general population.

SDG 9. Safer chemical processes contribute to more resilient and secure industrial infrastructure, reducing the likelihood of industrial accidents and their associated economic and environmental impacts.

SDG 12. By minimizing risks in production processes, this principle promotes safer, more sustainable industrial practices that prevent environmental and health hazards.

It is vital that disasters like Bhopal and others already mentioned do not happen again. This requires improvements not only in the design and monitoring of processes but also in policy, quality assurance, and a holistic effort (Bowonder et al., 2005).

Each of these green chemistry principles strengthens the sustainability of industrial processes, reduces environmental impacts, and enhances public health, aligning with SDG objectives. Many SDGs and principles overlap, so there is some repetition.

The prevention principle of green chemistry is paramount in pesticide production because it focuses on avoiding the creation of hazardous substances from the outset, thereby reducing environmental pollution, protecting non-target species, and safeguarding human health. Conventional pesticides often result in contamination of soil, water, and air, while also posing risks to beneficial organisms like pollinators and natural predators. By prioritizing prevention, the need for harmful chemicals is minimized, reducing exposure to toxic substances and encouraging the development of sustainable pest management practices, such as biopesticides and integrated pest management (IPM). This approach not only enhances environmental and human safety but also reduces costs associated with pollution control, remediation, and healthcare, making it a crucial strategy for long-term sustainability in agriculture.

Sustainable development goal (SDG) 12, which promotes responsible consumption and production, is critical in pesticide production as it encourages the use of sustainable practices that minimize environmental harm and health risks. Traditional pesticides often lead to soil and water contamination, harm to non-target species, and adverse effects on human health. By adhering to SDG 12, pesticide production can shift toward more sustainable alternatives, such as biopesticides and integrated pest management (IPM), which reduce reliance on hazardous chemicals. This approach not only supports environmental preservation and biodiversity but also aligns with the principles of waste reduction, resource efficiency, and long-term agricultural sustainability, ultimately leading to reduced costs in pollution control, remediation, and healthcare.

7 Conclusion

There is a synergy in the combination of SDGs and the principles of green chemistry when devising a roadmap for the development of environmentally friendly pesticides. Pesticides are one of the elements, and ideally the last line in pest management, where a series of procedures that center in control, not eradication of pests, preventive practices, monitoring, and scaling control procedures: mechanical, biological, and chemical control only when the previous steps have proven insufficient.

It seems a paradox that the best way to use environmentally friendly pesticides is to use them very sparingly under integrated systems, especially IPM. That way is in alignment with both SDGs and green chemistry principles, particularly responsible production and consumption (SDG 12) and the prevention of green chemistry principle.

Natural products are an area that requires more research to obtain effective pesticides with low environmental impact.

References

AbuQamar, S. F., El-Saadony, M. T., Alkafaas, S. S., Elsalahaty, M. I., Elkafas, S. S., Mathew, B. T., Aljasmi, A. N., Alhammadi, H. S., Salem, H. M., Abd El-Mageed, T. A., Zaghloul, R. A., Mosa, W. F. A., Ahmed, A. E., Elrys, A. S., Saad, A. M., Alsaeed, F. A., & El-Tarabily, K. A. (2024). Ecological impacts and management strategies of pesticide pollution on aquatic life and human beings. *Marine Pollution Bulletin, 206*, 116613. https://doi.org/10.1016/j.marpolbul.2024.116613

Angon, P. B., Mondal, S., Jahan, I., Datto, M., Antu, U. B., Ayshi, F. J., & Islam, M. S. (2023). Integrated pest management (IPM) in agriculture and its role in maintaining ecological balance and biodiversity. *Advances in Agriculture, 2023*(1), 5546373. https://doi.org/10.1155/2023/5546373

Arévalo-Jaramillo, P., Idrobo, A., Salcedo, L., Cabrera, A., Vintimilla, A., Carrión, M., & Bailon-Moscoso, N. (2019). Biochemical and genotoxic effects in women exposed to pesticides in Southern Ecuador. *Environmental Science and Pollution Research, 26*(24), 24911–24921. https://doi.org/10.1007/s11356-019-05725-7

Arrobo, M. J. C. (2023). Participación comunitaria en la percepción de la contaminación por agroquímicos en la Laguna de Cube. *Revista Criterios, 30*(2), Article 2. https://doi.org/10.31948/rev.criterios/30.2-art2

Barzman, M., Bàrberi, P., Birch, A. N. E., Boonekamp, P., Dachbrodt-Saaydeh, S., Graf, B., Hommel, B., Jensen, J. E., Kiss, J., Kudsk, P., Lamichhane, J. R., Messéan, A., Moonen, A.-C., Ratnadass, A., Ricci, P., Sarah, J.-L., & Sattin, M. (2015). Eight principles of integrated pest management. *Agronomy for Sustainable Development, 35*(4), 1199–1215. https://doi.org/10.1007/s13593-015-0327-9

Batra, V., Avinash, Kumar, A., & Soti, N. (2024). Achieving zero hunger under sustainable development goals concerning organic agriculture. *Current Agriculture Research Journal, 12*(1), 219–230. https://doi.org/10.12944/CARJ.12.1.18

Benelli, G., Lucchi, A., Thomson, D., & Ioriatti, C. (2019). Sex pheromone aerosol devices for mating disruption: Challenges for a brighter future. *Insects, 10*(10), Article 10. https://doi.org/10.3390/insects10100308

Berestetskiy, A. (2023). Modern approaches for the development of new herbicides based on natural compounds. *Plants, 12*(2), Article 2. https://doi.org/10.3390/plants12020234

Bloomfield, J. P., Williams, R. J., Gooddy, D. C., Cape, J. N., & Guha, P. (2006). Impacts of climate change on the fate and behaviour of pesticides in surface and groundwater—A UK perspective. *Science of the Total Environment, 369*(1), 163–177. https://doi.org/10.1016/j.scitotenv.2006.05.019

Bowonder, B., Kasperson, J. X., & Kasperson, R. E. (2005). Avoiding future Bhopals. In *Social contours of risk*. Routledge.

Burbano, J., Cruz, I., Colina-Márquez, J., López-Vásquez, A., & Machuca, F. (2008). Evaluation of zinc oxide-based photocatalytic degradation of a commercial pesticide. *Journal of Advanced Oxidation Technologies, 11*(1), 49–55. https://doi.org/10.1515/jaots-2008-0106

Carson, R., Lear, L., & Wilson, E. O. (2002). *Silent spring* (Anniversary ed.). Mariner Books Classics.

Chatterjee, S., Bag, S., Biswal, D., Sarkar Paria, D., Bandyopadhyay, R., Sarkar, B., Mandal, A., & Dangar, T. K. (2023). Neem-based products as potential eco-friendly mosquito control agents over conventional eco-toxic chemical pesticides-A review. *Acta Tropica, 240*, 106858. https://doi.org/10.1016/j.actatropica.2023.106858

Damalas, C. A. (2021). Farmers' intention to reduce pesticide use: The role of perceived risk of loss in the model of the planned behavior theory. *Environmental Science and Pollution Research, 28*(26), 35278–35285. https://doi.org/10.1007/s11356-021-13183-3

Damalas, C. A., & Koutroubas, S. D. (2020). Botanical pesticides for eco-friendly pest management. In *Pesticides in crop production* (pp. 181–193). Wiley. https://doi.org/10.1002/9781119432241.ch10

Deepa, T. V., Lakshmi, G., Lakshmi, P. S., & Sreekanth, S. K. (2011). Ecological effects of pesticides. In *Pesticides in the modern world—Pesticides use and management* (pp. 327–336). IntechOpen. https://doi.org/10.5772/950

Deguine, J.-P., Aubertot, J.-N., Flor, R. J., Lescourret, F., Wyckhuys, K. A. G., & Ratnadass, A. (2021). Integrated pest management: Good intentions, hard realities. A review. *Agronomy for Sustainable Development, 41*(3), 38. https://doi.org/10.1007/s13593-021-00689-w

Delcour, I., Spanoghe, P., & Uyttendaele, M. (2015). Literature review: Impact of climate change on pesticide use. *Food Research International, 68*, 7–15. https://doi.org/10.1016/j.foodres.2014.09.030

Devineni, G., Lewer, J. M., & Kostal, J. (2023). In silico design vectoring for safer pesticides. In *Sustainable agricultural practices and product design* (Vol. 1449, pp. 11–30). American Chemical Society. https://doi.org/10.1021/bk-2023-1449.ch002

Ejeta, S. Y., & Imae, T. (2021). Photodegradation of pollutant pesticide by oxidized graphitic carbon nitride catalysts. *Journal of Photochemistry and Photobiology A: Chemistry, 404*, 112955. https://doi.org/10.1016/j.jphotochem.2020.112955

Elizalde, L., Arbetman, M., Arnan, X., Eggleton, P., Leal, I. R., Lescano, M. N., Saez, A., Werenkraut, V., & Pirk, G. I. (2020). The ecosystem services provided by social insects: Traits, management tools and knowledge gaps. *Biological Reviews, 95*(5), 1418–1441. https://doi.org/10.1111/brv.12616

Fenibo, E. O., Ijoma, G. N., & Matambo, T. (2021). Biopesticides in sustainable agriculture: A critical sustainable development driver governed by green chemistry principles. *Frontiers in Sustainable Food Systems, 5*, 619058. https://doi.org/10.3389/fsufs.2021.619058

Fenibo, E. O., Ijoma, G. N., Nurmahomed, W., & Matambo, T. (2022). The potential and green chemistry attributes of biopesticides for sustainable agriculture. *Sustainability, 14*(21), Article 21. https://doi.org/10.3390/su142114417

Gatto, M. P., Cabella, R., & Gherardi, M. (2016). Climate change: The potential impact on occupational exposure to pesticides. *Annali Dell'Istituto Superiore Di Sanità, 52*(3), Article 3.

Gomes, H. D. O., Menezes, J. M. C., da Costa, J. G. M., Coutinho, H. D. M., Teixeira, R. N. P., & do Nascimento, R. F. (2020). A socio-environmental perspective on pesticide use and food production. *Ecotoxicology and Environmental Safety, 197*, 110627. https://doi.org/10.1016/j.ecoenv.2020.110627

Harwood, R. R. (1990). A history of sustainable agriculture. In *Sustainable agricultural systems*. CRC Press.

Hayat, S., Ahmad, A., Ahmad, H., Hayat, K., Khan, M. A., & Runan, T. (2022). Garlic, from medicinal herb to possible plant bioprotectant: A review. *Scientia Horticulturae, 304*, 111296. https://doi.org/10.1016/j.scienta.2022.111296

Jørgensen, P. S., Folke, C., Henriksson, P. J. G., Malmros, K., Troell, M., & Zorzet, A. (2020). Coevolutionary governance of antibiotic and pesticide resistance. *Trends in Ecology & Evolution, 35*(6), 484–494. https://doi.org/10.1016/j.tree.2020.01.011

Karimi, H., Mahdavi, S., Asgari Lajayer, B., Moghiseh, E., Rajput, V. D., Minkina, T., & Astatkie, T. (2022). Insights on the bioremediation technologies for pesticide-contaminated soils. *Environmental Geochemistry and Health, 44*(4), 1329–1354. https://doi.org/10.1007/s10653-021-01081-z

Kate, A., Sahu, L. K., Pandey, J., Mishra, M., & Sharma, P. K. (2022). Green catalysis for chemical transformation: The need for the sustainable development. *Current Research in Green and Sustainable Chemistry, 5*, 100248. https://doi.org/10.1016/j.crgsc.2021.100248

Kaur, N., Khunger, A., Wallen, S. L., Kaushik, A., Chaudhary, G. R., & Varma, R. S. (2021). Advanced green analytical chemistry for environmental pesticide detection. *Current Opinion in Green and Sustainable Chemistry, 30*, 100488. https://doi.org/10.1016/j.cogsc.2021.100488

Khan, N., Ray, R. L., Sargani, G. R., Ihtisham, M., Khayyam, M., & Ismail, S. (2021). Current progress and future prospects of agriculture technology: Gateway to sustainable agriculture. *Sustainability, 13*(9), Article 9. https://doi.org/10.3390/su13094883

Liu, X., Cao, A., Yan, D., Ouyang, C., Wang, Q., & Li, Y. (2021). Overview of mechanisms and uses of biopesticides. *International Journal of Pest Management, 67*(1), 65–72. https://doi.org/10.1080/09670874.2019.1664789

Ma, C.-S., Zhang, W., Peng, Y., Zhao, F., Chang, X.-Q., Xing, K., Zhu, L., Ma, G., Yang, H.-P., & Rudolf, V. H. W. (2021). Climate warming promotes pesticide resistance through expanding overwintering range of a global pest. *Nature Communications, 12*(1), 5351. https://doi.org/10.1038/s41467-021-25505-7

Måren, I. E. (2019). Food systems for sustainable terrestrial ecosystems (SDG 15). *Food Ethics, 2*(2), 155–159. https://doi.org/10.1007/s41055-018-00032-2

Martinez, D. A., Gathorne-Hardy, A., & Smith, B. M. (2024). Impacts of polycultural cropping on crop yields and biodiversity: A systematic map protocol. *Ecological Solutions and Evidence, 5*(3), e12349. https://doi.org/10.1002/2688-8319.12349

Mehdizadeh, M., Mushtaq, W., Anusha Siddiqui, S., Ayadi, S., Kaur, P., Yeboah, S., Mazraedoost, S., K.A.AL-Taey, D., & Tampubolon, K. (2021). Herbicide residues in agroecosystems: Fate, detection, and effect on non-target plants. *Reviews in Agricultural Science, 9*, 157–167. https://doi.org/10.7831/ras.9.0_157

Mischko, W., Hirte, M., Roehrer, S., Engelhardt, H., Mehlmer, N., Minceva, M., & Brück, T. (2018). Modular biomanufacturing for a sustainable production of terpenoid-based insect deterrents. *Green Chemistry, 20*(11), 2637–2650. https://doi.org/10.1039/C8GC00434J

Musarurwa, H., & Tavengwa, N. T. (2020). Green aspects during synthesis, application and chromatographic analysis of chiral pesticides. *Trends in Environmental Analytical Chemistry, 27*, e00093. https://doi.org/10.1016/j.teac.2020.e00093

Najmi, A., Javed, S. A., Al Bratty, M., & Alhazmi, H. A. (2022). Modern approaches in the discovery and development of plant-based natural products and their analogues as potential therapeutic agents. *Molecules, 27*(2), Article 2. https://doi.org/10.3390/molecules27020349

Ortega, P., Salcedo, R., Sánchez, E., & Gil, E. (2023). Biopesticides as alternatives to reduce the use of copper in Spanish and Portuguese viticulture: Main trends in adoption. *European Journal of Agronomy, 151*, 126996. https://doi.org/10.1016/j.eja.2023.126996

Panis, C., Candiotto, L. Z. P., Gaboardi, S. C., Gurzenda, S., Cruz, J., Castro, M., & Lemos, B. (2022). Widespread pesticide contamination of drinking water and impact on cancer risk in Brazil. *Environment International, 165*, 107321. https://doi.org/10.1016/j.envint.2022.107321

Pistocchi, A., Dorati, C., Galimberti, F., Udias, A., Bopp, S., D'Andrimont, R., Catarino, R., & Schaefer, R. B. (2023). A screening study of the spatial distribution and cumulative toxicity of agricultural pesticides in the European Union's waters. *Frontiers in Environmental Science, 11*, 1101316. https://doi.org/10.3389/fenvs.2023.1101316

Pizzi, S., Caputo, A., Corvino, A., & Venturelli, A. (2020). Management research and the UN sustainable development goals (SDGs): A bibliometric investigation and systematic review. *Journal of Cleaner Production, 276*, 124033. https://doi.org/10.1016/j.jclepro.2020.124033

Pradyumna, A., & Chelaton, J. (2018). The endosulfan tragedy of Kasaragod: Health and ethics in non-health sector programs. In A. Mishra & K. Subbiah (Eds.), *Ethics in public health practice in India* (pp. 85–104). Springer. https://doi.org/10.1007/978-981-13-2450-5_5

Pratap, B., Tiwari, H., & Singh, R. S. (2024). Optimization of atrazine degradation process by Bacillus paramycoides strain in a batch system, and assessment of growth kinetics, bacterial toxicity, phytotoxicity and chlorophyll: Optimization and assessment of atrazine degradation process. *Indian Journal of Experimental Biology (IJEB), 62*(7), Article 07. https://doi.org/10.56042/ijeb.v62i07.12068

Prăvălie, R., Patriche, C., Borrelli, P., Panagos, P., Roşca, B., Dumitraşcu, M., Nita, I.-A., Săvulescu, I., Birsan, M.-V., & Bandoc, G. (2021). Arable lands under the pressure of multiple land degradation processes. A global perspective. *Environmental Research, 194*, 110697. https://doi.org/10.1016/j.envres.2020.110697

Raimi, M. O., Vivien, O. T., Alima, O., Akpojubaro, E. H., & Williams, E. A. (2021). Articulating the effect of pesticides use and sustainable development goals (SDGs): The science of improving lives through decision impacts (SSRN Scholarly Paper 3838210). *Research on World Agricultural Economy, 2*, 29–36. https://papers.ssrn.com/abstract=3838210

Rajendran, T. P., & Singh, D. (2016). Chapter 1—Insects and pests. In Omkar (Ed.), *Ecofriendly pest management for food security* (pp. 1–24). Academic Press. https://doi.org/10.1016/B978-0-12-803265-7.00001-4

Rathore, S. S., Babu, S., Shekhawat, K., Singh, R., Yadav, S. K., Singh, V. K., & Singh, C. (2022). Designing energy cum carbon-efficient environmentally clean production system for achieving green economy in agriculture. *Sustainable Energy Technologies and Assessments, 52*, 102190. https://doi.org/10.1016/j.seta.2022.102190

Ravula, A. R., & Yenugu, S. (2021). Pyrethroid based pesticides – Chemical and biological aspects. *Critical Reviews in Toxicology, 51*(2), 117–140. https://doi.org/10.1080/10408444.2021.1879007

Sadigov, R. (2022). Rapid growth of the world population and its socioeconomic results. *The Scientific World Journal, 2022*(1), 8110229. https://doi.org/10.1155/2022/8110229

Shaikh, M. A., Hadjikakou, M., Geyik, O., & Bryan, B. A. (2024). Assessing global agri-food system exceedance of national cropland limits for linking responsible consumption and production under SDG 12. *Ecological Economics, 215*, 107993. https://doi.org/10.1016/j.ecolecon.2023.107993

Siviter, H., Brown, M. J. F., & Leadbeater, E. (2018). Sulfoxaflor exposure reduces bumblebee reproductive success. *Nature, 561*(7721), 109–112. https://doi.org/10.1038/s41586-018-0430-6

Souto, A. L., Sylvestre, M., Tölke, E. D., Tavares, J. F., Barbosa-Filho, J. M., & Cebrián-Torrejón, G. (2021). Plant-derived pesticides as an alternative to pest management and sustainable agricultural production: Prospects, applications and challenges. *Molecules, 26*(16), Article 16. https://doi.org/10.3390/molecules26164835

Sparks, T. C., Sparks, J. M., & Duke, S. O. (2023). Natural product-based crop protection compounds—Origins and future prospects. *Journal of Agricultural and Food Chemistry, 71*(5), 2259–2269. https://doi.org/10.1021/acs.jafc.2c06938

Tewari, V. K., Pareek, C. M., Lal, G., Dhruw, L. K., & Singh, N. (2020). Image processing based real-time variable-rate chemical spraying system for disease control in paddy crop. *Artificial Intelligence in Agriculture, 4*, 21–30. https://doi.org/10.1016/j.aiia.2020.01.002

Tiwari, S., & Brahmpurkar, A. (2021). Green chemistry: A new trend in the chemical synthesis to prevent our earth. *Acta Scientific Pharmaceutical Sciences, 5*, 39–52. https://doi.org/10.31080/ASPS.2021.05.0726

Umapathi, R., Ghoreishian, S. M., Sonwal, S., Rani, G. M., & Huh, Y. S. (2022). Portable electrochemical sensing methodologies for on-site detection of pesticide residues in fruits and vegetables. *Coordination Chemistry Reviews, 453*, 214305. https://doi.org/10.1016/j.ccr.2021.214305

Urionabarrenetxea, E., Casás, C., Garcia-Velasco, N., Santos, M., Tarazona, J. V., & Soto, M. (2022). Predicting environmental concentrations and the potential risk of Plant Protection Products (PPP) on non-target soil organisms accounting for regional and landscape ecological variability in European soils. *Chemosphere, 303*, 135045. https://doi.org/10.1016/j.chemosphere.2022.135045

van der Werf, W., & Bianchi, F. (2022). Options for diversifying agricultural systems to reduce pesticide use: Can we learn from nature? *Outlook on Agriculture, 51*(1), 105–113. https://doi.org/10.1177/00307270221077442

van Dijk, J., Flerlage, H., Beijer, S., Slootweg, J. C., & van Wezel, A. P. (2022). Safe and sustainable by design: A computer-based approach to redesign chemicals for reduced environmental hazards. *Chemosphere, 296*, 134050. https://doi.org/10.1016/j.chemosphere.2022.134050

Wahab, S., Muzammil, K., Nasir, N., Khan, M. S., Ahmad, M. F., Khalid, M., Ahmad, W., Dawria, A., Reddy, L. K. V., & Busayli, A. M. (2022). Advancement and new trends in analysis of pesticide residues in food: A comprehensive review. *Plants, 11*(9), Article 9. https://doi.org/10.3390/plants11091106

Wee, S. Y., & Aris, A. Z. (2017). Ecological risk estimation of organophosphorus pesticides in riverine ecosystems. *Chemosphere, 188*, 575–581. https://doi.org/10.1016/j.chemosphere.2017.09.035

Westerhoven, A. C. V., Meijer, H. J. G., Seidl, M. F., & Kema, G. H. J. (2022). Uncontained spread of Fusarium wilt of banana threatens African food security. *PLoS Pathogens, 18*(9), e1010769. https://doi.org/10.1371/journal.ppat.1010769

Zhang, Z., Yan, X., Jones, K. C., Jiao, C., Sun, C., Liu, Y., Zhu, Y., Zhang, Q., Zhai, L., Shen, Z., & Chen, L. (2022). Pesticide risk constraints to achieving Sustainable Development Goals in China based on national modeling. *npj Clean Water, 5*(1), 1–10. https://doi.org/10.1038/s41545-022-00202-0

Zhao, F., He, J., Li, X., Bai, Y., Ying, Y., & Ping, J. (2020). Smart plant-wearable biosensor for in-situ pesticide analysis. *Biosensors and Bioelectronics, 170*, 112636. https://doi.org/10.1016/j.bios.2020.112636

Zhou, Y., Li, Y., & Liu, Y. (2020). The nexus between regional eco-environmental degradation and rural impoverishment in China. *Habitat International, 96*, 102086. https://doi.org/10.1016/j.habitatint.2019.102086

Zimmerman, J. B., Anastas, P. T., Erythropel, H. C., & Leitner, W. (2020). Designing for a green chemistry future. *Science, 367*(6476), 397–400. https://doi.org/10.1126/science.aay3060

Part III
Challenges

This part explores the intricate network of environmental dynamics, addressing transformative changes in pest dynamics driven by the use of pesticides and the effects of climate change.

The Nexus Between Pesticides, Climate Change, Carbon Sequestration, and Soil Health

V. Mageshwaran, O. T. Lawal, O. O. Osemwegie, R. Arutselvan, and A. F. Ologundudu

1 Introduction

Pesticides are essential in agriculture, as they target pests that can harm crop yields. They are categorized based on the types of organisms they control: insecticides, herbicides, fungicides, and nematicides (Khan et al., 2023). Each category has distinct chemical structures and mechanisms of action, which can influence not only the intended pests but also the health of the soil (Stenberg et al., 2021). Insecticides, including organophosphates, carbamates, pyrethroids, and neonicotinoids, are designed to disrupt the nervous systems of insects. These can be applied through foliar sprays, soil drenching, or seed treatments, depending on the specific pest and crop involved (Alsafran et al., 2022). Herbicides are utilized to manage weeds that compete with crops. Common examples include glyphosate, atrazine, 2,4-D, and

V. Mageshwaran (✉)
Microbial Technology Lab, ICAR-National Bureau of Agriculturally Important Microorganisms, Kushmaur, Mau, Uttar Pradesh, India
e-mail: mageshwaran.v@icar.gov.in

O. T. Lawal
Department of Medical Biochemistry, Federal University of Technology, Akure, Ondo State, Nigeria

O. O. Osemwegie
Department of Food Science and Microbiology, Landmark University, Omu Aran, Kwara State, Nigeria

R. Arutselvan
Regional Station, ICAR-Central Tuber Crops Research Institute (CTCRI), Dumuduma, Bhubaneswar, Odisha, India

A. F. Ologundudu
School of Computer, Mathematics and Natural Sciences, Morgan State University, Baltimore, MD, USA

B. R. Babaniyi, E. E. Babaniyi (eds.), *The Interplay of Pesticides and Climate Change*, https://doi.org/10.1007/978-3-031-81669-7_12

paraquat, which function by inhibiting processes like photosynthesis or hormonal regulation in plants (Cobb, 2022). Herbicides are typically applied either before or after the crops emerge, directly to the soil or onto the plants themselves. Fungicides are used to combat fungal diseases that can harm crops (Santra & Banerjee, 2020). Chemical groups such as azoles, strobilurins, copper-based fungicides, and sulfur-based fungicides target fungal pathogens by interfering with their cell membrane synthesis or respiration. These are usually applied as foliar sprays, soil drenches, or seed treatments. However, fungicides can also harm beneficial soil fungi, like mycorrhizal fungi, which are crucial for nutrient absorption in plants, thereby diminishing soil biodiversity and resilience. Nematicides are used to control nematodes, which are tiny worms that damage plant roots. There are two main types: fumigant nematicides, such as methyl bromide, and non-fumigant ones like aldicarb. These are typically applied to the soil before or during planting (Doran & Zeiss, 2000).

Soil health is a multifaceted concept that refers to the continued capacity of soil to function as a living ecosystem, which is critical for sustaining plants, animals, and humans. It encompasses physical, chemical, and biological properties, ensuring the soil remains productive, resilient, and capable of supporting a wide range of ecosystem services (Doran & Zeiss, 2000). Key components of soil health include soil organic matter, nutrient cycling, water retention, and the activity of soil microorganisms, all of which play a crucial role in maintaining soil structure and fertility. The soil's ability to support plant growth, regulate water flow, and filter pollutants depends on these interacting components, making soil health essential for both agricultural productivity and environmental sustainability (Delgado & Gómez, 2016).

Biological factors, particularly soil microorganisms, are integral to soil health as they contribute to organic matter decomposition, nutrient cycling, and the formation of soil aggregates. Microbial activity, especially from bacteria, fungi, and actinomycetes, enhances the soil's capacity to sequester carbon, thus playing a role in mitigating climate change (Lal, 2004). The balance of these microorganisms is vital for controlling soil-borne diseases, enhancing plant growth, and detoxifying harmful substances like pesticides. Therefore, maintaining soil health is crucial for reducing the negative impacts of both pesticides use and climate change (Swift et al., 2004; Lehmann et al., 2020a).

2 Pesticides and Their Applications

The application methods for pesticides vary widely and include foliar sprays, soil drenching, and seed treatments. Foliar sprays are aimed at the surface of the plants, while soil drenching allows systemic pesticides to be taken up by the roots (Zheng & Xu, 2023). Seed treatments offer protection in the early stages, but all these methods can result in pesticide residues building up in the soil (Ul Haq et al., 2020).

3 Impact of Pesticides on Soil Health

The effects of pesticide use on soil health are considerable. Pesticides, especially those with long-lasting effects, can contaminate the soil, harming soil microorganisms, disrupting nutrient cycling, and altering overall soil structure (Tripathi et al., 2020). Excessive use of pesticides can lead to a decline in microbial diversity, the development of herbicide resistance, and the contamination of water sources through leaching (Ali et al., 2021).

4 Pesticide Residues and Soil Contamination

The presence of persistent pesticide residues in soil raises significant concerns regarding long-term soil health. The accumulation of these residues can result in the following

(i) *Bioaccumulation*: Certain pesticides, particularly those that degrade slowly, can build up in the soil and enter the food chain through crops or soil organisms.
(ii) *Leaching and Water Contamination*: Persistent residues have the potential to leach into groundwater, leading to the contamination of water sources and negatively impacting aquatic ecosystems.
(iii) *Soil Contamination and Phytotoxicity*: Residual pesticides may hinder plant growth by disrupting root development or limiting nutrient availability, which can result in decreased crop productivity.

5 Effects on Soil Fauna

Soil fauna, such as earthworms, nematodes, and arthropods, play a vital role in soil aeration, organic matter breakdown, and nutrient cycling (Yatoo et al., 2020). However, exposure to certain pesticides can lead to:

(i) *Mortality and Population Decline*: Some pesticides, especially insecticides, are harmful to beneficial soil fauna, causing population declines and a reduction in soil biodiversity.
(ii) *Behavioral Changes*: Sublethal exposure to pesticides can alter the feeding, reproduction, and burrowing behaviors of soil organisms like earthworms, which further affect soil structure and fertility.
(iii) *Impaired Ecosystem Services*: The decline in soil fauna disrupts essential soil functions, such as nutrient cycling and organic matter decomposition, ultimately diminishing soil health over time.

6 Pesticide Impact on Soil Enzyme Activity

Soil enzymes, produced by soil microorganisms and plants, are crucial for various biochemical processes in the soil, including the breakdown of organic matter and nutrient cycling. The application of pesticides can inhibit enzyme activity, particularly:

(i) *Dehydrogenase*: This enzyme, vital for microbial respiration, is often suppressed by pesticide exposure, leading to a decrease in soil metabolic activity.
(ii) *Phosphatase and Urease*: These enzymes, which are responsible for the cycling of phosphorus and nitrogen, respectively, are also adversely affected by pesticide residues, resulting in nutrient imbalances.

7 Impact on Soil Microbial Communities

Soil microbial communities represent some of the most diverse and dynamic ecosystems on our planet, consisting of bacteria, fungi, archaea, protozoa, and various other microorganisms (Nkongolo & Narendrula-Kotha, 2020). These communities are vital for sustaining soil health and fertility, as they play key roles in organic matter decomposition, nutrient cycling (including nitrogen, phosphorus, and sulfur), carbon sequestration, and interactions between plants and microbes. However, the use of pesticides can significantly disrupt these fragile microbial ecosystems, resulting in numerous harmful effects on soil functionality (Bhattacharyya et al., 2022).

(i) *Reduction in Microbial Diversity*

Microbial diversity is essential for the resilience of soil ecosystems. A varied community of microorganisms ensures that multiple metabolic pathways operate effectively, even under stress. Unfortunately, pesticides can selectively eliminate or inhibit beneficial soil microorganisms, leading to a decrease in microbial diversity (Lahlali et al., 2021).

(ii) *Selective Toxicity*

Various classes of pesticides target specific organisms, but they often have unintended effects on non-target soil microorganisms. For instance, organophosphates, which are frequently used as insecticides, can be toxic to nitrogen-fixing bacteria like Rhizobium and Azotobacter (Zhao et al., 2022). These bacteria are crucial for converting atmospheric nitrogen into ammonia, a form of nitrogen that plants can absorb and use.

(iii) *Fungi and Mycorrhizae*

Certain pesticides, especially fungicides (such as azoles), can decrease the populations of beneficial soil fungi, including mycorrhizal fungi. Mycorrhizae establish symbiotic relationships with plant roots, enhancing the plant's nutrient uptake (Smith et al., 2020).

8 Alteration of Microbial Activity

Microbial activity in soil plays a crucial role in essential processes like nutrient cycling, organic matter decomposition, and maintaining soil structure. Pesticides can disrupt these processes by affecting the activity of microorganisms or by inhibiting the production of enzymes that are vital for these functions (Kirchman, 2018).

(i) *Enzyme Suppression*

Numerous soil microorganisms produce enzymes that are critical for nutrient cycling. For instance, dehydrogenase is involved in the oxidative processes that break down organic matter, while urease facilitates the conversion of urea to ammonia, a key step in the nitrogen cycle (Zhao et al., 2022).

(ii) *Nitrogen and Phosphorus Cycling*

The nitrogen cycle, which heavily relies on microbial activity, is particularly vulnerable to disruption by pesticides. For example, nitrifying bacteria, which convert ammonia to nitrite and nitrate (forms of nitrogen that plants can utilize), can be inhibited by herbicides and insecticides. Likewise, denitrifying bacteria, which convert nitrate to nitrogen gas (thus completing the nitrogen cycle), can also be adversely affected by pesticides. This disruption creates an imbalance in nitrogen availability, which is essential for plant growth (Wang et al., 2023).

(iii) *Organic Matter Decomposition*

Decomposer organisms such as bacteria and fungi are responsible for breaking down organic material in the soil, transforming it into simpler forms of organic carbon that can be utilized by plants and other soil organisms. Pesticides can disrupt this vital process (Chen et al., 2021).

9 Antibiotic Resistance

One of the most concerning effects of pesticide use in soil is the potential for fostering the growth of antibiotic-resistant bacteria. Certain pesticides, especially those with antimicrobial properties, can create selective pressure on soil bacteria, leading to the evolution and dissemination of resistance genes.

(i) *Cross-Resistance Between Pesticides and Antibiotics*

Studies have indicated that some bacteria exposed to pesticides develop resistance mechanisms that resemble those used against antibiotics. For instance, exposure to herbicides like glyphosate and insecticides such as chlorpyrifos has been associated with increased tolerance to specific antibiotics in soil bacteria. This occurs because pesticides and antibiotics may target similar cellular processes, allowing bacteria that evolve resistance to one to unintentionally gain resistance to the other (Kurenbach et al., 2017).

(ii) *Horizontal Gene Transfer*

Soil serves as a hotspot for horizontal gene transfer, the process through which bacteria share genetic material, including antibiotic resistance genes. The presence of pesticides can facilitate this process by creating stressful environmental conditions that heighten the chances of gene exchange. The spread of resistance genes within soil microbial communities not only jeopardizes soil health but also poses risks to human health, as these resistant bacteria may transfer their resistance to human pathogens (Martínez, 2017).

(iii) *Impact on Biocontrol Agents*

The emergence of pesticide resistance in soil bacteria can also influence the effectiveness of biocontrol agents utilized in agriculture. For example, bacteria that have developed resistance to fungicides or bactericides may outcompete beneficial organisms employed in biological pest control, diminishing the effectiveness of these sustainable pest management approaches (González et al., 2020).

(iv) *Impact on Nutrient Cycling*

Soil nutrient cycling is a vital process that ensures essential elements such as nitrogen, phosphorus, and carbon are available for plants and other organisms. This cycle is mainly driven by soil microorganisms, including bacteria, fungi, and archaea, which help convert nutrients into forms that plants can absorb and utilize. However, the use of pesticides in agriculture can disrupt these important biogeochemical cycles, leading to decreased soil fertility, lower crop yields, and long-term soil degradation. This section explores how pesticides interfere with the nitrogen, phosphorus, and carbon cycles, emphasizing the effects on soil health and plant growth.

(v) *Impact on Nitrogen-Fixing Bacteria*

Insecticides and fungicides have been found to decrease the population and activity of nitrogen-fixing bacteria in the soil, which leads to lower nitrogen inputs into the ecosystem. Pesticides such as organophosphates and carbamates can inhibit nitrogenase, the enzyme responsible for nitrogen fixation, by disrupting microbial metabolic processes. This suppression can diminish nitrogen availability for plants, especially in legume crops that depend on symbiotic nitrogen fixation.

10 Carbon Sequestration in Soil Under Climate Change Scenarios

When plants take up atmospheric carbon dioxide (CO_2) through photosynthesis and store it as organic carbon in the soil, this process is known as carbon sequestration in soil (Sadatshojaei et al., 2021). Because CO_2 is a greenhouse gas (GHG) that greatly contributes to global warming, this process helps moderate climate change by lowering its atmospheric concentration.

(i) *Photosynthesis and Plant Biomass*

During photosynthesis, plants take in CO_2 from the atmosphere and transform it into organic matter (OM). A certain amount of this organic matter is converted into plant biomass (leaf, stem, and root), which eventually breaks down and adds to the soil's organic carbon content (SOC) (Paul, 2016).

(ii) *Root Exudation*

Soil microorganisms absorb the organic substances released into the soil by the roots. Root exudation improves soil carbon storage by assisting in the development of stable soil organic matter (Panchal et al., 2022).

(iii) *Humification*

Humus is a stable soil of organic carbon that decomposing plant and animal wastes in the soil undergo microbial transformation into. It can stay in the soil for generations (Prescott & Vesterdal, 2021).

(iv) *Soil Aggregation*

Soil particles bind together to form aggregates, which protect organic carbon from decomposition. The formation of stable soil aggregates is crucial for long-term carbon storage (Mustafa et al., 2020).

(v) *Biochar Addition*

Carbon sequestration in soil can be improved by adding biochar, a long-lasting form of carbon made from pyrolyzing organic materials. Carbon may be stored in biochar for hundreds to thousands of years since it is not easily broken down (Rahman et al., 2020).

11 Types of Carbon Storage

(i) *Soil Organic Carbon (SOC)*: SOC, or soil organic carbon storage, is mostly made up of microbial biomass, root exudates, and degraded plant and animal matter. Climate, land use, and management techniques can all have an impact on SOC, which is a dynamic process (Paul, 2016).
(ii) *Particulate Organic Matter (POM)*: A portion of SOC that contains more labile and quickly decomposing plant material that has partially decomposed (Panchal et al., 2022).
(iii) *Humus*: A steady, dark-colored form of SOC that comes from organic matter being broken down by microbes. Long-term carbon storage and soil fertility are enhanced by humus (Paul, 2016).
(iv) *Soil Inorganic Carbon (SIC)*: It comprises the bicarbonates and carbonates that are produced when soils precipitate carbonates or when rocks weather chemically. SIC can live in soils for millennia and is more stable than SOC (Batool et al., 2024). Carbonates can precipitate in the soil in dry and semi-arid locations, which helps with long-term carbon storage (Abdelhak, 2022).

12 Roles of Organic Matter and Microorganisms in Carbon Sequestration

Microorganisms and organic matter are essential for soil carbon sequestration. Soil organic carbon (SOC) is mostly obtained from organic matter, which includes decomposed animal wastes, plant residues, and root exudates (Sadatshojaei et al., 2021). Carbon dioxide (CO_2) from the atmosphere is taken up by plants during photosynthetic processes and transformed into organic matter that is subsequently absorbed by the soil (Kirkham, 2016). This organic matter is a dynamic carbon reservoir that decomposes to release carbon back into the atmosphere or store it in the soil for extended periods of time.

The decomposition process depends on microorganisms, such as fungi, bacteria, and other soil microbes (Zifcakova, 2020). Organic matter is broken down into simpler compounds by them, some of which go on to produce stable forms of SOC, like humus (Paul, 2016). Along with improving soil structure, fertility, and water retention, this process stabilizes carbon in the soil (Zhang et al., 2019). Additionally, as a result of microbial activity, soil aggregates are formed, improving long-term carbon storage by physically preventing further organic carbon breakdown (Bhattacharyya et al., 2022). The carbon cycle in soils is largely dependent on the relationship between organic matter and microbes, which affects the equilibrium between sequestration and release of carbon and, in the end, helps to mitigate climate change.

13 Agricultural Practices That Enhance Sequestration

Agricultural practices that enhance soil carbon sequestration focus on increasing organic matter inputs and minimizing carbon losses (Hussain et al., 2021). Reducing tillage or adopting no-till practices minimizes soil disturbance, helping to retain organic matter and reduce carbon loss through erosion and decomposition (Hussain et al., 2021). Planting cover crops during off-seasons adds organic matter to the soil, improves soil structure, and enhances microbial activity, all of which contribute to carbon sequestration (Aziz, 2022). Diverse crop rotations, including legumes, improve soil health and increase carbon inputs through varied root structures and residues (Liu et al., 2022). Integrating trees and shrubs into agricultural systems increases biomass production, which contributes to long-term carbon storage in both soil and plant biomass (Meena et al., 2020). Applying compost, manure, and biochar boosts organic carbon levels in soils and improves microbial activity (Rahman et al., 2020).

14 Importance of Carbon Sequestration to Soil Health and Climate Change Mitigation

Carbon sequestration is vital for soil health and climate change mitigation. By capturing and storing atmospheric carbon in soils, it enhances soil structure, fertility, and water retention, leading to more productive and resilient agricultural systems (Meena et al., 2020). Increased soil organic carbon also supports diverse microbial communities, which are essential for nutrient cycling (Jiao et al., 2021). For climate change mitigation, carbon sequestration reduces greenhouse gas concentrations, helping to curb global warming (Fornara et al., 2011). Thus, soil carbon sequestration contributes to both environmental sustainability and food security, making it a key strategy in addressing climate change and promoting healthy ecosystems (Telo da Gama, 2023).

15 Challenges and Opportunities for Increasing Soil Carbon Storage

Increasing soil carbon storage faces challenges such as soil degradation, intensive agricultural practices that deplete organic matter, and climate change, which can accelerate carbon loss through higher temperatures and extreme weather events (Smith et al., 2020). Additionally, economic and knowledge barriers limit the adoption of carbon-sequestering practices (Buck & Palumbo-Compton, 2022).

Opportunities include adopting sustainable farming practices like no-till, cover cropping, and agroforestry, which enhance carbon storage (Dang et al., 2020). Advances in technology and better land management policies can support farmers in implementing these practices (Agidew & Singh, 2019). Carbon markets also present financial incentives for farmers to prioritize soil carbon sequestration, benefiting both agriculture and the environment (Gelardi et al., 2023).

16 Influence of Pesticides on Soil Organic Carbon Dynamics

Soil organic carbon (SOC) dynamics are principally affected by pesticides, which in many cases influence the SOC cycle causing unfavorable effects (Swarup et al., 2019). Pesticides are used to protect the crops from pests and diseases, but, in addition to their purpose of increasing yields, they also have effects on microbial soil communities responsible for organic matter degradation and carbon cycling (Meena et al., 2020). Pesticides and especially synthesized pesticides reduce soil biodiversity resulting in disrupting the equilibrium of synnemata that plays a huge part on residue decomposition and sustenance into soil organic carbon (Hussain et al., 2021).

This leads to a disruption of the natural humification—which is part of decomposition where organic matter transforms into stable soil carbon. It makes SOC formation decrease, lowers the potential of carbon storage by soils. (Hayes & Swift, 2020) Gradual destruction of microbial communities over time also results in loss of soil structure, hence degradation and carbon elution (Sang et al., 2021). Additionally, pesticide residues can sequester with soil organic matter binding up carbon making for an inaccessibility of resources to the microbial community adding further insult to below ground C depletion (Kaufman, 2020).

However, this is also the point at which environmentally friendly pest management techniques, such as integrated pest management (IPM), become useful in promoting healthy soils by reducing the need for pesticides and protecting microbial biodiversity (Angon et al., 2023). Reducing the pesticide input and improving natural soil processes to keep SOC at considerable levels would help in both maintaining soils health for secular demand of enhancing agricultural production and mitigating climate change through long-term carbon sequestration (Mueller et al., 2021).

17 Impact of Pesticides on Soil Microbial Activities

One of the major environmental concerns that pertains to pesticide use is that it negatively influences soil microbial activities which in turn affect stability and diversity of ecosystem as whole (Wołejko et al., 2020). Most synthetic pesticides are toxic and biocide to beneficial soil microorganisms that include bacteria, fungi or earthworms which significantly help maintain a healthy soil through the cycling of nutrients like nitrogen and carbon as well in breaking down organic matter (Tripathi et al., 2020). A drop in the population of such microorganisms may result in slower decomposition rates, limiting soil organic carbon sequestration and nutrient release.

Furthermore, pesticides have the ability to change the composition of the microbial community by suppressing sensitive species and favoring pesticide-resistant ones, creating an unbalanced ecosystem that adversely affects critical soil processes like soil aggregation, nitrogen fixation, and organic matter decomposition. Eventually, the decreased microbial activity can deteriorate soil fertility and health, making the soil more vulnerable to erosion (Vlaiculescu & Varrone, 2022).

These negative consequences can be mitigated by using sustainable techniques like integrated pest management (IPM), which reduces the need for pesticides and promote healthier microbial communities and improved soil efficiency (Angon et al., 2023).

18 Studies on the Combined Effects of Pesticides and Carbon Sequestration Practices

Various findings are found in studies examining the combined effects of carbon sequestration techniques and insecticides (Stanley et al., 2018). Pesticide use can thwart sustainable agricultural methods that improve soil carbon storage, such as no-till farming, cover crops, and organic additions (Nelson, 2023). Pesticides frequently decrease microbial variety and activity, which are essential for the breakdown of organic matter and the production of soil organic carbon (Parte et al., 2017). However, some research indicates that soil health and carbon storage can be enhanced by combining reduced pesticide usage with carbon-sequestering techniques, such as agroecological approaches (Johansson et al., 2022). In general, the secret to maximizing carbon sequestration efforts is to use as little pesticide as possible while implementing sustainable soil management techniques.

19 Mitigation Strategies to Balance Pesticide Use and Carbon Sequestration

Implementing integrated mitigation techniques that enhance soil health and manage pests is necessary to strike a balance between the use of pesticides and carbon sequestration. Integrated pest management (IPM) is a useful strategy that lessens the need for synthetic pesticides by combining chemical, mechanical, and biological pest control strategies (Baker et al., 2020). IPM reduces pesticide inputs while maintaining microbial variety and activity, which are critical for the breakdown of organic matter and the creation of soil organic carbon (SOC) (Angon et al., 2023).

Pest pressure is naturally reduced by crop rotation and diversification, which lowers the demand for pesticides. These techniques increase soil organic matter while shielding soil ecosystems from pesticide damage when paired with carbon-sequestering techniques like cover crops and no-till farming (Jaworski et al., 2023).

By limiting misuse of pesticides and focusing exclusively on impacted areas, precision agriculture can further optimize their administration while lessening their detrimental effects on soil bacteria (Raj et al., 2022). Additionally, maintaining microbial populations essential for carbon cycling can be achieved by the use of biopesticides, which are neutral to beneficial soil species (Meena et al., 2020).

Through encouraging microbial resilience, organic additions like compost and biochar not only increase soil organic carbon (SOC) but also act as a buffer against pesticide residues (Bolan et al., 2023). Farmers may maintain efficient pest control while increasing carbon sequestration and strengthening soil health and the environment's ability to slow down climate change by combining these tactics (Shah & Wu, 2019).

20 Pesticide Applications and Climate Change

The ecological interlink between pesticide applications and climate change dynamics and impacts is complex. Many studies have established the consistent looping influence of pesticides on environmental sustainability and climate change impact (Choudhury & Saha, 2020; Van Deynze et al., 2024). Pesticide applications are a product of the global need for food security and their persistent use has not been without high health and environmental concerns (Adil et al., 2023). However, growing consciousness and studies on the risk of pesticide applications in food safety, and air and groundwater pollution, all subject to climate influence affirms a connection. Persistent pesticide applications can influence climate change impacts through several interconnected ways, supporting the superior benefit of sustainable agricultural practices in climate change mitigation. The linkage between these two is more obvious in pest pressure within the ecological and agroecological context. In an era demanding biological and ecosystem resilience for favorable adaption to climate change impact, the entire pesticide lifecycle production, transportation, and end-use interplay with climate factors, contributing, often indirectly, to ecosystem disruption, biodiversity loss, increased pest pressure, decline in soil health, and rise in water pollution and greenhouse gas emissions (Delcour et al., 2015). Therefore, it is logical to characterize the relationship (Table 1) between pesticide industries, particularly its application and climate change determining the fate of environmental dynamics as a vicious cycle or pesticide-climate loop, encompassing soil types and agroclimate zones (Pérez-Lucas et al., 2024). Persistent application of pesticides releases toxic and greenhouse gases into the atmosphere exacerbating global warming and creating pest pressures, particularly loss of beneficial species (natural predators, pollinators, burrowers, and decomposers), and pesticide-resistant species (Siebielec et al., 2020; Zinyemba et al., 2020). This increased pesticide dosage, health risks, ecological signaling disruptions, ecotoxicity, soil microbial pressure, stressing agricultural systems, and pestilence. It also confirms that the entire lifecycle of pesticides including their disposal and degradation contributes to climate change in numerous ways even though studies are more focused on climate influences on the fate of applied pesticides than upstream of their cycle (Kaka et al., 2021; Martínez-Megías et al., 2023). A few pieces of information also exist on the consequences of the interplay of pesticide applications and climate change on the fate of environmental dynamics in the tropical compared to the Mediterranean parts of the world, emphasizing their unique ecosystems and agroecology (AbuQamar et al., 2024). Likewise, Tudi et al. (2021) noted that most agricultural hotspots worldwide are often vulnerable to climate variability and faced with heightened risks of pest pressure compared to other regions, suggesting increased use of pesticides.

The exacerbating challenges arising from climate-pesticides application feedback loops on the environment, particularly agricultural practices, have attracted solution efforts from farmers, governments, researchers, communities, and environmentalists to cover their health, environmental, economic, and ecosystem impacts.

Table 1 The summary of interplays between pesticide applications and climate change on the fate of environmental dynamics

Pesticide application	Climate change effect	Environmental challenges	Fate	Reference
Increases greenhouse gas (GHG) emissions	Pesticides' production, application, and degradation release greenhouse (CO_2, CH_4, and N_2O) gases, contributing to climate change.	Increases GHG, especially carbon emissions from energy-intensive production, into the atmosphere, contributing to global warming	Raises temperature, accelerates pest lifecycle, increases pesticide use, resistance, and breakdown, exacerbating climate change impacts	Adil et al. (2023)
Soil degradation and water pollution	Pesticides disrupt soil microbiota, reducing carbon sequestration capacity in soils, and alter the efficacy and environmental behavior of pesticides	Altered rainfall pattern leading to increased runoffs, leaching, eutrophication	Water pollution with lasting impacts on humans, animals, and agriculture, eutrophication, disruption of aquatic ecosystems and food chain, decline in soil health, and depletes natural carbon sinks	Kaka et al. (2021); Brüggemann et al. (2024)
Biodiversity loss	Non-target organisms (e.g., pollinators, natural predators, burrowers, decomposers, and beneficial insects) are affected, reducing biodiversity	Ecosystem imbalances, food chain disruptions, habitat loss, changes in species distribution and range, and reduced resilience to environmental changes	Decline in biodiversity, disruption of ecosystem functions and services, making the ecosystems more vulnerable to climate change impacts	Saha et al. (2019)
Pest pressure dynamics	Increases reliance on chemical controls at even higher doses, the potential for resistance	Shifts in pest populations and ranges; new pest introductions	Greater pest pressure, escalation of pesticide uses, creating a cycle of pesticide dependence	Tudi et al. (2021); Elhamalawy et al. (2024)
Land use change	Encourages land use changes	Deforestation, and habitat loss, leads to a reduction in natural sinks	Enhanced carbon footprint and loss of ecosystems critical for climate regulation, requiring agroecological practices for impact mitigation	Odewale et al. (2023); Van Deynze et al. (2024)
Increases human health risks	Increases exposure to pesticide residues in food and water	Higher temperatures and extreme weather events affecting pesticide behavior	Health risks from contaminated food and water are demonstrated in the high incidences of pesticide-related illnesses	Zinyemba et al. (2020); Adil et al. (2023)

Moreira et al. (2022) and Zhou et al. (2025) acknowledged the exigent need to break the cycle dynamic of climate-pesticide applications and their devasting impacts by strategizing to promote agricultural resilience (climate-resilient crops) and sustainable practices (advanced irrigation techniques, cover cropping, climate-smart, and precision agriculture), biopesticide applications, and integrated pest management approaches. According to Yahyah et al. (2024), these are crucial in promoting a promising agricultural and safe environmental future if combined with public awareness campaigns, education and training, funding support for research and development, and policies and legislation.

20.1 Climate Change and Its Global Impact on Soil Health

Even though complex, the multidimensional influence of climate change is critical in the fate of soil health and cannot be ignored in attaining sustainable agroecology and soil ecosystems. Climate change substantially affects the global soil's chemical, physical, and biological properties, contributing to varying regional land use policies, soil types, cropping, and management practices. This equally suggests that climate change can cause modifications of existing soils, altering their physicochemical profiles, fertility, and services (van der Meij et al., 2020). Furthermore, shifting climate patterns, particularly temperature, and precipitation, significantly determine soil health (resilience or vulnerability), impacting soil organic matter content, physicochemical attributes, carbon storage, biodiversity, productivity, and pathogen-antagonist balance (Tripathi et al., 2021). For instance, increased temperature and altered precipitation patterns exacerbate soil erosion, reduce soil organic matter, modify soil structure, and alter pedogenesis, resulting in soil vulnerability, and inability of the soil to support plant growth.

These climate-driven changes to soil ecosystem functions and services affect sustainable agriculture with significant implications for global food security, encouraging numerous innovations in traditional and emerging mitigation efforts among stakeholders worldwide (Issoufou et al., 2020). Even though the efforts may vary across regions, they focus more on integrating traditional cropping practices like crop rotation, cover cropping or mulching, conservation tillage, and agroforestry to build long-term soil ecosystem resilience and adaptation to climate change vagaries (AL-Falahi et al., 2023; Okoronkwo et al., 2024; Visscher et al., 2024). Similarly, research investments are now introducing time-tested emerging strategies like the sustainable use of soil amendments, microbial inoculants or consortia, and precision agriculture tools (drones, soil sensors, digital technologies, predictive modeling, nanotechnology, and molecular engineering) to complement integrated traditional approach to mitigating the impacts of climate change on soil health (Parra-López et al., 2024). These notwithstanding, exploring the benefits of emerging approaches to mitigating climate change's impacts on soil health and productivity requires more groundwork to surmount significant barriers to their widespread adoption. Therefore, it is important for the prospect of mitigation efforts to identify the barrier militating against the widespread adoption of advanced soil resilient-promoting strategies. These barriers may include the lack of infrastructure, training/literacy, digital and technological divides, policy frameworks, financial incentives, and scalability (Okoronkwo et al., 2024).

21 The Role of Pesticide Applications in Climate Change Impacts

Pesticide applications combined with intensive agriculture have revolutionized agricultural yield and productivity, facilitating food production for the growing global population. Despite being deliberately designed to target specific pests and minimize crop loss due to their rampaging impacts and transmission of pathogens, their persistent applications within the climate change context are not without environmental, ecological, and health consequences (Popp et al., 2013; Bonner & Alavanja, 2017). Climate change significantly influences pesticide applications and their outcomes, contributing to food safety and environmental concerns (Zinyemba et al., 2020; Tudi et al., 2021). Climate change enhances the survival and reproduction rates of mostly non-beneficial pests and weeds, altering pesticide and pest adaptation behaviors (Delcour et al., 2015). The entire pesticide value chain from production to applications and disposal is subject to climate influences, worsening climate change impacts on the environment, human health, and soil. The fate of pesticide spray is significantly affected by climate change factors, facilitating their spatial capacity to produce substantial amounts of volatile organic compounds (VOCs) or act as potent greenhouse gases when sprayed (Bhatti et al., 2024). Moreover, pesticide microbial degradation with its fossil-energy-consuming production and transportation contributes to GHG emissions, thinning the atmosphere's ozone layer and causing global warming.

22 Crop, Pests, and Weed Responses in Pesticidal and Climate Change Conditions

The interaction between pesticides and climate change creates feedback loops, impacting pest pressure, pesticide efficacy, soil carbon sequestration dynamics, physicochemical profiles, and nutrient bioavailability (Van Deynze et al., 2024). Also, the responses of crops, pests, and weeds to pesticide and climate change conditions are interconnected, differential, and complex (Ramesh et al., 2017; Heeb et al., 2019). Pesticide applications in crop fields were intended to promote crop growth and yield by limiting pest-related stress but such objectives may be derailed by climate change. Furthermore, elevated temperature and carbon dioxide have been linked to negative responses leading to increased pesticide residues on crops and soils (Hutton et al., 2024). This posed high risks to food quality, safety, and essential nutrients, expanded pest distribution and abundance, deteriorated soil health, and increased resistance (Saha et al., 2019; Ma et al., 2021). The risks of pesticide-climate change interplays underscore the need for proper monitoring and tracking of pesticide applications and challenges to innovate more efficient integrated management and sustainable agricultural practices coupled with emerging technologies as mitigating strategies.

23 Pesticides and Greenhouse Gas Emissions

Fossil fuel remains the most widely common energy source constituting the economic and technological mainstream energy in most parts of the world. Therefore, manufacturing and industrial production sectors, including pesticide, plastic, and other petrochemical (chemical fertilizers) production, are powered by fossil energy, contributing to their unhealthy role in exacerbating global warming and climate change impacts (Demeneix, 2020). These critical production sectors are fossil fuel-intensive with pesticide production demanding the highest energy in the agricultural sectors worldwide (Stull & Patz, 2024). They are regarded as major amplifiers of climate change, health risks, and environmental menace, escalating emissions (carbon and GHG). Furthermore, Fredric Bauer et al. (2022) and Wei et al. (2024) recognized that like plastic production, the pesticide production sector encourages the vicious cycle of pesticide-related gas emissions, complicating decarbonization (low carbon transition), and circular economy strategies. This supports the urgency of a systemic change in the approach to pesticide production, chemical feedstock supply, pesticide transportation, use, and demands. Fossil fuel-intensive production of pesticides poses severe challenges in climate scenarios, perpetuating a vicious cycle of climate-related agricultural stresses that increase dependence on more pesticides. However, growing investments in biopesticide production, renewable energies, biomass feedstock, recycling technologies, sustainable agroecological practices, and estate planning are foreseeable strategies that can reduce overreliance on pesticides (Bouri et al., 2023).

The application of some classes of agro-allied pesticides (Table 2) like insecticides (neonicotinoids, and organophosphates), herbicides (glyphosate, paraquat), and fungicides (Triazoles) has a significant environmental footprint and is critical in greenhouse gas emission dynamics (Bouri et al., 2023). Many pesticides contribute substantially to GHG emissions from their upstream production line to downstream applications, promoting declines in soil health, biodiversity, and ecosystem services. Accumulated residues of pesticides that inadvertently penetrate the soil cause affect soil organic matter and microbial activity, facilitating pesticide degradation and emissions of GHG like N_2O and CO_2 (Pahalvi et al., 2021). Additionally, many organochlorine classes of pesticides including sulfuryl fluoride, methyl bromide, and chloropicrin constitute emerging pollutants, acting directly as potent GHGs with ozone-depleting and ecotoxicological attributes (Hughes et al., 2011; Baweja et al., 2020).

Table 2 Interplay of pesticide applications with greenhouse gas emissions and relative mitigation strategies

Class pesticides	Brands	Impact	GHG emitted	Stage in value chain	Mitigation approach	Reference
Insecticides	Chlorpyrifos. Imidacloprid, Thiamethoxam, Diazinon, Chloropicrin, Malathion, Permethrin, Deltamethrin	Toxic to non-target species; contributes to pollinator decline	N_2O, CO_2, CH_4	Production, application	Integrated pest management (IPM), use of biological control	Heimpel et al. (2013); Cech et al. (2022)
Herbicides	Glyphosate, Paraquat, Atrazine, Dicamba, 24-dichloro phenoxy acetic acid, Bensulfuron-methyl	Soil degradation; disrupts soil microbial activities; and increases weed resistance	N_2O, CO_2, CH_4	Production, application	Agroecology, precision agriculture, use of microbial inoculants, buffer zones, no-tilling system, cover cropping, crop rotation	Shi et al. (2020); Cech et al. (2022)
Fungicides	Tebuconazole, Chlorothalonil, Azoxystrobin, Propiconazole, Copper-based fungicides	Disruption of soil microbial communities; reduced	CO_2, CH_4, N_2O	Production	Adoption of climate-smart pest management strategies	Heimpel et al. (2013); Cech et al. (2022)
General/ multi-action pesticides	Various synthetic pesticides like Chlorpyrifos and Mancozeb	Alter microbial community, disrupting nutrient cycling, soil microbial activity	CO_2, N_2O	Production, transportation, degradation, application	Biopesticides adoption, promotion of organic agricultural practices	Audsley et al. (2009); Shi et al. (2020)

24 The Role of Soil Microorganisms in Mitigating Pesticide Contamination and Climate Change

24.1 Soil Health Indicators and Assessment Methods

Soil health is a fundamental component of sustainable agricultural systems and environmental conservation, playing a vital role in mitigating both pesticide contamination and the effects of climate change. Various biological, chemical, and physical indicators are used to assess soil health, each reflecting different aspects of soil functionality. Biological indicators, such as microbial biomass, enzymatic activity, and the presence of specific soil microorganisms, provide insights into the soil's biological activity and its potential to degrade pesticides (Croft & Van De Baan, 1988). Chemical indicators, like soil organic matter, nutrient availability, and pH, are crucial for understanding how soil can buffer against external contaminants, including pesticide residues. Physical indicators, such as soil structure, bulk density, and water infiltration rates, help assess the soil's ability to sustain plant growth while also influencing carbon sequestration and mitigation of greenhouse gases. The integration of these indicators into comprehensive soil health assessments is necessary to fully understand the impact of pesticides and climate factors on soil ecosystems (Lehmann et al., 2020b; Croft & Van De Baan, 1988).

To effectively assess soil health, a variety of methods have been developed, ranging from traditional laboratory analyses to more modern techniques that involve remote sensing and molecular tools. Standard methods for biological assessments include microbial respiration tests and measurements of soil enzyme activities, which provide direct indicators of microbial function in pesticide degradation. Additionally, advances in molecular biology have enabled the use of DNA-based techniques to monitor microbial community structure and function (Smith et al., 2008). Chemical assessments often rely on soil nutrient testing, carbon and nitrogen ratios, and pH measurements, while physical health can be evaluated using soil texture analysis, aggregate stability tests, and measurements of porosity. These methods are increasingly being combined with digital technologies, allowing for more accurate, site-specific assessments of soil conditions and their response to both pesticide application and climate variability (Guo, 2021).

Incorporating soil health indicators into management strategies is essential for both mitigating pesticide contamination and adapting to climate change. Effective soil health management can enhance the ability of soils to sequester carbon, buffer against extreme climate conditions, and degrade pesticide residues (Reganold & Wachter, 2016). Techniques such as the use of cover crops, organic amendments, and reduced tillage are known to improve microbial activity and soil structure, leading to increased resilience against pesticide accumulation and climate impacts. Moreover, monitoring soil health over time allows for the detection of early signs of degradation, offering opportunities for timely intervention. This proactive approach can ensure that soils remain a vital resource for sustainable agriculture, even in the face of the dual challenges posed by pesticide use and a changing climate (de Vries et al., 2018; Dalal et al., 2011).

Reducing the impact of pesticides on soil health requires implementing several sustainable practices that strike a balance between effective pest control and environmental conservation. Integrated pest management (IPM) integrates biological control, cultural practices, and careful pesticide application to limit the use of chemicals. This comprehensive strategy helps lower pesticide residues in the soil while supporting natural pest predators.

Biopesticides, which come from natural sources like plants, bacteria, and fungi, provide an environmentally friendly alternative to synthetic chemicals. They break down more quickly and are less harmful to soil organisms, helping to maintain soil biodiversity.

For soils that are already contaminated with pesticide residues, remediation techniques such as bioremediation (utilizing microorganisms to decompose pesticides) and phytoremediation (employing plants to absorb and detoxify chemicals) can help restore soil health. These approaches assist in cleaning up pesticide-laden soils and reducing environmental harm.

Organic farming enhances soil health by steering clear of synthetic pesticides and employing natural pest control methods, including biological agents and crop rotation. Organic practices boost soil biodiversity, improve soil structure, and promote long-term fertility without relying on chemical inputs.

24.2 Sustainable Soil Health Management Practices

Sustainable soil health management is a critical component in mitigating both pesticide contamination and the adverse effects of climate change. One of the foundational approaches involves enhancing the biological diversity of soil ecosystems, particularly through the promotion of beneficial microorganisms that can degrade harmful pesticide residues. These microorganisms, including bacteria, fungi, and actinomycetes, play a significant role in breaking down chemical compounds into less harmful substances, thereby reducing the toxicity of pesticides in agricultural systems (Mustapha et al., 2018). Moreover, the application of organic amendments such as compost and biochar has been shown to improve soil structure, increase microbial biomass, and enhance the overall resilience of soils to both chemical pollutants and climate variability (Lai, 2007).

In addition to microbial degradation, soil health management practices that promote carbon sequestration are increasingly recognized as essential strategies for addressing climate change. No-till farming, cover cropping, and crop rotation are examples of agronomic practices that not only reduce soil erosion and improve moisture retention but also enhance soil organic carbon levels. These practices provide a dual benefit by mitigating greenhouse gas emissions while improving soil fertility and biodiversity (Paustian et al., 2016). Furthermore, biochar, a form of charcoal used as a soil amendment, has gained attention for its ability to sequester carbon in soils for long periods, while simultaneously providing habitat for microorganisms that aid in the breakdown of pesticide residues (Jeffery et al., 2011).

Integrating these sustainable soil management strategies into agricultural systems not only enhances the resilience of soils to the impacts of climate change but also plays a pivotal role in reducing pesticide contamination. For example, the use of biofertilizers and microbial inoculants like *Trichoderma* spp. can increase nutrient availability, suppress soil-borne pathogens, and degrade pesticides more efficiently (Renjith et al., 2020). Combined with conservation practices such as contour farming and terracing, these approaches create a more sustainable agricultural ecosystem that protects soil health, enhances crop productivity, and contributes to long-term environmental sustainability (Pretty, 2008).

24.3 Case Studies of Policies and Management Recommendations for Pesticide Application on Soil Health

Several case studies have demonstrated the importance of regulatory frameworks and management practices in promoting soil health while mitigating pesticide contamination. For example, the European Union's pesticide regulation emphasizes the need for sustainable use of pesticides, focusing on reducing their adverse impacts on soil biodiversity and water systems. In the United States, the integrated pest management (IPM) program has been a successful approach in encouraging the use of biopesticides and reducing synthetic pesticide applications to preserve soil health (Tahat et al., 2020). Additionally, Australia's National Pesticide Risk Management Framework promotes responsible pesticide use and encourages the adoption of best practices such as crop rotation, cover cropping, and reduced tillage, all of which support microbial diversity and soil organic carbon sequestration. These policies are critical in balancing the need for pest control with the protection of soil ecosystems in the face of climate change (Romeh, 2018).

25 Case Studies and Real-World Applications

The increasing threats posed by climate change have led to the adoption of more resilient agricultural practices, including integrated pest management (IPM) and climate-smart agriculture (CSA). IPM emphasizes reducing reliance on chemical pesticides by using biological control agents, cultural practices, and resistant crop varieties. These methods help mitigate the environmental impact of pesticides while enhancing the resilience of ecosystems. In combination with CSA, which focuses on improving productivity, adaptation, and mitigation to climate change, these approaches offer sustainable solutions (Pretty & Bharucha, 2014). For instance, in regions vulnerable to drought and temperature fluctuations, CSA strategies like drought-tolerant crop varieties, soil moisture management, and agroforestry have demonstrated positive outcomes in pest control and climate adaptation.

Several case studies underscore the success of IPM and CSA in reducing pesticide use while maintaining crop productivity. One such example comes from Sub-Saharan Africa, where smallholder farmers have integrated push-pull strategies with climate-resilient cropping systems, reducing pest pressure and improving soil health. Similarly, in Southeast Asia, rice farmers have implemented IPM techniques alongside water management practices, enhancing both pest control and carbon sequestration in paddy fields (Pretty & Bharucha, 2014). The examples presented in Table 3 highlight the potential of integrating pest management with climate-adaptive strategies to establish a resilient and sustainable agricultural system.

Table 3 Real-world examples of the implementation of IPM and CSA across various regions to enhance agricultural sustainability

Case study location	Integrated pest management (IPM) strategy	Climate-smart agriculture (CSA) strategy	Outcome	Reference
Sub-Saharan Africa	Push-pull method for controlling maize stemborers	Drought-tolerant maize varieties and intercropping	Reduced pest pressure and improved crop resilience	Khan et al. (2014)
India	Biological control of sugarcane pests using *Trichogramma chilonis*	Soil moisture conservation and drip irrigation	Enhanced pest control and water use efficiency	Geetha et al. (2009)
China	Ecological engineering for managing rice planthoppers	Alternate wetting and drying irrigation techniques	Increased pest suppression and reduced methane emissions	Lu et al. (2015)
Brazil	Use of parasitoids to control sugarcane borers	Conservation tillage and cover cropping	Improved pest control and soil health	Bezerra et al. (2021)
Kenya	Biological control of fruit flies in mango orchards	Agroforestry and organic mulching	Reduced pest damage and enhanced soil carbon storage	Kibira et al. (2015)
Mexico	Biological control of coffee borer beetles with entomopathogens	Shade-grown coffee systems and soil management	Reduced pest infestation and increased biodiversity	Oliva et al. (2020)
United States	Trap cropping for cucumber beetle control in cucurbits	Crop rotation and no-till farming	Improved pest management and reduced soil erosion	Shelton and Badenes-Perez (2006)
Ethiopia	Use of botanicals for controlling the pests	Conservation agriculture and reduced tillage	Reduced pest impact and improved soil structure	Derso (2020)

26 Future Perspectives and Research Needs

26.1 Emerging Trends and Technologies in Pesticide Use, Soil Health, and Carbon Sequestration

As the environmental consequences of conventional pesticide use become more evident, there is a growing focus on developing eco-friendly alternatives that minimize soil degradation and promote carbon sequestration (Malik et al., 2017). Recent innovations include bio-based pesticides and integrated pest management (IPM) systems that enhance soil biodiversity and organic matter retention. Additionally, advanced technologies such as precision agriculture are enabling more targeted pesticide applications, reducing overuse and enhancing soil health (Jacquet et al., 2022).

26.2 Knowledge Gaps and Areas for Further Research

While considerable progress has been made in understanding the interactions between pesticide use and soil health, significant knowledge gaps remain. There is limited research on the long-term impacts of novel bio-pesticides on soil microbial communities and their ability to contribute to carbon sequestration. Future studies are needed to explore these dynamics under various environmental conditions and farming systems (Bijilaxmi Devi et al., 2018; Rengalakshmi et al., 2018).

26.3 Potential Impacts of Future Climate Scenarios on Soil Health and Pesticide Dynamics

Climate change is projected to alter rainfall patterns, temperature regimes, and soil moisture content, all of which can influence pesticide efficacy and persistence. In future climate scenarios, these changes could lead to increased volatilization of certain chemicals and altered degradation pathways, potentially exacerbating soil degradation (Tudi et al., 2021). Moreover, shifts in pest populations may require changes in pesticide use patterns, further complicating soil and ecosystem health management (Delcour et al., 2015).

26.4 Innovative Approaches for Sustainable Soil Management

To mitigate the negative impact of pesticides, use on soil health, there is a need for sustainable soil management strategies that integrate biological, chemical, and technological approaches (Powlson et al., 2011). Practices such as organic amendments, crop rotations, and the use of biochar are being explored for their potential to improve soil structure, enhance microbial diversity, and increase carbon sequestration. Combining these practices with smart technology can optimize soil health while reducing dependency on chemical inputs (Lehmann et al., 2011).

27 Conclusions

Pesticides are essential in modern agriculture, helping to protect crops from pests and ensuring food security. However, their extensive use has serious consequences for soil health. The application of different types of pesticides—such as insecticides, herbicides, fungicides, and nematicides—can result in harmful residues accumulating in the soil, disrupting the fragile balance of soil ecosystems. While they are effective in controlling pests, these chemicals can also harm non-target organisms, leading to a decline in soil biodiversity, changes in nutrient cycling, and a decrease in overall soil fertility. Although pesticides are vital for agricultural productivity, we must not overlook their effects on soil health. Transitioning to sustainable pest management practices is essential for maintaining soil fertility, protecting biodiversity, and ensuring the long-term sustainability of agricultural systems. By adopting strategies like integrated pest management (IPM), biopesticides, soil remediation, and organic farming, we can reduce the adverse impacts of pesticides on soil health and protect the foundation of food production for future generations.

References

Abdelhak, M. (2022). Soil improvement in arid and semiarid regions for sustainable development. In *Natural resources conservation and advances for sustainability* (pp. 73–90). Elsevier.

AbuQamar, S. F., El-Saadony, M. T., Alkafaas, S. S., Elsalahaty, M. I., Elkafas, S. S., et al. (2024). Ecological impacts and management strategies of pesticide pollution on aquatic life and human beings. *Marine Pollution Bulletin, 206*, 116613. https://doi.org/10.1016/j.marpolbul.2024.116613

Adil, M., Abbas, G., Khan, R. N., & Abbas, F. (2023). Impact of climate change on environmental fate and ecological effects of pesticides. In S. A. Bandh (Ed.), *Strategizing agricultural management for climate change mitigation and adaptation*. Springer. https://doi.org/10.1007/978-3-031-32789-6_15

Agidew, A. M. A., & Singh, K. N. (2019). Factors affecting the adoption of sustainable land management practices at farm level in the northeastern highlands of Ethiopia: The Teleyayen subwatershed case study. *Journal of Environmental Pollution and Management, 2*(1), 103–115.

AL-Falahi, A. H., Barry, S., Gebrechorkos, S. H., Spank, U., & Bernhofer, C. (2023). Potential of traditional adaptation measures in mitigating the impact of climate change. *Sustainability, 15*(21), 15442. https://doi.org/10.3390/su152115442

Ali, S., Ullah, M. I., Sajjad, A., Shakeel, Q., & Hussain, A. (2021). Environmental and health effects of pesticide residues. In *Sustainable agriculture reviews 48: Pesticide occurrence, analysis and remediation. Vol. 2 Analysis* (pp. 311–336). Springer.

Alsafran, M., Rizwan, M., Usman, K., Saleem, M. H., & Al Jabri, H. (2022). Neonicotinoid insecticides in the environment: A critical review of their distribution, transport, fate, and toxic effects. *Journal of Environmental Chemical Engineering, 10*(5), 108485.

Angon, P. B., Mondal, S., Jahan, I., Datto, M., Antu, U. B., Ayshi, F. J., & Islam, M. S. (2023). Integrated pest management (IPM) in agriculture and its role in maintaining ecological balance and biodiversity. *Advances in Agriculture, 2023*(1), 5546373.

Audsley, E., Stacey, K. F., Parsons, D. J., & Williams, A. G. (2009). *Estimation of the greenhouse gas emissions from agricultural pesticide manufacture and use*. Cranfield University.

Aziz, K. (2022). *Effects of cover crops on nitrous oxide (N_2O) emissions in cereal cropping*. SLU.

Baker, B. P., Green, T. A., & Loker, A. J. (2020). Biological control and integrated pest management in organic and conventional systems. *Biological Control, 140*, 104095.

Batool, M., Cihacek, L. J., & Alghamdi, R. S. (2024). Soil inorganic carbon formation and the sequestration of secondary carbonates in global carbon pools: A review. *Soil Systems, 8*(1), 15.

Bauer, F., Nielsen, T. D., Nilsson, L. J., Palm, E., Ericsson, K., Fråne, A., & Cullen, J. (2022). Plastics and climate change—Breaking carbon lock-ins through three mitigation pathways. *One Earth, 5*(4), 361–376. https://doi.org/10.1016/j.oneear.2022.03.007

Baweja, P., Kumar, S., & Kumar, G. (2020). Fertilizers and pesticides: Their impact on soil health and environment. In *Soil health* (pp. 265–285). Springer.

Bezerra, J. I., Molter, A., Rafikov, M., & Frighetto, D. F. (2021). Biological control of the chaotic sugarcane borer parasitoid agroecosystem. *Ecological Modelling, 450*, 109564.

Bhattacharyya, S. S., Ros, G. H., Furtak, K., Iqbal, H. M., & Parra-Saldívar, R. (2022). Soil carbon sequestration–An interplay between soil microbial community and soil organic matter dynamics. *Science of the Total Environment, 815*, 152928.

Bhatti, U. A., Bhatti, M. A., Tang, H., Syam, M. S., Awwad, E. M., Sharaf, M., & Ghadi, Y. Y. (2024). Global production patterns: Understanding the relationship between greenhouse gas emissions, agriculture greening and climate variability. *Environmental Research, 245*, 118049. https://doi.org/10.1016/j.envres.2023.118049

Bijilaxmi Devi, Y., Meetei, T. T., & Kumari, N. (2018). Impact of pesticides on soil microbial diversity and enzymes: A review. *International Journal of Current Microbiology and Applied Sciences, 7*(6), 952–958.

Bolan, S., Sharma, S., Mukherjee, S., Kumar, M., Rao, C. S., Nataraj, K. C., et al. (2023). Biochar modulating soil biological health: A review. *Science of the Total Environment, 914*, 169585.

Bonner, M. R., & Alavanja, M. C. (2017). Pesticides, human health, and food security. *Food and Energy Security, 6*(3), 89–93.

Bouri, M., Arslan, K. S., & Şahin, F. (2023). Climate-smart pest management in sustainable agriculture: Promises and challenges. *Sustainability, 15*(5), 4592. https://doi.org/10.3390/su15054592

Brüggemann, M., Mayer, S., Brown, D., Terry, A., Rudiger, J., & Hoffmann, T. (2024). Measuring pesticides in the atmosphere: Current status, emerging trends and future perspectives. *Environmental Sciences Europe, 36*, 39. https://doi.org/10.1186/s12302-024-00870-4

Buck, H. J., & Palumbo-Compton, A. (2022). Soil carbon sequestration as a climate strategy: What do farmers think? *Biogeochemistry, 161*(1), 59–70.

Cech, R., Leisch, F., & Zaller, J. G. (2022). Pesticide use and associated greenhouse gas emissions in sugar beet, apples, and viticulture in Austria from 2000 to 2019. *Agriculture, 12*(6), 879. https://doi.org/10.3390/agriculture12060879

Chen, L., Zhang, S., Li, F., Wang, Z., Gao, G., & Tang, L. (2021). Effects of pesticides on soil enzymes: A review. *Environmental Science and Pollution Research, 28*(2), 2041–2057.

Choudhury, P. P., & Saha, S. (2020). Dynamics of pesticides under changing climatic scenario. *Environmental Monitoring and Assessment, 192*(Suppl 1), 814. https://doi.org/10.1007/s10661-020-08719-y

Cobb, A. H. (2022). *Herbicides and plant physiology*. Wiley.

Croft, B. A., & Van De Baan, H. E. (1988). Ecological and genetic factors influencing evolution of pesticide resistance in tetranychid and phytoseiid mites. *Experimental & Applied Acarology, 4*, 277–300.

Dalal, R. C., Allen, D. E., Chan, K. Y., & Singh, B. P. (2011). Soil organic matter, soil health and climate change. In B. Singh, A. Cowie, & K. Chan (Eds.), *Soil health and climate change* (Soil Biology) (Vol. 29). Springer.

Dang, Y. P., Page, K. L., Dalal, R. C., & Menzies, N. W. (2020). No-till farming systems for sustainable agriculture: An overview. In *No-till farming systems for sustainable agriculture: Challenges and opportunities* (pp. 3–20). Springer.

de Vries, F. T., Griffiths, R. I., Bailey, M., Craig, H., Girlanda, M., et al. (2018). Soil bacterial networks are less stable under drought than fungal networks. *Nature Communications, 9*(1), 3033.

Delcour, I., Spanoghe, P., & Uyttendaele, M. (2015). Literature review: Impact of climate change on pesticide use. *Food Research International, 68*, 7–15.

Delgado, A., & Gómez, J. A. (2016). The soil physical, chemical and biological properties. In F. Villalobos & E. Fereres (Eds.), *Principles of agronomy for sustainable agriculture*. Springer.

Demeneix, B. A. (2020). How fossil fuel-derived pesticides and plastics harm health, biodiversity, and the climate. *The Lancet Diabetes and Endocrinology, 8*(6), 462–464. https://doi.org/10.1016/S2213-8587(20)30116-9. PMID: 32445732; PMCID: PMC7239621.

Derso, C. (2020). Farmers' indigenous knowledge practice for crop protection: A case study of ant protection on Teff (*Eragrostis teff*) crop in North Wollo, Ethiopia. *Asian Journal of Research in Crop Science, 5*(1), 14–20.

Doran, J. W., & Zeiss, M. R. (2000). Soil health and sustainability: Managing the biotic component of soil quality. *Applied Soil Ecology, 15*(1), 3–11.

Elhamalawy, O., Bakr, A., & Eissa, F. (2024). Impact of pesticides on non-target invertebrates in agricultural ecosystems. *Pesticide Biochemistry and Physiology, 202*, 105974. https://doi.org/10.1016/j.pestbp.2024.105974

Fornara, D. A., Steinbeiss, S., McNamara, N. P., Gleixner, G., Oakley, S., Poulton, P. R., et al. (2011). Increases in soil organic carbon sequestration can reduce the global warming potential of long-term liming to permanent grassland. *Global Change Biology, 17*(5), 1925–1934.

Geetha, N., Shekinah, E., & Rakkiyappan, P. (2009). Comparative impact of release frequency of *Trichogramma chilonis* Ishii against Chilo *sacchariphagus indicus* (Kapur) in sugarcane. *Journal of Biological Control, 23*, 343–351.

Gelardi, D. L., Rath, D., & Kruger, C. E. (2023). Grounding United States policies and programs in soil carbon science: Strengths, limitations, and opportunities. *Frontiers in Sustainable Food Systems, 7*, 1188133.

González, A., King, A., & Yung, P. (2020). The impact of pesticide resistance on biocontrol agents in agriculture. *Ecological Applications, 30*(5), e02148.

Guo, M. (2021). Soil health assessment and management: Recent development in science and practices. *Soil Systems, 5*, 61.

Hayes, M. H., & Swift, R. S. (2020). Vindication of humic substances as a key component of organic matter in soil and water. *Advances in Agronomy, 163*, 1–37.

Heeb, L., Jenner, E., & Cock, M. J. W. (2019). Climate-smart pest management: Building resilience of farms and landscapes to changing pest threats. *Journal of Pest Science, 92*, 951–969. https://doi.org/10.1007/s10340-019-01083-y

Heimpel, G. E., Yang, Y., Hill, J. D., & Ragsdale, D. W. (2013). Environmental consequences of invasive species: Greenhouse gas emissions of insecticide use and the role of biological control in reducing emissions. *PLoS One, 8*(8), e72293.

Hughes, D. J., West, J. S., Atkins, S. D., Gladders, P., Jeger, M. J., & Fitt, B. D. (2011). Effects of disease control by fungicides on greenhouse gas emissions by UK arable crop production. *Pest Management Science, 67*(9), 1082–1092.

Hussain, S., Hussain, S., Guo, R., Sarwar, M., Ren, X., et al. (2021). Carbon sequestration to avoid soil degradation: A review on the role of conservation tillage. *Plants, 10*(10), 2001.

Hutton, S. J., Siddiqui, S., & Brander, S. M. (2024). Ecotoxicology challenges during climate change scenarios. In S. Siddiqui & S. M. Brander (Eds.), *Aquatic ecotoxicology*. Springer. https://doi.org/10.1007/978-3-031-53130-9_11

Issoufou, A. A., Soumana, I., Maman, G., Konate, S., & Mahamane, A. (2020). Dynamic relationship of traditional soil restoration practices and climate change adaptation in semi-arid Niger. *Heliyon, 6*(1), e03265.

Jacquet, F., Jeuffroy, M. H., Jouan, J., Le Cadre, E., Litrico, I., Malausa, T., Reboud, X., & Huyghe, C. (2022). Pesticide-free agriculture as a new paradigm for research. *Agronomy for Sustainable Development, 42*(1), 1–24.

Jaworski, C. C., Thomine, E., Rusch, A., Lavoir, A. V., Wang, S., & Desneux, N. (2023). Crop diversification to promote arthropod pest management: A review. *Agriculture Communications, 1*(1), 100004.

Jeffery, S., Verheijen, F. G. A., van der Velde, M., & Bastos, A. C. (2011). A quantitative review of the effects of biochar application to soils on crop productivity using meta-analysis. *Agriculture, Ecosystems & Environment, 144*(1), 175–187.

Jiao, S., Peng, Z., Qi, J., Gao, J., & Wei, G. (2021). Linking bacterial-fungal relationships to microbial diversity and soil nutrient cycling. *mSystems, 6*(2), 10–1128.

Johansson, E. L., Brogaard, S., & Brodin, L. (2022). Envisioning sustainable carbon sequestration in Swedish farmland. *Environmental Science & Policy, 135*, 16–25.

Kaka, H., Opute, P. A., & Maboeta, M. S. (2021). Potential impacts of climate change on the toxicity of pesticides towards earthworms. *Journal of Toxicology, 2021*, 8527991. https://doi.org/10.1155/2021/8527991. PMID: 34456999; PMCID: PMC8397574.

Kaufman, M. (2020). An organic-based food system: A voyage back and forward in time. In *Diet for a sustainable ecosystem: The science for recovering the health of the Chesapeake Bay and its people* (pp. 375–395). Springer.

Khan, Z., Midega, C., Pittchar, J., Pickett, J., & Bruce, T. (2014). Push—Pull technology: A conservation agriculture approach for integrated management of insect pests, weeds and soil health in Africa. *International Journal of Agricultural Sustainability, 9*(1), 162–170.

Khan, B. A., Nadeem, M. A., Nawaz, H., Amin, M. M., Abbasi, G. H., et al. (2023). Pesticides: Impacts on agriculture productivity, environment, and management strategies. In *Emerging contaminants and plants: Interactions, adaptations and remediation technologies* (pp. 109–134). Springer International Publishing.

Kibira, M., Affognon, H., Njehia, B., Muriithi, B., Mohamed, S., & Ekesi, S. (2015). Economic evaluation of integrated management of fruit fly in mango production in Embu County, Kenya. *African Journal of Agricultural and Resource Economics, 10*(4), 343–353.

Kirchman, D. L. (2018). *Processes in microbial ecology*. Oxford University Press.

Kirkham, M. B. (2016). *Elevated carbon dioxide: Impacts on soil and plant water relations*. CRC Press.

Kurenbach, B., Marjoshi, D., Amábile-Cuevas, C. F., Ferguson, G. C., Godsoe, W., Gibson, P., & Heinemann, J. A. (2017). Herbicide ingredients induce antibiotic resistance in bacterial communities. *Proceedings of the National Academy of Sciences, 114*(18), 4298–4303.

Lahlali, R., Ibrahim, D. S., Belabess, Z., Roni, M. Z. K., Radouane, N., Vicente, C. S., et al. (2021). High-throughput molecular technologies for unraveling the mystery of soil microbial community: Challenges and future prospects. *Heliyon, 7*(10), e08142.

Lai, R. (2007). Managing soils for food security and climate change. *Journal of Crop Improvement, 19*, 49–71.

Lal, R. (2004). Soil carbon sequestration impacts on global climate change and food security. *Science, 304*(5677), 1623–1627.

Lehmann, J., Rillig, M. C., Thies, J., Masiello, C. A., Hockaday, W. C., & Crowley, D. (2011). Biochar effects on soil biota – A review. *Soil Biology and Biochemistry, 43*(9), 1812–1836.

Lehmann, J., Hansel, C. M., Kaiser, C., Kleber, M., Maher, K., Manzoni, S., et al. (2020a). Persistence of soil organic carbon caused by functional complexity. *Nature Geoscience, 13*(8), 529–534.

Lehmann, J., Bossio, D. A., Kögel-Knabner, I., & Rillig, M. C. (2020b). The concept and future prospects of soil health. *Nature Reviews Earth and Environment, 1*, 544–553.

Liu, C., Plaza-Bonilla, D., Coulter, J. A., Kutcher, H. R., Beckie, H. J., et al. (2022). Diversifying crop rotations enhances agroecosystem services and resilience. *Advances in Agronomy, 173*, 299–335.
Lu, Z., Zhu, P., Gurr, G. M., Zheng, X., Chen, G., & Heong, K. L. (2015). Rice pest management by ecological engineering: A pioneering attempt in China. In K. Heong, J. Cheng, & M. Escalada (Eds.), *Rice planthoppers*. Springer.
Ma, C. S., Zhang, W., Peng, Y., Zhao, F., Chang, X. Q., Xing, K., et al. (2021). Climate warming promotes pesticide resistance through expanding overwintering range of a global pest. *Nature Communications, 12*(1), 5351.
Malik, Z., Ahmad, M., Abassi, G. H., Dawood, M., Hussain, A., & Jamil, M. (2017). Agrochemicals and soil microbes: Interaction for soil health. In M. Hashmi, V. Kumar, & A. Varma (Eds.), *Xenobiotics in the soil environment* (Soil Biology) (Vol. 49). Springer.
Martínez, J. L. (2017). Ecological aspects of antibiotic resistance genes transfer and spread. *Environmental Pollution, 220*, 1322–1329.
Martínez-Megías, C., Mentzel, S., Fuentes-Edfuf, Y., Moe, S. J., & Rico, A. (2023). Influence of climate change and pesticide use practices on the ecological risks of pesticides in a protected Mediterranean wetland: A Bayesian network approach. *Science of the Total Environment, 878*, 163018. https://doi.org/10.1016/j.scitotenv.2023.163018
Meena, R. S., Kumar, S., & Yadav, G. S. (2020). Soil carbon sequestration in crop production. In *Nutrient dynamics for sustainable crop production* (pp. 1–39). Springer.
Moreira, R. A., Cordero-de-Castro, A., Polo-Castellano, C., Pinto, T. J. S., Dias, M. A., Montagner, C. C., Espíndola, E. L. G., Araújo, C. V. M., & Blasco, J. (2022). Avoidance responses by Danio rerio reveal interactive effects of warming, pesticides and their mixtures. *Science of the Total Environment, 847*, 157525. https://doi.org/10.1016/j.scitotenv.2022.157525
Mueller, L., Eulenstein, F., Dronin, N. M., Mirschel, W., McKenzie, B. M., et al. (2021). Agricultural landscapes: History, status and challenges. In *Exploring and optimizing agricultural landscapes* (pp. 3–54). Springer.
Mustafa, A., Minggang, X., Shah, S. A. A., Abrar, M. M., Nan, S., et al. (2020). Soil aggregation and soil aggregate stability regulate organic carbon and nitrogen storage in a red soil of southern China. *Journal of Environmental Management, 270*, 110894.
Mustapha, M. U., Halimoon, N., Johari, W. L. W., & Abdul Shukor, M. Y. (2018). Soil microorganisms and their potential in pesticide biodegradation: A review. *Journal of Sustainable Agricultural Sciences, 44*(2), 39–61.
Nelson, P. N. (2023). Sustainable soil management in tropical agriculture. In P. N. Nelson (Ed.), *Sustainable soil management: Beyond food production* (p. 204). Cambridge Scholars Publishing.
Nkongolo, K. K., & Narendrula-Kotha, R. (2020). Advances in monitoring soil microbial community dynamic and function. *Journal of Applied Genetics, 61*(2), 249–263.
Odewale, S. A., Odekanle, E. L., & Fakinle, B. S. (2023). Global environmental sustainability and agrochemical use. In M. C. Ogwu & S. Chibueze Izah (Eds.), *One health implications of agrochemicals and their sustainable alternatives* (Sustainable Development and Biodiversity) (Vol. 34). Springer. https://doi.org/10.1007/978-981-99-3439-3_26
Okoronkwo, D. J., Ozioko, R. I., Ugwoke, R. U., Nwagbo, U. V., Nwobodo, C., Ugwu, C. H., Okoro, G. G., & Mbah, E. C. (2024). Climate smart agriculture? Adaptation strategies of traditional agriculture to climate change in sub-Saharan Africa. *Frontiers in Climate, 6*, 1272320.
Oliva, M., Rubio, K., Rivasplata, E., & Leiva, S. (2020). Integrated management of the coffee berry borer: A comparison of cultural, biological, and ethological control. *Entomological Research, 50*(11), 539–544.
Pahalvi, H. N., Rafiya, L., Rashid, S., Nisar, B., & Kamili, A. N. (2021). Chemical fertilizers and their impact on soil health. In G. H. Dar, R. A. Bhat, M. A. Mehmood, & K. R. Hakeem (Eds.), *Microbiota and biofertilizers* (Vol. 2). Springer. https://doi.org/10.1007/978-3-030-61010-4_1
Panchal, P., Preece, C., Peñuelas, J., & Giri, J. (2022). Soil carbon sequestration by root exudates. *Trends in Plant Science, 27*(8), 749–757.

Parra-López, C., Abdallah, S. B., Garcia-Garcia, G., Hassoun, A., Sánchez-Zamora, P., Trollman, H., Jagtap, S., & Carmona-Torres, C. (2024). Integrating digital technologies in agriculture for climate change adaptation and mitigation: State of the art and future perspectives. *Computers and Electronics in Agriculture, 226*, 109412. https://doi.org/10.1016/j.compag.2024.109412
Parte, S. G., Mohekar, A. D., & Kharat, A. S. (2017). Microbial degradation of pesticide: A review. *African Journal of Microbiology Research, 11*(24), 992–1012.
Paul, E. A. (2016). The nature and dynamics of soil organic matter: Plant inputs, microbial transformations, and organic matter stabilization. *Soil Biology and Biochemistry, 98*, 109–126.
Paustian, K., Lehmann, J., Ogle, S., Reay, D., Robertson, G. P., & Smith, P. (2016). Climate-smart soils. *Nature, 532*(7597), 49–57.
Pérez-Lucas, G., Navarro, G., & Navarro, S. (2024). Adapting agriculture and pesticide use in Mediterranean regions under climate change scenarios: A comprehensive review. *European Journal of Agronomy, 161*, 127337. https://doi.org/10.1016/j.eja.2024.127337
Popp, J., Pető, K., & Nagy, J. (2013). Pesticide productivity and food security. A review. *Agronomy for Sustainable Development, 33*, 243–255. https://doi.org/10.1007/s13593-012-0105-x
Powlson, D. S., Gregory, P. J., Whalley, W. R., Quinton, J. N., Hopkins, D. W., Whitmore, A. P., & Goulding, K. W. (2011). Soil management in relation to sustainable agriculture and ecosystem services. *Food Policy, 36*, 72–87.
Prescott, C. E., & Vesterdal, L. (2021). Decomposition and transformations along the continuum from litter to soil organic matter in forest soils. *Forest Ecology and Management, 498*, 119522.
Pretty, J. (2008). Agricultural sustainability: Concepts, principles and evidence. *Philosophical Transactions of the Royal Society B: Biological Sciences, 363*(1491), 447–465.
Pretty, J., & Bharucha, Z. P. (2014). Sustainable intensification in agricultural systems. *Annals of Botany, 114*(8), 1571–1596.
Rahman, G. M., Rahman, M. M., Alam, M. S., Kamal, M. Z., Mashuk, H. A., Datta, R., & Meena, R. S. (2020). Biochar and organic amendments for sustainable soil carbon and soil health. In *Carbon and nitrogen cycling in soil* (pp. 45–85). Springer.
Raj, E. F. I., Appadurai, M., & Athiappan, K. (2022). Precision farming in modern agriculture. In *Smart agriculture automation using advanced technologies: Data analytics and machine learning, cloud architecture, automation and IoT* (pp. 61–87). Springer Singapore.
Ramesh, K., Matloob, A., Aslam, F., Florentine, S. K., & Chauhan, B. S. (2017). Weeds in a changing climate: Vulnerabilities, consequences, and implications for future weed management. *Frontiers in Plant Science, 8*, 95.
Reganold, J., & Wachter, J. (2016). Organic agriculture in the twenty-first century. *Nature Plants, 2*, 15221.
Rengalakshmi, R., Manjula, M., Prabavathy, V. R., Jegan, S., & Selvamukilan, B. (2018). Building bioeconomy in agriculture: Harnessing soil microbes for sustaining ecosystem services. In W. Leal Filho, D. Pociovălişteanu, P. Borges de Brito, & I. Borges de Lima (Eds.), *Towards a sustainable bioeconomy: Principles, challenges and perspectives* (World Sustainability Series). Springer.
Renjith, P. S., Sheetal, K. R., Kumar, S., Choudhary, J., & Prasad, S. (2020). Microbial and biotechnological approaches in the production of biofertilizer. In A. Singh, S. Srivastava, D. Rathore, & D. Pant (Eds.), *Environmental microbiology and biotechnology*. Springer.
Romeh, A. A. (2018). Integrated pest management for sustainable agriculture. In A. Negm & M. Abu-hashim (Eds.), *Sustainability of agricultural environment in Egypt: Part II* (The Handbook of Environmental Chemistry) (Vol. 77). Springer.
Sadatshojaei, E., Wood, D. A., & Rahimpour, M. R. (2021). Potential and challenges of carbon sequestration in soils. In *Applied soil chemistry* (pp. 1–21). Wiley.
Saha, A., Ghosh, R. K., & Basak, B. B. (2019). Fate and behavior of pesticides and their effect on soil biological properties under climate change scenario. In R. Meena, S. Kumar, J. Bohra, & M. Jat (Eds.), *Sustainable management of soil and environment*. Springer. https://doi.org/10.1007/978-981-13-8832-3_8

Sang, Y., Yu, W., He, L., Wang, Z., Ma, F., Jiao, W., & Gu, Q. (2021). Sustainable remediation of lube oil-contaminated soil by low temperature indirect thermal desorption: Removal behaviors of contaminants, physicochemical properties change and microbial community recolonization in soils. *Environmental Pollution, 287*, 117599.

Santra, H. K., & Banerjee, D. (2020). Natural products as fungicide and their role in crop protection. In *Natural bioactive products in sustainable agriculture* (pp. 131–219). Springer.

Shah, F., & Wu, W. (2019). Soil and crop management strategies to ensure higher crop productivity within sustainable environments. *Sustainability, 11*(5), 1485.

Shelton, A. M., & Badenes-Perez, F. R. (2006). Concepts and applications of trap cropping in pest management. *Annual Review of Entomology, 51*(1), 285–308.

Shi, L., Guo, Y., Ning, J., Lou, S., & Hou, F. (2020). Herbicide applications increase greenhouse gas emissions of alfalfa pasture in the inland arid region of northwest China. *PeerJ, 8*, e9231. https://doi.org/10.7717/peerj.9231. PMID: 32509469; PMCID: PMC7255331.

Siebielec, S., Siebielec, G., Klimkowicz-Pawlas, A., Gałązka, A., Grządziel, J., & Stuczyński, T. (2020). Impact of water stress on microbial community and activity in sandy and loamy soils. *Agronomy, 10*(9), 1429. https://doi.org/10.3390/agronomy10091429

Smith, P., Martino, D., Cai, Z., Gwary, D., Janzen, H., Kumar, P., McCarl, B., Ogle, S., O'Mara, F., et al. (2008). Greenhouse gas mitigation in agriculture. *Philosophical Transactions of the Royal Society of London. Series B Biological Sciences, 363*(1492), 789–813.

Smith, P., Calvin, K., Nkem, J., Campbell, D., Cherubini, F., Grassi, G., et al. (2020). Which practices co-deliver food security, climate change mitigation and adaptation, and combat land degradation and desertification? *Global Change Biology, 26*(3), 1532–1575.

Stanley, P. L., Rowntree, J. E., Beede, D. K., DeLonge, M. S., & Hamm, M. W. (2018). Impacts of soil carbon sequestration on life cycle greenhouse gas emissions in Midwestern USA beef finishing systems. *Agricultural Systems, 162*, 249–258.

Stenberg, J. A., Sundh, I., Becher, P. G., Björkman, C., Dubey, M., Egan, P. A., et al. (2021). When is it biological control? A framework of definitions, mechanisms, and classifications. *Journal of Pest Science, 94*(3), 665–676.

Stull, V. J., & Patz, J. A. (2024). Agriculture policy. In *Climate change and public health* (pp. 292–318). Oxford Academic Press.

Swarup, A., Manna, M. C., & Singh, G. B. (2019). Impact of land use and management practices on organic carbon dynamics in soils of India. In *Global climate change and tropical ecosystems* (pp. 261–281). CRC Press.

Swift, M. J., Izac, A. M. N., & van Noordwijk, M. (2004). Biodiversity and ecosystem services in agricultural landscapes—Are we asking the right questions? *Agriculture, Ecosystems & Environment, 104*(1), 113–134.

Tahat, M. M., Alananbeh, K. M., Othman, Y. A., & Leskovar, D. I. (2020). Soil health and sustainable agriculture. *Sustainability, 12*(12), 4859.

Telo da Gama, J. (2023). The role of soils in sustainability, climate change, and ecosystem services: Challenges and opportunities. *Ecologies, 4*(3), 552–567.

Tripathi, S., Srivastava, P., Devi, R. S., & Bhadouria, R. (2020). Influence of synthetic fertilizers and pesticides on soil health and soil microbiology. In *Agrochemicals detection, treatment and remediation* (pp. 25–54). Butterworth-Heinemann.

Tripathi, A., Pandey, V., & Ranjan, M. R. (2021). Climate change and its impact on soil properties. In D. K. Choudhary, A. Mishra, & A. Varma (Eds.), *Climate change and the microbiome* (Soil Biology) (Vol. 63). Springer. https://doi.org/10.1007/978-3-030-76863-8_7

Tudi, M., Daniel Ruan, H., Wang, L., Lyu, J., Sadler, R., Connell, D., Chu, C., & Phung, D. T. (2021). Agriculture development, pesticide application and its impact on the environment. *International Journal of Environmental Research and Public Health, 18*(3), 1112. https://doi.org/10.3390/ijerph18031112

Ul Haq, I., Sarwar, M. K., Faraz, A., & Latif, M. Z. (2020). Synthetic chemicals: Major component of plant disease management. In *Plant disease management strategies for sustainable agriculture through traditional and modern approaches* (pp. 53–81). Springer.

Van der Meij, W. M., Temme, A. J., Wallinga, J., & Sommer, M. (2020). Modeling soil and landscape evolution – The effect of rainfall and land-use change on soil and landscape patterns. *The Soil, 6*(2), 337–358.

Van Deynze, B., Swinton, S. M., Hennessy, D. A., Haddad, N. M., & Ries, L. (2024). Insecticides, more than herbicides, land use, and climate, are associated with declines in butterfly species richness and abundance in the American Midwest. *PLoS One, 19*(6), e0304319. https://doi.org/10.1371/journal.pone.0304319

Visscher, A. M., Vanek, S., Huaraca, J., Mendoza, J., Ccanto, R., et al. (2024). Traditional soil fertility management ameliorates climate change impacts on traditional Andean crops within smallholder farming systems. *Science of the Total Environment, 912*, 168725.

Vlaiculescu, A., & Varrone, C. (2022). Sustainable and eco-friendly alternatives to reduce the use of pesticides. In *Pesticides in the natural environment* (pp. 329–364). Elsevier.

Wang, X., Yu, J., Shen, W., Zhou, P., & Zhu, L. (2023). Impact of pesticide application on nitrification and denitrification processes in agricultural soils. *Journal of Agricultural and Food Chemistry, 71*(3), 225–236.

Wei, X. F., Yang, W., & Hedenqvist, M. S. (2024). Plastic pollution amplified by a warming climate. *Nature Communications, 15*, 2052. https://doi.org/10.1038/s41467-024-46127-9

Wołejko, E., Jabłońska-Trypuć, A., Wydro, U., Butarewicz, A., & Łozowicka, B. (2020). Soil biological activity as an indicator of soil pollution with pesticides – A review. *Applied Soil Ecology, 147*, 103356.

Yahyah, H., Kameri-Mbote, P., & Kibugi, R. (2024). Implications of pesticide use regulation on soil sustainability in Uganda. *Soil Security, 16*, 100133. https://doi.org/10.1016/j.soisec.2024.100133

Yatoo, A. M., Ali, S., Hamid, S., Hassan, B., Zaheen, Z., Ali, M. N., et al. (2020). Role of soil biota and associated threats. In *Bioremediation and biotechnology, Vol 4: Techniques for noxious substances remediation* (pp. 143–165). Springer.

Zhang, W., Li, S., Xu, Y., Liu, X., et al. (2019). Advances in research on relationships between soil pore structure and soil microenvironment and organic carbon turnover. *Journal of Soil and Water Conservation, 33*(4), 1–9.

Zhao, Y., Fan, W., Liu, Z., & Liu, J. (2022). Effects of urease and dehydrogenase on soil microbial activity under pesticide stress. *Soil Biology and Biochemistry, 163*, 108429.

Zheng, J., & Xu, Y. (2023). A review: Development of plant protection methods and advances in pesticide application technology in agro-forestry production. *Agriculture, 13*(11), 2165.

Zhou, W., Li, M., & Achal, V. (2025). A comprehensive review on environmental and human health impacts of chemical pesticide usage. *Emerging Contaminants, 11*(1), 100410. https://doi.org/10.1016/j.emcon.2024.100410

Zifcakova, L. (2020). Factors affecting soil microbial processes. In *Carbon and nitrogen cycling in soil* (pp. 439–461). Springer.

Zinyemba, C., Archer, E., & Rother, H. A. (2020). Climate change, pesticides and health: Considering the risks and opportunities of adaptation for Zimbabwean smallholder cotton growers. *International Journal of Environmental Research and Public Health, 18*(1), 121. https://doi.org/10.3390/ijerph18010121

Pesticides and Microbiome Shifts

Uchechukwu Chinwe Nebo, Amarachi Nchekwube Ugwuja, Ayoigbala Monioluwa Lawal, and Stanley Ozoemena Agbo

1 Introduction to Microbiome

Environmental and agricultural sustainability are essential for preserving biodiversity, maintaining plant and animal health, ensuring ecosystem stability, and supporting human well-being. With increasing human impacts on the climate and environment, adopting environmentally sustainable practices is critical, not only to protect natural ecosystems but also to sustain a growing global population. Since soil is a cornerstone of agriculture and food production, a deeper understanding of soil microbiomes can significantly enhance future agricultural methods and applications (Tosi et al., 2020).

The microbiome refers to a diverse community of microorganisms, such as bacteria, fungi, viruses, and archaea, or the collective genetic material of these organisms, present in various environments like soil, water, and the bodies of plants and animals. Zhu et al. (2023) define the microbiome as including all microbes, such as archaea, bacteria, fungi, protists, and viruses that are associated with an animal or plant host or inhabit specific environments like soil and water. These microorganisms play vital roles in ecosystem health by recycling nutrients, supporting plant health, aiding decomposition, purifying water, and enabling bioremediation. Beyond their functions in soil organic matter decomposition, carbon cycling, and nutrient

U. C. Nebo (✉) · A. M. Lawal
Department of Microbiology, Federal University Oye-Ekiti, Oye-Ekiti, Ekiti State, Nigeria
e-mail: uchechukwu.nebo@fuoye.edu.ng

A. N. Ugwuja
Department of Biology, Adeyemi Federal University of Education Ondo, Ondo City, Ondo State, Nigeria

S. O. Agbo
Environmental Toxicology Unit, Department of Animal and Environmental Biology, Federal University Oye-Ekiti, Oye-Ekiti, Ekiti State, Nigeria

B. R. Babaniyi, E. E. Babaniyi (eds.), *The Interplay of Pesticides and Climate Change*, https://doi.org/10.1007/978-3-031-81669-7_13

mobilization, saprotrophic fungi contribute to soil structure by secreting extracellular compounds and binding soil particles through their hyphal networks (Bergmann et al., 2016).

Research into microbiomes spans fields such as medical science, environmental studies, and agriculture. Microbiome studies offer new perspectives for medical treatments (Baquero & Nombela, 2012), while engineered microbiomes replace toxic chemicals, enhancing environmental health. The use of microbiomes in food processing and agriculture is a rapidly growing area of research. Historically, research focused primarily on microorganisms associated with disease, while beneficial microorganisms that play key roles in ecosystem functions and offer advantages to other organisms, plants, and animals were often overlooked (Berg et al., 2016).

The term "microbiome" merges "micro," meaning small, with "biome," referring to a specific community of organisms in a distinct environment. It describes a unique microbial community within a defined habitat that possesses particular physicochemical characteristics, influencing microbial behavior and interactions (Whipps et al., 1988) (Fig. 1).

Microbes interact with each other in diverse ways, forming symbiotic relationships that greatly influence their fitness, population dynamics, and functionality within the microbiome (Banerjee et al., 2018). These interactions may occur within a single species or between different species, genera, families, or even broader taxonomic groups. Interactions can be positive, as in mutualism, synergism, or commensalism; negative, such as in amensalism (including predation, parasitism, antagonism, or competition); or neutral, with no notable impact on the involved species' fitness or functions.

The advancement of genomic and molecular techniques, including second- and third-generation sequencing and metabolomics, has led to significant breakthroughs in understanding microbial diversity and interactions. These tools have enriched our

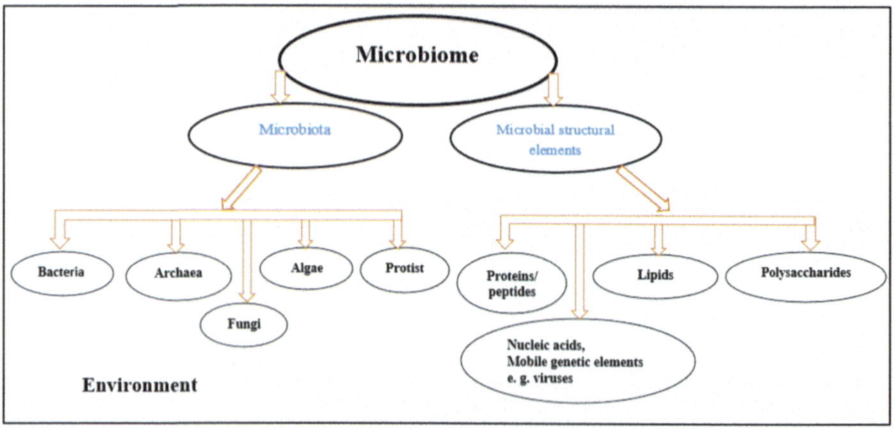

Fig. 1 Schematic composition of microbiome. (Modified from Berg et al. (2020))

comprehension of microbial community complexity and dynamics, revealing new details about microbial functions and interactions across diverse environments (Zhu et al., 2023). Research shows that microbes enhance plant growth and productivity by mediating stress tolerance and supporting nutrient uptake, and they are crucial to nearly all biogeochemical cycles (Falkowski et al., 2008; Müller et al., 2016).

The stability and health of environments, as well as the organisms within them, depend heavily on soil microbial communities. Any imbalance in the soil microbiome can lead to widespread ecological effects (Oluwadara & Olubukola, 2021). Recently, attention has grown around global soil microbiome shifts caused by climate change, which impacts soil health, nutrient cycling, and broader ecosystem processes as climate factors reshape microbial composition and function in soils worldwide (Naylor et al., 2020).

Agricultural practices also affect soil conditions and microbial communities. Some practices may harm the soil microbiome, while others can boost the diversity, structure, and abundance of beneficial microbes, thereby promoting healthier soils. Additionally, pesticide use can disrupt microbial communities in soil, plants, animals, and even humans, as exposure may unintentionally affect beneficial microorganisms. Such microbiome shifts can alter many essential functions, often with far-reaching environmental consequences.

2 Impact of Pesticides on Soil Microbiome

Pesticides are chemical agents intended to eliminate or control pests, thereby safeguarding crops, supporting public health, and managing ecosystems. They may be naturally derived or chemically synthesized and are applied across various sectors to manage different types of pests. By targeting specific organisms, pesticides help boost agricultural productivity, curb vector-borne diseases, and maintain ecosystem balance in designated areas. However, their application brings concerns, particularly regarding environmental impact, human health, and the emergence of pest resistance (Zhan et al., 2020; Bhatt et al., 2021; Zhang et al., 2021). As a result, sustainable use and regulatory measures are essential to maximize their benefits while minimizing associated risks. The different classes of pesticides are illustrated in Fig. 2 (Table 1).

Historically, various organic and inorganic chemicals, including saltwater, were used to control pests before the advent of dichloro-diphenyl-trichloroethane (DDT) (Abubakar et al., 2020). DDT contributed to increased food production and extended the shelf life of food products, leading to a surge in its global demand and paving the way for synthesizing additional chemical pesticides. Today, farmers frequently apply pesticides to increase crop yields, particularly to address global food shortages. This practice helps protect crops from pests and diseases, supporting higher agricultural productivity. However, this widespread use has led to environmental pollution and a shift in the disease burden due to various health risks, including respiratory issues, immune disruptions, and neurological disorders (Ohlander et al.,

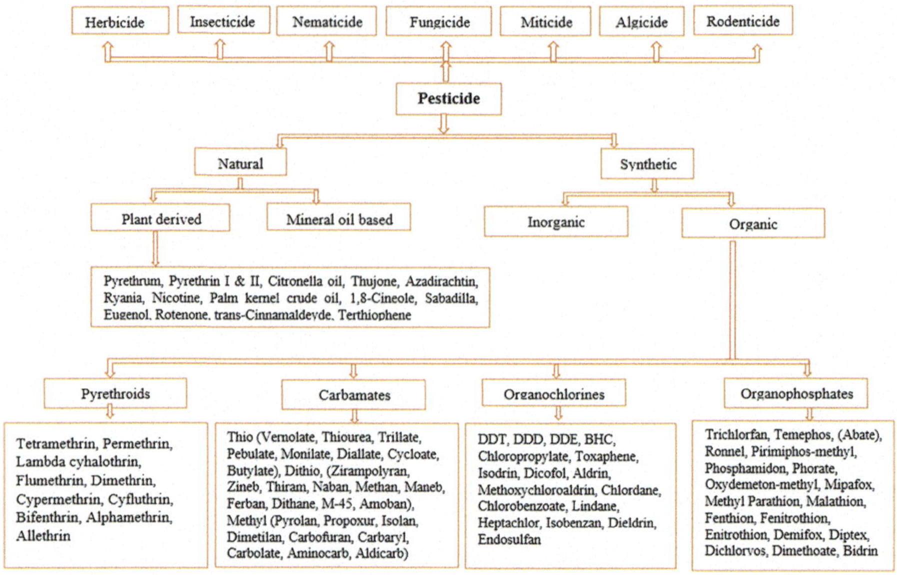

Fig. 2 Classification of pesticides. (Source: Jayaraj et al. (2016), Hassaan & El Nemr (2020), Malhotra et al. (2021), Souto et al. (2021), Parra-Arroyo et al. (2022), Pathak et al. (2022)

Table 1 Chemical structures of commonly used pesticides

Name	Structure	Name	Structure
DDT (Dichlorodiphenyltrichloroethane)	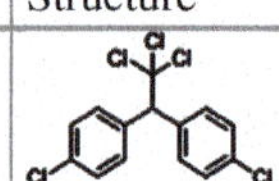	Lindane	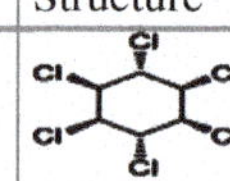
DDD (Dichlorodiphenyldichloroethane)	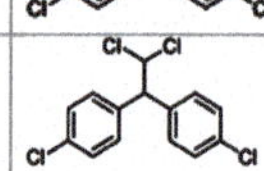	HCH	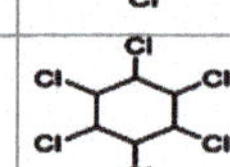
DDE (Dichlorodiphenyldichloroethylene)	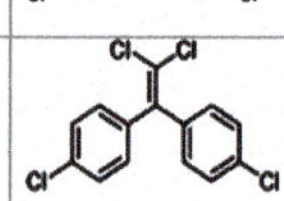	Chlordecone	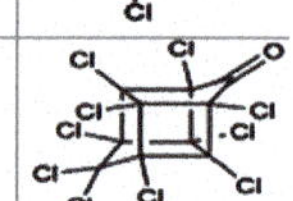
Dieldrin	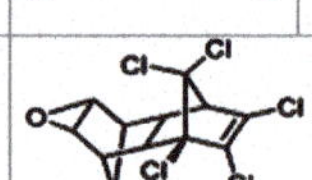	Toxaphene	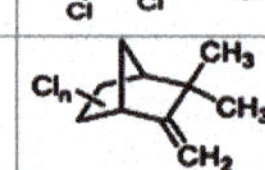
Aldrin	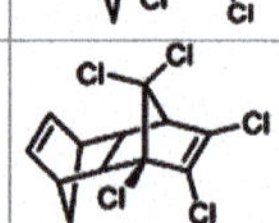	Mirex	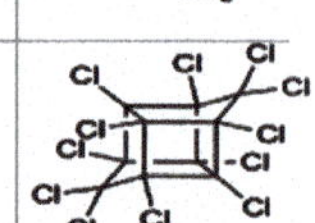
Endrin	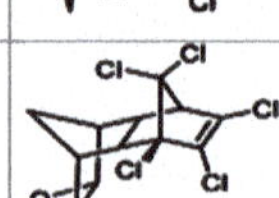	Endosulfan	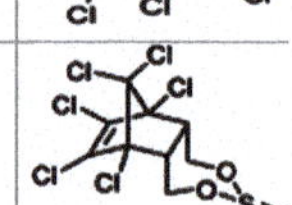
Heptachlor	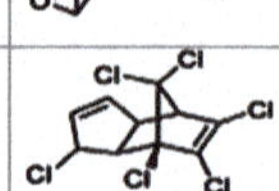	Chlordane	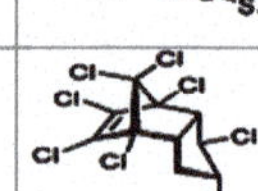

Source: Pathak et al. (2022)

2020; Vermeulen et al., 2020). Pesticides can also harm soil microbial diversity, essential for nutrient cycling, organic matter breakdown, and plant growth, which may result in degraded soil health and reduced productivity.

Research indicates that despite the ban on numerous harmful pesticide ingredients, chemical pesticides continue to contribute significantly to environmental pollution in Europe. This ongoing use poses risks to biodiversity and public health, affecting pollinators, soil microorganisms, and natural pest controllers, all of which are vital to food security (Piselli, 2023). Although the European Union (EU) Directive on the Sustainable Use of Pesticides was adopted in 2009, pesticide sales in Europe have remained relatively steady over the past decade, showing the reliance on chemical pesticides in industrial agriculture and highlighting challenges in transitioning to sustainable practices (Destoumieux-Garzón et al., 2018).

Pesticides are inherently harmful, often impacting non-target organisms and ecosystems. They can contaminate soil and harm beneficial insects, plants, and microorganisms. High levels of pesticide use can impair critical ecological functions, reducing biodiversity and ecosystem resilience. Excessive pesticide use can also damage beneficial soil microbes. Dr. Elaine Ingham, a soil scientist, explains that the degradation of soil organisms due to pesticides parallels the adverse effects of overusing antibiotics on human health. For example, various herbicides inhibit soil bacteria and fungi that contribute to nitrogen fixation and other soil processes, directly impacting soil health and plant nutrition (Santos & Flores, 1995; Arias & Fabra, 1993; Singh & Singh, 1989; Kelley & South, 1978).

Modern pesticides are highly potent, allowing smaller quantities to be effective, yet they can still cause significant ecological harm (Schulz et al., 2021). According to Walder et al. (2022), pesticide residues in arable soils have been shown to affect soil microbiomes, impacting microbial diversity and the abundance of nitrogen cycle-related functional genes. These effects vary depending on the pesticide compound, although certain microbial taxa may adapt to metabolize pesticides, using them as sources of carbon, nitrogen, or energy. However, this shift does not negate the detrimental impact on biological nitrogen fixation, an essential process for soil fertility.

A reliable indicator for predicting the environmental impact of pesticides is toxicity-weighted use, which assesses potential effects on organisms, considering differences in pesticide properties, application methods, and exposure conditions (Wolfram et al., 2019). Enzymatic activities are also vital, as they drive essential biochemical reactions. Pesticides can disrupt these activities and other biological processes (Antonious, 2003). Concerns over pesticide contamination have spurred interest in bio-pesticides as safer alternatives. Still, even natural pesticides like azadirachtin may harm soil fungi and nitrifiers at recommended doses, with biocidal effects becoming evident at higher levels (Gopal et al., 2007).

3 Consequences for Plant Health and Disease Resistance

Pesticides can indeed benefit agriculture by controlling pests, like caterpillars that consume crops, potentially resulting in higher yields and better-quality produce, which can enhance farmers' revenues. For instance, Chibu et al. (2002) found that applying chitosan early in development promoted growth and increased seed production in rice and soybeans. Pesticides target specific organisms; however, it is essential to recognize their growing biological and physiological effects on crops themselves (Pathak et al., 2022). Over time, these chemicals can disrupt normal growth patterns, influencing seed size, biomass, and storage in plant tissues (Jan et al., 2012). Research has shown that glyphosate, for example, reduces leaf dry matter in *Phaseolus vulgaris* (Brecke & Duke, 1980), and higher doses can lower germination rates, dry weight, and root length in Vigna radiata (Basantani et al., 2011).

While agrochemicals provide benefits, they carry risks, especially due to phytotoxicity, where high chemical exposure can harm plants. Young, chemically treated seeds often face higher concentrations than mature plants, increasing the likelihood of damage. Pesticides are generally toxic to target and non-target species alike, and pests can develop resistance, which has been documented with chemicals like DDT (Pathak et al., 2022). Pesticide effects on plant physiology are well documented; studies by Fuerst and Norman (1991) and Sharples et al. (1997) emphasize that accumulated pesticides can disturb vital plant processes. For instance, chlortoluron disrupts the photosynthetic electron transport chain, while uracil-type herbicides inhibit the Hill reaction and damage photosystem II, impairing photosynthesis (Barry et al., 1990). Fungicides can reduce chlorophyll content, affecting overall photosynthetic efficiency (Tort & Turkyilmaz, 2003). Additionally, herbicides are linked to necrosis, stunting, chlorosis, and other symptoms, and prolonged pesticide use may even reduce vegetation diversity (Donald, 2004).

Beyond physical growth, pesticides influence plant defense mechanisms, often causing oxidative stress by generating reactive oxygen species (ROS), which can limit growth and photosynthetic ability. In response, plants activate their antioxidative defenses, including both non-enzymatic antioxidants and antioxidative enzymes (Xia et al., 2009; Sharma et al., 2015). However, oxidative stress can reduce plant proteins, chlorophyll levels, and overall photosynthetic performance (Xia et al., 2006).

Glyphosate-based herbicides (GBHs), specifically, have broader implications on crops and their environment. Martinez et al. (2018) highlight that GBHs impair the shikimic acid pathway in glyphosate-sensitive crops, weakening their natural defenses against pathogens. Even glyphosate-resistant crops, though genetically modified to tolerate glyphosate, may experience weakened disease resistance. GBHs also alter microbial communities in the rhizosphere, fostering harmful phytopathogens by inhibiting beneficial microbes. Additionally, they can reduce nutrient metal uptake in crops, further compromising resistance to diseases. These adverse effects often interact, amplifying the detrimental impact on crop health and resilience (Fig. 3).

Fig. 3 Molecular structure of glyphosate

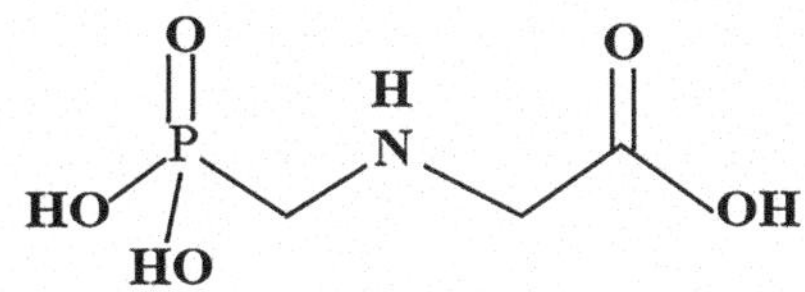

Glyphosate inhibits the enzyme 5-enolpyruvylshikimate-3-phosphate synthase (EPSPS), a key enzyme in the shikimic acid pathway, essential for synthesizing aromatic amino acids like tryptophan, phenylalanine, and tyrosine (Caetano et al., 2012; Martinez et al., 2018). Blocking EPSPS disrupts the production of these amino acids, vital for plant growth and survival, ultimately leading to plant death. Inhibition of this pathway also halts the production of other essential aromatic compounds involved in plant metabolism, including lignin, pigments, and hormones. Consequently, amino acid deficiency affects all plant tissues, disrupting protein synthesis and metabolic processes, which contributes to plant death. This interruption also weakens plants' natural defenses, such as the production of defensive metabolites and lignification, making them more susceptible to pathogens (Keen et al., 1982; Sharon et al., 1992; Liu et al., 1997). This vulnerability can occur even at sublethal glyphosate doses, where impaired protective metabolite production and weakened lignification reduce disease resistance.

Larson et al. (2006) demonstrated the reduction in disease resistance with glyphosate-based herbicides (GBHs) in previously resistant crops. Their study on a glyphosate-resistant sugar beet cultivar, "B4RR," revealed that treatment with Roundup Weather MAX at 0.84 kg a.e. ha^{-1} increased susceptibility to Rhizoctonia solani infection, resulting in heightened disease damage from Rhizoctonia root rot. Untreated plants, by contrast, maintained resistance. The researchers noted that this vulnerability arose from plant-specific responses, as in vitro assays showed no direct impact of glyphosate on Rhizoctonia solani growth. Instead, elevated shikimate levels—a marker of EPSPS disruption—were observed in glyphosate-treated plants, particularly outside the root zone. This shikimate accumulation, documented at intervals up to 14 days post-treatment, was absent in untreated controls and likely impaired the plants' defense mechanisms, leaving them more susceptible to infection. The metabolic disruption from glyphosate exposure weakens the plant's innate pathogen defenses, accounting for increased disease vulnerability in glyphosate-treated sugar beets.

While herbicides target plants, they can also be toxic to bacteria. Bacteria, like plants, have developed resistance mechanisms against herbicides, such as reduced membrane permeability, altered metabolic pathways, or efflux pumps to remove toxic substances (Miller et al., 2022). Resistance to glyphosate in bacteria has been observed via efflux through membrane transporters and mutations in the aroA gene, which plays a role in aromatic amino acid synthesis (Hertel et al., 2021). Additionally, studies have shown that herbicides like dicamba, 2,4-dichlorophenoxyacetic acid, and glyphosate can alter antimicrobial resistance in bacteria, including Salmonella enterica Typhimurium and *Escherichia coli* (Kurenbach et al., 2015). Field studies on citrus greening treatments found that prolonged glyphosate use correlated with penicillin resistance in glyphosate-resistant bacteria, raising concerns about potential cross-resistance between herbicides and antibiotics (Shin et al., 2016).

4 Microbiome-Mediated Pesticide Degradation

Biodegradation is a key process in reducing pesticide pollution and promoting environmental sustainability. Traditional physical and chemical methods for pesticide removal often create toxic by-products and are costly, while bioremediation offers a more eco-friendly alternative (Desisa et al., 2022). Bioremediation utilizes natural processes involving plants, algae, fungi, and bacteria to break down toxins in a cost-effective and sustainable manner. Common bioremediation techniques include phytoremediation, microalgae bioremediation, mycoremediation, and bacterial degradation (Singh et al., 2020).

Microorganisms, particularly bacteria and fungi, play essential roles in breaking down pesticides into less toxic forms. Through processes like biotransformation, bacterial communities can metabolize persistent pesticides either as nutrient sources or through co-metabolism (Kumar et al., 2020). Co-metabolism involves converting compounds without directly benefiting microbial growth, which can effectively degrade resistant pesticides. Additionally, consortia of microbial populations are generally more efficient in degrading pesticides than individual species, accelerating bioremediation and enhancing soil health (Kumar et al., 2020).

Numerous microorganisms have been identified for their pesticide-degrading capabilities. Bacterial strains like *Pseudomonas, Bacillus*, and *Streptomyces*, as well as fungal strains like *Aspergillus* and *Penicillium*, degrade pesticides and convert them into non-toxic compounds. *Actinomycetes* also contribute to pesticide degradation in soil environments, while some algae have been effective in breaking down specific pesticides (Kafilzadeh et al., 2015; Bueno et al., 2010).

For instance, *Aspergillus species* can degrade atrazine, producing metabolites such as deethylatrazine and deisopropylatrazine (Kaufman & Blake, 1970). Bacterial degradation of atrazine generates 2-hydroxyatrazine, with *Mucilaginibacter* identified as a significant aerobic degrader (Douglass et al., 2017). Other studies have shown *Burkholderia species* to be effective in degrading triazole fungicides under tebuconazole exposure (Walder et al., 2022; Han et al., 2021). Additional genera like *Polaromonas, Fusarium*, and *Acidobacteria* are also known for their role in pesticide degradation (Jiang et al., 2018; Xie et al., 2011).

However, pesticide residues can impact microbial diversity differently, often reducing bacterial diversity due to the toxic effects on certain sensitive taxa, such as *Skermanella* and *Pedomicrobium* (Ezeokoli et al., 2020; Yu et al., 2020). These effects contrast with an observed positive effect on fungal diversity, potentially due to the ecological niches created by pesticide residues (Walder et al., 2022).

Efforts to enhance bioremediation efficacy have demonstrated promising results. Sangeetha et al. (2016) reported successful chlorpyrifos degradation by *Aspergillus fumigatus* and *Aspergillus niger*, achieving degradation rates up to 93.6 and 82.2%, respectively. Pathak et al. (2022) further identified bacterial strains like *Alcaligenes, Flavobacterium*, and *Acinetobacter* as capable of degrading endosulfan, illustrating the broad applicability of microbial degradation in managing pesticide pollution.

Biodegradation, thus, represents a sustainable approach to pesticide remediation, leveraging microbial diversity to reduce environmental impacts and support ecosystem health. Future research and metagenomic analyses can further illuminate the role of biodegradation in shaping soil microbial communities and improving agricultural sustainability.

5 Mechanisms and Environmental Benefits

Advanced techniques like bioaugmentation, biostimulation, and natural attenuation are increasingly used to improve pesticide biodegradability in contaminated environments. Bioaugmentation involves introducing specialized, potent bacteria into a site to enhance degradation processes. Biostimulation, on the other hand, relies on adding nutrients or other substances to support indigenous microbial populations. Meanwhile, natural attenuation leverages native microbial species already present in the contaminated site to carry out biodegradation, reducing external intervention (Pathak et al., 2022).

Microbial bioremediation is a promising, cost-effective, and eco-friendly method for removing pollutants from the environment. This approach uses microorganisms, primarily bacteria, to break down pesticides into smaller, non-toxic compounds, thus avoiding secondary pollution. Studies, including those by Kafilzadeh et al. (2015) highlight the effectiveness of microorganisms in reducing pesticide contamination in soil and water. Bacteria play a significant role in these processes, breaking down complex organic molecules via mineralization and co-metabolism, converting them into less harmful intermediates (Huang et al., 2018; Mai et al., 2001; Boivin et al., 2005; Arora et al., 2012; Ye et al., 2018).

Microorganisms employ various metabolic pathways to break down pesticides, which include mitochondrial energy metabolism, fatty acid and lipid metabolism, amino acid metabolism, as well as oxidative and hydrolytic reactions. Some bacteria also utilize methylation pathways to alter the structure of pesticides, aiding in their breakdown and mineralization (Guerrero et al., 2023). These pathways facilitate the transformation of harmful chemicals into stable end-products, demonstrating the potential of microbial metabolism in sustainable pollution control.

Mechanisms of Biodegradation

- *Mineralization*: Mineralization and co-metabolism are key microbial processes in the biodegradation of pesticides. Mineralization involves the conversion of complex organic molecules into inorganic substances, such as carbon dioxide, water, and simple minerals. This process is crucial for pesticide degradation because it leads to the complete breakdown of harmful chemicals into non-toxic, inorganic end-products. Through mineralization, bacteria can fully degrade pesticides, ensuring that the contaminants are transformed into substances that are harmless to the environment and living organisms.

Co-metabolism, on the other hand, allows microorganisms to degrade compounds that are not directly used as growth substrates. In this process, bacteria and fungi metabolize one substance while simultaneously breaking down another, even if the second compound cannot be used as a primary energy or carbon source. Co-metabolism enhances the overall degradation of various pollutants, including pesticides, by enabling microorganisms to degrade compounds that would otherwise be resistant to breakdown. This ability allows for a more comprehensive and effective breakdown of harmful substances, facilitating the removal of pesticide residues from contaminated environments (Zhang et al., 2010). Together, these processes support a sustainable, efficient approach to pesticide biodegradation, helping mitigate the environmental impact of pesticide pollution and promoting ecological balance.

- *Enzymatic degradation*: Microorganisms play a key role in breaking down pesticide molecules through the production of specialized enzymes, such as esterases, hydrolases, and oxygenases. These enzymes target specific bonds within the pesticide structures, facilitating their conversion into simpler, non-toxic compounds. Many pesticides share structural similarities with natural substances, allowing certain microorganisms to identify and degrade them efficiently (Van Eerd et al., 2003, Ortiz-Hernández et al., 2011). By using pesticides as a source of nutrients, these microbes contribute to the breakdown of harmful chemicals into harmless by-products. Both fungi and bacteria are involved in these processes, producing enzymes either inside their cells or releasing them into their environment, including hydrolases, peroxidases, and oxygenases (Bansal, 2012). Hydrolases are particularly effective at breaking down various types of pesticide molecules, such as esters, peptide bonds, carbon-halide bonds, ureas, and thioesters, and they typically operate without the need for redox cofactors, making them well-suited for use in bioremediation efforts. Among the hydrolases, esterases are notable for their ability to hydrolyze a broad range of compounds, including carboxylic esters, amides, and phosphate esters. In the reactions they catalyze, esterases break down ester-based substrates into alcohols and acids, thus playing a significant role in the detoxification of pesticides in the environment.

$$R = O\text{-}OCH_3 + H_2O\ R = O\text{-}OH + CH_3OH$$

Oxidases, a subclass of oxidoreductases, are enzymes that facilitate the transfer of electrons from one molecule (the electron donor or reductant) to another (the electron acceptor or oxidant) (Scott et al., 2008). Many oxidoreductases require additional cofactors to act as electron donors, acceptors, or both. These enzymes are significant in bioremediation, where they catalyze redox reactions using molecular oxygen (O_2) as the electron acceptor. In such reactions, oxygen is reduced to form water (H_2O) or hydrogen peroxide (H_2O_2) (Ortiz-Hernández et al., 2013).

The primary environmental advantage of biodegradation lies in its ability to promote long-term sustainability. By breaking down harmful pesticides, this process helps prevent the accumulation of toxic substances in ecosystems, ultimately leading to healthier soils and water sources. Through the use of targeted microbial strains, researchers and environmentalists can employ bioremediation techniques to effectively address pesticide contamination in various environments. These specialized microbes possess the capability to degrade harmful chemicals, offering an eco-friendly and efficient method to reduce pollution and restore affected areas.

6 Interplay Between Pesticides and Aquatic Microbiomes

Herbicides, being specifically designed to target and kill plants, can have devastating effects on aquatic plants when pesticides contaminate water sources (Aktar et al., 2009). These agricultural chemicals are highly persistent in soil, and once applied to fields, they can seep into the ground and eventually reach groundwater at detectable levels. Once in the groundwater, pesticides can continue to migrate over long distances, contaminating a wide area. This unpredictability in the fate of hazardous pesticides means their impacts often extend far beyond the original application site, making decontamination a complex and widespread challenge (Kumar et al., 2020).

Pesticides typically reach surface water through runoff from treated soil and plants, causing widespread contamination. Poorly managed agricultural practices are a major contributor to this pesticide pollution. Hydrology transport models are often used to study how these chemicals move and behave in rivers and streams. As early as the 1970s, researchers were conducting quantitative analyses of pesticide runoff to predict how much would reach surface waters (Hogan et al., 1973). There are several routes through which pesticides contaminate water: they can drift during spraying, percolate through the soil, be carried away by runoff, or be spilled, whether accidentally or negligently. Additionally, soil erosion can transport pesticides to water bodies (Papendick et al., 1986).

The extent to which pesticides contaminate water depends on various factors, including the pesticide's solubility in water, the distance from the application site to nearby water bodies, weather conditions, soil type, the presence of crops, and the application method used (Pedersen, 1997). A comprehensive study by the United States Geological Survey (USGS) in the 1990s revealed disturbing findings: over 90% of water and fish samples from streams in major US river basins contained one or more pesticides, often several at a time (Kole et al., 2001). Pesticides were found in all samples from rivers influenced by agricultural or urban land use, with 99% of urban stream samples contaminated (Bortleson & Davis, 1987–1995). In many cases, pesticide levels in urban streams exceeded safety guidelines for aquatic life. Herbicides like 2,4-D, diuron, and prometon, as well as insecticides such as chlorpyrifos and diazinon, were commonly found in both surface and groundwater (USGS, 1998). Alarmingly, in the Puget Sound Basin, 23 pesticides, including 17 herbicides, were detected, with diazinon and diuron exceeding recommended safety levels for aquatic ecosystems (Bevans et al., 1998; Fenelon, 1998). These findings underscore the pervasive environmental damage caused by pesticide runoff and the significant risks posed to aquatic ecosystems and water quality.

The movement of pesticides via runoff and drift can transport them to distant aquatic environments, other fields, grazing areas, human settlements, and undeveloped regions. Further complications arise from inadequate production, storage, transport, and disposal practices (Tashkent, 1998). Additionally, the repeated use of pesticides over time leads to pest resistance, while also harming non-target species, which can ultimately contribute to the resurgence of pests (Damalas & Eleftherohorinos, 2011). To mitigate these issues, alternatives such as integrated pest management and sustainable agricultural practices like polyculture offer effective solutions that reduce reliance on harmful chemical applications.

7 Restoring Microbiome Health Post-Pesticide Exposure: Approaches and Challenges

The two primary approaches for pesticide remediation are in situ and ex situ bioremediation. In situ bioremediation involves treating hazardous materials directly at the contaminated site, while ex situ bioremediation requires removing the pollutants from the site for treatment at an offsite location (Giri et al., 2021).

In situ *bioremediation* works by stimulating microbial activity at the contaminated site through the addition of microbes and nutrients, and by optimizing environmental conditions. This technique treats pollutants onsite, without the need for excavation, minimizing disruption to the soil structure. In situ bioremediation relies on natural biological processes to degrade organic contaminants, breaking them down into non-toxic by-products such as carbon dioxide and water. It is a cost-effective, low-maintenance, and environmentally friendly approach to detoxifying polluted areas. A case study by Seech et al. (2008) used the cycled DARAMEND treatment for in situ remediation of dieldrin-contaminated soil in coastal Florida in 2004, where 2600 tons of soil were treated at an operational cost of around 12.5 USD per cubic yard.

Ex situ *bioremediation*, on the other hand, involves excavating contaminated soil from the polluted site and transporting it to a separate location for treatment. The effectiveness and cost of ex situ methods depend on various factors, including the type of pollutant, pollution depth, geographical location, and degree of contamination. Common ex situ techniques include land farming and composting, where the polluted materials are rehabilitated at designated offsite locations. Ex situ methods tend to be more expensive than in situ techniques due to the additional requirements for excavation and transportation (Jorgensen, 2007). Both in situ and ex situ methods' biodegradation rates depend on the microbial populations involved and the specific environmental conditions (Jorgensen, 2007). In situ bioremediation is often preferred for restoring polluted soils and water ecosystems due to its lower cost and minimal disruption to the environment. The feasibility of either approach depends onsite-specific conditions, including the level and toxicity of the pesticides present and the microbial communities in place.

Bioremediation is recognized as a highly cost-effective alternative to traditional pesticide removal methods, with estimated savings of around 65–85%. For example, incineration of contaminated soil can cost between 250 and 500 USD per ton, while biotechnological methods cost only 40–70 USD per ton (Eykelbosh, 2014). Microbe-assisted treatments are about one-third cheaper than conventional soil remediation techniques (Onwurah et al., 2007). The costs of bio-treatment, incineration, and landfill disposal typically range from 50 to 130 USD/m^3, 300 to 1000 USD/m^3, and 200 to 300 USD/m^3, respectively (Asadi et al., 2012). Despite its advantages, a key limitation of bioremediation is maintaining optimal conditions for microbial growth, and the degradation process can be slow and time-consuming.

8 Conclusion

The relationship between pesticides and microbiomes has profound implications for both soil health and environmental sustainability. While pesticides are highly effective in controlling pests, they can disrupt the balance of microbial communities, leading to negative consequences for nutrient cycling, plant vitality, and overall ecological stability. The persistence of pesticides in the environment poses long-term risks, particularly in agricultural systems where healthy soil microbiomes are crucial for maintaining ecosystem functions.

Despite these challenges, bioremediation techniques offer promising solutions. The use of microbial communities to degrade pesticides is a cost-effective and environmentally friendly approach. However, for this method to be more effective, it requires continued refinement and research to optimize the conditions that support microbial growth and enhance the rate of degradation. Advances in biotechnology and genetic engineering offer great potential to develop resilient microbial strains and improve treatment strategies. These innovations can help manage pesticide contamination more efficiently and sustainably. Moving forward, the integration of these cutting-edge techniques will be essential for restoring and preserving the health of ecosystems affected by pesticide use, paving the way for more sustainable agricultural practices.

References

Abubakar, Y., Tijjani, H., Egbuna, C., Adetunji, C. O., Kala, S., Kryeziu, T. L., et al. (2020). Pesticides, history, and classification. In C. Egbuna & B. Sawicka (Eds.), *Natural remedies for pest, disease and weed control* (pp. 29–42). Academic Press.

Aktar, M. W., Sengupta, D., & Chowdhury, A. (2009). Impact of pesticides use in agriculture: Their benefits and hazards. *Interdisciplinary Toxicology, 2*(1), 1–12.

Antonious, G. F. (2003). Impact of soil management and two botanical insecticides on urease and invertase activity. *Journal of Environmental Science and Health Part B, 38*, 479–488.

Arias, R. N., & Fabra, P. A. (1993). Effects of 2,4-dichlorophenoxyacetic acid on *Rhizobium* sp. growth and characterization of its transport. *Toxicology Letters, 68*, 267–273.

Arora, P. K., Sasikala, C., & Ramana, C. V. (2012). Degradation of chlorinated nitroaromatic compounds. *Applied Microbiology and Biotechnology, 93*, 2265–2277.

Asadi, A., Zinatizadeh, A. A. L., & Sumathi, S. (2012). Simultaneous removal of carbon and nutrients from an industrial estate wastewater in a single up-flow aerobic/anoxic sludge bed (UAASB) bioreactor. *Water Resources, 46*(15), 4587–4598.

Banerjee, S., Schlaeppi, K., & van der Heijden, M. G. A. (2018). Keystone taxa as drivers of microbiome structure and functioning. *Nature Reviews Microbiology, 16*, 567–576.

Bansal, O. P. (2012). Degradation of pesticides. In S. Hamir, Rathore, & L. M. L. Nollet (Eds.), *Pesticides evaluation of environmental pollution* (pp. 47–77). CRC Press.

Baquero, F., & Nombela, C. (2012). The microbiome as a human organ. *Clinical Microbiology and Infection, 18*, 2–4.

Barry, P., Young, A. J., & Britton, G. (1990). Photodestruction of pigments in higher plants by herbicide action: I. The effect of DCMU (diuron) on isolated chloroplasts. *Journal of Experimental Botany, 41*, 123–129.

Basantani, M., Srivastava, A., & Sen, S. (2011). Elevated antioxidant response and induction of tau-class glutathione S-transferase after glyphosate treatment in Vigna radiata (L.) Wilczek. *Pesticide Biochemistry and Physiology, 99*, 111–117.

Berg, G., Rybakova, D., Grube, M., & Köberl, M. (2016). The plant microbiome explored: Implications for experimental botany. *Journal of Experimental Botany, 67*, 995–1002.

Berg, G., Rybakova, D., Fischer, D., Cernava, T., Vergès, M. C., Charles, T., Chen, X., Cocolin, L., Eversole, K., Corra, G. H., Kazou, M., Kinke, L., Lange, L., Lima, N., Loy, A., Macklin, A., Emmanuelle Maguin, E., Tim Mauchline, T., McClure, R., Mitter, B., Ryan, M., Sarand, I., Smidt, H., Schelkle, B., Roume, H., Kiran, G. S., Selvin, J., de Correa Souza, R. S., Overbeek, L., Singh, B. K., Wagner, M., Walsh, A., Sessitsch, A., & Schloter, M. (2020). Microbiome definition re-visited: Old concepts and new challenges. *Microbiome, 8*, 103.

Bergmann, J., Verbruggen, E., Heinze, J., Xiang, D., Chen, B., Joshi, J., et al. (2016). The interplay between soil structure, roots, and microbiota as a determinant of plant-soil feedback. *Ecology and Evolution, 6*, 7633–7644.

Bevans, H. E., Lico, M. S., & Lawrence, S. J. (1998). *Water quality in the Las Vegas Valley area and the Carson and Truckee River basins, Nevada and California, 1992–96* (p. 1170). USGS.U.S. Geological Survey Circular.

Bhatt, P., Joshi, T., Bhatt, K., Zhang, W., Huang, Y., & Chen, S. (2021). Binding interaction of glyphosate with glyphosate oxidoreductase and C–P lyase: Molecular docking and molecular dynamics simulation studies. *Journal of Hazardous Materials, 409*, 124927.

Boivin, A., Amellal, S., Schiavon, M., & Van Genuchten, M. T. (2005). 2, 4-Dichlorophenoxyacetic Acid 2, 4-D sorption and degradation dynamics in three agricultural soils. *Environmental Pollution, 138*, 92–99.

Bortleson, G., & Davis, D. (1987–1995). *U.S. Geological Survey & Washington State Department of Ecology*. Pesticides in selected small streams in the Puget Sound Basin; pp. 1–4.

Brecke, B. J., & Duke, W. B. (1980). Effect of glyphosate on intact bean plants (Phaseolus vulgaris L.) and isolated cells. *Plant Physiology, 66*, 656–659.

Bueno, A., Diez, J. J., & Fernández, M. M. (2010). Ophiostomatoid fungi transported by Ips sexdentatus (Coleoptera; Scolytidae) in Pinus pinaster in NW Spain. *Silva Fennica, 44*(3), 387–397.

Caetano, M. S., Ramalho, T. C., Botrel, D. F., da Cunha, E. F. F., & de Mello, W. C. (2012). Understanding the inactivation process of organophosphorus herbicides: A DFT study of glyphosate metallic complexes with Zn^{2+}, Ca^{2+}, Mg^{2+}, Cu^{2+}, Co^{3+}, Fe^{3+}, Cr^{3+}, and Al_3. *International Journal of Quantum Chemistry, 112*, 2752–2762.

Chibu, H., Shibayama, H., & Arima, S. (2002). Effects of chitosan application on the shoot growth of rice and soybean. *Japan Journal of Crop Science, 71*, 206–211.

Damalas, C. A., & Eleftherohorinos, I. G. (2011). Pesticide exposure, safety issues, and risk assessment indicators. *International Journal of Environmental Research and Public Health, 8*(12), 1402–1419.

Desisa, B., Getahun, A., & Muleta, D. (2022). Advances in biological treatment technologies for some emerging pesticides. In S. Siddiqui, M. K. Meghvansi, & K. K. Chaudhary (Eds.), *Pesticides bioremediation*. Springer. https://doi.org/10.1007/978-3-030-97000-0_10

Destoumieux-Garzón, D., et al. (2018). The One Health concept: 10 years old and a long road ahead. *Frontiers in Veterinary Science, 5* https://doi.org/10.3389/fvets.2018.00014.

Donald, P. F. (2004). Biodiversity impacts of some agricultural commodity production systems. *Conservation Biology, 18*, 17–37.

Douglass, J. F., Radosevich, M., & Tuovinen, O. H. (2017). Microbial attenuation of atrazine in agricultural soils: Biometer assays, bacterial taxonomic diversity, and catabolic genes. *Chemosphere, 176*, 352–360.

Eykelbosh, A. (2014). *Short-and long-term health impacts of marine and terrestrial oil spills*. Vancouver Coastal Health. Retrieved 1st October, 2024.

Ezeokoli, O. T., Bezuidenhout, C. C., Maboeta, M. S., Khasa, D. P., & Adeleke, R. A. (2020). Structural and functional differentiation of bacterial communities in post-coal mining reclamation soils of South Africa: Bioindicators of soil ecosystem restoration. *Scientific Reports, 10*, 1759.

Falkowski, P. G., Fenchel, T., & Delong, E. F. (2008). The microbial engines that drive Earth's biogeochemical cycles. *Science, 320*, 1034–1039.

Fenelon, J. M. (1998). *Water quality in the White River Basin, Indiana, 1992–96* (p. 1150). USGS. U.S. Geological Survey Circular.

Fuerst, E. P., & Norman, M. A. (1991). Interactions of herbicides with photosynthetic electron transport. *Weed Science, 39*, 458–464.

Giri, M., Bista, G., Singh, P. K. & Pandey, R. (2021). Climate change vulnerability assessment of urban informal settlers in Nepal, a least developed country. Journal of Cleaner Production, 307:405.

Gopal, M., Gupta, A., Arunachalam, V., & Magu, S. P. (2007). Impact of azadirachtin, an insecticidal allelochemical from neem on soil microflora, enzyme and respiratory activities. *Bioresource Technology, 98*, 3154–3158.

Guerrero Ramírez, J. R., Ibarra Muñoz, L. A., Balagurusamy, N., Frías Ramírez, J. E., Alfaro Hernández, L., & Carrillo Campos, J. (2023). Microbiology and biochemistry of pesticides biodegradation. *International Journal of Molecular Science, 24*(21), 15969.

Han, L., Kong, X., Xu, M., & Nie, J. (2021). Repeated exposure to fungicide tebuconazole alters the degradation characteristics, soil microbial community and functional profiles. *Environmental Pollution, 287*, 117660.

Hassaan, M. A., & El Nemr, A. (2020). Pesticides pollution: Classifications, human health impact, extraction and treatment techniques. *Egyptian Journal of Aquatic Resource, 46*, 207–220.

Hertel, R., Gibhardt, J., Martienssen, M., Kuhn, R., & Commichau, F. M. (2021). Molecular mechanisms underlying glyphosate resistance in bacteria. *Environmental Microbiology, 23*, 2891–2905.

Hogan, C. M., Patmore, L., Latshaw, G., Seidman, H, et al. (1973). Computer modeling of pesticide transport in soil for five instrumented watersheds, U.S. Environmental Protection Agency Southeast Water laboratory, Athens, Ga. by ESL Inc., Sunnyvale, California. States of Jersey (2007), Environmental protection and pesticide use Archived. 25 August 2006 at the Wayback Machine. Retrieved on 10 October 2007.

Huang, Y., Xiao, L., Li, F., Xiao, M., Lin, D., Long, X., & Wu, Z. (2018). Microbial degradation of pesticide residues and an emphasis on the degradation of Cypermethrin and 3-phenoxy benzoic acid: A review. *Molecules, 23*(9), 2313.

Jan, S., Parween, T., & Siddiqi, T. (2012). Effect of gamma radiation on morphological, biochemical, and physiological aspects of plants and plant products. *Environmental Reviews, 20*, 17–39.

Jayaraj, R., Megha, P., & Sreedev, P. (2016). Organochlorine pesticides, their toxic effects on living organisms and their fate in the environment. *Interdisciplinary Toxicology, 9*, 90–100.

Jiang, B., Jin, N., Xing, Y., Su, Y., & Zhang, D. (2018). Unraveling uncultivable pesticide degraders via stable isotope probing (SIP). *Critical Reviews in Biotechnology, 38*, 1025–1048.

Jorgensen, K. S. (2007). In situ bioremediation. *Advanced. Applied Microbiology, 61*, 285–305.

Kafilzadeh, F. (2015). Distribution and sources of polycyclic aromatic hydrocarbons in water and sediments of the Soltan Abad River, Iran. Egyptian Journal of Aquatic Research, 41(3):227–231.

Kaufman, D., & Blake, J. (1970). Degradation of atrazine by soil fungi. *Soil Biology and Biochemistry, 2*, 73–80.

Keen, N. T., Holliday, M. J., & Yoshikawa, M. (1982). Effects of glyphosate on glyceollin production and the expression of resistance to *Phytophthora megasperma* f. sp. glycinea in soybean. *Phytopathology, 72*, 1467–1470.

Kelley, W. D., & South, D. B. (1978). *In vitro effects of selected herbicides on growth and mycorrhizal fungi.* Weed Sci. Soc. America Meeting. Auburn University, Auburn, Alabama: 38.

Kole, R. K., Banerjee, H., & Bhattacharyya, A. (2001). Monitoring of market fish samples for Endosulfan and Hexachlorocyclohexane residues in and around Calcutta. *Bulletin of Environmental Contamination and Toxicology, 67*, 554–559.

Kumar, M., Yadav, A. N., Saxena, R., Paul, D., & Tomar, R. S. (2020). Biodiversity of pesticides degrading microbial communities and their environmental impact. *Biocatalysis and Agricultural Biotechnology, 31*, 101883.

Kurenbach, B., Marjoshi, D., Amábile-Cuevas, C. F., Ferguson, G. C., Godsoe, W., Gibson, P., & Heinemann, J. A. (2015). Sublethal exposure to commercial formulations of the herbicides Dicamba, 2,4-Dichlorophenoxyacetic acid, and glyphosate cause changes in antibiotic susceptibility in *Escherichia coli* and *Salmonella enterica* serovar Typhimurium. *MBio, 6*, e00009–15.

Larson, R. L., Hill, A., Fenwick, A., Kniss, A. R., Hanson, L. E. A., & Miller, S. D. (2006). Influence of glyphosate on *Rhizoctonia* and *Fusarium* root rot in sugar beet. *Pest Management Science, 62*, 1182–1192.

Liu, L. Z. K., Punja, Z. K., & Rahe, J. E. (1997). Altered root exudation and suppression of induced lignification as mechanisms of predisposition by glyphosate of bean roots (*Phaseolus vulgaris* L.) to colonization by *Pythium* spp. *Physiological and Molecular Plant Pathology, 51*, 111–127.

Mai, P., Jacobsen, O. S., & Aamand, J. (2001). Mineralization and co-metabolic degradation of phenoxyalkanoic acid herbicides by a pure bacterial culture isolated from an aquifer. *Applied Microbiology and Biotechnology, 56*, 486–490.

Malhotra, H., Kaur, S., & Phale, P. S. (2021). Conserved metabolic and evolutionary themes in microbial degradation of carbamate pesticides. *Frontiers in Microbiology, 12*, 648868.

Martinez, D. A., Loening, U. E., & Graham, M. C. (2018). Impacts of glyphosate-based herbicides on disease resistance and health of crops: A review. *Environmental Science Europe, 30*, 2.

Miller, S. A., Ferreira, J. P., & LeJeune, J. T. (2022). Antimicrobial use and resistance in plant agriculture: A one health perspective. *Agriculture, 12*(2), 289.

Müller, D. B., Vogel, C., Bai, Y., & Vorholt, J. A. (2016). The plant microbiota: Systems-level insights and perspectives. *Annual Review of Genetics, 50*, 211–234.

Naylor, D., Sadler, N., Bhattacharjee, A., Graham, E. B., Anderton, C. R., McClure, R., Lipton, M., Hofmockel, K. S., & Jansson, J. K. (2020). Soil microbiomes under climate change and implications for carbon cycling. *Annual Review of Environment and Resources, 45*(1), 29–59.

Ohlander, J., Fuhrimann, S., Basinas, I., Cherrie, J. W., Galea, K. S., Povey, A. C., Tongeren, M. V., Harding, A. H., Jones, K., Vermeulen, R., & Kromhout, H. (2020). Systematic review of methods used to assess exposure to pesticides in occupational epidemiology studies, 1993–2017. *Occupational and Environmental Medicine, 77*, 357–367.

Oluwadara, P. O., & Olubukola, O. B. (2021). Resident rhizosphere microbiome's ecological dynamics and conservation: Towards achieving the envisioned sustainable development goals, a review. *International Soil and Water Conservation Research, 9*(1), 127–142.

Onwurah, I. N. E., Ogugua, V. N., Onyike, N. B., Ochonogor, A. E., & Otitoju, O. F. (2007). Crude oil spills in the environment, effects and some innovative clean-up biotechnologies. *International Journal of Environmental Research, 1*(4), 307–320.

Ortiz-Hernández, M. L., Sánchez-Salinas, E., Olvera-Velona, A., & Folch-Mallol, J. L. (2011). Pesticides in the environment: Impacts and its biodegradation as a strategy for residues treatment. In M. Stoytcheva (Ed.), *Pesticides-formulations, effects, fate* (pp. 551–574). InTech.

Ortiz-Hernández, M. L., Sánchez-Salinas, E., Castrejón Godínez, M. L., Dantan González, E., Popoca, U., & Elida, C. (2013). Mechanisms and strategies for pesticide biodegradation: Opportunity for waste, soils and water cleaning. *Revista Internacional de Contaminación Ambiental, 29*, 85–104.

Papendick, R. I., Elliott, L. F., & Dahlgren, R. B. (1986). Environmental consequences of modern production agriculture: How can alternative agriculture address these issues and concerns? *American Journal of Alternative Agriculture, 1*(1), 3–10.

Parra-Arroyo, L., González-González, R. B., Castillo-Zacarías, C., Martínez, E. M. M., Sosa-Hernández, J. E., Bilal, M., et al. (2022). Highly hazardous pesticides and related pollutants: Toxicological, regulatory, and analytical aspects. *Science of the Total Environment, 807*, 151879.

Pathak, V. M., Verma, V. K., Rawat, B. S., Kaur, B., Babu, N., Sharma, A., Dewali, S., Yadav, M., Kumari, R., Singh, S., Mohapatra, A., Pandey, V., Rana, N., & Cunill, J. M. (2022). Current status of pesticide effects on environment, human health and it's eco-friendly management as bioremediation: A comprehensive review. *Frontiers in Microbiology, 13*, 962619.

Pedersen, T. L. (1997, June). *Pesticide residues in drinking water.* extoxnet.orst.edu. Retrieved on 15th September, 2007.

Piselli, V. (2023). *Pesticides: What are the risks to our health and to the environment.* European Environment Agency.

Sangeetha, M., Kanimozhi, K., Panneerselvam, A., & Kumar, R. S. (2016). Biodegradation of pesticide using fungi isolated from Paddy fields of Thanjavur District, India. *International Journal of Current Microbiology and Applied Science, 5*, 348–354.

Santos, A. and Flores, M. (1995). Effects of glyphosate on nitrogen fixation of free-living heterotrophic bacteria. Letters in Applied Microbiology, 20(20):349–352.

Souto, A.L., Sylvestre, N., Tölke, E. D., Tavares, J. F., Barbosa-Filho, J. M. and Cebrián-Torrejón, G. (2021). Plant-derived pesticides as an alternative to pest management and sustainable agricultural production: prospects, applications and challenges. Molecules, 10;26(16):4835.

Schulz, R., Bub, S., Petschick, L. L., Stehle, S., & Wolfram, J. (2021). Applied pesticide toxicity shifts toward plants and invertebrates, even in GM crops. *Science, 372*(6537), 81–84.

Scott, C., Pandey, G., Hartley, C. J., Jackson, C. J., Cheesman, M. J., Taylor, M. C., Pandey, R., Khurana, J. L., Teese, M., Coppin, C. W., Weir, K. M., Jain, R. K., Lal, R., Russell, R. J., & Oakeshott, J. G. (2008). The enzymatic basis for pesticide bioremediation. *Indian Journal of Microbiology, 48*, 65–79.

Seech, A., Bolanos, K., Hill, D., & Molin, J. (2008). In situ bioremediation of pesticides in soil and groundwater. *Remediation Journal, 19*(1), 87–98.

Sharma, I., Bhardwaj, R., & Pati, P. K. (2015). Exogenous application of 28 homobrassinolide modulates the dynamics of salt and pesticides induced stress responses in an elite rice variety Pusa Basmati-1. *Journal of Plant Growth Regulation, 34*, 509–518.

Sharon, A., Amsellem, Z., & Gressel, J. (1992). Glyphosate suppression of an elicited defense response. *Plant Physiology, 98*, 654–659.

Sharples, C. R., Hull, M. R., & Cobb, A. H. (1997). Growth and photosynthetic characteristics of two biotypes of the weed black-grass (Alopecurus myosuroidesHuds.) resistant and susceptible to the herbicide chlorotoluron. *Annals of Botany, 79*, 455–461.

Shin, K., Ascunce, M. S., Narouei-Khandan, H. A., Sun, X., Jones, D., Kolawole, O. O., Goss, E. M., & van Bruggen, A. H. C. (2016). Effects and side effects of penicillin injection in huanglongbing affected grapefruit trees. *Crop Protection, 90*, 106–116.

Singh, J. B., & Singh, S. (1989). Effect of 2,4-dichlorophenoxyacetic acid and ma leichydrazide on growth of blue-green algae (cyanobacteria) *Anabaena doliolum* and *Anacystis nidulans*. *Science and Culture, 55*, 459–460.

Singh, D. V., Ali, R., Kulsum, M., & Bhat, R. A. (2020). Ecofriendly approaches for remediation of pesticides in contaminated environs. In R. Bhat, K. Hakeem, & N. S. Al-Saud (Eds.), *Bioremediation and biotechnology*. Springer. https://doi.org/10.1007/978-3-030-46075-4_8

Tashkent. (1998). Part 75. Conditions and provisions for developing a national strategy for biodiversity conservation Archived 13 October 2007 at the Wayback Machine. Biodiversity Conservation National Strategy and Action Plan of Republic of Uzbekistan. Prepared by the National Biodiversity Strategy Project Steering Committee with the Financial Assistance of The Global Environmental Facility (GEF) and Technical Assistance of United Nations Development Programme (UNDP). Retrieved on 17 September, 2007.

Tort, N., & Turkyilmaz, B. (2003). Physiological effects of captan fungicide on pepper (Capsicum annuum L.) plant. *Pakistan Journal of Biological Sciences, 6*, 2026–2029.

Tosi, M., Mitter, E. K., Gaiero, J., & Dunfield, K. (2020). It takes three to tango: The importance of microbes, host plant, and soil management to elucidate manipulation strategies for the plant microbiome. *Canadian Journal of Microbiology, 66*(7), 413–433.

Van Eerd, L. L., Hoagland, R. E., Zablotowicz, R. M., & Hall, J. C. (2003). Pesticide metabolism in plants and microorganisms. *Weed Science, 51*, 472–495.

Vermeulen, R., Schymanski, E. L., Barabási, A. L., & Miller, G. W. (2020). The exposome and health: Where chemistry meets biology. *Science, 367*, 392–396.

Walder, F., Schmid, M. W., Riedo, J., Valzano-Held, A. Y., Banerjee, S., Büchi, L., Bucheli, T. D., & van der Heijden, M. G. A. (2022). Soil microbiome signatures are associated with pesticide residues in arable landscapes. *Soil Biology and Biochemistry, 174*, 108830.

Whipps, J., Lewis, K., & Cooke, R. (1988). Mycoparasitism and plant disease control. In M. Burge (Ed.), *Fungi biol control syst* (Vol. 1988, pp. 161–187). Manchester University Press.

Wolfram, J., Stehle, S., Bub, S., Petschick, L. L., & Schulz, R. (2019). Insecticide risk in US surface waters: Drivers and spatiotemporal modeling. *Environmental Science and Technology, 53*, 12071–12080.

Xia, X. J., Huang, Y. Y., Wang, L., Huang, L. F., Yu, Y. L., Zhou, Y. H., et al. (2006). Pesticides-induced depression of photosynthesis was alleviated by 24 epibrassinolide pretreatment in *Cucumis sativus* L. *Pesticide Biochemistry and Physiology, 86*, 42–48.

Xia, X. J., Zhang, Y., Wu, J. X., Wang, J. T., Zhou, Y. H., Shi, K., et al. (2009). Brassinosteroids promote metabolism of pesticides in cucumber. *Journal of Agricultural and Food Chemistry, 57*, 8406–8413.

Xie, S., Sun, W., Luo, C., & Cupples, A. M. (2011). Novel aerobic benzene degrading microorganisms identified in three soils by stable isotope probing. *Biodegradation, 22*, 71–81.

Ye, X., Dong, F., & Lei, X. (2018). Microbial resources and ecology-microbial degradation of pesticides. *Natural Resources Conservation and Research, 1*. https://doi.org/10.24294/nrcr.v1i1.242

Yu, B., Chen, Z., Lu, X., Huang, Y., Zhou, Y., Zhang, Q., Wang, D., & Li, J. (2020). Effects on soil microbial community after exposure to neonicotinoid insecticides thiamethoxam and dinotefuran. *Science of the Total Environment, 725*, 138328.

Zhan, H., Huang, Y., Lin, Z., Bhatt, P., & Chen, S. (2020). New insights into the microbial degradation and catalytic mechanism of synthetic pyrethroids. *Environmental Research, 182*, 109138.

Zhang, C., Jia, L., Wang, S. H., Qu, J., Li, K., Xu, L. L., Shi, Y. H., & Han, Y. C. (2010). Biodegradation of Beta-Cypermethrin by two Serratia Spp. with different cell surface hydrophobicity. *Bioresource Technology, 101*, 3423–3429.

Zhang, W., Pang, S., Lin, Z., Mishra, S., Bhatt, P., & Chen, S. (2021). Biotransformation of perfluoroalkyl acid precursors from various environmental systems: Advances and perspectives. *Environmental Pollution, 272*, 115908.

Zhu, Y. G., Zhu, D., Rillig, M. C., Yang, Y., Chu, H., Chen, Q. L., Penuelas, J., Cui, H. L., & Gillings, M. (2023). Ecosystem microbiome science. *mLife, 2*(1), 2–10.

Emerging Pests and Disease Vectors

Prity Das, Rakesh Das, Manish Kumar Gautam, and Sandip Mondal

1 Introduction

Climate change chiefly indicates the large-scale deviation of the weather conditions and the temperature of the earth (Pielke Jr, 2004). Though climate change can be the consequence of natural causes, currently it is evidently mostly driven by mankind, which is leading us to global warming, the greenhouse effect, acidification of the ocean, extreme change in weather, melting of glaciers, etc. The greenhouse effect generally represents the entrapment of heat by greenhouse gases (GHGs) such as carbon dioxide, methane, and carbon monoxide in the atmosphere, causing the elevation of the average surface temperature, that is, global warming. Global worming directly and indirectly effects the melting of glacier and polar ice, increased sea level, heat waves, drought, etc. (Kweku et al., 2018). The change in climate is due to several factors, but it can broadly be classified in two types based on origin:

(a) Anthropogenic activities such as overexploitation of land, translocation of spices, air pollution due to aviation, shipping (marine pollutant), groundwater exploitation, ozone layer depletion, overconsumption of fossil fuels livestock farming, etc. (Zheng et al., 2021).
(b) Natural processes: they are not as harmful and frequent as anthropogenesis but cause pollution, for example, volcanic eruptions, solar radiation fluctuations, and orbital change (Bryant, 1997).

P. Das · R. Das
School of Pharmaceutical Technology, School of Health and Medical Sciences, Adamas University, Kolkata, India

M. K. Gautam
Faculty of Pharmaceutical Science, Assam Down Town University, Guwahati, Assam, India

S. Mondal (✉)
Amity Institute of Pharmacy, Amity University Kolkata, Kolkata, West Bengal, India

B. R. Babaniyi, E. E. Babaniyi (eds.), *The Interplay of Pesticides and Climate Change*, https://doi.org/10.1007/978-3-031-81669-7_14

The distortion in climate has given rise to many global issues like global warming, floods, droughts, severe storms, rising sea levels, loss of spices due to a lack of adequate environment and climate, scarcity of food, and an increase in pests and diseases (Hitz & Smith, 2004). Pests can be described as the living organisms that have the capability to cause negative impact for the human being and be present where they are not wanted. Fundamentally, there are pathogens present in the pest; when they come into contact with a human being, they spread them and cause disease (Bos & Parlevliet, 1995). Vector-borne disease is a disease that is carried by a vector. These vectors are living agents, such as different types of arthropods (mosquitoes, fleas, ticks), that carry and transmit the infectious pathogen, which can be a parasite, microbe, or bacteria (Müller et al., 2019).

The increase in pests is leading to an increase in vector-borne diseases like dengue, malaria, yellow fever, and zika virus. Basically, pests are any unwanted living organism that causes harmful effects for human beings. Pests have a tendency to spread deadly diseases by carrying infectious pathogens within themselves. This variation in disease is widely known as vector-borne disease (Organization, 2014; Caminade et al., 2019; Skendžić et al., 2021).

The significance of studying the emerging pests and vector-borne disease comes under global health and well-being, as it is not only affecting health conditions but also the economics and ecological system of the earth (Morales-Hojas, 2017).

Given the direct and indirect effects of pests on our lives, it becomes necessary to control them. Fortunately, pesticides are now easily accessible and readily available. Pesticide is a chemical compound that can kill pests (plants, rodents, fungi, insects, etc.) and reduce the invasion of vector-borne diseases and loss of agricultural products. The relationship between use of pesticides and climate change is directly proportional and connected to each other, as the more the climate is changing, the more the pests are increasing in number, thus the need and use of pesticides are increasing. On the other hand, the more pesticides are used, the more GHGs are produced, resulting in global warming, that is, climate change (Matthews et al., 2014; Delcour et al., 2015).

In this chapter, universal issue of emerging pests and vector diseases, are going to be discussed. The primary objectives of this chapter are as follows:

1. How the pests are affecting our daily lives.
2. Understanding of the dynamics of the disease due to pests.
3. Identification and monitoring of the emerging threats for the futuristic benefits.
4. Development of the control and prevention strategies.
5. Understanding the impact of climate as well as environmental changes.

2 Historical Background

2.1 *Historical Outbreaks*

The presence of vector-borne disease due to pests has been the cause of the suffering of the population by becoming epidemic throughout the centuries. The following list (Table 1) highlights a few significant historical outbreaks:

2.2 *Insights Gained from Previous Occurrences*

In most cases, the spread or outbreak of vector-borne diseases is caused by people changing places. In the fourteenth to nineteenth centuries, colonization and travel were the primary causes of the outbreak. Today, in addition to the impact of travel and climate change, urbanization is also a significant contributing factor. Whenever a country or region is being affected due to the epidermis, the economy is also being affected. Not only the mortality rate is increasing, but also the public health and ecosystem are being affected, leaving a long-term imbalance.

Following a thorough analysis of the causes of the outbreak and the historical context of the diseases, the governing body, the Audi Healthcare Organization, has established a set of guidelines that individuals should adhere to in order to prevent vector-borne diseases (Githeko et al., 2000). Current methodologies have taken into account the following changes:

1. Awareness among the people.
2. Setting strict rules and the regulations.
3. Surveillance and monitoring of the places to control the pests and the outbreak of disease.
4. A proper management system to control the vector-borne diseases.
5. Use of vaccination to prevent.
6. Research and analysis for futuristic approach of prevention and modification of the present methodologies.

3 Pests and Vector-Borne Diseases

In order to comprehend the far-reaching effects of pests and vector-borne illnesses on ecology, economics, and human health, research into these topics is essential. Public health planning and protection, dealing with chronic health disorders, and controlling co-infections and other diseases that pests may spread all rely on these kinds of research from a health standpoint. In order to create health plans and interventions that work, this information is crucial. Significant loss of production, higher

Table 1 Historical outbreaks

Year	Disease	Vector	Pest	Region	Symptoms	Reference
1346–1953	The black death	Fleas (*Xenopsylla cheopis*)	Rats	Europe, Asia, North America	Fever, coughing with blood, abdomen muscle pain, chills, headache, fatigue	Benedictow (2004)
1492	Malaria	*Anopheles* mosquitoes	Mosquitoes	America	Fever with shivering, chills, fatigue, muscle pain	Guerra (1993)
1793	Yellow fever	*Aedes aegypti* mosquitoes	Mosquitoes	Philadelphia USA	Yellow skin, fever, chills, fatigue, headache, loss of appetite	Kornfeld (1984)
1878	Yellow fever	*Aedes aegypti* mosquitoes	Mosquitoes	Mississippi valley	Yellow skin, fever, chills, fatigue, headache, loss of appetite	Nuwer (2015)
1897–1898	Malaria	*Anopheles* mosquitoes	Mosquitoes	Colonial Africa	Fever with shivering, chills, fatigue, muscle pain	Webb Jr (2011)
1918–1919	Malaria	*Anopheles* mosquitoes	Mosquitoes	Europe eastern front and Italy	Fever with shivering, chills, fatigue, muscle pain	Brabin (2014)
1930	Malaria	*Anopheles* mosquitoes	Mosquitoes	South east Asia	Fever with shivering, chills, fatigue, muscle pain	Brabin (2014)

(continued)

Table 1 (continued)

Year	Disease	Vector	Pest	Region	Symptoms	Reference
1950s	Malaria	*Anopheles* mosquitoes	Mosquitoes	Globally	Fever with shivering, chills, fatigue, muscle pain	Hamoudi and Sachs (1999)
2004	Chikungunya	*Aedes* mosquitoes	Mosquitoes	Indian ocean	Joint pain headache and skin rash	Schuffenecker et al. (2006)
2015	Zika virus	*Aedes aegypti* mosquitoes	Mosquitoes	Latin America and Caribbean	Fever rack headache joint and muscle pain red eyes	Pettersson et al. (2016)
2017	Plague	Fleas (*Xenopsylla cheopis*)	Rats	Madagascar	Fever, coughing with blood, abdomen muscle pain, chills, headache, fatigue	Parra-Rojas and Hernandez-Vargas (2018)
2019	Dengue	*Aedes aegypti* and *Aedes albooictus* mosquitoes	Mosquitoes	Asia Latin America Caribbean	Headache, muscle bone and joint pain, nausea vomiting, pain behind the eyes	Tsheten et al. (2021)

healthcare expenditures, and severe effects on agriculture are some of the economic impacts of pests and vector-borne illnesses. Trade and tourism are also affected by these problems, which have an impact on the stability of the global economy. Finally, pests have an impact on ecological monitoring, biodiversity protection, and interactions within ecosystems. Preserving ecological equilibrium requires management of invading species. In sum, these factors show why thorough research on pests and vector-borne diseases is crucial.

3.1 How Pests Cause Vector-Borne Diseases

Broadly speaking, there are two types of vector-borne disease transmission from pests to living organisms (such as humans): Mechanical and biological processes. As part of the biological process, when a pest or vector feeds on the blood of an infected host, it consumes the pathogens. The pathogen undergoes adaptation and multiplication as it enters the vector, eventually reaching an infectious stage. While this is happening, the vector and the pathogen are interacting physically and biochemically. When a pathogen reaches maturity, its vector may pass it on to a new host; this process continues throughout the lifespan of the vector, which makes it infectious. Malaria is a good illustration of this because mosquitoes are the carriers of the disease (Chandrasegaran et al., 2020). As part of the mechanical process, the vector is not harboring the virus inside; rather, it is the external carrier of the infectious agent that spreads it to other organisms via passive or direct contact. The pathogen within the vector does not undergo any modifications or multiplication throughout this phase. This is exemplified by the house fly, which carries several germs on its body and may thus transmit illnesses (Khamesipour et al., 2018). These two processes demonstrate how important vectors are in the spread of illness.

3.2 Types of Pests Causing Health Issues

Vector-borne illnesses are disseminated by several pests (Table 2) that present considerable health hazards worldwide. Significant categories include insects such as fleas, ticks, mosquitoes, bed bugs, cockroaches, and flies, which transmit illnesses like malaria, dengue, Lyme disease, and typhus. Rodents such as rats and mice are implicated in the transmission of illnesses like leptospirosis and plague. Arachnids, encompassing spiders and mites, are implicated in illnesses like scabies. Nematodes, such as roundworms and filarial worms, induce parasitic illnesses, including filariasis. Lice, encompassing head, body, and pubic varieties, spread illnesses including typhus. Birds and bats, including pigeons, may transmit illnesses such as rabies, but snails and mollusks can harbor parasitic infections like schistosomiasis. Effective pest management and sanitation are crucial for averting the transmission of these vector-borne illnesses.

3.3 Types of Emerging Pests

Emerging pests indicate that specific kinds of pests have increased in number over time and have developed resistance to the traditional pest control method, resulting in more harmful effects. These pests can be classified into three types based on this emerging population category. They are, respectively, insects, rodents, and weeds

Table 2 Summary of vector-borne diseases and their treatment

Disease name	Causative pest	Pathogen	Symptoms (effect on the body)	Treatment	References
Malaria	Mosquitos	Protozoa of the genus *Plasmodium*	Fever with shivering, chills, fatigue, muscle pain	Administration of antibiotics and antiparasitic	White (1996)
Dengue	Mosquitos	*Dengue virus* (*Flavivirus*)	Headache; muscle, bone and joint pain; nausea; vomiting; pain behind the eyes	Analgesic	Wiwanitkit (2010)
Yellow fever	Mosquitos	*Yellow fever virus* (*Arbovirus*)	Yellow skin, fever, chills' fatigue, headache, loss of appetite	Oral rehydration therapy	Monath (2008)
Zika virus	Mosquitos	*Zika virus* (*Flavivirus*)	Fever, rack headache, joint and muscle pain, red eyes	Antipyretic, analgesia, rest, fluids (no specific treatment)	Musso and Gubler (2016)
West Nile virus	Mosquitos	*West Nile virus* (*Flavivirus*)	Fever, fatigue, red spots blister on the skin, headache, sore throat	Analgesic, antihistamine	Petersen et al. (2013)
Chikungunya	Mosquitos	*Chikungunya virus* (CHIKV)	Joint pain, headache and skin rash	Analgesics, nonsteroidal anti-inflammatory agents	de Lamballerie et al. (2009)
Lyme disease	Tick	*Borrelia burgdorferi* (bacteria)	Stiffness and pain in joints, fever, fatigue, bull's eye pattern rash	Nonsteroidal anti-inflammatory agents, antibiotics	Bratton et al. (2008)
Rocky mountain spotted fever	Tick	*Rickettsia rickettsii* (bacteria)	Fever, muscle pain, nausea, eye redness, rashes	Antibiotics	Dantas-Torres (2007)
Ehrlichiosis	Tick	*Ehrlichia* spp. (bacteria)	Fever, muscle pain, severe headache	Antibiotics (most preferred doxycycline)	Blanco and Oteo (2002)
Anaplasmosis	Tick	*Anaplasma phagocytophilum* (bacteria)	Fever headache, malaise	Antibiotics (most preferred doxycycline)	Bakken and Dumler (2015)

(continued)

Table 2 (continued)

Disease name	Causative pest	Pathogen	Symptoms (effect on the body)	Treatment	References
Bubonic plague	Fleas	*Yersinia pestis (bacteria)*	Muscle pain, cough with blood, diarrhea, bleeding, fever	Antibiotics	Nikiforov et al. (2016)
Murine typhus	Flea of the rodent	*Rickettsia typhi* (bacteria)	Fever, body ache, headache, loss of appetite	Antibiotics (most preferred doxycycline)	Tsioutis et al. (2017)
Cat scratch fever	Infected cats (infected by fleas)	*Bartonella henselae* (bacteria)	Fever, joint pain, swollen lymph	Antibiotics	Greer and Keefer (1951)
Typhoid	Flies	*Salmonella typhi* (bacteria)	Muscle pain, nausea, constipation, fever headache. Loss of appetite	Antibiotics, fluid	Nicoll (1917)
Cholera	Flies	*Vibrio cholerae* (bacteria)	Nausea, severe diarrhea, abdomen pain, dehydration	Antibiotics, rehydration, IV fluid	Fotedar (2001)
Trypanosomiasis	Flies	Protozoan of the genus *Trypanosoma*	Recurrent fever, swollen bump, chills, headache	Fexinidazole, NECT, pentamidine, suramin	Lambrecht (1964)
Food poisoning	Cockroaches, houseflies	*E. coli, Salmonella typhi, Staphylococcus, Shigella*	Pain in abdomen, dehydration, bloating, diarrhea, headache, weakness	Oral electrolytes, activated charcoal, metoclopramide, azithromycin	Donkor (2020)
Gastroenteritis	Cockroaches	Norovirus, calicivirus, rotavirus, astrovirus, adenovirus	Diarrhea, abdominal pain, fever, muscle pain loss of appetite	Drinking clear fluid with electrolytes	Adeleke et al. (2012)
Leptospirosis	Rats and mice	*Leptospira* sp.	Jaundice, red eyes, diarrhea, rash, chills, fever	Doxycycline, azithromycin, ceftriaxone, ampicillin	Priya et al. (2007)
Hantavirus	Rats and mice	Orthohantavirus	Fever, fatigue, muscle pain, nausea, diarrhea, dizziness	No specific treatment available	Mir (2010)
Lymphocytic choriomeningitis	Rats and mice	*Lymphocytic choriomeningitis virus*	Anorexia, nausea, vomiting, photophobia, malaise, myalgia	No specific treatment available	Bonthius et al. (2002)
Salmonellosis	Rats and mice	*Salmonella* sp.	Diarrhea, fever, chills, headache, rash, lethargy, blood in urine or stool	Antibiotics such as ampicillin, fluoroquinolone, ciprofloxacin, ceftriaxone	Lapuz et al. (2008)

Disease name	Causative pest	Pathogen	Symptoms (effect on the body)	Treatment	References
Necrosis	Spider	Sphingomyelinase D (component of the venom of *Loxosceles* spp.)	Headache, nausea, vomiting, increased sweating, blister, itching or rash	NSAIDS, antihistamines, antibiotics	Baniardalani et al. (2020)
Scabies	*Sarcoptes scabiei*	Mite's presence triggers a significant immune response	Itchy skin, rash, sores	Permethrin cream, benzyl benzoate lotion, sulfur ointment, lidane lotion	Hicks and Elston (2009)
Ascariasis	*Ascaris lumricoides*	*Ascaris lumricoides*	Fever, abdominal pain, bloating, weight loss, anorexia, nausea	Mebendazole, albendazole, or any other anthelminthic drugs	Lamberton and Jourdan (2015)
Hookworm infection	Hookworm	*Ancylostoma duodenale* and *Necator americanus* (Parasites)	Diarrhea, anemia, weight loss, abdominal pain, coughing	Mebendazole, albendazole or any other anthelminthic drugs	Brooker et al. (2004)
Filariasis	Mosquitoes	*Wuchereria bancrofti*	Fever, skin exfoliation, inguinal pain, genital swelling, lymphadenopathy	Diethylcarbamazine	Sasa (1976)
Cryptococcosis	Cats	*Cryptococcus neoformans*	Headache, fever, neck pain, nausea, cough, chest pain	Flucytosine and liposomal amphotericin B fluconazole	Malik et al. (1992)
Parrot fever	Birds	*Chlamydia psittaci*	Fever, headache, muscle pains, cough, pneumonia	Antibiotics such as doxycycline, azithromycin, erythromycin	Fever (2003)
Ebola	Fruit bats	*Ebola virus*	Diarrhea, fever, vomiting, chest pain, abdominal pain, eye irritation, hiccups	Two monoclonal antibodies such as Inmazeb and Ebanga	Olivero et al. (2020)
Marburg	Fruit bats	*Marburg virus* (*Filoviridae*)	Fever, chills, rash, jaundice, severe weight loss, hemorrhaging	No specific treatment, patients are generally treated in ICU	Amman et al. (2015)
Schistosomiasis	Blood flukes (Schistosoma)	Blood flukes (*Schistosoma*)	Fever, rashes, cough, diarrhea, tummy pain, blood in urine and stool	Praziquantel	Leshem et al. (2008)

(continued)

Table 2 (continued)

Disease name	Causative pest	Pathogen	Symptoms (effect on the body)	Treatment	References
Trench fever	Body lice	*Bartonella quintana*	Fever, weakness, dizziness, conjunctival injection, severe back and leg pains	Antibiotics such doxycycline and ceftriaxone	Anstead (2016)
Relapsing fever	Ticks and lice	Spirochetes belonging to *Borrelia* sp.	Fever chills, malaise, arthralgias, myalgias, nausea, diarrhea	Antibiotics such as doxycycline, tetracycline, or azithromycin	Cutler (2015)
Epidemic typhus	Body lice	*Rickettsia prowazekii*	Fever and chills, headache, rashes, muscle ache, rapid breathing	Antibiotic doxycycline and chloramphenicol	Bechah et al. (2008)

(new species). These types of pests are mostly found in the crop-vegetating and infested areas. As these places can be sources of food for them, but for human beings, it is becoming the origin of many unwanted diseases and also affecting crop health and the economy by corrupting the food material, affecting ecology by spreading disease while being carriers of it.

3.3.1 Insect Pests

The emerging insect pests are those species of insects that have grown in number in large range due to climate change on earth, accidental introduction of pests to the new environment, habitat destruction of those species, monoculture farming, etc.

In the present scenario, many insects can be seen creating a worldwide problem by attacking the crops, which is affecting the economy rapidly and significantly.

A very few of the incidents that are significantly threatening to the agriculture of the region are given below in a manner of table (Table 3).

Table 3 Emerging pests, their regions, and affected vegetation

Emerging pest	Scientific name	Region	Type of vegetation being affected	Reference
Fall armyworm	*Spodoptera frugiperda*	Native America (origin), Africa, Asia, Australia	Maize, rice, sorghum	Day et al. (2017)
Spotted lanternfly	*Lycorma delicatula*	Native to Asia, invasive in USA	Fruit crops, ornamental plants, hard wood trees	Barringer and Ciafré (2020)
Brown marmorated stink bug	*Halyomorpha halys*	Native to east Asia, invasive in north America and Europe	Fruits and vegetable (wide variety)	Leskey and Nielsen (2018)
Asian long horned beetle	*Anoplophora glabripennis*	Native to Asia, invasive in north America and Europe	Kills hard wood trees	Haack et al. (2010)
Emerald ash borer	*Agrilus planipennis*	Native to Asia, invasive to north America	Destroys ash trees	Dang et al. (2022)
Tomato leafminer	*Phthorimaea absoluta*	Native to south America, Invasive to Europe, Africa, Asia	Tomato productions are in threat	Zhang et al. (2021)
Desert locust	*Schistocerca gregaria*	Africa, middle east, south Asia	Massive crop destruction in a short amount of time	Lecoq (2005)

3.3.2 Rodent Pests

Rodent specifies rat and mouse species, which are now an increasing threat in agricultural places as well as food storage new areas along with context. It is expanding its range based on the food availability while destroying the food for the human population and spreading various vector-borne diseases. It is not only seen in countryside regions but also in urban areas. It has the capability to damage and destroy the agricultural land and also the wetland ecosystem, which causes the depletion of agricultural crop quality and also affects the ecological balance of the earth. This rodent issue is an uprising matter because it has developed resistance against the traditional rodent controlling method, making it less effective.

In the present state, there are many species of rodents, but the black rats (*Rattus rattus*) are most able to adapt to urban environments while resisting the common control methods, which is now an emerging threat to urbanization. However, nutrient rodents are affecting the ecosystem and agricultural activity of America. Similarly, Gambian pouched rats are potential agricultural pests as well as competitors to native species in Africa. Though the invasion of rodents is not as emerging as the insects, it still has the potential to affect the economy and ecology in the same manner (Stenseth et al., 2003; Brown et al., 2010).

3.3.3 Invasive Plant Species

Invasive plant species are those that can outcompete native species through rapid growth, reproduction, and dispersal mechanisms in a new environment, while also adapting to various environmental conditions. It impacts the ecology of the region by reducing the biodiversity of the native plant species alongside altering the ecosystem function and natural habitat structure of the wild life. In agriculture, it reduces the desired crop yield while increasing the cost of weed control. Consequently, the region's economy suffers as the property value in heavily infested areas declines. In some cases, it has a tendency to cause allergies or dermatological issues in humans. As a result, the invasive new plant species is becoming an emerging pest in certain regions, particularly tropical and subtropical areas. Examples of these include water hyacinth, garlic mustard, honeysuckle, buckthorn, and common ivy (Chun et al., 2010; Iacarella et al., 2015).

3.4 Disease Vectors

As discussed, earlier vectors are those pests that are the carriers and transmitters of the pathogen that can cause ailment in the human population. There are several types of vectors that are responsible for spreading disease, but the most common insect vectors are mosquitoes, ticks, and flies.

3.4.1 Mosquitoes

Mosquitoes are one of those insects that are not only becoming large in population in the suburban and rural side but also in urban areas, making them a non-specific region emerging pest worldwide. It causes many vector-borne diseases that affect the patient severely, such as malaria, dengue, chikungunya, zika fever, and West Nile fever. These diseases have common symptoms of muscle pain, headache, fever, and fatigue, which leave the patient weak for certain time duration. The natural habitat of mosquitoes is stagnant water bodies, so it is suggested to not let any type of standing water remain collected in an area for more than 5–7 days (Tolle, 2009).

3.4.2 Ticks

Ticks are the parasites that live on the host body and consume the warm blood of the host as sustenance. They tend to infect humans by transmitting bacteria, protozoa, and viruses. At the moment, Lyme disease is in the USA. The latest rise was reported in 2023, causing nationwide health problems (Mead et al., 2024).

In addition to this, the Powassan virus is taking a toll. Headache, vomiting, and weakness are the primary symptoms. There is no proper vaccine or treatment available for this particular ailment, making it an uprising threat to health. Sometimes it has no symptoms. If left untreated, it can lead to brain swelling, which ultimately results in the patient's death (Hermance & Thangamani, 2017). Babesiosis, another tick-borne disease, has spread to over 10 states in the northeast and Midwest. Though with the combination drug therapy of the antibiotics and antiparasitic, it can be controlled and treated (Krause, 2019).

3.4.3 Other Vectors

There are many more vectors present in the environment aside from mosquitoes and ticks, such as flies and lice, which can also cause diseases, as mentioned in earlier section. Trypanosomiasis, or sleeping sickness, is caused by infected testes flies (a protozoan pathogen) causing recurrent fever in the patient (Lambrecht, 1964).

Lice also actively participate in the spread of vector-borne diseases like epidemic typhus and trench fever, resulting in fever at first, then fatigue and weakness in the body (Bechah et al., 2008; Anstead, 2016).

4 Factors Contributing to the Emergence of New Pests and Vectors

As we previously discussed, the importance of emerging pests and vectors in the world cannot be denied. The population, economy, ecology, health condition, and environment are all impacted, resulting in a lasting impact on the region. Various factors, either directly or indirectly, impact the emergence (Fig. 1), and we will discuss them below:

4.1 Climate Change

The principal cause of the change in climate is global warming, which is elevating the average temperature of the earth, resulting in the range and reproduction expansion of the pests. The warmer the temperature, the more hospitable the condition for the pest to survive and also accelerates the life cycle of the pest, leading to more generation per year comparatively before the global warming era. The increased average temperature also alters the precipitation rate worldwide, which alters the overwintering condition of a few regions, leading to a higher population of the pest. The change in precipitation is responsible for extreme weather conditions such as droughts, floods, and storms, which are not only disrupting the normal ecosystem but also causing pest outbreaks, resulting in invasion of pests in new regions (Skendžić et al., 2021).

Climate change also elevates the carbon dioxide level universally, which is changing the plant chemistry by making the crops more susceptible to a handful of pests. The nutrition quality is also being altered in the crops due to changes in

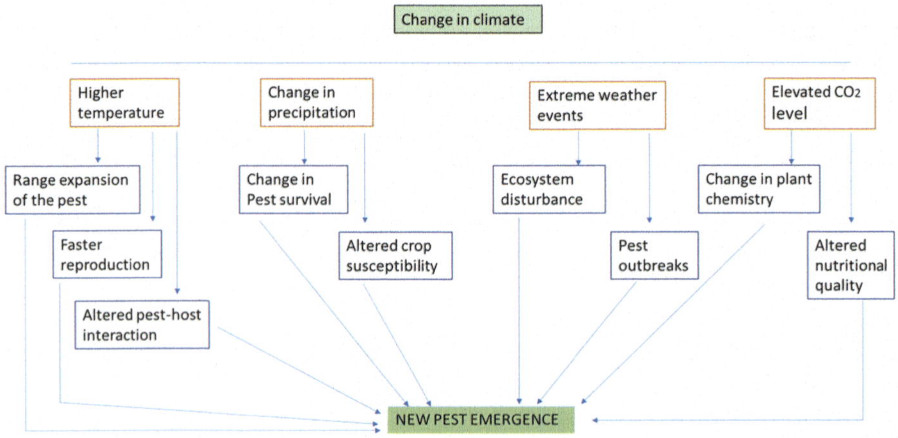

Fig. 1 How the change in climate is causing pest emergence

climate, which sometimes makes them more prone to pests. The impact on changing the synchrony of the pest host depends on the specific relationship between the pest and the host.

The indirect effects are through ecosystem changes, where the altered habitat of the pest decreases the natural enemies of the particular pest species. The natural pest control mechanism is also under question due to changes in biodiversity. The human adaptation strategy for the altered crop choices and cultivation practices is also opening the emergence of new pests and vectors (Forrest, 2016; El-Sayed & Kamel, 2020).

4.2 *Urbanization and Habitat Changes*

Urbanization and deforestation are internally related. The start of the urbanization was by cutting down the trees in the forest areas. As we know, forests are the natural habitat of pests and vectors. Therefore, the reduction in forest areas is leading to a scarcity of food for these pests and vectors, which in turn is driving them to urban areas in search of food.

As the agricultural crop field is filled with food components, it makes it easier for the pests to attack and consume. Deforestation destroys not only the food but also the habitat, enabling pests to explore and infiltrate new areas, where they can succumb to infected pathogens and act as carriers of diseases.

We, as humans, are accountable for the pest invasion in our habitat, as we have destroyed it to meet our basic needs (Schmitt & Burghardt, 2021; Guegan et al., 2023).

4.3 *Agricultural Practices*

The change in agricultural practices is due to the change in environment and change in demand. But the traditional all-cultivating method was not well in every way. Monoculture, one of the traditional agricultural practices, negatively impacts the environment.

Monoculture is single crop culturing in a particular region, and though it takes comparatively less effort than multiculturing, it affects the soil profile and the environment negatively, resulting in a greater number of pest emergences. The culture of one crop makes it more susceptible to pests and vectors by providing the same type of nutrients, which makes it easier for them to attain food (Altieri et al., 1984).

As we all know, farming involves numerous requirements to achieve the desired crop production, one of which is the use of pesticides. Pesticides help in the prevention of unwanted pests, which can damage production and delay the supply of crops. Observations show that recurrent use of a similar type of pesticide leads to pest resistance, rendering the pesticide ineffective. This is a contributing factor to the pests' resistance to traditional pest control methods. So, to prevent the reoccurring pest attacks on the crops, different pest-control methods are being suggested (Davies et al., 2012; Delcour et al., 2015).

5 Spread of Invasive Pests and Vector-Borne Diseases Due to Climate Change

Pest invasion in a region can be attributed to several factors. Climate change, including shifts in temperature, precipitation patterns, and extreme weather events, can alter habitats and make them more suitable for pests. Human activities such as global trade, travel, tourism, habitat modification, and agricultural practices can introduce pests into new regions. Ecological factors like ecosystem disturbance, niche availability, and resource abundance also create conditions favorable for pest establishment. Additionally, the biological characteristics of pests, including rapid reproduction, adaptability, and the absence of natural predators, contribute to their spread. Lastly, a lack of natural control, such as effective dispersal mechanisms, a lack of competing species, or the absence of environmental stressors, further facilitates pest invasion.

5.1 *Possible Reasons for Invasion*

There are multiple reasons for the invasion of pests in a particular region, which sometimes work in a combination manner, creating a complex scenario. We can broadly categorize it in several ways, including the following:

- Climate Change: The elevated temperature due to global warming alters the ecosystem of nature while providing better conditions for the reproduction of pests and vector-borne diseases (El-Sayed & Kamel, 2020).
- Human Activity: Global trade, travel, and tourism are the accidental transport of pests, which cause uprisings of the pest invasion. Deforestation and agricultural practices alter the habitat, paving the way for pest invasions in other regions of their native origin (Guegan et al., 2023).
- Ecological Factor: Ecosystem disturbance due to natural and human activities, niche availability in different regions, and food resources in agricultural fields provide the optimum conditions for the invasion (Altieri et al., 1984).
- Biological characteristics of the pest: Rapid reproduction, adaptability to the different environment, and quick spread contribute to the invasion (Loucks, 1977).
- Others: Lack of natural control such as scarcity of natural enemies and competing species; lag phase between the introduction of the pest to the new region and invasion; genetic diversity provides the time period to reproduce without being noticed (Bajwa et al., 2020).

5.2 *Reasons for Their Undesirability*

Pests' undesirability is deeply rooted in various unwanted impacts on human activities, ecology, and so on. The reasons for the pests' inexpediency are listed below.

- Economic Impact: Loss of crop yield, infrastructure damage, pest management costs, and restrictions on trade create the reason for not being wanted as it is affecting the economic condition of a particular region.
- Environmental and health impacts: loss of native species and biodiversity, alteration of ecological function and services, habitat destruction of native species, vector-borne diseases to humans and animals, allergen production (cockroach), toxin production (weed), etc. Thus, the invasion of pests has a bad impact on the environment and the population's health, resulting in the undesirability.
- Social and Cultural Impact: It can store food and reduce historically significant structures, as well as the beauty of the landscape.
- Others: Psychological impact by causing stress and anxiety, competition of resources with the pest, significantly affecting the lives of the affected area, etc.

The pest is inconvenient and affects the lives of those around it. As a result, pest control and management are essential (Haddi et al., 2020).

5.3 *Impact on Population and Health*

Pest- and vector-borne disease invasions have a significant impact on population and health, broadly categorized into two types: direct and indirect. Directly it can impacts the food security and by affecting the food it can increase disease outbreaks. Due to the numerous outbreaks the demand of the medical devices and health care is also uprising day by day. The loss of biodiversity due to invasion cannot be denied. Thus, the social impact is clearly another negative aftereffect of invasion. In case of indirect impacts, the economical side-effects, elevated community disruption, poor nutrition in the population and increased long-Term public health challenges are causing undeniable damage. The key impacts on population and health are outlined below:

- Increased disease: When the invasion of pests is rising, then the number of vectors is also increasing. Therefore, more vector-borne diseases outbreak among the the population.
- Reduced food security: Due to the pest attack, there is a shortage of food causing malnutrition in the population and increased food prices in agriculturally dependent regions or countries like India.
- Ecological degradation refers to the loss of biodiversity resulting from the destruction of native species' habitats, water contamination by invasive pests, and the introduction of pathogens via vectors, all of which degrade the natural ecological balance of that region.

- The social impact of the increasing number of vector-borne diseases and widespread crop failures creates distress among people, leading to a psychological burden that negatively impacts the population's social health.
- Demand for medical services: With the outbreak of disease, the demand of the local healthcare system as well as registered physicians increases, which leads to a shortage of medical supplies and facilities.
- Economical effect: The destruction of agricultural products leads to the loss of the agricultural community, causing economic distress. Also, due to the inadequate medical services (as they are in high demand), the cost of the healthcare facility is increasing rapidly, which is impacting the economic condition of the population.
- Community disruption: The cost of the medical facility creating a wedge between the economically weaker section and the rich. The multiple outbreaks of disease sometimes compel people to migrate from one place to another, resulting in resource competition and overcrowding.
- Poor Nutrition: The destruction of crops and livestock reduces the food security and availability, leading to poor nutrition available for most of the population. Stunted growth and development delays are the consequences of such a situation.
- Long-term public health challenge: Overusing antibiotics for vector-borne diseases causes antibiotic resistance (Mondal et al., 2019), making the disease difficult to control (Savary et al., 2012; Hassaan & El Nemr, 2020; Hosseinzadeh-Bandbafha et al., 2022).

5.4 Long-Term Effects

Pest infestations and vector-borne illnesses have significant and enduring effects on several facets of society and the environment. The long-term repercussions are evident in health, social structures, ecosystems, future generations, and economic stability. Figure 2 depicts the interrelated repercussions, emphasizing persistent health problems, ecological disturbance, cultural degradation, economic decline, and obstacles encountered by future generations due to these invasions (Banks, 2004; Holmes et al., 2009; Soliman et al., 2010; Mazza et al., 2014) .

5.5 Number of Average Cases Reported in Past Years

According to reports from the WHO and CDC in 2023, the number of cases affected by vector-borne diseases were as follows: Malaria affected 249 million individuals (Lubanga et al., 2024), Dengue impacted 300 million, Typhoid ranged between 11 to 21 million cases, Yellow Fever affected 200,000 people, Plague impacted 3000, Chikungunya had 450,000 cases, Cholera affected 535,000, Hantavirus ranged between 150,000 to 200,000 cases, and Lyme Disease impacted 63,000 individuals (CDC, 2024).

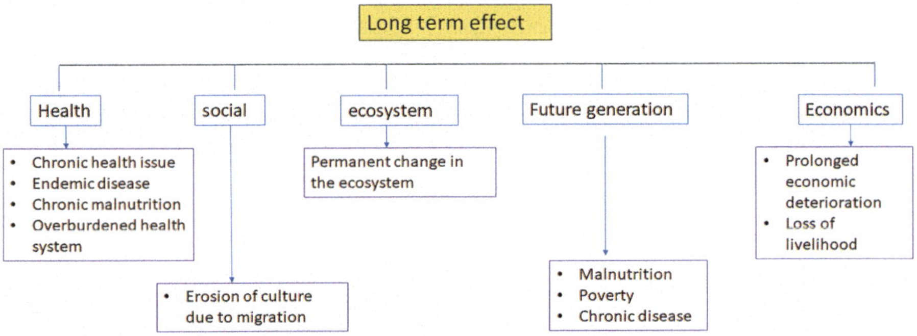

Fig. 2 Long-term effects of the pest invasion and vector-borne disease

6 Management and Control Strategies

6.1 Monitoring and Surveillance

- Importance of monitoring

When there is a monitoring system, it is availing of early-stage detection of the pest. This aids in safeguarding the product from damage and enables us to manage the pest before it poses a threat to the food material.

Controlled usage of pesticides: As they are monitoring the pest and vectors, they are aware of their whereabouts. Sodhi farmers apply pesticides precisely, aiding in the control of pesticide resistance by avoiding overuse in the field.

Public health protection: It involves surveillance and monitoring to aid in pest control, which in turn helps control disease. As a result, food safety increases.

Economical factor: It involves targeting the use of pesticides and enhancing food safety, thereby saving lives. Increasing pest control through monitoring and violence tends to lead to the importation of more products from outside the region, resulting in increased market profits.

Rules and regulations: The pest control method must follow the standard rules given by the official government authorities, and all the methods and the amount of pesticide usage, etc., should be abided by law.

Environmental factor: The excess use of pesticides and other chemical pest control methods always has a negative impact on the environment, so when there is a monitoring and surveillance system to control the invasion of pests and vectors, it gives a positive output (Petrovskii et al., 2014; Poland & Rassati, 2019; Preti et al., 2021) (Fig. 3).

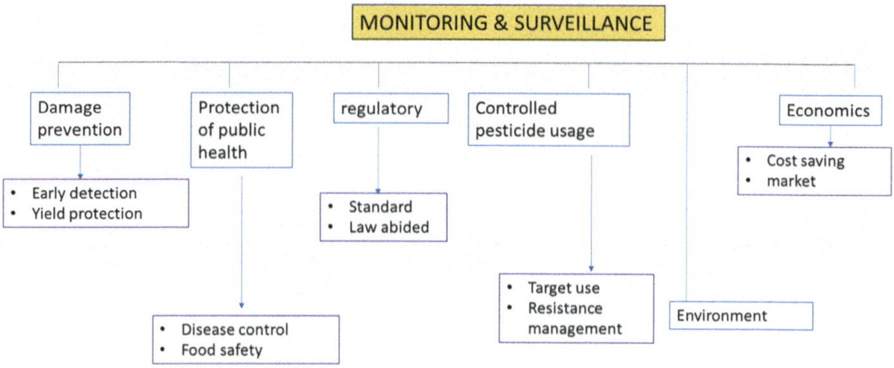

Fig. 3 Monitoring and surveillance for the pest control

- Early-stage detection

One of the major components of the monitoring and surveillance system is only detection, which simplifies its implementation in the field. There are several methods that are used for early detection; some of them are listed below:

Visual inspection: routine scouting, trap monitoring, etc., are done to visually inspect the presence of pests in the field.

Trapping: it can be divided into three types: pheromone trap, light trap, and sticky trap. In pheromone traps, the pheromones of the pest are used to attract the pest and trap them. Whereas in a sticky trap, the best sticks on the trap and becomes unable to move. In light traps, only the nocturnal type of insect is trapped by using the source of light (e.g., moths, beetles).

Remote inspection: the use of satellites and drones to detect the best where the control system is far from the field is known as remote inspection.

Acoustic inspection: here listening devices and vibrating sensors are used to detect the presence of the pest in the field.

Sampling test: the soil and the plants are sent to test as a sampling test to detect the presence of the pest around the yield area.

Molecular inspection: PCR, LAMP (loop-mediated isothermal amplification), and DNA barcoding are used to detect the pest.

Aside from these, there are many other detection methods, such as chemical analysis, climate and weather models, predictive models, etc. (Azfar et al., 2015; Lima et al., 2020; Nagar & Sharma, 2020; Elakya & Manoranjitham, 2022).

Early Detection Methods for Pests and Diseases

Early detection of pests and diseases is crucial for effective agricultural management. Various methods can be employed to identify and monitor potential threats to crops. These include the following:

- *Sample Inspection*: Examining soil and plant specimens for signs of infestation or infection.

- *Acoustic Inspection*: Using listening devices and vibration sensors to detect pest activity.
- *Molecular Inspection*: Employing techniques like PCR and DNA barcoding to identify pathogens at the molecular level.
- *Remote Inspection*: Utilizing satellites and drones for aerial surveillance and visual inspection of crops.
- *Trapping*: Employing pheromone traps, light traps, and sticky traps to capture specific pests.

By combining these approaches, early detection can help prevent significant damage to crops and control pest populations effectively.

6.2 Pest Surveillance and Management Strategies

Pest management is the foundation of controlling pests and prevention is a copy of thorough monitoring and surveillance that give the earliest sign of pest invasion. By applying these measures, the damage caused by pests in the crops is reduced thus increasing production and quality of foods crops.

6.2.1 Pest Surveillance

Pest surveillance is essentially the process of monitoring pest populations and recording their activities over time. The essential elements of pest surveillance include detecting the presence of pests, tracking the population's changes over time, monitoring their activities, and modifying these activities to facilitate easy control over the population and their attacks in the field. Controlling the test is the most effective way to prevent the spread of waste, vectors, and diseases, ensuring the food quality remain high. So, it also affects the economic prospects (Poland & Rassati, 2019).

6.2.2 Presence of Specific Pests in Crop Production

Table 4 presents the presence of specific pests in crop production.

6.2.3 Pest Control/Management Strategy

One of the major ways of controlling pests or the strategy to manage pests is having a variation in the agricultural field. As discussed in Sect. 4.3, if there is a culture of a single crop (monoculture) in the field, then it tends to attract a similar kind of pest. When there is no variation in crop, it helps in the growth of the pest via easy

Table 4 Impact of various pests on crop production and associated health issues

Pest name	Affected crops	Reason	Impact on production	Health issues	References
Tomato leaf miner	Tomato	It has host plant preference toward the Solanaceae family, as the chemical constituents help in their survival	Leaf fall at premature state, cosmetic damage to the plant, desiccation	Skin itching	Zhang et al. (2021)
Boll weevil	Cotton	As it consumes cotton as their food and place to lay eggs (in case of female species)	Fiber yield reduction, degradation of fiber grade, decreased production overall	None in human	Bottger et al. (1964)
Colorado potato beetle	Potato	As it has specific adaptation and fondness toward potatoes, which help them in their growth and survival	Damage to the leaf, yield and product loss, pesticide resistance	None in human	Alyokhin (2009)
Cabbage looper	Cabbage	For the larval feeding	Stunted growth, head aborted buds	None in human	Badenes-Pérez (2019)
Rice weevil	Rice	As they tend to feed on the root part of rice plants (behavioral condition),	Hollowed out product grains, degradation of quality, product loss, increase fungal growth	None in human	Chen et al. (2005)
Grape phylloxera	Grapevines	As they tend to feed on the root part of rice plants (behavioral condition)	Damaged root, necrotic spots on the plant	None in human	Powell (2008)
Codling moth	Apple and pear	The apple and pear provide the optimum environment for their growth	Quality degradation, damaged crop production, not easy to detect	None in human	Lacey et al. (2008)
Corn earworm	Corn	Corn provides ideal condition for the growth of the larva	Mold & fungal growth on the plant, product yield loss	None in human	Zuber and Lillehoj (1979)
Citrus greening	Citrus fruit	As the HLB disease pathogen easily affects the citrus plant, they have a preference for citrus fruit plants	death of the plan, degraded product quality	None in human	Ghosh et al. (2018)
Peach tree borer	Peach	The bark of the peach tree and the cambium layer are ideal for the borer larva as habitat	Yield reduction, death of the plant	None in human	Neiswander and Blackburn (1943)

adaptation to that particular crop. Thus, the tendency of invasion of pests increases with the possible outbreaks of disease and food shortage.

The pest control is being used in the agricultural for the very long time. The controlling methods are also developing with the time. This is the precise reason why the methods are divided into two types: traditional and modern pest controlling methods. In traditional methods, burning and mulching, trap cropping, handpicking of the pests, crop rotation, and companion plantation are done. But in the modern methods, integrated pest management (IPM), genetic modification, use of insect growth regulator, sterile insect technique, and use of biopesticide and drone can be seen.

6.3 Biological Control

The biological control depends on two types of living organisms: natural predators and natural pathogens.

Natural predators are those living species that are capable of haunting and survive on consumption of the prey. In the realm of agriculture, we employ natural predators as a substitute for pesticides, enabling them to manage pest populations and prevent the contamination of products. For this management strategy, the companion vegetation is planted alongside the agricultural products so that the natural mites and parasitic wasps for caterpillars, etc. (Gerson, 2014).

Natural pathogens, another type of biological pest control method, have the ability to infect insects and pests. Some naturally occurring bacteria, fungi, and viruses are there that can infect the insect and pest and act as a biological pest management strategy (Roy & Cottrell, 2008).

6.4 Chemical Control

Pesticides are chemical compounds that have the capability to kill the pests and vectors (which tend to cause harmful effects to humans, animals, or any living being). Using pesticides is one of the chemical control methods of pest management (Matthews et al., 2014). Figure 4 illustrates the classification of pesticides.

The significance of pesticides is inevitable. It not only controls the pest but also the vector-borne diseases, which helps in the public health management. Since foods are not destroyed by the festival due to the pesticide usage, the economic benefit is also there. The food quality remains intact. Though it has some potential risks and limits of usage, the support it provides for modern agriculture is undeniable (Leong et al., 2020).

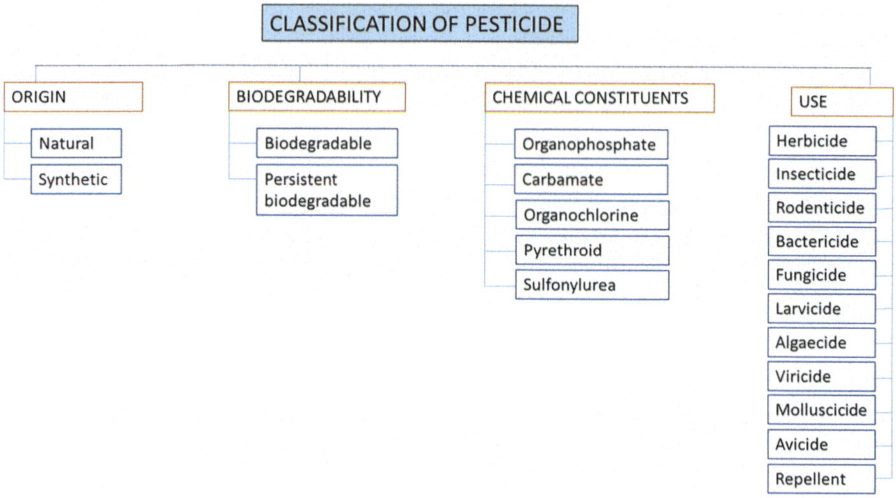

Fig. 4 Classification of the pesticide

- *Mechanism of pesticide*

There are a number of ways in which pesticides are able to suppress insect populations. Neurotoxins, which hinder the transmission of nerve signals in various ways—as acetylcholinesterase (ACHE) inhibitors, sodium ion modulators, or GABA antagonists—make up a common mechanism. A different approach focuses on the respiratory system; for example, some pesticides work as fumigants or block the electron transport chain in the mitochondria, reducing the pest's energy production capacity. In the same way as metabolic disruptors impede vital biochemical reactions, hormone blockers, such as growth regulators, upset the delicate hormonal equilibrium required for maturation and procreation. Finally, the pest's physiological processes are further hindered by signal transduction disruptors, which impact intracellular communication. By interfering with essential biological functions, each mechanism offers a unique approach to pest management or elimination (Casida, 2009, 2017; Das, 2013; Lushchak et al., 2018; Li et al., 2020).

Inhibitors of acetylcholinesterase (ACHE) prevent the enzyme from doing its regular job of degrading the synaptic cleft neurotransmitter acetylcholine. Acetylcholine builds up in the synapses of neurons when an ACHE inhibitor is present because it blocks its breakdown. Due to the inability to inhibit the overstimulation of acetylcholine receptors, this accumulation causes the ongoing propagation of nerve impulses. Because of the disruption of normal nerve function and muscle control, the pest becomes paralyzed and dies as a result. Because of the cumulative effect on the central nervous system, ACHE inhibitors are a powerful neurotoxic chemical class that is used for pest management (Main, 1979).

Pyrethroids and DDT are both effective insecticides because they work by binding to sodium channels in the bugs' nerve cells. The correct closure of sodium channels following the transmission of a nerve impulse is hindered by these substances' binding to them. This results in persistent nerve signaling because sodium keeps flowing into the nerve cells. The pest is finally overwhelmed by the hyperexcitation of its neural system caused by this prolonged and uncontrolled signaling. The hyperexcitation causes a loss of motor control and, eventually, death. Pyrethroids and DDT are efficient neurotoxins because they disturb the natural nerve activity of pests through this technique (Soderlund, 2012).

- *Controlling collateral damage*
  - The goal of targeted application is to lessen the damage that pesticides cause.
  - Rules and regulations include adhering to the law and limiting the use of pesticides to prevent any negative consequences.
  - Integrated pest management: using the pest control method in the combination of mechanical, chemical, biological, and traditional methods to balance the ecological system.
  - Selective pesticide: The prestige side, which has low toxicity and has less impact on the human animal and the ecosystem, can be a beneficial alternative.
  - Education and technique: The unique pesticide applicator can apply the pesticides at the optimal time and place without affecting the unwanted zones.
  - Maintaining a buffer zone is crucial when applying pesticides. It helps with soil profiling and keeps the water body pollutant-free (Rehman et al., 2021; Deguine et al., 2023; Kebe et al., 2023).

6.5 Biopesticides

Biopesticides are living organisms that have the ability to control the pest and its invasion. It is one of the alternatives to chemical pesticides, which does not leave a negative impact on the environment as much as pesticides (Sudakin, 2003). Figure 5 depicts the classification of biopesticides.

There are many advantages of biopesticides over conventional chemical pesticides; some of them are listed below:

(a) Less toxic.
(b) There is more diversity in the modes of action, which reduces the tendency to resist.
(c) As they are living organisms, it is biodegradable in nature.
(d) Integrated pest management includes this as one of its key elements.
(e) Sustainable pest management system.
(f) More compatible with organic farming.
(g) Lower health risks.
(h) Environment friendly.
(i) Increase food safety.
(j) Easily approved by the authorities, etc. (Kumar et al., 2021).

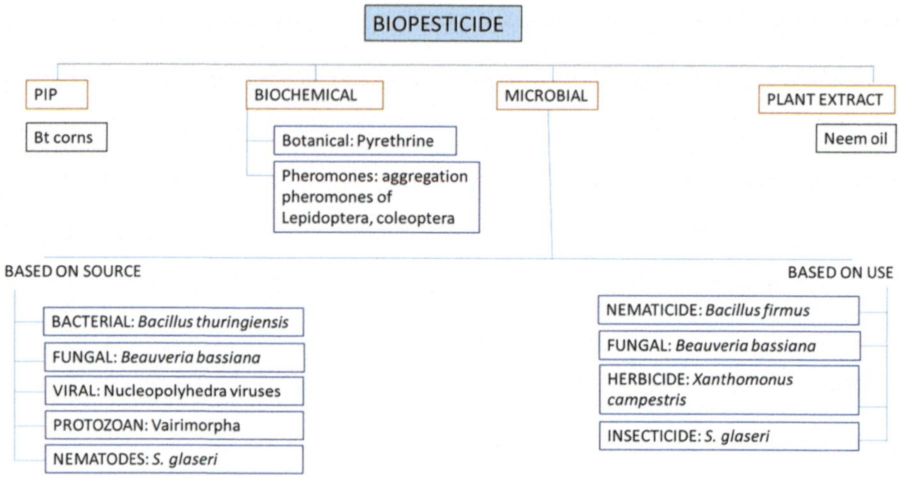

Fig. 5 Classification of biopesticide

6.6 *Integrated Pest Management (IPM)*

IPM, or integrated pest management, is a modern, sustainable approach to pest control where instead of only chemical methods, other methods such as physical, biological, and chemical methods are used in an integrated manner. It not only helps in better pest control and monitoring but is also less toxic, more environmentally friendly, and focuses on the long-term solution (Jørgensen et al., 2014).

At first, the best monitoring and surveillance are done to identify the type of pest and their activity. Then the action threshold is set based on economic conditions to prevent the excess use of pesticides. Agricultural practices are also focused on decreasing the livelihood of the pest and vector outbreaks in the field. Physical control and mechanical control take place by manually working on the field, such as handpicking and soil preparation. Natural predators and the application of biopesticides are examples of biological methods in action. In the case of chemical pesticides, targeted application, reduced usage, and pesticide selection are important factors to maintain the health of the crops.

In the current scenario, it has turned out to be a great alternative to control the pests and the vector alongside the disease outbreak (Wearing, 1988; Allen & Rajotte, 1990; Binns & Nyrop, 1992).

6.7 *Policy and Regulation*

Every country has specific boards and regulatory authorities that oversee the management and surveillance of pest control. In India, there is the Central Insecticides Board & Registration Committee (CIBRC) (Kumar et al., 2019), in Europe, the

European Food Safety Authority (EFSA), and in the US Environmental Protection Agency (EPA) for all kinds of evolution and approval (Authority et al., 2018).

There are several kinds of rules to follow for the ethical use of pest control in crop production. Here are a few examples of these rules:

- Registration approval of pesticide.
- The product is appropriately labeled, ensuring it contains all the necessary information. (For example, safety use, limitation, direction of usage, etc.) pesticide.
- Promotion of the integrated pest management.
- Training and education for the farmer or agriculture committee.
- Not using the banned pesticides.
- The provision of appropriate licenses and certificates is necessary for the marketing and production of chemical components needed for agricultural purposes.
- Maintenance of the buffer zone.
- The regional pest monitoring system is designed for the surveillance of pests and vectors.
- Quarantine regulation for the prevention of the spread of pest invasion.
- Environment protection regulations must be followed.
- Public health protection rules and regulations must be considered.
- When importing a crop, it must meet all international standards and receive approval.

The pest control and the surveillance are abided by a complex structure of rules and regulations for the protection of the respective population, environment, and economic needs. Compliance with these rules and regulations ensures the safety, quality, and efficacy of the pest control method (Foster, 1995; Waterfield & Zilberman, 2012; Lefebvre et al., 2015).

7 Impact on Human Health

Invasions of pests and vectors are dangerous to both crops and human life and thereby threatening the productivity of an economy. Such invasions contribute to the emergence of vector-borne diseases with direct consequences on the health of populations and which reinforce, in addition to psychological and social problems.

7.1 Direct Health Impacts

As previously mentioned, the invasion of pests and the emergence of vectors in the environment have led to the occurrence of numerous vector-borne diseases. The emergence of the invasive vector has put human health at risk. Many diseases such as malaria, dengue, chikungunya, and hantavirus have taken over (Daraban et al., 2023).

7.2 *Psychological and Social Effects*

Every time a pest or vector invasion causes a disease outbreak, the population experiences unavoidable social distress.

For example, whenever a pest invasion takes place, people become anxious due to the sudden outbreak of the disease. Secondly, even in the absence of an outbreak, the intrusion of pests raises concerns about the safety of food. Not only does food safety come into question, but the scarcity of food also leads to stress among people.

This entire problem due to the emergence of pests and vectors leads to psychological distress among the people, and this psychological distress sometimes leads to the migration of the people from rural or suburban areas to urban areas, causing overcrowding and competition for the resources, which results in social uncertainty and psychological effects.

Thus, the vicious cycle of psychological and social issues keeps on repeating and causing distress among the people due to the emergence and invasion of pests and vectors (Mankad, 2016; Khan et al., 2019).

8 Economic Impact

As discussed before, not only the environment and public health, but the economic state of the region also suffers losses. The damage to the crops not only increases the demand in the market but also provides signs of the scarcity of food. The imported goods (crops) do not meet the internal standard, resulting in a loss. In addition to this, the cost-effectiveness of pesticides and other pest control methods contributes significantly to the economy. Damage to crops not only affects the region but also the agricultural community and farmers, leading to significant economic disadvantages for the population.

When an outbreak of disease occurs due to pest invasion, the cost of public health care also rises, necessitating the provision of medical services and the training of medical practitioners. Regional healthcare costs encounter numerous challenges (Soliman et al., 2010; Waterfield & Zilberman, 2012).

9 Future Prospects

- Research and development: The new surge in the disease's effects on human health is still not completely known. Similarly, the detection of vector-borne diseases is not always straightforward, as the absence of symptoms can make identification challenging. So, the research and development on the treatment and the pest activities, the adaptation of the pests, and the variation in disease are needed to be studied and properly identified. The pest control methods need to be developed in a more cost-effective manner, for example, pulsation burners, insect gustatory (Ke & Chen-Zhu, 2023; Linyun et al., 2024).

- Public awareness and education are crucial, as people worldwide must be aware of the emergence of pests and the outbreak of diseases. The maintenance of proper sanitation and taking a few measurements at home can even make a big difference in the spread of the disease. The more people will be educated in this matter, the more the prevention of the disease is possible (Siddiqi, 2023; Sitar & Rusu, 2023).
- Adaptation and mitigation strategies: The agricultural community must maintain a proper management strategy to mitigate disease outbreaks, pest invasions, and the emergence of pests and their vectors. According to the situation, the adaptation technique will vary. IPM in the current scenario is a beneficial alternative, but there are many more places to modify and do better in these mitigation strategies (Bouri et al., 2023; Chiriloaie-Palade & Georgescu, 2023; Grigorieva et al., 2023).

10 Research Gaps

- The rapid adaptation mechanism of the pests to the different environments needs to be studied more so that the prevention of this problem can be prevented from the very beginning (McCulloch & Waters, 2023).
- The understanding of vector-borne diseases is that when there are no symptoms, it can be difficult to detect the disease and initiate treatment, potentially leading to a serious condition for the affected individual (Mantlo & Haley, 2023).
- The population and the invasion of pests and vectors are affected by climate change, but we need to carefully consider how climate change contributes to the establishment of new invaded species in completely different geographical areas (Ioriatti et al., 2023).
- More sustainable control methods to manage the pests and vectors (Bouri et al., 2023).
- We require more sustainable approaches to combat pesticide resistance (Manzoor et al., 2024).

11 Conclusion

Last but not least, climate change is making an already difficult situation much worse for farmers and public health workers dealing with newly emerged pests and disease vectors. In addition to wreaking havoc on food storage and production, these pests help transmit illnesses carried by vectors, such as yellow fever, dengue, and malaria. These pests are becoming more common and adaptable as a result of ongoing climate change, thus a comprehensive strategy to control them is required.

Overuse of pesticides has exacerbated the greenhouse effect and degraded the ecosystem, despite their usefulness in managing pests. Because of this, more eco-friendly and long-term solutions are required, and Integrated Pest Management (IPM) and biopesticides fit the bill. Reducing the prevalence of pests and illnesses relies heavily on public education, cleanliness, and awareness. To further our understanding of pest dynamics, refine our detection methods, and create efficient management measures, ongoing research and development are of the utmost importance.

Improving the diagnosis of asymptomatic vector-borne illnesses, filling knowledge gaps regarding pest adaption mechanisms, and creating long-term strategies to control pest populations and pesticide resistance are all critical areas to focus on moving forward. Protecting agricultural output, human health, and the environment in the face of a climate-change-shaped uncertain future may be achieved through the adoption of adaptable solutions and the promotion of innovation in pest management.

References

Adeleke, M., Akatah, H., Hassan, A., Sam-Wobo, S., Famodimu, T., Olatunde, G., & Mafiana, C. (2012). Implication of cockroaches as vectors of gastrointestinal parasites in parts of Osogbo, southwestern Nigeria. *Munis Entomology & Zoology, 7*, 1106–1110.

Allen, W., & Rajotte, E. (1990). The changing role of extension entomology in the IPM era. *Annual Review of Entomology, 35*, 379–397.

Altieri, M. A., Letourneau, D. K., & Risch, S. J. (1984). Vegetation diversity and insect pest outbreaks. *Critical Reviews in Plant Sciences, 2*, 131–169.

Alyokhin, A. (2009). Colorado potato beetle management on potatoes: Current challenges and future prospects. *Fruit Veg Cereal Sci Biotech, 3*, 10–19.

Amman, B. R., Jones, M. E., Sealy, T. K., Uebelhoer, L. S., Schuh, A. J., Bird, B. H., Coleman-McCray, J. D., Martin, B. E., Nichol, S. T., & Towner, J. S. (2015). Oral shedding of Marburg virus in experimentally infected Egyptian fruit bats (Rousettus aegyptiacus). *Journal of Wildlife Diseases, 51*, 113–124.

Anstead, G. M. (2016). The centenary of the discovery of trench fever, an emerging infectious disease of World War 1. *The Lancet Infectious Diseases, 16*, e164–e172.

Authority, E. F. S., Arena, M., Auteri, D., Barmaz, S., Brancato, A., Brocca, D., Bura, L., Carrasco Cabrera, L., Chiusolo, A., & Court Marques, D. (2018). Peer review of the pesticide risk assessment of the active substance spinosad. *EFSA Journal, 16*, e05252.

Azfar, S., Nadeem, A., & Basit, A. (2015). Pest detection and control techniques using wireless sensor network: A review. *Journal of Entomology and Zoology Studies, 3*, 92–99.

Badenes-Pérez, F. R. (2019). Trap crops and insectary plants in the order Brassicales. *Annals of the Entomological Society of America, 112*, 318–329.

Bajwa, A. A., Farooq, M., Al-Sadi, A. M., Nawaz, A., Jabran, K., & Siddique, K. H. (2020). Impact of climate change on biology and management of wheat pests. *Crop Protection, 137*, 105304.

Bakken, J. S., & Dumler, J. S. (2015). Human granulocytic anaplasmosis. *Infectious Disease Clinics, 29*, 341–355.

Baniardalani, M., Saghafipour, A., Kababian, M., & Abai, M. R. (2020). Cutaneous necrosis following brown recluse spider bite. *Iranian Journal of Dermatology, 23*, 40–42.

Banks, J. E. (2004). Divided culture: Integrating agriculture and conservation biology. *Frontiers in Ecology and the Environment, 2*, 537–545.

Barringer, L., & Ciafré, C. M. (2020). Worldwide feeding host plants of spotted lanternfly, with significant additions from North America. *Environmental Entomology, 49*, 999–1011.

Bechah, Y., Capo, C., Mege, J.-L., & Raoult, D. (2008). Epidemic typhus. *The Lancet Infectious Diseases, 8*, 417–426.

Benedictow, O. J. (2004). *The Black Death, 1346–1353: The Complete History*. Boydell Press.

Binns, M., & Nyrop, J. (1992). Sampling insect populations for the purpose of IPM decision making. *Annual Review of Entomology, 37*, 427–453.

Blanco, J., & Oteo, J. (2002). Human granulocytic ehrlichiosis in Europe. *Clinical Microbiology and Infection, 8*, 763–772.

Bonthius, D. J., Mahoney, J., Buchmeier, M. J., Karacay, B., & Taggard, D. (2002). Critical role for glial cells in the propagation and spread of lymphocytic choriomeningitis virus in the developing rat brain. *Journal of Virology, 76*, 6618–6635.

Bos, L., & Parlevliet, J. (1995). Concepts and terminology on plant/pest relationships: Toward consensus in plant pathology and crop protection. *Annual Review of Phytopathology, 33*, 69–102.

Bottger, G., Cross, W. H., Gunderson, W., & Wene, G. (1964). Recent research on the boll weevil in northern Sonora, Mexico, and the thurberia weevil in Arizona. *Journal of Economic Entomology, 57*, 286–290.

Bouri, M., Arslan, K. S., & Şahin, F. (2023). Climate-smart pest management in sustainable agriculture: Promises and challenges. *Sustainability, 15*, 4592.

Brabin, B. J. (2014). Malaria's contribution to World War One–the unexpected adversary. *Malaria Journal, 13*, 497.

Bratton, R. L., Whiteside, J. W., Hovan, M. J., Engle, R. L., & Edwards, F. D. (2008). Diagnosis and treatment of Lyme disease. In *Mayo Clinic Proceedings* (Vol. 83, pp. 566–571). Elsevier.

Brooker, S., Bethony, J., & Hotez, P. J. (2004). Human hookworm infection in the 21st century. *Advances in Parasitology, 58*, 197.

Brown, P. R., Singleton, G. R., Pech, R. P., Hinds, L. A., & Krebs, C. J. (2010). Rodent outbreaks in Australia: Mouse plagues in cereal crops. In *Rodent outbreaks: Ecology and impacts* (Vol. 225). International Rice Research Institute (IRRI).

Bryant, E. (1997). *Climate process and change*. Cambridge University Press.

Caminade, C., McIntyre, K. M., & Jones, A. E. (2019). Impact of recent and future climate change on vector-borne diseases. *Annals of the New York Academy of Sciences, 1436*, 157–173.

Casida, J. E. (2009). Pest toxicology: The primary mechanisms of pesticide action. *Chemical Research in Toxicology, 22*, 609–619.

Casida, J. E. (2017). Pesticide interactions: Mechanisms, benefits, and risks. *Journal of Agricultural and Food Chemistry, 65*, 4553–4561.

CDC. (2024). *About vector-borne diseases*. Vector-Borne Diseases.

Chandrasegaran, K., Lahondère, C., Escobar, L. E., & Vinauger, C. (2020). Linking mosquito ecology, traits, behavior, and disease transmission. *Trends in Parasitology, 36*, 393–403.

Chen, H., Chen, Z., & Zhou, Y. (2005). Rice water weevil (Coleoptera: Curculionidae) in mainland China: Invasion, spread and control. *Crop Protection, 24*, 695–702.

Chiriloaie-Palade, A., & Georgescu, R.-G. (2023). Current approaches to prevention and integrated pest control on maize crop in the context of climate change–mini review. *Romanian Journal for Plant Protection, 16*.

Chun, Y. J., Van Kleunen, M., & Dawson, W. (2010). The role of enemy release, tolerance and resistance in plant invasions: Linking damage to performance. *Ecology Letters, 13*, 937–946.

Cutler, S. J. (2015). Relapsing fever borreliae: A global review. *Clinics in Laboratory Medicine, 35*, 847–865.

Dang, Y., Wei, K., Wang, X., Duan, J. J., Jennings, D. E., & Poland, T. M. (2022). Introduced plants induce outbreaks of a native pest and facilitate invasion in the plants' native range: Evidence from the emerald ash borer. *Journal of Ecology, 110*, 593–604.

Dantas-Torres, F. (2007). Rocky Mountain spotted fever. *The Lancet Infectious Diseases, 7*, 724–732.

Daraban, G. M., Hlihor, R.-M., & Suteu, D. (2023). Pesticides vs. biopesticides: From pest management to toxicity and impacts on the environment and human health. *Toxics, 11*, 983.
Das, S. K. (2013). Mode of action of pesticides and the novel trends-a critical review. *International Research Journal of Agriculture and Soil Science, 3*(11), 393–401.
Davies, T., Field, L., & Williamson, M. (2012). The re-emergence of the bed bug as a nuisance pest: Implications of resistance to the pyrethroid insecticides. *Medical and Veterinary Entomology, 26*, 241–254.
Day, R., Abrahams, P., Bateman, M., Beale, T., Clottey, V., Cock, M., Colmenarez, Y., Corniani, N., Early, R., & Godwin, J. (2017). Fall armyworm: Impacts and implications for Africa. *Outlooks on Pest Management, 28*, 196–201.
de Lamballerie, X., Ninove, L., & Charrel, R. N. (2009). Antiviral treatment of chikungunya virus infection. *Infectious Disorders-Drug Targets (Formerly Current Drug Targets-Infectious Disorders), 9*, 101–104.
Deguine, J.-P., Aubertot, J.-N., Bellon, S., Côte, F., Lauri, P.-E., Lescourret, F., Ratnadass, A., Scopel, E., Andrieu, N., & Bàrberi, P. (2023). Agroecological crop protection for sustainable agriculture. *Advances in Agronomy, 178*, 1–59.
Delcour, I., Spanoghe, P., & Uyttendaele, M. (2015). Literature review: Impact of climate change on pesticide use. *Food Research International, 68*, 7–15.
Donkor, E. S. (2020). Cockroaches and food-borne pathogens. *Environmental Health Insights, 14*, 1178630220913365.
Elakya, R., & Manoranjitham, T. (2022). A novel approach for early detection of disease and pest attack in food crop: A review. In *Advanced computational paradigms and hybrid intelligent computing: Proceedings of ICACCP* (Vol. 2021, pp. 93–101). Springer.
El-Sayed, A., & Kamel, M. (2020). Climatic changes and their role in emergence and re-emergence of diseases. *Environmental Science and Pollution Research, 27*, 22336–22352.
Fever, P. (2003). *Psittacosis/Avian Chlamydiosis*.
Forrest, J. R. (2016). Complex responses of insect phenology to climate change. *Current Opinion in Insect Science, 17*, 49–54.
Foster, J. (1995). Government policy strategies relating to the enhanced implementation of integrated pest management: A review. *Australian Journal of Environmental Management, 2*, 234–244.
Fotedar, R. (2001). Vector potential of houseflies (Musca domestica) in the transmission of Vibrio cholerae in India. *Acta Tropica, 78*, 31–34.
Gerson, U. (2014). Pest control by mites (Acari): Present and future. *Acarologia, 54*, 371–394.
Ghosh, D., Motghare, M., & Gowda, S. (2018). Citrus greening: Overview of the most severe disease of citrus. *Advanced Agricultural Research and Technology Journal, 2*, 83–100.
Githeko, A. K., Lindsay, S. W., Confalonieri, U. E., & Patz, J. A. (2000). Climate change and vector-borne diseases: A regional analysis. *Bulletin of the World Health Organization, 78*, 1136–1147.
Greer, W. E., & Keefer, C. S. (1951). Cat-scratch fever: A disease entity. *New England Journal of Medicine, 244*, 545–548.
Grigorieva, E., Livenets, A., & Stelmakh, E. (2023). Adaptation of agriculture to climate change: A scoping review. *Climate, 11*, 202.
Guegan, J.-F., de Thoisy, B., Gómez-Gallego, M., & Jactel, H. (2023). World forests, global change, and emerging pests and pathogens. *Current Opinion in Environmental Sustainability, 61*, 101266.
Guerra, F. (1993). The European-American exchange. *History and Philosophy of the Life Sciences, 15*, 313–327.
Haack, R. A., Hérard, F., Sun, J., & Turgeon, J. J. (2010). Managing invasive populations of Asian longhorned beetle and citrus longhorned beetle: A worldwide perspective. *Annual Review of Entomology, 55*, 521–546.

Haddi, K., Turchen, L. M., Viteri Jumbo, L. O., Guedes, R. N., Pereira, E. J., Aguiar, R. W., & Oliveira, E. E. (2020). Rethinking biorational insecticides for pest management: Unintended effects and consequences. *Pest Management Science, 76*, 2286–2293.

Hamoudi, A., & Sachs, J. D. (1999). *The changing global distribution of malaria: A review*. CID working paper series.

Hassaan, M. A., & El Nemr, A. (2020). Pesticides pollution: Classifications, human health impact, extraction and treatment techniques. *The Egyptian Journal of Aquatic Research, 46*, 207–220.

Hermance, M. E., & Thangamani, S. (2017). Powassan virus: An emerging arbovirus of public health concern in North America. *Vector-Borne and Zoonotic Diseases, 17*, 453–462.

Hicks, M. I., & Elston, D. M. (2009). Scabies. *Dermatologic Therapy, 22*, 279–292.

Hitz, S., & Smith, J. (2004). Estimating global impacts from climate change. *Global Environmental Change, 14*, 201–218.

Holmes, T. P., Aukema, J. E., Von Holle, B., Liebhold, A., & Sills, E. (2009). Economic impacts of invasive species in forests: Past, present, and future. *Annals of the New York Academy of Sciences, 1162*, 18–38.

Hosseinzadeh-Bandbafha, H., Kiehbadroudinezhad, M., Khanali, M., & Taghizadehghasab, A. (2022). Emerging risks to plant health. In *Biodiversity, functional ecosystems and sustainable food production* (pp. 41–72). Springer.

Iacarella, J. C., Mankiewicz, P. S., & Ricciardi, A. (2015). Negative competitive effects of invasive plants change with time since invasion. *Ecosphere, 6*, 1–14.

Ioriatti, C., Anfora, G., Bagnoli, B., Benelli, G., & Lucchi, A. (2023). A review of history and geographical distribution of grapevine moths in Italian vineyards in light of climate change: Looking backward to face the future. *Crop Protection, 173*, 106375.

Jørgensen, L. N., Hovmøller, M. S., Hansen, J. G., Lassen, P., Clark, B., Bayles, R., Rodemann, B., Flath, K., Jahn, M., & Goral, T. (2014). IPM strategies and their dilemmas including an introduction to www. eurowheat. org. *Journal of Integrative Agriculture, 13*, 265–281.

Ke, Y., & Chen-Zhu, W. (2023). Progress in insect gustatory research and the application of related principles to pest control. *Chinese Journal of Applied Entomology, 60*, 486.

Kebe, A. A., Hameed, S., Farooq, M. S., Sufyan, A., Malook, M. B., Awais, S., Riaz, M., Waseem, M., Amjad, U., & Abbas, N. (2023). Enhancing crop protection and yield through precision agriculture and integrated pest management: A comprehensive review. *Asian Journal of Research in Crop Science, 8*, 443–453.

Khamesipour, F., Lankarani, K. B., Honarvar, B., & Kwenti, T. E. (2018). A systematic review of human pathogens carried by the housefly (Musca domestica L.). *BMC Public Health, 18*, 1–15.

Khan, N., Kennedy, A., Cotton, J., & Brumby, S. (2019). A pest to mental health? Exploring the link between exposure to agrichemicals in farmers and mental health. *International Journal of Environmental Research and Public Health, 16*, 1327.

Kornfeld, E. (1984). Crisis in the capital: The cultural significance of Philadelphia's great yellow fever epidemic. *Pennsylvania History: A Journal of Mid-Atlantic Studies, 51*, 189–205.

Krause, P. J. (2019). Human babesiosis. *International Journal for Parasitology, 49*, 165–174.

Kumar, K. K., Sridhar, J., Murali-Baskaran, R. K., Senthil-Nathan, S., Kaushal, P., Dara, S. K., & Arthurs, S. (2019). Microbial biopesticides for insect pest management in India: Current status and future prospects. *Journal of Invertebrate Pathology, 165*, 74–81.

Kumar, J., Ramlal, A., Mallick, D., & Mishra, V. (2021). An overview of some biopesticides and their importance in plant protection for commercial acceptance. *Plants, 10*, 1185.

Kweku, D. W., Bismark, O., Maxwell, A., Desmond, K. A., Danso, K. B., Oti-Mensah, E. A., Quachie, A. T., & Adormaa, B. B. (2018). Greenhouse effect: Greenhouse gases and their impact on global warming. *Journal of Scientific Research and Reports, 17*, 1–9.

Lacey, L. A., Thomson, D., Vincent, C., & Arthurs, S. P. (2008). Codling moth granulovirus: A comprehensive review. *Biocontrol Science and Technology, 18*, 639–663.

Lamberton, P. H., & Jourdan, P. M. (2015). Human ascariasis: Diagnostics update. *Current Tropical Medicine Reports, 2*, 189–200.

Lambrecht, F. L. (1964). Aspects of evolution and ecology of tsetse flies and trypanosomiasis in prehistoric African environment. *The Journal of African History, 5*, 1–24.
Lapuz, R., Tani, H., Sasai, K., Shirota, K., Katoh, H., & Baba, E. (2008). The role of roof rats (Rattus rattus) in the spread of Salmonella Enteritidis and S. Infantis contamination in layer farms in eastern Japan. *Epidemiology and Infection, 136*, 1235–1243.
Lecoq, M. (2005). Desert locust management: From ecology to anthropology. *Journal of Orthoptera Research, 14*, 179–186.
Lefebvre, M., Langrell, S. R., & Gomez-y-Paloma, S. (2015). Incentives and policies for integrated pest management in Europe: A review. *Agronomy for Sustainable Development, 35*, 27–45.
Leong, W.-H., Teh, S.-Y., Hossain, M. M., Nadarajaw, T., Zabidi-Hussin, Z., Chin, S.-Y., Lai, K.-S., & Lim, S.-H. E. (2020). Application, monitoring and adverse effects in pesticide use: The importance of reinforcement of Good Agricultural Practices (GAPs). *Journal of Environmental Management, 260*, 109987.
Leshem, E., Maor, Y., Meltzer, E., Assous, M., & Schwartz, E. (2008). Acute schistosomiasis outbreak: Clinical features and economic impact. *Clinical Infectious Diseases, 47*, 1499–1506.
Leskey, T. C., & Nielsen, A. L. (2018). Impact of the invasive brown marmorated stink bug in North America and Europe: History, biology, ecology, and management. *Annual Review of Entomology, 63*, 599–618.
Li, X., Yang, X., Zheng, X., Bai, M., & Hu, D. (2020). Review on structures of pesticide targets. *International Journal of Molecular Sciences, 21*, 7144.
Lima, M. C. F., de Almeida Leandro, M. E. D., Valero, C., Coronel, L. C. P., & Bazzo, C. O. G. (2020). Automatic detection and monitoring of insect pests—A review. *Agriculture, 10*, 161.
Linyun, X., Aiqi, Z., Jing, J., & Bing, Y. (2024). Research progress of pulsation burner and its application in pest control. *Journal of Chinese Agricultural Mechanization, 45*, 51.
Loucks, O. L. (1977). Emergence of research on agro-ecosystems. *Annual Review of Ecology and Systematics, 8*, 173–192.
Lubanga, A. F., Bwanali, A. N., Moyo, C. S., Chumbi, G. D., & Matola, Y. M. (2024). Beyond RTS'S Malaria vaccine piloting to adoption and historic introduction in sub-Saharan Africa: A new hope in the fight against the vector-borne disease. *Frontiers in Tropical Diseases, 5*, 1387078.
Lushchak, V. I., Matviishyn, T. M., Husak, V. V., Storey, J. M., & Storey, K. B. (2018). Pesticide toxicity: A mechanistic approach. *EXCLI Journal, 17*, 1101.
Main, A. (1979). Mode of action of anticholinesterases. *Pharmacology & Therapeutics, 6*, 579–628.
Malik, R., Wigney, D., Muir, D., Gregory, D., & Love, D. (1992). Cryptococcosis in cats: Clinical and mycological assessment of 29 cases and evaluation of treatment using orally administered fluconazole. *Journal of Medical and Veterinary Mycology, 30*, 133–144.
Mankad, A. (2016). Psychological influences on biosecurity control and farmer decision-making. A review. *Agronomy for Sustainable Development, 36*, 40.
Mantlo, E. K., & Haley, N. J. (2023). Heartland virus: An evolving story of an emerging zoonotic and vector-borne disease. *Zoonotic Diseases, 3*, 188–202.
Manzoor, S., Nabi, S. U., Rather, T. R., Gani, G., Mir, Z. A., Wani, A. W., Ali, S., Tyagi, A., & Manzar, N. (2024). Advancing crop disease resistance through genome editing: A promising approach for enhancing agricultural production. *Frontiers in Genome Editing, 6*, 1399051.
Matthews, G., Bateman, R., & Miller, P. (2014). *Pesticide application methods*. John Wiley & Sons.
Mazza, G., Tricarico, E., Genovesi, P., & Gherardi, F. (2014). Biological invaders are threats to human health: An overview. *Ethology Ecology & Evolution, 26*, 112–129.
McCulloch, G. A., & Waters, J. M. (2023). Rapid adaptation in a fast-changing world: Emerging insights from insect genomics. *Global Change Biology, 29*, 943–954.
Mead, P., Hinckley, A., & Kugeler, K. (2024). Lyme disease surveillance and epidemiology in the United States: A historical perspective. *The Journal of Infectious Diseases, 230*, S11–S17.
Mir, M. A. (2010). Hantaviruses. *Clinics in Laboratory Medicine, 30*, 67–91.
Monath, T. P. (2008). Treatment of yellow fever. *Antiviral Research, 78*, 116–124.

Mondal, S., Jiang, J., Li, Y., & Ouyang, G. (2019). Carbon and tin-based polyacrylonitrile hybrid architecture solid phase microextraction fiber for the detection and quantification of antibiotic compounds in aqueous environmental systems. *Molecules, 24*, 1670.
Morales-Hojas, R. (2017). Molecular ecology of insect pests of agricultural importance: The case of aphids. *Ecological Entomology, 42*, 18–27.
Müller, R., Reuss, F., Kendrovski, V., & Montag, D. (2019). Vector-borne diseases. In *Biodiversity and health in the face of climate change* (pp. 67–90). Springer.
Musso, D., & Gubler, D. J. (2016). Zika virus. *Clinical Microbiology Reviews, 29*, 487–524.
Nagar, H., & Sharma, R. (2020). A comprehensive survey on pest detection techniques using image processing. In *2020 4th International Conference on Intelligent Computing and Control Systems (ICICCS)* (pp. 43–48). IEEE.
Neiswander, R., & Blackburn, N. (1943). *Recent developments in peach insect control.*
Nicoll, W. (1917). Flies and typhoid. *Epidemiology and Infection, 15*, 505–526.
Nikiforov, V. V., Gao, H., Zhou, L., & Anisimov, A. (2016). Plague: Clinics, diagnosis and treatment. In *Yersinia pestis: Retrospective and perspective* (pp. 293–312). Springer.
Nuwer, D. S. (2015). *Plague among the magnolias: The 1878 yellow fever epidemic in Mississippi.* University of Alabama Press.
Olivero, J., Fa, J. E., Farfán, M. Á., Márquez, A. L., Real, R., Juste, F. J., Leendertz, S. A., & Nasi, R. (2020). Human activities link fruit bat presence to Ebola virus disease outbreaks. *Mammal Review, 50*, 1–10.
Organization, W. H. (2014). *Vector-borne diseases*. WHO Regional Office for South-East Asia.
Parra-Rojas, C., & Hernandez-Vargas, E. A. (2018). The 2017 plague outbreak in Madagascar: Data descriptions and epidemic modelling. *Epidemics, 25*, 20–25.
Petersen, L. R., Brault, A. C., & Nasci, R. S. (2013). West Nile virus: Review of the literature. *JAMA, 310*, 308–315.
Petrovskii, S., Petrovskaya, N., & Bearup, D. (2014). Multiscale approach to pest insect monitoring: Random walks, pattern formation, synchronization, and networks. *Physics of Life Reviews, 11*, 467–525.
Pettersson, J. H.-O., Eldholm, V., Seligman, S. J., Lundkvist, Å., Falconar, A. K., Gaunt, M. W., Musso, D., Nougairède, A., Charrel, R., & Gould, E. A. (2016). How did Zika virus emerge in the Pacific Islands and Latin America? *MBio, 7*. https://doi.org/10.1128/mbio.01239-01216
Pielke, R. A., Jr. (2004). What is climate change? *Energy & Environment, 15*, 515–520.
Poland, T. M., & Rassati, D. (2019). Improved biosecurity surveillance of non-native forest insects: A review of current methods. *Journal of Pest Science, 92*, 37–49.
Powell, K. S. (2008). Grape phylloxera: An overview. In *Root feeders: An ecosystem approach* (pp. 96–114). CABI.
Preti, M., Verheggen, F., & Angeli, S. (2021). Insect pest monitoring with camera-equipped traps: Strengths and limitations. *Journal of Pest Science, 94*, 203–217.
Priya, C., Hoogendijk, K., Berg, M., Rathinam, S., Ahmed, A., Muthukkaruppan, V., & Hartskeerl, R. (2007). Field rats form a major infection source of leptospirosis in and around Madurai, India. *Journal of Postgraduate Medicine, 53*, 236–240.
Rehman, A., Feng, J., Qunyi, T., Korma, S. A., Assadpour, E., Usman, M., Han, W., & Jafari, S. M. (2021). Pesticide-loaded colloidal nanodelivery systems; preparation, characterization, and applications. *Advances in Colloid and Interface Science, 298*, 102552.
Roy, H. E., & Cottrell, T. E. (2008). Forgotten natural enemies: Interactions between coccinellids and insect-parasitic fungi. *European Journal of Entomology, 105*, 391.
Sasa, M. (1976). *Human filariasis: A global survey of epidemiology and control.* University Park Press.
Savary, S., Horgan, F., Willocquet, L., & Heong, K. L. (2012). A review of principles for sustainable pest management in rice. *Crop Protection, 32*, 54–63.
Schmitt, L., & Burghardt, K. T. (2021). Urbanization as a disrupter and facilitator of insect herbivore behaviors and life cycles. *Current Opinion in Insect Science, 45*, 97–105.

Schuffenecker, I., Iteman, I., Michault, A., Murri, S., Frangeul, L., Vaney, M.-C., Lavenir, R., Pardigon, N., Reynes, J.-M., & Pettinelli, F. (2006). Genome microevolution of chikungunya viruses causing the Indian Ocean outbreak. *PLoS Medicine, 3*, e263.

Siddiqi, Z. (2023). A review of advances in pest control products and services in urban pest management. In *Urban pest management: An environmental perspective* (pp. 50–61). CAB International.

Sitar, G.-M., & Rusu, A. S. (2023). The impact of environmental educational programs in promoting insects conservation awareness: A scoping review. *Journal of Educational Sciences, 24*, 74–92.

Skendžić, S., Zovko, M., Živković, I. P., Lešić, V., & Lemić, D. (2021). The impact of climate change on agricultural insect pests. *Insects, 12*, 440.

Soderlund, D. M. (2012). Molecular mechanisms of pyrethroid insecticide neurotoxicity: Recent advances. *Archives of Toxicology, 86*, 165–181.

Soliman, T., Mourits, M., Lansink, A. O., & Van der Werf, W. (2010). Economic impact assessment in pest risk analysis. *Crop Protection, 29*, 517–524.

Stenseth, N. C., Leirs, H., Skonhoft, A., Davis, S. A., Pech, R. P., Andreassen, H. P., Singleton, G. R., Lima, M., Machang'u, R. S., & Makundi, R. H. (2003). Mice, rats, and people: The bio-economics of agricultural rodent pests. *Frontiers in Ecology and the Environment, 1*, 367–375.

Sudakin, D. L. (2003). Biopesticides. *Toxicological Reviews, 22*, 83–90.

Tolle, M. A. (2009). Mosquito-borne diseases. *Current Problems in Pediatric and Adolescent Health Care, 39*, 97–140.

Tsheten, T., Gray, D. J., Clements, A. C., & Wangdi, K. (2021). Epidemiology and challenges of dengue surveillance in the WHO South-East Asia Region. *Transactions of the Royal Society of Tropical Medicine and Hygiene, 115*, 583–599.

Tsioutis, C., Zafeiri, M., Avramopoulos, A., Prousali, E., Miligkos, M., & Karageorgos, S. A. (2017). Clinical and laboratory characteristics, epidemiology, and outcomes of murine typhus: A systematic review. *Acta Tropica, 166*, 16–24.

Waterfield, G., & Zilberman, D. (2012). Pest management in food systems: An economic perspective. *Annual Review of Environment and Resources, 37*, 223–245.

Wearing, C. (1988). *Evaluating the IPM implementation process* (Vol. 33, p. 17).

Webb, J. L., Jr. (2011). Malaria in africa. History compass. *Annual Review of Entomology, 9*, 162–170.

White, N. J. (1996). The treatment of malaria. *New England Journal of Medicine, 335*, 800–806.

Wiwanitkit, V. (2010). Dengue fever: Diagnosis and treatment. *Expert Review of Anti-Infective Therapy, 8*, 841–845.

Zhang, G. F., Xian, X. Q., Zhang, Y. B., Liu, W. X., Liu, H., Feng, X. D., Ma, D. Y., Wang, Y. S., Gao, Y. H., Zhang, R., & Li, Q. H. (2021). Outbreak of the South American tomato leafminer, Tuta absoluta, in the Chinese mainland: Geographic and potential host range expansion. *Pest Management Science, 77*, 5475–5488.

Zheng, K., Tan, L., Sun, Y., Wu, Y., Duan, Z., Xu, Y., & Gao, C. (2021). Impacts of climate change and anthropogenic activities on vegetation change: Evidence from typical areas in China. *Ecological Indicators, 126*, 107648.

Zuber, M., & Lillehoj, E. (1979). Status of the aflatoxin problem in corn. *Journal of Environmental Quality, 8*, 1–5.

Dynamic Shift in Plant Biodiversity by Herbicides Application

Oni Ayodeji Christopher, Daramola Oluwatosin Olaoluwa, Ewuola Akinola Ayoola, and Aliu Folorunso Patrick

1 Introduction

Contemporary patterns and shifts in plant dynamism have been observed as ubiquitous and an existential threat to plant biodiversity. Plant biodiversity comprises the representative components of our rich natural resources. Gradual and drastic degradation of natural vegetation have been shown to change the pattern of crops' physiological and agronomical attributes, as well as their evolutionary factors. The introgression of invasive crop species is a common occurrence, which is of major concern to environmental conservationists. These changes in temporal and spatial distribution of ecological vegetation are highly influenced by biological, environmental, and anthropological factors (human activities), which singly or in combination are of significant importance. The significance of the isolated effects of human activities is massive in relation to the globalization, industrialization, trade, and development of societies (Olorode, 2007). The pollution of natural genetic resources of agroecological areas for agricultural purposes is highly implicated in the loss of such resources. There are many cases of extinct, threatened, or endangered plant species that once contributed to the richness of crop biodiversity.

The establishment of invasive and alien plant species is one of the main reasons for loss of plant biodiversity, constituting an apparent threat to the existence of endangered species, combined with the devastating effect of extensive human activities in trade, industrialization, and globalization. It was reported in Nigeria, due to

O. Ayodeji Christopher (✉) · A. Folorunso Patrick
Bioresource Development Centre, National Biotechnology Research and Development Agency, Abuja, Nigeria

D. Oluwatosin Olaoluwa
Department of Chemistry, Federal University of Technology, Akure, Ondo State, Nigeria

E. Akinola Ayoola
Department of Chemistry, Osun State College of Technology, Esa Oke, Nigeria

B. R. Babaniyi, E. E. Babaniyi (eds.), *The Interplay of Pesticides and Climate Change*, https://doi.org/10.1007/978-3-031-81669-7_15

overexploitation, that the natural ecosystems and forests remaining are restricted to conserved areas, like national parks, botanical gardens, and forest reserves, which have been rapidly decimated by deforestation, agricultural cultivation, and other forms of land encroachment (Borokini, 2011). In Borokini's survey (*ibid*) to determine the general assessment and impact of invasiveness on the field gene bank in southwest Nigeria, the results revealed that 25 notable invasive plant species had successfully displaced tree plants like *Theobroma cacao*, *Irvingia gabonensis*, *Mangifera indica*, and *Cassia spp.* by 18% of the total land cover in the field gene bank. The aggressiveness of these species is attributed to their ability to easily form canopies, which increases receptivity to intercept light and effectively increases their competitiveness for space and nutrients.

From a conservational point of view, weeds are essential components of agroecosystem biodiversity. The possibility of coexistent and ecological interactions between crops and weeds depends on diverse morphophysiological traits in the plant community. This promotes the idea of a neutral weed community that promotes simultaneous growth and development of plants without significant detrimental effects (Esposito et al., 2023).

Weed communities, in general, play an essential dual role in agriculture—as an alternative repository of beneficial or parasitic hosts to crop plants, as well as a diverse bioresources entity (Gharde et al., 2018). Consequently, the need to develop elaborate and innovative management schemes to balance dynamism in plant biodiversity, while mitigating the adverse impacts of weeds on agricultural activities, is necessary. The proponents of neutral weed community encourage the coexistence of weeds and crops, considering their interactions as crucial ecological factors. These contribute enormous benefits to agriculture in terms of pollination regulation, control of soil erosion, and improvement of crop quality. Moreover, their cooperation promotes the transition to desirable weed complexes.

It is a known fact that chemical methods of weed control destabilize crop-weed communities by eliminating diverse plants that could survive adverse competitive effects of the induced toxic environments, limiting the diversity and abundance of plant bioresources by favoring the growth of certain groups of crops. This paper examines the adverse effect of herbicides on plant biodiversity and consequently the trend towards plant dynamic shift.

2 Existence of Weed Neutral Community

The following are examples of neutral plant communities. In France, speedwell, *Veronica spp.*, common chickweed, *Stellaria media* L, cut-leaf geranium, *Geranium dissectum* L, and catchweed bedstraw, *Galium aparine* L, contribute significantly to the growth of winter cereal crops. In Italy, weed communities of corn chamomile, *Anthemis arvensis* L, California burclover, *Medicago polymorpha* L, scarlet pimpernel, *Anagallis arvensis* L, night-flowering catchfly, *Silene noctiflora* L, *Euphorbia spp.*, field violet, *Viola arvensis*, common lambs quarters, *Chenopodium album* L and *Polygonum spp.*, grow simultaneously with winter wheat, *Triticum aestivum* L, and barley, *Hordeum vulgare* L (Adeux et al., 2019; Esposito et al., 2023). The coexistence of wild buckwheat, *Polygonum convolvulus* L and *Fallopia convolvulus* L, prostrate knotweed, *Polygonum aviculare* L, false cleavers, *Galium spurium* L, and field forget-me-not, *Myosotis arvensis* L, with autumn-sown cereals is reported as well in Sweden. In an arid region of India, the coexistent field of annual leguminous weed, *Indigofera cordifolia*, positively improved the growth parameters and yield in millet, *Pennisetum glaucum* L, and sesame, *Sesamum indicum* L, by contributing 19.8% and 22.4% higher respectively in mixed plots compared with weeded plots. In the same area, weeds like Arabian-primrose, *Arnebia hispidissima* and *Spermacoce articularis*, and leafy vegetable feather cockscomb, *Celosia argentea*, increased the growth parameters and yield of millet, while redroot pigweed, *Amaranthus retroflexus*, established a symbiotic relationship with maize, as reported by Gholamhoseini et al. (2013).

Different methods are employed in controlling or managing noxious weeds in agricultural crop production, these are presented in Fig. 1 below:

The intensity of herbicidal use negatively impacts the abundance and diversity of weed species. The increase in the rate of application of herbicides to control weeds is gradually becoming an integral component of agricultural practice as a proven cheap and easy method for controlling weeds in order to ensure higher crop yields. Adverse situations have arisen such as herbicide drift to nontargeted plants, herbicide chemical retention in the soil, tolerance, and susceptibility of plant among others are commonly reported and linked to heavy agricultural activities. The growing concerns over herbicidal impacts on the dynamic shifts in plant vegetation among ecologists, environmental conservationists, and plant biodiversity–oriented individuals are becoming astronomically high. Besides the global effects of climate change, effects from the interaction of chemical components of herbicides with plants have become evident in crop genetic erosion, physiological transformation, and differences in biochemical constituents and ultimately admixture of vegetative

Fig. 1 Combination of Weed management practices. (Source: Pontes et al., 2022)

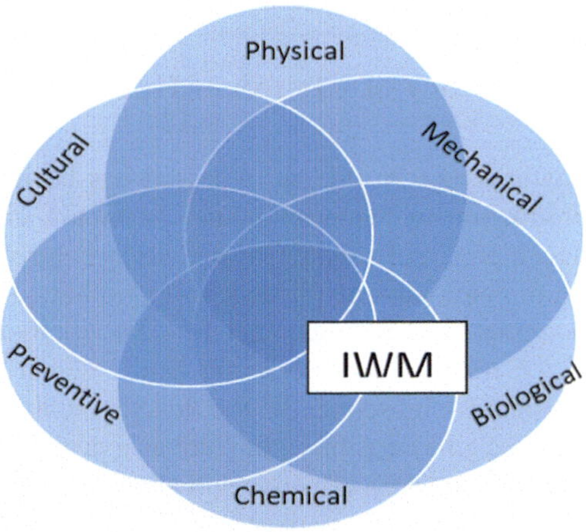

pollution. Additionally, the use of herbicides has led to dilution in vegetative patterns, partly due to accumulative degradation of native plant species and natural selection pressure with an increase in introgression of invasive crop species that are either comparatively tolerant or opportunistic in nature. These significantly change the terrain of plant diversity and result in an alteration of plants' ecosystems.

In recent developments, the trend in managing the effect of climate change and increasing levels of carbon dioxide is a massive pointer to resurgent biological constraints in crop production to meet the demand to feed the teeming global population. This illustrates the drastic shift in weed response, insect adaptive patterns, and distribution of disease occurrence within agronomic sectors. Among other biological parameters, crop losses due to weed management strategy were effectively reduced by 75%. Chemical herbicides are the most economical and widely used method of controlling weeds globally, because of significant reduction in production cost and energy, thus imposing intense pressure on their chemical formulation, potency, and safety. Cumulatively, climate change effects influence the plant biodiversity dynamics shift that has occurred in crop composition, and abundance, along with biological and physiological differences, especially in weed-crop–herbicide resistant situations (Ziska, 2020; Nath et al., 2024).

The development and adoption of better weed management practices, aimed at reducing weed dominance while still promoting diversity, are justified as holistic strategies for providing healthier competitive growth. The application of some chemical herbicides as a means of weed control, sparing some weed species while negatively affecting others, favors the dominance of a particular morphological group of weeds such as grasses, broadleaf weeds, sedges, and spiderwort over others (Ayodele et al., 2024). Similarly, the pressure to manage composition and intensity in chemical weed control is becoming steeper in sustaining crop production despite the climate change effects of increased atmospheric CO_2 concentration, erratic rainfall patterns, higher temperature, and drought and flooding occurrences. All of these elements are implicated in a gradual shift of weed-crop dynamics and competitiveness, as corroborated in many scientific studies.

3 History of Chemical Herbicide Development

Post–Second World War research has been tailored towards intensified industrial production of heavy and fine chemicals, an integral part of which included the development of suitable herbicides to control weeds in agricultural production. The prominent outcome yielded, among others, the production and development of two potential compounds, 2,4-D (2,4 dichlorophenoxyacetic acid) and 2,3,4-T (2,4,5-trichlorophenoxyacetic acid), which are selective broadleaf weed–control chemicals. Subsequently, the development of glyphosate (with a toxic dose as low as 1–2 kg/ha) was brought forth as a promising candidate for chemical weed herbicide. It acts as an effective signal compound that disrupts protein production in the biochemical pathway of plants. However, both weeds and crops are eradicated by its application. A binary effect on the environment was thus created in combining genetic engineering with glyphosate herbicides—the development of genetically modified/transgenic crops tolerant to glyphosate chemical are conspicuous in soybean, canola, cotton, maize/corn, and sugar-beet (Gianessi, 2016; Duke, 2018; Nandula, 2010).

Herbicides are natural or synthetic chemicals applied at low concentrations to control, suppress, or eradicate the growth and development of unwanted noxious plants. As of 2001, herbicides had the largest share world-wide among pesticides, with about 267 active ingredients directed towards crops in the following proportion: 5% for cotton, 6% for fruits and nuts, 8% for rice, 17% for cereals, 19% for maize/corn, and 20% for oil-seeds and soybeans. About 30 or more chemical compounds are nonselective, while 130 are employed as selective herbicides, exhibiting about 16 different modes of action (Shekhawat & Singh, 2024). The jump in the production and application of synthetic herbicides was an estimated 50-fold increase to about 2.3 million tons. Approximately 75% of the utilization of all global herbicides is being pushed to the developing countries to boost their agricultural production, accompanied by high rates of indiscriminate doses. Prominent families of herbicides are phenoxy and benzoic acids such as 2,4-D tyriazines (atrazine), urea (diuron) and chloroacetanilides (alachlor) (Kughur, 2012). Herbicides could be systemic or nonsystemic (contact) in modes of action.

According to De Souza et al. (2022), these are the advantages of herbicide application:

1. Less dependence on labor; inexpensive
2. Efficient and not subjected to seasonal fluctuations
3. Minimal or zero tillage system required
4. Control of vegetatively propagated weeds
5. Tailored towards eliminating weeds only

3.1 Types of Herbicides

1. Based on composition
2. Time of application
3. Location of action

Classification based on specificity divides herbicides into two types: *selective*, which affect a specific class of plants, and *nonselective*, which are broad spectrum, containing chemicals with burning effects. The timing of herbicides application can be classified as being *preemergence* (i.e., before) and *postemergence* (i.e., after). The use of butachlor is recommended as a preemergence herbicide, effective for weed control in okra production. This ensures minimum tilling of the soil. The effect of preemergence herbicides is non-lasting in perennial crops especially during the seedling period. Therefore, practices such as combination of aggressive post- and preemergence herbicides are commonly employed to combat weed problems. For example, a combination of glyphosate and paraquat is used for land preparation as an alternative to manual bush clearing. There is a substantial claim of an exponential increase in the use of herbicide to control weed in cocoyam production in Nigeria. The common active herbicide groups are atrazine, S-metolachlor + atrazine, glyphosate, and paraquat (Adewumi et al., 2024; Ekeleme et al., 2020).

The general mechanisms employed by chemical herbicides vary based on the specific modes of action of their principal active compounds. Some of the mechanisms of action are targeted for specific reactions and are itemized below according to Rehman et al. (2023):

1. Photosynthetic Inhibition
2. Amino Acid Synthesis Inhibition
3. Seedling Growth Inhibition
4. Cell Membrane Disruption
5. Lipid Synthesis Inhibition
6. Pigment Inhibition

Additionally, the biochemical and physiological processes initiated by chemical herbicides in plants are highlighted by Nath et al. (2024), as given in Fig. 2 below;

Nontarget site mechanisms follow the pattern of characteristic changes occurring before an herbicide's lethal dose arrived at the targeted site. These could be from increased metabolism or differential translocation rates (Alcántara-de la Cruz et al., 2021; Nandula et al., 2013). Target-site mechanisms are molecularly oriented towards overexpression of gene-encoding enzymes of interest or substitution of nucleotide sequences (Mendes et al., 2020).

3.2 *Mode of Action in Designing Herbicide Ingredients*

The active participation of proteins in plant biochemical and physiological pathways has made them potential targets for the development of herbicides. This includes processes such as photosynthesis, cell division and growth, cellular metabolism, and material transportation. Concerns in the development of herbicides based on mode of action are often characterized by (i) longer development period, (ii) high rate of failure in development, (iii) high developmental cost.

The success rates of employing mode of action as a means of herbicide development were limited to a small number of classes during the past two decades, which came with the introduction of transgenic crops—those genetically modified to resist the effects of herbicides. Developing novel herbicide through mode of action faces stringent quality control measures to assure a safe product with no environmental and toxicological effects.

Current efforts and scientific works are geared towards developing herbicides with new modes of action inhibition targets such as homogentisate solanesyltransferase (HST), fatty acid thioesterase (FAT), solanyl diphosphate synthase, among others. These efforts also involve the alteration of active natural (organic) products

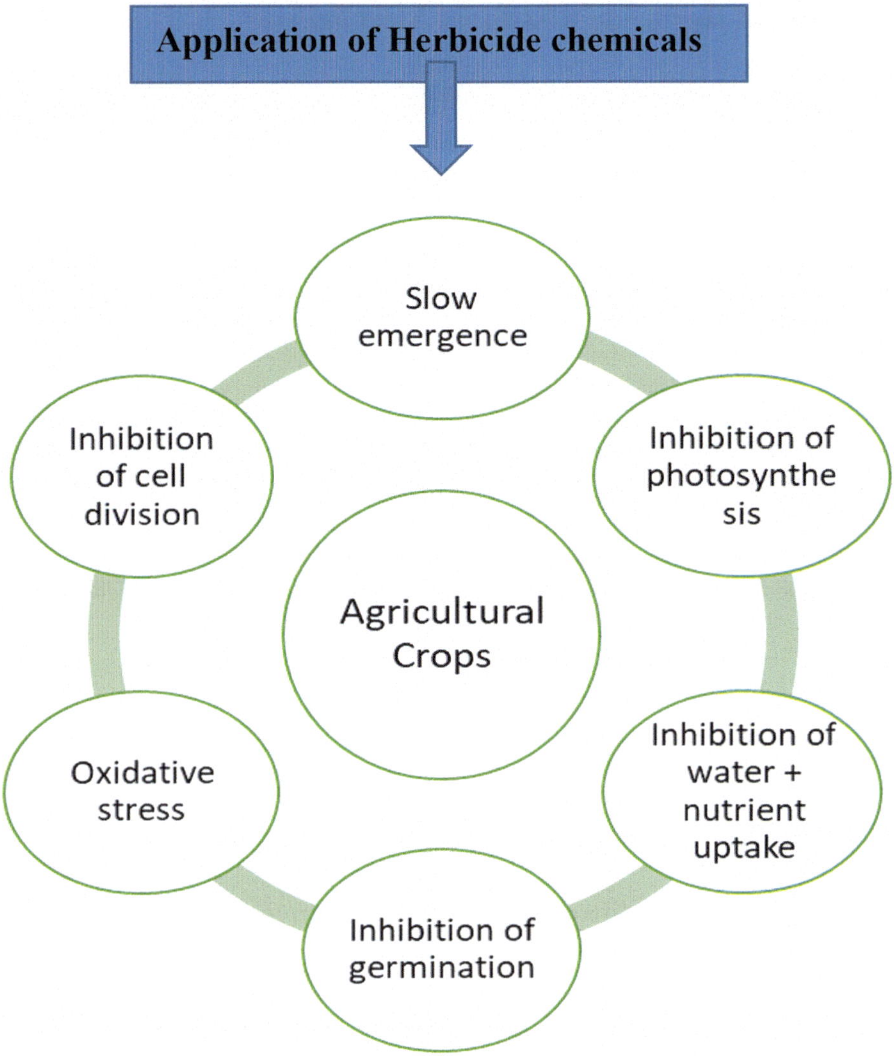

Fig. 2 The effect of herbicides on biochemical and physiological processes in plants

like enzymes, metabolites, and phytotoxins from biological entities such as gostatin, cornexistin, AAL-toxin romidepsin, spliceostatin C from *Burkholderia rinojensis*, targeting different molecular sites of action (Duke & Dayan, 2022). Employment of advanced techniques in molecular omics, bioinformatics, computational biology, and chemistry present possible opportunities for designing, synthesis, screening, and development of targeted ingredients with molecular precision (He et al., 2022; Kahlau et al., 2020; Liu & Lu, 2016).

4 Cases of Herbicide Resistance Mechanisms

Resistance cases in plants could be:

(a) *Simple*: When a species offers resistance to the same herbicide
(b) *Cross*: When there is resistance to initially susceptible application
(c) *Multiple*: When weed biotypes offer resistance to herbicides of different MoAs without past record of use, especially in grasses, to ALS and ACCase inhibitors (Mendes et al., 2020)

There are more cases becoming global concerns of weed biotypes being resistant and some even multi-resistant to alternate use of herbicides with different modes of action. The trend of developing new herbicides or transgenic-resistant crops is quite challenging, as the mechanism of action for each seems to follow the same biochemical pathways. These biochemical and genetic mutations in both weeds and crops are the lifeform's evasion of the targeted actions of the engineered herbicides. The proposed solutions are:

(i) Increasing the concentration of glyphosate herbicides
(ii) Reducing chemical drifts and volatility in application of older herbicides (e.g., Dicamba)
(iii) Mixtures of different herbicides
(iv) New formulations (Strom et al., 2019; Egan et al., 2014)

The pressure of advancements in herbicide chemicals development and weed control strategies intensifies widespread herbicide resistance, implicating climate change's indirect effects and rising CO_2 concentration.

5 Genetically Modified Herbicide Resistance Crops and Their Mechanisms

The introduction of genetically modified herbicide-resistant crops such as soybean, maize, and cotton offers many advantages by allowing for increased application of broad-spectrum nonselective herbicides, such as glyphosate or ammonium–glufosinate. It also allows for increased selectivity of herbicides on a particular weed of interest, encouraging the combination of different chemical groups with diverse mechanisms of action to control resistant weed biotypes. Genetic modification for this purpose also facilitates adoption of minimum or zero tillage system and ultimately reduces the residual effects of persistent chemicals in soil.

The negative impacts of GM herbicide-resistant crops are: (i) promoting selection of acquired resistance in weed biotypes, (ii) introduction of ecosystem imbalances among weed communities, and (iii) mixture of different herbicides, which encourages antagonistic interactions that negate the intended actions.

Glyphosate-resistant GM crops are developed to resist the action of glyphosate by inactivating the binding EPSPs enzymes with phosphoenolpyruvate substrate in the biosynthetic pathway of essential aromatic amino acids. Even though it is systemic, the possibility of being used as pre-emergence herbicide is high, when combined with other herbicides. Glyphosate is used to control ALS inhibitors resistant weeds. Many studies suggested that the broad-spectrum effect of glyphosate is not that potent in controlling some resistant or tolerant weed species, so combination of other herbicides or other control methods becomes necessary to ensuring efficient use. The case of resistance in chemical groups of imidazolinones is made possible through the introduction of the *csr1-2* gene to *Arabidopsis thaliana*, which encodes the subunit of ALS enzyme into its genome. In the absence of this gene, herbicide-induced ALS enzymes disrupt the synthesis of branched amino acids, leading to death in plants.

Crops enabled with ammonium-glufosinate (AG) can express the phosphinothricin acetyltransferance (*pat*) gene from bacterium *Streptomyces viridochromogenes*, which is inserted by bio-ballistic transformation. AG acts in the balance of reactive oxygen species (ROS) to inhibit synthesis of glutamine thereby disrupting the photorespiratory pathway. Other forms of GM herbicide resistance are linked to the 4-hydroxyphenylpyruvate dioxygenase inhibitor, dicamba, and 2, 4-dichlorophenoxyacetic acid through insertion of foreign genes encoding a protein that degrades herbicide chemicals into nontoxic inactive compounds (De Aguiar et al., 2022).

The dichotomy between the proponents and antagonists of GM herbicide–resistant crops affecting the rate of herbicide application still varies in the research, but ecological changes causing biodiversity shifts in the population of weed communities or herbicide-resistant biotypes is an inevitable reality. Upon repeated use, volunteered crops or weeds exposed to heavy doses of herbicides exhibit some degree of resistance as indicated in cases of herbicide-resistant weeds to ALS, EPSPS, ACCase, and PSII inhibitors in *Conyza spp.*, *Digitaria insularis*, *Euphorbia heterophylla*, *Echinochloa spp.*, and *Eleusine indica* (Heap, 2021; Mendes et al., 2020).

Strategies to reduce the risk of herbicide resistance evolution and biodiversity shifts should be put in place to combat this menace. They are highlighted as follows:

1. Use chemicals of different mechanisms of action
2. Perform rotation of chemicals with diverse MoAs
3. Limit the number of herbicide applications on the same family of crop
4. Perform a mixture or sequential application of herbicides
5. Use chemicals with lower selection pressure
6. Rotate crop/weed control methods
7. Apply herbicide at recommended doses (Mendes et al., 2020)

The classes of herbicide chemicals, their mechanisms of action, effects, and resistance in crops are presented in Tables 1 and 2.

Current changes in wind velocity, rainfall patterns, and extreme thermal effects are directly linked to changes in the climate, a significant proportion of which invariably results from application of herbicide chemicals. The incidence of

Table 1 Classes of herbicide chemicals, mechanisms of action, and effects on crops

(a)

Chemical classes	*Mechanism of action*	*Herbicides*	*Time of Application*	*Translocation*	*Effects on crops*
Aryloxyphenoxy propionate	Acetyl CoA carboxylase (ACC) inhibitors	Fenoxaprop, Fluazifop-P, Cyhalofop-butyl, Diclofop, Quizalofop-P and Clodinafop,	Post-emergence	Phloem	Inhibit synthesis of fatty acids and destroy the membrane structure
Carboxylic acid		Picloram, Fluroxypyr, Clopyralid, Triclopyr,			
Pyrimidinylthio-benzoate		Pyrithiobac			
Sulfonylamino-carbonyltriazolinone		Flucarbazonesodum, Propoxycarbazone			
Phenoxy	Synthetic auxins	2,4-D, MCPA, 2,4-DB, 2,4-DP, MCPB, MCPP	Post-emergence	Xylem and phloem	Interrupt cell growth in newly forming leaves and stem, cell wall plasticity and nucleic acid metabolism
Sulfonylurea		Chlorsulfuron, Sulfometuron, Mesosulfuron, Metsulfuron, etc			
Triazolopyrimidine		Cloransulammethyl, Diclosulam, Flumetsulam etc			
Dinitroaniline	Microtubule assembly inhibitors	Benefn, Trifuralin, Prodiamine, Pendimethalin etc			Inhibit cell division in plant roots
Imidazolinone	Acetolactate (ALS) synthase inhibitors	Imazapic, Imazamethabenz, Imazapyr, Imazamox etc	Pre and Post-emergence	Xylem and phloem	Inhibit production of three essential branched-chain amino acids
Pyridine		Dithiopyr, Thiazopyr			
None		DCPA			
Cyclohexanedione		Sethoxydim, Clethodim, Tralkoxydim			

Benzoic acid		Dicamba			
Quinaline carboxcylic acid		Quinclorac			
(b)					
Chemical classes	*Herbicides*	*Mechanism of action*	*Time of Application*	*Translocation*	*Effects on crops*
Phenyl-carbamate	Desmedipham, Phenmediphm	Photosynthetic inhibitors at Photosystem II Site			Block electron transport at second stable electron acceptor of photosystem II which interrupt photosynthesis and energy production
Thiocarbamates	Butylate, Cycloate, Prosulfocarb, Vernolate	Lipid synthesis inhibitors			Reduce cuticular wax formation and inhibit shoot-growth
Organophosphates Isopropylamine	Glyphosate	EPSP (5-enolypyruvylshikimate 3-phosphate), Synthase Inhibitors	Post-emergence	Xylem and phloem	Inhibit the synthesis of aromatic amino acids
Pyridazinone	Pyrazon				
Triazine	Atrazine, Ametryn, Propazine, Simazine		Pre-emergence	Xylem	
Triazinone	Hexazinone, Metribuzin				
Benzothiadiazole	Bentazon		Pre-emergence	Non-translocated (contact)	
Urea	Diuron, Fluometuron		Pre-emergence	Xylem	

(continued)

Table 1 (continued)

Bipyridylium	Diquat, Paraquat, Gramoxone		Post-emergence	Non-translocated (contact)	
None	Glufosinate	Glutamine-synthesis inhibitors			Accumulate huge ammonia and destroy plant cells
Triazoles	Amitrole	Carotenoid biosynthesis inhibitors/ pigment synthesis inhibitors	Pre-emergence	Xylem	Damage the chlorophyll pigments
Isoxazolidinnes	Clomazone	Carotenoid synthesis inhibitors		Xylem	
Diphenyleth-rs	Acifuorfen, Lactofen, Oxyfuorfen	Cell membrane disruptors or PPO inhibitors			Disrupt cell membrane
Oxadiazoles	Oxadiazon				
Pyridazinone	Norfurazon				

Sources: Oliveira Júnior and Bacarin (2011), Heap (2021), Parven et al. (2024), Adesina et al. (2024)

Table 2 Ten active chemical ingredients with the number of resistances in crops

Active Chemicals	Mode of Action	Cases of Weed Resistance	Targeted Crops	Number of Weed Species
Glyphosate	Inhibitors of EPSP synthase	North and South Americas, Oceania, Europe, Asia and South Africa	Wheat, cereals, canola, spring barley, lentils, peas, grapes, corn, chickpeas, cotton, soybean, alfalfa, sorghum, beans, rice, coffee, lime, almonds, pumpkin, squash, apples,	57
2-4-D	Synthetic Auxin (Plant cell growth disruptor)	North and South Americas, Oceania, Europe and Asia	Wheat, cereals, barley, corn, pastures, roadsides, soybean, oats, turf, sugarcane, rice	25
Paraquat	Inhibitor of photosynthesis at photosystem I	North and South America, Oceania, Europe, Africa and Asia	Wheat, pears, chestnut, peaches, sugarcane, vegetables, corn, cotton, soybean, beans, almonds, coffee, rubber, sweet potato, taro, alfalfa	31
Atrazine	Inhibitor of photosynthesis at photosystem II	North and South Americas, Oceania, Europe, Asia and Africa	Wheat, canola, corn, sugar beets, cotton, soybean, cotton, tomatoes, sorghum, vegetables, blueberries, carrots, asparagus	66
Tribenuron-methyl	Inhibitors of acetolactate synthase	North America, New Zealand, Europe, Asia and South Africa	Wheat, canola, corn, sunflower, lentils, barley, alfalfa, fallow, beans, cereals, grapes, rapeseed, peas, chickpeas, pastures	45
Imazamox	Inhibitors of acetolactate synthase	North and South Americas, Australia, Europe, Asia and South Africa	Wheat, sunflower, canola, barley, lentils, chickpeas, rice, rapeseed, alfalfa, grapes, sugar beets, sorghum, tomatoes, corn, cotton, soybean	40
Metsulfuron-methyl	Inhibitors of acetolactate synthase	North and South America, Australia, Europe, Asia (Iran and Malaysia) and South Africa	Wheat, sunflower, canola, barley, lentils, chickpeas, rice, grape, peas, grass seed, oil palm, corn, cotton, soybean, cereals	39
Imazethapyr	Inhibitors of acetolactate synthase	North and South America, Australia, Europe and Asia	Wheat, canola, corn, sunflower, soybean, cotton, onions, lentils, alfalfa, rice, cereals, peas, chickpeas, lettuce, cabbage, peanut	44

(continued)

Table 2 (continued)

Active Chemicals	Mode of Action	Cases of Weed Resistance	Targeted Crops	Number of Weed Species
Simazine	Inhibitor of photosynthesis at photosystem II	North America (USA), Australia, Europe and Asia	Vegetables, canola, barley, grapes, wheat, corn, sugar beets, potatoes, blueberries, wheat	31
Fenoxaprop-P-ethyl	Inhibitors of fat synthesis acetyl coA carboxylase inhibitors (ACCase inhibitors)	North and South America, Oceania, Europe, Africa and Asia	Wheat, sunflower, canola, lentils, peas, sugar beets, chickpeas, pulses, flax, spring barley, corn, cotton, soybean, beans	33

Source: Heap (2023): International Herbicide-Resistant Weed Database
Some other active herbicides exhibiting inhibition against acetolactate synthase, such as Bensulfuron-methyl, Thifensulfuron-methyl, Mesosulfuron-methyl, Iodosulfuron-methyl sodium, and Chlorsulfuron, were all used to control 29–38 weed population across the Americas, Europe, Oceania, Asia, and South Africa

herbicide persistence and efficacies in plants affect weed-plant demography and distribution. The physiological and phenological shifts in crop-herbicide interaction modifies the mechanism of the targeted modes of action. Furthermore, the joint influence of rising CO_2 and modifying the photosynthetic activities of plants are manifested in stomatal characteristics, leaf thickness, root-to-shoot ratio, etc., which directly affect the nutrient uptake and translocation processes of crops, thus impacting herbicide assimilation processes. For example, it was noted in *Parthenium hysterophorus* L, a noxious weed, that CO_2-induced biomass increases improved the recovery of a plant after glyphosate application and also increased the resistance of multiple-resistant strains of jungle-rice (*Echinochloa colona* L) to cyhalofop-butyl (Bajwa et al., 2019; Refatti et al., 2019).

The incidence of herbicide resistance has been growing since the 1970s, but as of 2020, it was suggested that almost 50 species had undergone evolved resistance to glyphosate (Ziska, 2020). As of writing, reports have provided over 500 cases of weed-herbicide–resistant species by site-of-action interactions worldwide, as identified by Weed Science Society of America, who opined that 23 out of the 26 distinct sites of action exhibited resistance to 167 herbicide chemicals. In another report, the incidence of herbicide resistance dated back as far as 1970s, estimating the situation at present at about 218 herbicide-resistant crop cases and counting. This phenomenon anchored on evolutionary processes of interaction among plants, chemicals, and their environment. Factors implicated in this process were the genetic basis of resistance in plants, characteristics of herbicides, intensity of use, and biology of weed species (Shekhawat & Singh, 2024). In America's Corn Belt, multiple weed-herbicide resistances are found in water hemp, *Amaranthus tuberculatus* (Moq.), giant ragweed, *Ambrosia trifida* L, velvet leaf, *Abutilon theophrasti*, and morning glory, *Ipomea* spp. The ever-increasing occurrence of herbicide resistance in weeds is conclusively traceable to these parameters:

(i) Continuous application of chemicals possessing the same mode of action
(ii) Disparity, in slow rate of adoption of genetically modified crops especially engineered to be resistant to herbicide, especially glyphosate
(iii) Genetic resistance mutation mechanisms in plants
(iv) Overuse of a smaller number of herbicides develops herbicide-tolerant crops, which simultaneously favors tolerant weed population

The cases of using of herbicides to control invasive non-native plants by landowners were massive in the early Twenty-First Century in countries like Australia, Canada, and the United States. The seedling performances of established viable seeds in a bank are negatively impacted with prior application of nonselective herbicides because of their persistence and toxicity. A study that examined herbicide effects on a seed bank of native and non-native species indicated poor seedling emergence rates and lower biomass qualities even at recommended application rates or lower, affecting both mono- and dicot- plant species (Wagner & Nelson, 2014).

Investigation revealed a dynamic shift in weed diversity, species growth rates, and the structures of weed communities under the application of glyphosate and dicamba herbicides for an 8-year period. An initial reduction in weed population growth rates was observed, however, weed communities compensated for this trade-off by increasing their reproductive functions and thus their survival rates (Oreja et al., 2021). Also, through the mechanism of seed dormancy, some weeds escape the destructive effects of preemergence herbicide application (Obiazi et al., 2024). It was demonstrated by Nath et al. (2017) that pendimethalin, a widely used dinitroaniline herbicide, can reduce the biodiversity of weeds significantly, but this is generally not the case, as, on the contrary, the loss in weed biodiversity is not limited to chemical weed management. General methods of weeding significantly modify weed community composition and elimination of minor populations. Indeed, this was supported by Ayodele et al. (2024)—control of weeds significantly alters the weed flora composition in okra plots, allowing for the emergence of some new weed species and decreasing overall weed diversity.

Factors influencing emergence of resistance in crops are (i) selection pressure from regular use of the same agricultural practices, (ii) genetic variability, and (iii) evolutionary resistance changes.

The movement of herbicide resistance traits is governed by gene drifts as a result of the followings.

Gene flow through seed or pollen could be responsible for transfer of a resistant gene from a mother plant to its descendants or crosses between close/wild relatives and weeds in adjacent fields.

The attributes of weeds that promote weed survival and acquired resistance are (i) whether they are self- or cross-pollinated, (ii) whether their seeds are produced in large quantity and easily dispersed by wind or insects, and (iii) whether their seeds exhibit dormancy for a longer period of time (Beckert & Dessaux, 2016).

6 Notable Herbicide Resistance in Weeds

The cases of herbicide resistance in the USA, Australia, Canada, and Brazil are specifically noted in genetically modified (GMO) crops that are glyphosate-resistant crops programmed for a single use of glyphosate, resulting in increasing cases of glyphosate-resistant weeds. Whereas, in Europe, resistance has been directly linked to a low diversification of chemical products in herbicide used. This is exemplified in several cases of wide weed biotypes exhibiting lateral resistance to acetyl-CoA carboxylase (ACCase)–inhibiting herbicides, acetolactate synthethase (ALS)–inhibiting herbicides, and photosynthesis at photosystems I&II–inhibiting herbicides.

Moreover, notable weed species identified and reported to have developed herbicide resistance include *Conyza* species, against flazasulfuron, iodosulfuron, and penoxsulam in vineyards, olive, and citrus farms. *Lolium* species evolved against glyphosate herbicides. *Avena* spp., *Bromus* spp., *Digitaria sanguinalis*, *Panicum dichotomiflorum*, and *Echinochloa crusgalli* are resistant against ALS-inhibiting herbicides across European countries (Montull & Torra, 2023). Additionally, some active ingredients of herbicides have more cases of resistant weed species globally; however, most of these cases are associated with monoculture farming systems.

Extensive reports show that a representative tree species in Brazil, *Dipteryx alata*, is highly sensitive to the recommended dose of Nicosulfuron, an ALS-inhibiting herbicide, which causes a series of reactions involving oxidative stress and alteration of its photosynthetic ability. This underscores the impact of ALS-inhibiting herbicides, which pose a serious threat to forest biodiversity and conservation of vulnerable plant species. Oxidative damage has also been observed in rice, *Oryza sativa*, maize, *Zea mays*, and weed seedlings, *Arabidopsis thaliana*, when exposed to 50 µg/L, 80 mg, and 2.5 µg/L of different brands of ALS-inhibiting herbicides. In these cases, there has been a clear indication of elevation in oxidative enzymes such as peroxidases, superoxidase, catalase, and malondialdehyde (Silva et al., 2020).

Reports of resurgence of ALS-inhibiting, herbicide-tolerant weeds are well documented in the evolution of tolerant weedy red rice under intensive and continuous application of ALS-inhibiting herbicides. Mutant and resistant species possess an ALS-resistant gene with 0.17% out-crossing incidence. This can tolerate high doses of ALS-inhibiting herbicides compared to sensitive species. Samples from fields with such history had at least 20% resistant offspring. Similar incidents of resistant gene transfer also occurred in wild turnip (*Brassica rapa*) weed (Bourdineauda, 2020). The potential for weeds to evolve, their acquired herbicide resistance, and their tolerance is motivated by a survival mechanism, using hybridization to escape vulnerability across geo-ecological zones. This is coupled with their abilities to withstand environmental stresses posed by climate change effects and other external factors, as shown in Table 3.

Table 3 Common resistant weeds in soybean, cereals, and cotton crops with their families

Broad leaf weed species	Family	Inhibitor	Narrow leaf - Grasses	Family	Inhibitor	Sedge	Family	Inhibitor
Euphorbia heterophylla L.	Euphorbiaceae	A[a], DE[a,b]	*Digitaria horizontalis W*	Poaceae	*B*[a], *D*[a,b]	*Cyperus*	Cyperaceae	A[b]
Euphorbia hirta L.			*Paspalum scrobiculatum*			*rotundus (L)*		A[b]
Euphorbia graminea J			*L.*			*Cyperus*		
Phyllanthus amarus Sc			*Panicum maximum J.*			*difformis (L)*		
Gomphrena celosoides M			*Axonopus compressus S.*			*Cyperus*		
Acalypha indica L			*Pennisetum spp*			*esculentus (L)*		
			Echinochloa spp.		*C*, *A*[b]	*Kyllinga* spp.		
Amaranthus spinosus L	Amaranthaceae		*Eleusine indica G.*		*B*[a]	*Fimbristylis*		
Tridax procumbens L.			*Cynodon dactylon L.*			*littoralis G.*		
Bidens pilosa L.		A[a]	*Cynodon plectostadyus*					
Alternanthera sessilis L.			*K.*					
Alternanthera brasiliensis			*Imperata cylindrica L.*					
Amaranthus spp		A, F, DE[a,c]	*Imperata cylindrica L.*					
			Brachiaria deflexa Sc.					
Chromolaena odorata L.	Asteraceae		*Rottboellia*					
Tithonia diversifolia He		A[a,c]	*cochinchinensis*					
Ageratum conyzoides L.			*Leersia hexandra Sw.*					
Aspilla africana P.			*Leptochloa caerulescens*					
Ludwigia spp.			*St.*					
			Oryza longistaminata A.					
Ipomea triloba L.			*Sacciolepsis africana C.*					
Ipomoea involucrata P.	Convolvulaceae		*Chloris pilosa Sc.*		*D*[a]			
Ipomoea aquatica F			*Urochloa palntaginea*		B[a]			
Merremia kentrocaulos			*Lolium perenne*		AB[a,b], D[a]			
Spigelia anthelmia L	Loganiaceae							
Commelina benghalensis	Commelinaceae							

(continued)

Table 3 (continued)

Broad leaf weed species	Family	Inhibitor	Narrow leaf - Grasses	Family	Inhibitor	Sedge	Family	Inhibitor
Striga spp.	Scrophulariaceae							
Mimosa pudica L. *Calopogonium mucunoides* *Centrosema pubescens B* *Senna obtusifolia L.* *Mucuna pruriens L.*	Fabaceae							
Malvastrum coromandelianum L.	Malvaceae							
Oldenlandia corymbosa L. *Mitracapus villosus Sw*								
Talinum triangulare J.	Rubiaceae							
Boerhavia diffusa L.	Portulacaceae							
Schrankia leptocarpa	Nyctaginaceae							
Mimosa pudica L *Mimosa invisa Mart* *Leucaena leucocephala*	Mimosaceae							
Ruthalicia eglandulosa	Cucurbitaceae							
Passiflora foetida L. *Adenia cissampeloides*	Passifloraceae							

Azolla pinnata R.								
Ludwigia spp.								
Nymphaea lotus L.								
Sphenoclea zeylanica Ga								
Parthenium hysterophorus		A[a]						
Conyza spp		D[a,b], A, F[c], CG[a]						
Sagittaria spp		A[b], F[b]						

Source: Daramola et al. (2020a, 2020b), Ayodele et al. (2024), Alagbo et al. (2022)

Note: ALS Inhibitor-A, ACC Inhibitors-B, Auxin mimetizer-C, D-EPSPS Inhibitors, E-PROTOX Inhibitors, F-PSII, and G-PSI Inhibitors

[a]Soybean

[b]Cereal crops

[c]Cotton

7 Conclusion

The negative impacts of herbicide application in terms of their intensity and resistance-inducing abilities in crops should be well managed and strategic to maintain stable crop compositions and abundance, sustain rich plant biodiversity, and minimize losses of plant bioresources. The adoption of chemical herbicides should be incorporated with other methods of weed control management strategies in sustainable crop production, thus protecting the environment from the hazardous effects of chemical pollution.

References

Adesina, G. O., Adelasoye, K. A., Akinjide, B. I., Abiola, S. O., & Adeniji, A. A. (2024). Assessment of pesticides residue in selected arable farm lands in Ogbomoso South local government area of Oyo state, Nigeria. *Agricultural Science and Technology, 16*(2), 75–85., Trakia University, Bulgaria ISSN 1313-8820 (print) ISSN 1314-412X. http://www.agriscitech.eu. https://doi.org/10.15547/ast.2024.02.020

Adeux, G., Munier-Jolain, N., Meunier, D., Farcy, P., Carlesi, S., Barberi, P., & Cordeau, S. (2019). Diversified grain-based cropping systems provide long-term weed control while limiting herbicide use and yield losses. INRA and Springer-Verlag France SAS, part of Springer Nature. *Agronomy for Sustainable Development, 39*, 42. https://doi.org/10.1007/s13593-019-0587-x

Adewumi, A. D., Olaniyi, O. W., & Adebayo, O. O. (2024). Cost-benefit analysis of selected pre-emergence herbicides for weed control in Cocoyam (*Xanthosoma Sagittifolium L.*) production. *Open Journal of Agricultural Science*. ISSN: 2734-214X, Article Ref. No.: OJAS0501001-538, *5*(1), 1–9. https://doi.org/10.52417/ojas.v5i1.538

Alagbo, O. O., Akinyemiju, O. A., & Chauhan, B. S. (2022). Weed management in rain-fed lowland rice ecology in Nigeria – challenges and opportunities. *Weed Technology, 36*, 583–591. https://doi.org/10.1017/wet.2022.57

Alcántara-de la Cruz, R., Silva, A. G., & Mendes, K. F. (2021). Absorption, translocation, and metabolism studies of herbicides in weeds and crops. In K. F. Mendes (Ed.), *Radioisotopes in weed research* (pp. 127–154). CRC Press.

Aney Parven, Islam Md Meftaul, Kadiyala Venkateswarlu, Saianand Gopalan, Mallavarapu Megharaj. (2024). Pre-emergence herbicides widely used in urban and farmland soils: fate, and potential human and environmental health risks. Environ Geochem Health: 46:132. https://doi.org/10.1007/s10653-024-01907-6.

Ayodele, O. P., Aluko, O. A., Amosun, J. O., & Udemba, I. O. (2024). Assessing the phytosociological characteristics of weed complex in okra field under different control strategies. *Plant Protection News, 107*(1), 16–23. All-Russian Institute of Plant Protection (St. Petersburg). Agronomy. https://doi.org/10.31993/2308-6459-2024-107-1-16427

Bajwa, A. A., Wang, H., Chauhan, B. S., & Adkins, S. W. (2019). Effect of elevated carbon dioxide concentration on growth, productivity and glyphosate response of parthenium weed (*Parthenium hysterophorus L.*). *Pest Management Science, 75*, 2934–2941.

Beckert, M., & Dessaux, Y. (2016). *Effects of herbicide-tolerant crop cultivation: Investigating the durability of a weed management tool*. Springer Nature. ISBN 978-94-024-1007-5 (eBook). https://doi.org/10.1007/978-94-024-1007-5

Borokini, T. I. (2011). Invasive alien plant species in Nigeria and their effects on biodiversity conservation. *National Centre for Genetic Resources and Biotechnology, P.M.B 5382, Moor*

Plantation, Ibadan, Nigeria. Open Access Journal - Tropical Conservation Science, 4(1), 103–110.

Bourdineauda, J.-P. (2020). Toxicity of the herbicides used on herbicide-tolerant crops, and societal consequences of their use in France. *Drug and Chemical Toxicology*. https://doi.org/10.1080/01480545.2020.177078

Daramola, O. S., Adigun, J. A., & Adeyemi, O. R. (2020a). Efficacy and economics of integrated weed management in chilli pepper (*Capsicum frutescens L.*). *Journal of Crop Improvement*. https://doi.org/10.1080/15427528.2020.1795770

Daramola, O. S., Adigun, J. A., & Olorunmaiye, P. M. (2020b). Review article challenges of weed management in rice for food security in Africa: A review. *Agricultura Tropica et Subtropica, 53*(3), 107–115. https://doi.org/10.2478/ats-2020-0011

de Aguiar, A. C. M., da Silva, A. A., Mendes, K. F., & da Costa Lima, A. (2022). Genetically modifed crops resistant to herbicides and weed control. In K. F. Mendes & A. A. da Silva (Eds.), *Applied weed and herbicide science* (p. 255). Springer Nature Switzerland AG. https://doi.org/10.1007/978-3-031-01938-8_8

de Souza, W. M., Paiva, M. C. G., Zaidan, Ú. R., Mendes, K. F., & de Freitas, F. C. L. (2022). Parameters of the phytosociological survey to evaluate the abundance, distribution, and diversity of the weed community. In K. F. Mendes & A. A. da Silva (Eds.), *Applied weed and herbicide science*. Springer Nature Switzerland AG. https://doi.org/10.1007/978-3-031-01938-8_3

Duke, S. O. (2018). The history and current status of glyphosate. *Pest Management Science, 74*, 1027–1034.

Duke, S. O., & Dayan, F. E. (2022). The search for new herbicide mechanisms of action: Is there a 'holy grail'? Perspective. *Pest Management Science, 78*, 1303–1313. www.soci.org. Society of Chemical Industry. https://doi.org/10.1002/ps.6726

Egan, J. F., Barlow, K. M., & Mortensen, D. A. (2014). A meta-analysis on the effects of 2, 4-D and dicamba drift on soybean and cotton. *Weed Science, 62*, 193–206.

Ekeleme, F., Dixon, A., Atser, G., Hauser, S., Chikoye, D., Olorunmaiye, P., Olojede, A., Korie, S., & Weller, S. (2020). Screening pre-emergence herbicides for weed control in cassava. *Weed Technology*. https://doi.org/10.1017/wet.2020.26

Esposito M, Westbrook AS, Maggio A, Cirillo V, DiTommaso A (2023) Neutral weed communities: the intersection between crop productivity, biodiversity, and weed ecosystem services. Weed Sci. 71:301–311. https://doi.org/10.1017/wsc.2023.27.

Gianessi, L. P. (2016). The increasing importance of herbicides in worldwide crop production. *Pest Management Science, 69*, 1099–1105.

Gholamhoseini, Majid., Aghaalikhani, Majid., Seyed Majid Mirlatifi, Seyed Ali Mohammad Modarres Sanavy. (2013). Weeds — Friend or foe? Increasing forage yield and decreasing nitrate leaching on a corn forage farm infested by redroot pigweed. Agriculture, ecosystems & environment. 179:151–162, 12p; ref: 3/4p.

He, B., Hu, Y., Wang, W., Yan, W., & Ye, Y. (2022). The progress towards novel herbicide modes of action and targeted herbicide development. *Agronomy, 12*, 2792. https://doi.org/10.3390/agronomy12112792

Heap, I. A. (2021). *The international survey of herbicide resistant weeds*. http://www.weedscience.org/Pages/fIter.aspx. Accessed 27 Mar 2021.

Heap, I. (2023). *The international survey of herbicide resistant weeds*. Available online: http://www.weedscience.org/Account/FAQ.Aspx#cite.

Kahlau, S., Schroder, F., Freigang, J., Laber, B., Lange, G., Passon, D., Kleessen, S., Lohse, M., Schulz, A., & von Koskull-Doring, P. (2020). Aclonifen targets solanesyl diphosphate synthase, representing a novel mode of action for herbicides. *Pest Management Science, 76*, 3377–3388.

Kughur, P. G. (2012). The effects of herbicides on crop production and environment in Makurdi Local Government Area of Benue State, Nigeria. *Journal of Sustainable Development in Africa, 14*(4) ISSN: 1520-5509. Clarion University of Pennsylvania, Clarion, Pennsylvania.

Liu, M., & Lu, S. (2016). Plastoquinone and ubiquinone in plants: Biosynthesis, physiological function and metabolic engineering. *Frontiers in Plant Science, 7*, 1898.

Mendes, R. R., Takano, H. K., & Leal, J. F. (2020). Evolution of EPSPS double mutation imparting glyphosate resistance in wild poinsettia (*Euphorbia heterophylla* L.). *PLoS One, 15*, e0238818. https://doi.org/10.1371/journal.pone.0238818

Montull, J. M., & Torra, J. (2023). Herbicide resistance is increasing in Spain: Concomitant management and prevention. *Plants, 12*, 469.

Nandula, V. K. (2010). *Herbicide resistance: Definitions and concepts. Glyphosate resistance in crops and weeds: History, development, and management* (pp. 35–43). Wiley.

Nandula, V. K., Ray, J. D., & Ribeiro, D. N. (2013). Glyphosate resistance in tall waterhemp (*Amaranthus tuberculatus*) from Mississippi is due to both altered target-site and non-target site mechanisms. *Weed Science, 61*(3), 374–383. https://doi.org/10.1614/WS-D-12-00155.1

Nath, C. P., Dubey, R. P., Sharma, A. R., Hazra, K. K., Kumar, N., & Singh, S. S. (2017). Evaluation of new generation post-emergence herbicides in Chickpea (Cicer arietinum L.). *National Academy Science Letters*. https://doi.org/10.1007/s40009-017-0604-z

Nath, C. P., Singh, R. G., Choudhary, V. K., Datta, D., Nandan, R., & Singh, S. S. (2024). Challenges and alternatives of herbicide-based weed management. *Agronomy, 14*, 126. https://doi.org/10.3390/agronomy14010126

Obiazi, C. C., Okogu, G. O., Ugbo, O., Alama, S. I., Uwabor, C. S., Akpomedaye, J. F., & Otite, F. O. (2024). Combined use of butachlor and cowpea for weed management in okra production. *International Journal of Phytology Research, 4*(2), 08-12 ISSN NO: 2583–08-12 ISSN NO: 0635.

Oliveira Júnior, R. S., & Bacarin, M. A. (2011). Absorção e translocação de herbicidas. In R. S. Oliveira Júnior, J. Constantin, & M. H. Inoue (Eds.), *Biologia e Manejo de Plantas Daninhas* (pp. 215–242). Omnipax.

Olorode, O. (2007). Biodiversity, globalization and poverty. *African Journal of Traditional, Complementary and Alternative Medicines, 4*(4), 532–540. ISSN 0189-6016.

Oreja, F. H., Inman, M. D., Jordan, D. L., & Leon, R. G. (2021). Population growth rates of weed species in response to herbicide programme intensity and their impact on weed community. *Weed Research*. https://doi.org/10.1111/wre.12509

Pontes Junior, V. B., da Silva, A. A., D'Antonino, L., Mendes, K. F., & de Paula Medeiros, B. A. (2022). Methods of control and integrated management of weeds in agriculture. In K. F. Mendes & A. A. da Silva (Eds.), *Applied weed and herbicide science*. Springer Nature Switzerland AG. https://doi.org/10.1007/978-3-031-01938-8_4

Refatti, J. P., de Avila, L. A., Camargo, E. R., Ziska, L. H., Oliveira, C., Salas-Perez, R., Rouse, C. E., & Roma-Burgos, N. (2019). High CO_2 and temperature increase resistance to cyhalofop-butyl in multiple-resistant *Echinochloa colona*. *Frontiers in Plant Science, 10*, 529.

Rehman, A. D., Ayyub, M., & Zahir, Z. A. (2023). The trend towards suppressing most troublesome weeds for sustainable production of wheat. In *Food and agriculture on social, economic and environmental linkages* (p. 74). Iksad Publications.

Shekhawat, D., & Singh, A. (2024). Trends in process and application of herbicides: A mini review. *International Journal of Agrochemistry, 10*(2), 1–6p.

Silva, F. B., Costa, A. C., Müller, C., Nascimento, K. T., Batista, P. F., Vital, R. G., Megguer, C. A., Jakelaitis, A., & Domingos, M. (2020). *Dipteryx alata*, a tree native to the Brazilian Cerrado, is sensitive to the herbicide nicosulfuron. *Ecotoxicology, 29*(2), 217–225.

Strom, S. A., Gonzini, L. C., Mitsdarfer, C., Davis, A. S., Riechers, D. E., & Hager, A. G. (2019). Characterization of multiple herbicide–resistant waterhemp (*Amaranthus tuberculatus*) populations from Illinois to VLCFA-inhibiting herbicides. *Weed Science, 67*, 369–379.

Wagner, V., & Nelson, C. R. (2014). Herbicides can negatively affect seed performance in native plants. *Restoration Ecology, 22*(3), 288–291.

Yogita Gharde, P.K. Singh and P.K. Gupta. Crop-weed competition and yield loss due to weeds in India. 2018. https://www.researchgate.net/publication/362156380.

Ziska, L. H. (2020). *Climate change and the herbicide paradigm: Visiting the future*. Environmental Health Science, Mailman School of Public Health, Columbia University.

Immedicable Effects of Organic Pesticides on Aquatic Environment

Mukul Machhindra Barwant, Shalagha A. B. Sharma, Debasree Lodh, Dipanwita Chaudhuri Sil, and Odangowei Inetiminebi Ogidi

1 Introduction

Pesticides play a significant influence in the health sector in addition to agriculture. The World Health Organization (WHO) has published the most widely used and successful methods for using pesticides to prevent malaria-carrying insects from spreading (Islam et al., 2022). Organic pesticides, often hailed as a safer alternative to synthetic chemicals, have become increasingly popular in agricultural practices due to their perceived environmental benefits. However, the widespread adoption of these substances has raised concerns regarding their impact on aquatic ecosystems (Ogidi et al., 2024b). Unlike synthetic pesticides, organic variants are typically derived from natural sources, such as plants, bacteria, and minerals, and are generally considered less toxic to non-target organisms. Nevertheless, recent studies suggest that organic pesticides may not be as benign as once thought, particularly concerning their long-term effects on aquatic environments. These days, there are many different kinds of hazards to human existence due to the declining condition of the environment. These days, food to satisfy hunger, water to slake thirst, and air to breathe are all dangerous. The everyday risks of human life have been demonstrated by infectious diseases. There are many dangerous substances all around us,

M. M. Barwant (✉)
Department of Botany, Sanjivani Rural Education Society's, Sanjivani Arts Commerce and Science College Kopargaon, Ahmednagar, Maharashtra, India

S. A. B. Sharma
School of Biological Engineering, SIET, Meerut, Uttar Pradesh, India

D. Lodh · D. C. Sil
Department of Botany, Holy Cross College, Agartala, Tripura, India

O. I. Ogidi
Department of Biochemistry, Faculty of Basic Medical Sciences, Bayelsa Medical University, Yenagoa, Bayelsa State, Nigeria

B. R. Babaniyi, E. E. Babaniyi (eds.), *The Interplay of Pesticides and Climate Change*, https://doi.org/10.1007/978-3-031-81669-7_16

and breathing, eating, or being exposed to them in any way can have a major negative impact on one's health. Because arsenic is a hazardous element, living has become extremely difficult in locations where soil, water, and airborne concentrations of the element are higher (Iyiola et al., 2023). Particularly in groundwater, since the majority of people on the planet, especially in nations like Brazil, Bangladesh, China, India, Nepal, Thailand, United States, Canada, and England, depend on groundwater for drinking. Natural sources are typically to blame for groundwater contamination with arsenic. Anthropogenic sources of environmental arsenic pollution include mining and the smelting of minerals containing arsenic. Depending on the prevailing physicochemical circumstances of the environment, arsenic can exist in the environment in many forms, such as –III, 0, +III, and +V. Drinking water that is high in arsenic continuously can lead to a variety of chronic illnesses in people, including keratosis, pigmentation, nausea, and, most significantly, cancer. It is discovered that the harmful effects of inorganic arsenic compounds outweigh those of organic versions (Sarkar & Paul, 2016). One of the main obstacles in the face of this population growth is plant diseases, particularly those caused by soil-borne pathogens that result in a large loss of yield (Ogidi, 2023). Moreover, the commonly employed management techniques are not practicable in terms of the environment. Thus, the necessity of the hour is sustainable agriculture. We have covered the impact of soil pathogens on soil health in this chapter, as well as the necessity of employing sustainable agriculture to successfully control it (Debnath et al., 2020).

The immedicable effects of organic pesticides on aquatic ecosystems are becoming a growing area of concern among environmental scientists and policymakers. Aquatic systems, including rivers, lakes, and wetlands, are particularly vulnerable to pesticide contamination due to runoff from agricultural fields, leaching through soil, and direct application. These pesticides can enter water bodies and affect aquatic organisms, including fish, amphibians, and invertebrates, by disrupting their physiological processes, altering reproductive behaviors, and reducing biodiversity. Additionally, the bioaccumulation of certain organic pesticides in the food chain poses a significant threat to the health of higher trophic levels, including humans The Intergovernmental Panel on Climate Change (IPCC) of India predicts that among the most sensitive effects of climate change will be water shortages brought on by glacier retreat and concomitant sea level rise. However, India, a country with a vast population, relies heavily on agriculture, which is especially susceptible to water and climatic changes. As a result, adverse changes in the quantity and quality of water will endanger both food security and the country's economy. A shift in the temperature of the atmosphere also modifies the radiation balance, which throws off the rate of evaporation, the pattern of precipitation, etc.

As a result, regular occurrences of both floods and droughts disturb normal hydrological cycles. These natural disasters have been occurring more often in India's remote regions in recent years (Thokchom, 2020). The largest concern facing the entire planet is the growing population and the availability of food for everyone. In 2017, there were 821 million hungry people worldwide, or one in nine people. In Asia, there were 515 million hungry people, while in Africa, there were

256.5 million (Debnath et al., 2020). The environment and individuals are both greatly harmed by practically all pesticides, especially when applied incorrectly. Pesticide metabolites damage ecosystems and potential human health due to their presence in rivers, sediment, and the atmosphere, which are believed to be largely caused by the extensive use and discharge of pesticides. The destruction of the environment caused by pesticides is a major problem in many parts of the world, including Asia, Africa, Eastern Europe, Latin America, and the Middle East. Organochlorine pesticide (OCP) residues can enter aquatic habitats by a variety of pathways, including drainage water discharges, industry effluents, runoff from non-point sources, and wet/dry degradation. Pesticides can have extremely negative effects on several environmental compartments like water, soil, and other living things since they are toxic to particular specific groups of living things (Islam et al., 2022).

The interdependent arrangement of water, sediment, and biota that sustains human life is known as the aquatic environment. The degradation of aquatic environments caused by the accumulation of toxic inorganic and organic contaminants has the potential to have far-reaching, detrimental consequences on human health. Of all the dangerous pollutants, pesticides are among the most concerning. It is common practice to use pesticides in order to keep insects out of crops grown for human consumption and to clear homes and yards of disease-carrying animals such as rats, mice, ticks, and mosquitoes. The South Asian region is home to more than 30% of the world's top 10 countries that use pesticides.

Organochlorine pesticides (OCPs) such as dichloro-diphenyl-trichloroethane (DDT), aldrin, dieldrin, and polychlorinated biphenyls (PCBs) are among the most hazardous pesticides used today. Among environmental components like soil, water, fish tissues, and other aquatic biotas, OCPs can remain for over 40 years. More than half a million tonnes of OCPs in various nations pose a risk to both human and environmental health. The bioaccumulation and biomagnification of persistent pesticides contribute to the estimated 200,000 yearly incidences of severe pesticide toxicity, which affect 1–5 million people globally.

Toxic effects of pesticides in watery habitats include death, infertility, weak immune systems, and a host of other fish health issues including malignancies, tumors, lesions, and excess slime on scales and gills. Both acute and chronic effects, including skin and eye irritation, nausea, vertigo, and headaches, as well as diabetes, asthma, and cancer, can be caused by pesticide poisoning in humans. From both an Asian and global viewpoint, this review traces the historical development of organic pesticide levels and dispersion throughout several environmental compartments (Islam et al., 2022). What he calls "nutritional literacy" includes "physical, economic, and social access to a balanced diet, clean drinking water, environmental hygiene, primary health care, and nutrition."

The issues here are numerous and include dwindling land resources, increased productivity demands, reliable irrigation water sources, improved technology, farmer well-being, and, above all, environmental safety and economic stability. Due to the high cost of production, farming becomes demotivating, which exacerbates the issue of migration to the city and causes the village's life to stagnate for all eternity. Therefore, the primary goal of sustainable farming is to increase both large-scale and small-scale farm output while also satisfying the growing demand for food and farmers' own needs by utilizing locally accessible inputs and safeguarding the ecosystem (Debnath et al., 2020).

Particularly valuable natural resources include fisheries and aquatic resources, which include lakes, ponds, rivers, streams, and seas. Concern over the impact of a growing human population and human activities on aquatic ecosystems has been associated with the rise in awareness of fisheries and aquatic systems. One significant class of toxic substances that have been connected to human exposure and have a negative influence on aquatic habitats are pesticides. Because of the world's growing population and its constant need for more food, pesticide use is gradually rising worldwide (Islam et al., 2022). Despite the "organic" label, these pesticides can persist in the environment and may undergo transformation into more toxic compounds. The chronic exposure of aquatic organisms to sub-lethal concentrations of organic pesticides can lead to physiological stress, behavioral changes, and increased susceptibility to diseases. Moreover, the interaction of organic pesticides with other environmental stressors, such as climate change, pollution, and habitat destruction, can exacerbate their detrimental effects on aquatic ecosystems (Ogidi & Akpan, 2022; Ogidi et al., 2024b).

The aquatic environment is contaminated with chemicals that humans are exposed to through various means, including the use of organic pesticides for controlling aquatic weeds, insects, and garbage fish, runoff and percolation from agricultural areas, drift from aerial and land applications, and the discharge of industrial waste waters. The majority of the United States' surface waterways are generally contaminated with chlorinated hydrocarbon pesticides, according to chemical and biological data. The pesticide 2,4-D has contaminated ground and surface waterways in certain, limited instances. These organic pesticides are indirectly consumed by fish that is edible and are transferred to humans through drinking water. The use of organic pesticides to control aquatic weeds, insects, and garbage fish, runoff and percolation from agricultural areas, drift from aerial and land applications, and the discharge of industrial waste waters are some of the ways that humans are exposed to chemicals that contaminate the aquatic environment. Chemical and biological studies indicate that chlorinated hydrocarbon insecticides are typically present in most surface rivers in the United States. In a few isolated cases, the chemical 2,4-D has contaminated surface and ground streams. Fish that are edible indirectly eat these organic pesticides, which are then passed on to people through drinking water.

Organic matter is not rapidly removed via chemical precipitation, absorptionAQ1, retention and release by biologic systems, hydrolysis, adsorption on solid surfaces, and chemical and biologic destruction. pesticides, according to the vast majority of research, from a body of water. By imparting unpleasant flavors and odors, organic pesticides degrade the quality of drinking water. Threshold taste and odor levels can be utilized to develop drinking water standards with the required safety factors, since the concentrations of various organic thiophosphates and 2,4-D causing tastes and odors are well below subtoxic levels by factors ranging from 173,000 to 1113,000. At the current amounts thought to be present in surface and

Table 1 The different classes of pesticides and their mode of action

Pesticide class	Description	Toxicity	Mode of action	Reference
Rotenone (insecticides)	Produced mostly from Cubé resin, a root extract of *Lonchocarpus utilis* and *Lonchocarpus urucu,* and plant species of the genus Derris plant or Lonchocarpus (Leguminosae)	Extremely poisonous to fish, pigs, and animals with cold blood; used as a pesticide for beetles and caterpillars that devour leaves. Irritation of the skin and mucous membranes may result from direct touch	Complex I in the ETS is inhibited by rotenone, and blockage at this site may reduce energy output and cause toxicity in the entire animal. The slow-developing toxicity of these chemicals includes lethargy, paralysis, and death as a result of the disturbance of energy metabolism and the ensuing depletion of ATP	Isman (2006), Cabizza et al. (2004) and Bomford and Isman (1996)
Azadirachtin	*Azadirachta indica*, a member of the Meliaceae family, has become known for its extreme potency as a biopesticide. About 40% of neem tree seeds are composed of up of oil, with azadirachtin serving as the main active ingredient and the source of neem's insecticidal properties	Azadirachtin inhibits and eliminates a wide variety of caterpillar, thrips, and whitefly species	Azadirachtin prevents the prothoracic gland from producing and releasing moulting hormones, or ecdysteroids, which causes incomplete ecdysis in developing insects	Isman (2006)
Ryania	They are formed of *Ryania speciosa's* grounded stem	Extremely poisonous to citrus thrips and fruit moths	The stomach poison *Ryania* acts gradually. It causes insects to cease feeding shortly after consumption, although not producing instantaneous knockout paralysis. Regarding the precise mechanism of action in insect systems, not much information has been provided. Ryania is piperonyl butoxide (PBO) that is synergized successfully, and it is said to work best in heated weather	Isman (2006) and Dimetry et al. (1993)
Fluoroacetate	Numerous plants in South Africa and Australia produce fluoroacetate, which serves a significant use for the plants as a natural insecticide	It is extremely toxic to rodents and other mammals	First, it is transformed into fluoroacetyl-CoA, and then fluorocic acid is produced. The enzyme that transforms citric acid into cis-aconitic acid is inhibited by this structural counterpart, which also causes the citric acid to stop producing energy. Calcium is sequestered by the accumulating citric acid. Glutamic acid is consequently decreased along with A-ketoglutaric acid. The organism is negatively impacted by these changes. Due to the significant roles that calcium plays as an impulse mediator and glutamic acid as a transmitter in what are known as glutaminergic synapses, the nervous system is susceptible to these alterations. Moreover, it is extremely detrimental to stop aerobic energy generation	Horowitz and Isaaca (2004)

groundwaters, organic pesticides do not seem to pose any chronic or acute health risks to humans (Faust, 1964).

2 Overview of Organic Pesticides

Pesticides are organically manufactured compounds used in a variety of agricultural applications to avoid the propagation of diseases, weeds, and pests among plants (Sawyer et al., 2023; Ogidi et al., 2024c). Pesticides can be categorized in an array of ways. One general approach is by the type of pests they manage, such as herbicides, insecticides, fungicides, rodenticides, and nematicides. Using pesticides has several benefits, such as increased food production and quality and reduced insect-borne illness. However, it has raised concerns related to potential harm to the environment, particularly to water supplies (Sharma et al., 2019). Herbicides are chemical weed killers that are mostly used to restrict plant growth. Agronomic facilities, residential gardens, and agricultural sites employ insecticides to manage insects. Application of fungicides can occur either prior to or subsequent to the fungus infecting the plant species. Their purpose is to stop fungal infections in plants or seeds (Mahmood et al., 2016; Ogidi et al., 2024a, b, c).

Through the utilization of compounds that kill, mitigate, or repel pests, pesticides can be grouped based on how they impact them. The details of some of the most frequently used pesticides are shown Table 1.

3 Pathways of Pesticide Entry into Aquatic Environments

3.1 Atmospheric Deposition

Atmospheric deposition (AD) is a process that moves materials and contaminants from the atmosphere to the surface of urban catchments through precipitation or deposition in dry weather. Consequently, atmospheric deposition consists of both wet and dry components. Wet components lead to surface runoff pollution, whereas dry components may remove compounds and chemicals from catchment surfaces, potentially leading to surface runoff pollution (Marsalek et al., 2008). Air pollutants, both natural and artificial, are introduced to metropolitan areas through emissions and the movement of air from local, regional, and faraway sources. Numerous sources and atmospheric transport scales are responsible for the wide variation in chemical composition and rates of atmospheric deposition in both space and time (Rocher et al., 2004).

Among the substances carried by the air are those belonging to the class of volatile organic compounds (VOCs). Paints, fuels, fumigants, insecticides, solvents, and chemical precursors are just a few examples of the many items that include volatile organic compounds (VOCs). They can also form as a result of urban activity, like

Table 2 Different categories and classes of agriculture runoff with their mode of action

Category	Classes	Mode of action	References
Insecticides	Organophosphates, carbamates, pyrethroids, organochlorines, and neonicotinoids	Neurotoxic, bioaccumulates, and biomagnifies	Leroux et al. (1992)
Herbicides	Phosphonates, chlorophenoxy herbicides, bipyridyl herbicides	Neurotoxin to specific stages of insect during development	Jespers and De Waard (1995)
Fungicides	Thiocarbamates triazoles strobilurins	Prevent fungal spore formation and stop plant diseases	Jespers and De Waard (1995)

trihalomethanes, which are created when water is disinfected. Due to this, VOCs are abundant in urban areas and can enter surface and groundwater sources via many pathways, one of which is rainwater (Lopes, 1998). Twelve of the 129 volatile organic compound (VOC) priority pollutants identified by the United States Environmental Protection Agency (EPA) as having very high toxicity were included in the NURP (Naryn urban resilience program) investigation of urban runoff quality. This category includes three BTEX compounds and all of their derivatives, including chlorinated benzenes: benzene, ethylbenzene, and toluene. Only six samples per site and four chemicals in two to six of the 28 cities studied contained volatile organic compounds (VOCs) in the limited NURP sampling. The main sources of volatile organic compounds (VOCs) in stormwater include automobile fuels (either as particles and exhaust gases from vehicles or as spills from petrol stations), atmospheric transport and deposition, and industrial discharges (Lopes, 1998).

3.2 *Industrial Discharges*

Pollutants from industrial operations are known to infiltrate aquatic bodies through two different channels (Duke & Chung, 1995; Ellis, 1986). The first is through atmospheric emissions of pollutants, while the other is through surface runoff from industrial land.

Industrial activity categories clearly influence the amount of pollution that industries contribute to stormwater, and these contributions will vary as new technology for pollution management and industrial production are introduced and market needs change the nature of these activities. Most industrial sites have some commonalities, such as the following: operating vehicles, washing and storing goods and materials, managing waste (including wastewater pretreatment), and operating outdoor equipment (Duke & Chung, 1995). Transportation on industrial sites involves a diverse range of vehicles, including large industrial trucks and equipment, passenger autos, and more, all of which contribute to pollution emissions. Industrial contaminants that end up in stormwater include the following: micropollutants follow conventional pollutants. Microplastics may end up in stormwater from industrial sites due to a variety of manufacturing processes, such as those involved in making rubber, chemicals, paints, and varnishes; making personal care product plastic pellets; using plastic media in abrasive blasting; and so on (Horton et al., 2017).

Even though the stormwater retention ponds servicing commercial and industrial sectors produced the highest amounts of microplastics, stormwater was not shown to be the primary vector of microplastics from industrial activities (Liu et al., 2019).

3.3 Agriculture Runoff

Increasing the application of agrochemicals such fertilizers, fungicides, herbicides, and nematicides is one way to increase crop production (Ogidi & Akpan, 2023; Ogidi et al., 2023). According to Vandergragt et al. (2020), the world's population grew from 6.8 billion in 2009 to 7 billion in 2012. These agrochemicals are used to reduce the number of pests and increase output, which help provide enough food for everyone on the planet. Sufficient food supplies are necessary in developing nations, where 1.02 billion people or 15% of the population are undernourished and 1.3 billion people, or 19% of the population live on an inadequate diet (Ogidi & Izah, 2024). Nonetheless, a multitude of hazardous compounds seriously contaminate freshwater and terrestrial ecosystems, chiefly the agricultural sector's use of pesticides and fertilizers (Schwarzenbach et al., 2006) a problem that has worldwide implications (Table 2).

3.4 Urban Runoff

The water quality of these rivers may be influenced by any surface that is in contact with precipitation or storm discharge. Rain and precipitation can affect a variety of urban structures, including crash barriers, lampposts, sculptures, and buildings (Van Buren et al., 2000). Clay and concrete roofing materials are recognized for their capacity to elevate the pH of discharge (Lee & Bak, 2012). Runoff Cu and Zn are considerably influenced by metal building corrosion rust, which is visible on roofs and facades. The concentration of Cd, Pb, and Zn in the moisture-laden flows of a combined sewer system that serves a densely populated suburban district in central Paris was primarily attributed to roof runoff, as demonstrated by Gromaire et al. (2001). This was attributed to the easily corroded nature of roof cover materials. The primary locations where lead was found were in the slate roofing, window frames, and ventilation apertures of ancient structures. Zinc sheeting, which is employed for roofing and gutters, was the primary source of zinc and cadmium, which are minor components of zinc products. Another significant source of zinc is galvanized steel, which is employed on a variety of surfaces in the urban built environment, including lampposts, collision barriers, and building envelopes.

Robert-Sainte et al. (2009) discovered that the average annual Zn content in discharge from galvanized steel panels is 3081 μg/L. Just to emphasize this point. In addition, Steiner and Boller et al. (2002) predicted that roof copper could account for up to 50% of the copper utilized in urban drainage systems. The median value of the discharge from copper panels was 1905 μg/L, spanning 20 rain events. An average Cu concentration of 3090 μg/L was obtained from three replicate panel samples that were conducted for six rain events, as per Müller et al. (2019). There is evidence to suggest that the increasing prevalence of Corten weathering steel on urban surfaces, including sculptures and structures, contributes nickel to stormwater runoff. In order to safeguard structures and buildings from adverse weather conditions, atmospheric deposition, and stormwater, a diverse array of inorganic and organic materials have been integrated into the design of the building envelope. Many of these coatings, including biocides and paints, may contain hazardous substances and leach into stormwater. Urban runoff may come from new possible sources as a result of ongoing research and development of novel materials, paint formulas, surface coatings, and other materials, as well as from their uses in urban environments. Workable control solutions include material substitutes, policy controls, and locally applied aquatic body treatment procedures.

4 Toxicological Effects on Aquatic Organisms

Fish, amphibians, plants, microbes, invertebrates, and additional kinds of species constitute the aquatic ecosystem. Pesticides have both direct and indirect effects on these beings. The former entails changes to an organism's internal physiology (Singh & Mandal, 2013; Acosta-Sánchez et al., 2020). For instance, the immediate effect of pesticides on water flea exposure is their mortality, which can also be considered an indirect effect. Furthermore, the algae's proliferation could experience a significant increase in response to the removal of grazing pressure. Herbicides, particularly glyphosate, are implemented globally to regulate aquatic and terrestrial plants. It is also recognized that their use has had a detrimental impact on non-target aquatic organisms as a result of the significant increase in their prevalence in recent years (Paul et al., 2014). At its inception, the primary purpose of a process was to affect the flora (Larson, 2019). Nevertheless, in recent years, a number of articles have been published that emphasize the negative consequences of non-target organisms, some of which are fatal or sub-lethal (Preston, 2002; Khalid et al., 2020). Growth, morphological/behavioral changes, and a survival rate are all physical indicators of the exposed organisms. Reproductive performance, which encompasses sexual maturity, the time required to release the first brood, the time required for eggs to mature, fertility, and variations in reproduction characteristics, is frequently employed to quantify sub-lethal responses.

Furthermore, an array of biochemical markers, such as blood glucose levels, steroid metabolism, lipid peroxidation, AChE activity, as well as disruption of metabolic pathways, can be utilized to assess the toxicity in animals that have been exposed.

The evaluation of sub-lethal effects on arthropods has been done in numerous research using two direct measurements of growth: body weight and length. Drying organisms, which are sampled for 48 hours at an average temperature of 60 °C, yield simple dry weight (D'Andrea et al., 2020). The physical, biological, and chemical components of the marine environment are all intimately interacted with by fish, making them an integral part of the aquatic ecosystem. Because they are a source of nutrition for other marine animals and birds, they play a crucial role in the marine food chain. A large body of research has established a correlation between pesticide toxicity and declining fish populations (Salam et al., 2020; El-Shebly & El-Kad, 2008).

5 Sub-lethal and Chronic Effects

Sub-lethal and chronic effects of pesticides on aquatic organisms can have devastating impacts on their health and survival. These effects include impaired feeding behavior, reduced reproductive success, developmental abnormalities, and endocrine disruptions. Studies have shown that even at low concentrations, pesticides can disrupt the delicate balance of aquatic ecosystems and pose long-term threats to biodiversity (Kaur & Kaur, 2017).

5.1 Behavioral Changes

The sub-lethal effects of organic pesticides on aquatic organisms extend beyond altered behavior to encompass reproductive and developmental impairments. Female fish exposed to these compounds have exhibited reduced spawning success, decreased egg viability, and abnormal larval development, all of which can disrupt population dynamics and threaten species survival (Redshaw et al., 2019).

5.2 Reproductive and Developmental Effects

The reproductive and developmental effects of organic pesticides on aquatic organisms are a growing concern in environmental research. Exposure to these compounds has been linked to decreased fertility, increased embryo mortality, and developmental abnormalities in a wide range of species, including fish, amphibians, and invertebrates. These effects can have cascading impacts on population dynamics, biodiversity, and ecosystem stability (Van den Brink et al., 2018).

5.3 *Endocrine Disruptions*

One of the most insidious effects of organic pesticides on aquatic organisms is their ability to disrupt endocrine function. These compounds can mimic or interfere with hormones in the endocrine system, leading to altered reproductive behaviors, disrupted growth patterns, and impaired immune function. Endocrine disruptions can have long-lasting effects on individual organisms and populations, with implications for ecosystem health and resilience (Orton et al., 2020).

6 Risk Assessment and Management

Pesticides are also quite persistent even when they are of biological origin. Organic pesticides when introduced into water bodies through agricultural runoff account for a large percentage of the pollution due to unnecessary pesticide application on agricultural crops. Risk assessment can be defined as the formulation and evaluation of risk management strategies, including the mechanism of hazard identification, defining dose effect relationship, risk communication, exposure and risk characterization.

6.1 *Hazard Identification*

Hazard identification is the assessment of the pesticide's properties, which includes its chemical composition, toxicity, and persistence in the environment. Whereas organic pesticides, such as those derived from pyrethrin and neem, are less toxic to humans, it has been established through research that they do have damaging impacts on aquatic species like fish and amphibians-invertebrates (Chen et al., 2019). For example, pyrethrins are very toxic to fish, exhibiting neurotoxic effects in terms of their swimming, feeding, and reproduction even at the recommended concentrations (Mogren & Lundgren, 2016; Van Wijngaarden et al., 2005). In addition, other aquatic invertebrates such as crustaceans (*Daphnia magna*) are also vulnerable, though reduced survival and reproduction have been linked with neem oil and pyrethrins (Anderson, 2018). For example, some of the amphibians are highly sensitive to chemicals as their skin is permeable, and due to this development and reproduction are affected by pesticides such as azadirachtin and rotenone. Moreover, azadirachtin affects hormonal regulation in frogs, leading to malformations and reductions in survival in frogs (Hayes et al., 2006).

Based on identification and management of hazards, it would be understood that knowing the means through which these pesticides enter the aquatic systems via runoff, leaching, or aerial drift is prime importance for hazards (Giddings et al., 2001; Mian & Mulla, 1992; Solomon et al., 2001).

6.2 Dose–Response Assessment

The harmful effects of organic pesticides on non-target aquatic species are assessed, and their maximum harmful levels are determined in dose–response evaluation. The pesticides may cause significant hazards even though they originate from natural sources, especially at high doses. Toxicological studies, including LD50 and LC50, are considered crucial in assessing the pesticide-related risks (Meyer et al., 2013).

Exposure of pesticides over an extended period of time effects aquatic animals such as change in behavior, success in reproduction, and deformity in development which are important in assessing the dose–response relationship, as they can lead to population declines over time (Sánchez-Bayo & Tennekes, 2017). Even biodegradable pesticides such as spinosad and azadirachtin can be toxic to some aquatic species, during prolonged exposure (Goulson, 2014). For example, even in low quantities azadirachtin (active compound in neem-based pesticides) has been proven to affect developing fish and amphibians in which it alters their endocrine systems leading to abnormal growth and reproduction (Dhillon & Sharma, 2004). It also has been noticed that, accumulation of sub-lethal doses of azadirachtin interferes with metamorphosis in amphibians which disrupt enzymatic pathways required for prompt completion of metamorphosis causing malformations (Hayes et al., 2006).

6.3 Exposure Assessment

In exposure assessment, one of the major areas is understanding the pathway, timing, and the level of exposure by non-target aquatic organisms to organic pesticides. This explores how pesticides enter the environment, how often, for how long, and in what quantities. The water solubility, soil behavior, and degradation are some of the main determinants of the pesticide behavior. In case of such easily soluble pesticides as neem oil derivatives in high distribution systems, the exposure risks are higher (Baskaran et al., 2018).

Regions with significant agricultural activity are typically associated with increases in pesticide exposure in the surrounding water bodies (Schafer et al., 2019). Prolonged exposure, even at low concentrations, may lead to chronic toxicity as well as bioaccumulation. The other important pathways are soil leaching into groundwater, agricultural runoff, and fall out of pesticides applied during crop protection activities (Giddings et al., 2001).

To estimate exposure levels, environmental monitoring (measurement of pesticide concentrations in water, sediments, and aquatic organisms) and modeling (distribution of pesticides based on application rates, environmental conditions and chemical properties) approaches are often used. These techniques evaluate both short- and long-term exposure (Giddings et al., 2001). It has been observed that pesticide exposure is higher in areas closer to agricultural fields or in low-flow regions of a water body where pesticides are prone to deposition (Van Wijngaarden et al., 2005).

6.4 *Risk Characterization*

Organic pesticides risk characterization deals with the description of the risks that are encountered by aquatic ecosystems, through interlinking data on hazard, dose–response, and exposure assessment. This process is carried out by determining the risk quotient and differentiating between acute and chronic risks according to their levels of exposure (Solomon et al., 2001). Acute risks include instances where pesticide concentrations exceed toxicity thresholds, for example, upon pesticide runoff following heavy rainfall events that can directly affect non-target species, especially during sensitive life stages, such as fish larvae (Van Wijngaarden et al., 2005). Chronic risks result from an extended period of exposure to lower concentrations of pesticides, resulting in sub-lethal effects such as reproductive or developmental problems. For instance, the azadirachtin from neem has been well known to influence an amphibian's growth pattern and influence their reproductive system in the long run (Hayes et al., 2006).

Risk characterization also considers bioaccumulation, in which pesticides persist in organisms' tissues and magnify up the food chain and pose greater risks to higher trophic levels (Giddings et al., 2001). The outcome of this step is prediction of the potential risks under varying environmental scenarios and is often applied to making decisions regarding use and management interventions with respect to pesticides in vulnerable ecosystems with species that are highly sensitive to exposure to pesticides (Richardson et al., 2016).

6.5 *Mitigation Strategies*

The impact of organic pesticide on the aquatic ecosystems can be minimized through better agriculture practice, advancement of technology, and by following strict rule of law. Best management practices such as buffer strips and constructed wetlands reduce pesticides and enhance microbial degradation of pesticides (Vymazal, 2011). Integrated pest management (IPM), which is the integration of biological control with non-chemical methods, minimizes the use of pesticides (Gurr et al., 2016). Precision agriculture and drone technology also optimize pesticide application (Zhang & Kovacs, 2012; Zhang et al., 2018). Policies like the European Union Water Framework Directive and the U.S. Environmental Protection Agency's National Pollutant Discharge Elimination System govern pesticide discharge (European Parliament, 2000; EPA, 2011). Public awareness and bioremediation, using plants and microorganisms, further promote sustainable practices (Ogidi & Akpan, 2024; Pretty et al., 2011; Anwar et al., 2020).

7 Regulatory Framework and Policies

7.1 International Regulations

Organic pesticides, though less toxic when compared to those made synthetically, still affect aquatic systems. Treaties such as the Stockholm Convention limit the persistent organic pollutants that accumulate in water (Stockholm Convention, 2004). The Rotterdam Convention on the other hand controls the international trade in pesticides and exports of these substances must be agreed (FAO, 1998). American laws such as FIFRA and the Clean Water Act focus on pesticide safe use with no harm to aquatic ecosystems (EPA, 2011). Various treaties aim to help people reduce pesticide wastes (Secretariat of the Basel Convention, 1992) but Codex Alimentarius establishes no residue excesses (FAO & WHO, 1963). Strategic approach and International Chemicals Management (SAICM) and Convention on Biological Diversity (CDB) focus on the safe use of chemicals and preservation of the environment (SAICM, 2006; Secretariat of the CBD, 1992). The GHS also aims at the precise labeling of pesticides in easy understandable terms to the ordinary person (UNECE, 2003), while the United Nations Sustainable Development Goals (SDGs) aim to deal with water pollution by the year 2030 (United Nations, 2015). There are also Community Level Integrated Water Resource Management in which the European Union Water Framework Directive and the FAO's Code of Conduct deal with also promotion of citizens about the dangers and disadvantages of pesticides (FAO 2014; Ogidi & Richard-Nwachukwu, 2024).

7.2 National and Regional Policies

The Insecticides Act, 1968 controls all pesticides in India, organic pesticides included; however, organic pesticides have less stringent scrutiny compared to synthetic pesticide products, which may become a great risk to aquatic environments. The National Policy on IPM promotes biopesticides but has no policy to control runoff into the water bodies (Sharma et al., 2021). National Water Policy, 2012, shows importance on controlling water pollution, but it does not mention risks of organic pesticide (Ministry of Water Resources, 2012). Biological Diversity Act, 2002, protects biodiversity, but it also overlooks impacts of pesticide runoff in aquatic ecosystems (Biodiversity Authority of India, 2020).

In Northern India, Punjab and Haryana practice organic farming to reduce pesticide use and monitor water quality (Singh et al., 2020). In the Western region, Maharashtra promotes bio pesticides, while Gujarat monitors the pesticide's drift and runoff in the river systems (Patil et al., 2021). Kerala promotes organic farming near sensitive ecosystems and regularly monitors for pesticide contamination, and Tamil Nadu puts emphasis on guidelines that decrease pesticide runoff (Rajesh & Soman, 2021). West Bengal promotes integrated pest management through training farmers and monitoring the pesticide levels in the Ganges. In Assam, sustainable tea cultivation is designed to minimize pesticides and protect water bodies (Dutta et al., 2020).

7.3 Future Directions in Pesticide Regulation

New approach to pesticide regulations in the future should look at both the benefits and the dangers of the risks associated with synthetic and organic pesticides such as monocrotophos through toxicity and persistence risk assessments (Rajan et al., 2023; Pérez et al., 2022). Improved biosensors and satellite imaging monitoring system and currently available innovative green biopesticides should be used in order to minimize the impacts on water ecosystems (Ghosh & Sharma, 2020; Lee et al., 2022). Regulations must encourage development, education, and funding of safer options for ecological agriculture (Singh et al., 2021).

8 Conclusion

An aquatic environment is an ecosystem that survives freely without interference of human but due to the human activity organic pesticide release in aquatic body and which will directly and indirectly impact living world. It will be impactful various toxic impact, increases pollution, water borne diseases water quality will decreases. So, it is crucial to critically evaluate the environmental safety of organic pesticides, particularly in the context of aquatic environments.

References

Acosta-Sánchez, A., Soto-Garita, C., Masís-Mora, M., Cambronero-Heinrichs, J. C., & Rodríguez-Rodríguez, C. E. (2020). Impaired pesticide removal and detoxification by biomixtures during the simulated pesticide application cycle of a tropical agricultural system. *Ecotoxicology and Environmental Safety, 195*(110460), 110460.

Anderson, B. S., Phillips, B. M., Voorhees, J. P., Deng, X., Geraci, J., Worcester, K., & Tjeerdema, R. S. (2018). Changing patterns in water toxicity associated with current use pesticides in three California agriculture regions. *Integrated environmental assessment and management, 14*(2), 270–281.

Anwar, T., Ranjha, M. H., & Javed, N. (2020). Efficacy of bio-pesticides against larvae of Trogoderma granarium (Everts)(Coleoptera: Dermestidae) on maize. Pakistan Journal of Agricultural Sciences, 57(2).

Baskaran, R., Lee, J., & Yang, S. G. (2018). Clinical development of photodynamic agents and therapeutic applications. *Biomaterials research, 22*(1), 25.

Biodiversity Authority of India (2020), http://www.nbaindia.org/

Boller, M. A., Müller, M., Österlund, H., Nordqvist, K., & Marsalek, J. (2002). Viklander building surface materials as sources of micropollutants in building runoff: A pilot study. *Water Science and Technology, 46*, 190–197.

Bomford, M. K., & Isman, M. B. (1996). Desensitization of fifth instar Spodoptera litura (Lepidoptera: Noctuidae) to azadirachtin and neem. *Entomologia Experimentalis et Applicata, 81*, 307–313.

Cabizza, M., Angioni, A., Melis, M., Cabras, M., Tuberoso, C. V., & Cabras, P. (2004). Rotenone and rotenoids in cubè resins, formulations, and residues on olives. *Journal of Agricultural and Food Chemistry, 52*(2), 288–293.

Chen, K., Wang, J., Pang, J., Cao, Y., Xiong, Y., Li, X., ... & Lin, D. (2019). MMDetection: Open mmlab detection toolbox and benchmark. arXiv preprint arXiv:1906.07155.

Convention, R. (n.d.). Rotterdam convention – Home page. Pic.int. Retrieved September 21, 2024, from http://www.pic.int

D'Andrea, M. F., Letourneau, G., Rousseau, A. N., & Brodeur, J. C. (2020). Sensitivity analysis of the pesticide in water calculator model for applications in the Pampa region of Argentina. *The Science of the Total Environment, 698*(134232), 134232.

Debnath, D., Chakraborty, S., & Deb, G. (2020). Sustainable farming: The oldest but smartest soldier against hardest enemy; The soil pathogens. *Current Research and Innovations in Plant Pathology, 17*.

Debnath, B., Majumdar, M., Bhowmik, M., Bhowmik, K. L., Debnath, A., & Roy, D. N. (2020). The effective adsorption of tetracycline onto zirconia nanoparticles synthesized by novel microbial green technology. *Journal of environmental management, 261*, 110235.

Dhillon, M. K., & Sharma, P. D. (2004). Studies on biology and behavior of Earias vittella (Lepidoptera: Noctuidae) for mechanisms of resistance in different cotton genotypes. *Crop Protection (Guildford, Surrey), 23*(3), 235–241.

Dimetry, N. Z., Amer, S., & Reda, A. S. (1993). Biological activity of 2 neem seed kernel extracts against the 2-spotted spider mite Tetranychus urticae. *Journal of Applied Entomol, 116*, 308–312.

Directive – 2000/60 – EN – Water framework directive – EUR-Lex. (n.d.). Europa.Eu. Retrieved September 21, 2024, from https://eur-lex.europa.eu/legal-content/EN/TXT/?uri=celex%3A32000L0060

Duke, D. L., & Chung, Y. J. (1995). Industrial stormwater pollution prevention: Effectiveness and limitations of source controls in the transportation industry. *Waste Management, 15*, 543–558.

Dutta, P., Banerjee, A., & Ghosh, S. (2020). Pesticide contamination in the Sundarbans wetland ecosystem. *Environmental Science and Technology Review, 13*(7), 509–518.

Ellis, J. B. (1986). Pollutional aspects of urban runoff. In *Urban runoff pollution* (pp. 1–38). Springer.

El-Shebly, A., & El-Kad, M. (2008). Effects of glyphosate herbicide on serum growth hormone (GH) levels and muscle protein content in Nile tilapia (Oreochromis Niloticus L). *Oreochromis Niloticus L. Research Journal of Fisheries and Hydrobiology, 3*, 84–88.

Faust, S. D. (1964). Pollution of the water environment by organic pesticides. *Clinical Pharmacology and Therapeutics, 5*(6part1), 677–686.

Gromaire, M. C., Garnaud, S., & Saad, M. (2001). Chebbo contribution of different sources to the pollution of wet weather flows in combined sewers. *Water Research, 35*(2), 521–533.

Gurr, G. M., Wratten, S. D., & Snyder, W. E. (2016). *Biological control: Ecology and applications*. Cambridge University Press.
Hayes, T. B., Case, P., & Chui, S. (2006). Pesticide effects on amphibian development: Azadirachtin and its impact on hormonal regulation. *Biology of Amphibians, 22*(1), 23–29.
Home. (n.d.-a). Codexalimentarius.org. Retrieved September 21, 2024, from http://www.codexalimentarius.org
Home. (n.d.-b). FAOHome. Retrieved September 21, 2024, from http://www.fao.org
Home. (n.d.-c). Convention on Biological Diversity. Retrieved September 21, 2024, from https://www.cbd.int
Horowitz, A. R., & Isaaca, I. (2004). Bio-rational insecticides- mechanisms, selectivity and importance in pest management programs. In A. R. Horowitz & I. Isaaca (Eds.), *Insect pest management- field and protected crops* (pp. 1–28). Springer.
Horton, A., Walton, A., Spurgeon, D., & Lahive, E. (2017). Svendsen microplastics in freshwater and terrestrial environments: Evaluating the current understanding to identify the knowledge gaps and future research priorities. *Science of the Total Environment, 586*, 127–141.
Implementation of the Biological Diversity Act. 2002: An assessment. (2020). Biodiversity Authority of India.
Islam, M. A., Amin, S. M. N., Rahman, M. A., Juraimi, A. S., Uddin, M. K., Brown, C. L., & Arshad, A. (2022). Chronic effects of organic pesticides on the aquatic environment and human health: A review. *Environmental Nanotechnology Monitoring & Management, 18*(100740), 100740.
Isman, M. B. (2006). Botanical insecticides, deterrents, and repellents in modern agriculture and an increasingly regulated world. *Annual Review of Entomology, 51*(1), 45–66.
Iyiola, A. O., Kolawole, A. S., Ajayi, F. O., Ogidi, O. I., & Ogwu, M. C. (2023). Sustainable water use and management for agricultural transformation in Africa. In *Water resources management for rural development: Challenges and mitigation* (Vol. 1, pp. 287–300). Published by Elsevier. https://doi.org/10.1016/B978-0-443-18778-0.00001-5
Jespers, A. B. K., & De Waard, M. A. (1995). Effect of fenpiclonil on phosphorylation of glucose in *Fusarium sulphureum*. *Pesticide Science, 44*(2), 167–175.
Kattwinkel, M., Liess, M., & Beketov, M. A. (2015). Ecological impacts of pesticides and pesticide mixtures on aquatic ecosystems. *Aquatic Toxicology, 164*, 189–197.
Kaur, P., & Kaur, S. (2017). Sub lethal effects of pesticides on fish health. *International Journal of Fisheries and Aquatic Studies, 5*(5), 229–235.
Khalid, S., Shahid, M., Murtaza, B., Bibi, I., Natasha, A., Naeem, M., & Niazi, N. K. (2020). A critical review of different factors governing the fate of pesticides in soil under biochar application. *The Science of the Total Environment, 711*(134645), 134645.
Larson, S. J. (2019). *Pesticides in surface waters: Distribution, trends, and governing factors*. CRC Press.
Leaner, J. J., & Mason, R. P. (2001). The effect of thiolate organic compounds on methylmercury accumulation and redistribution in sheepshead minnows, Cyprinodon variegatus. *Environmental Toxicology and Chemistry, 20*(7), 1557–1563.
Lee, J. Y., & Bak, G. (2012). Han quality of roof-harvested rainwater – Comparison of different roofing materials. *Environmental Pollution, 162*, 422–429.
Leroux, P., Lanen, C., & Fritz, R. (1992). Similarities in the antifungal activities of fenpiclonil, iprodione and tolclofos-methyl against *Botrytis cinerea* and *Fusarium nivale*. *Pesticide Science, 36*(3), 255–261. https://doi.org/10.1002/ps.2780360312
Liu, F., Olesen, K. B., & Borregaard, A. R. (2019). Vollertsen microplastics in urban and highway stormwater retention ponds. *Science of the Total Environment, 671*, 992–1000.
Lopes, T. J. (1998). Bender nonpoint sources of volatile organic compounds in urban areas – Relative importance of land surfaces and air. *Environmental Pollution, 101*, 221–230.
Mahmood, I., Imadi, S. R., Shazadi, K., & Gul, A. (2016). Hakeem effects of pesticides on environment. In K. Hakeem, M. Akhtar, & S. Abdullah (Eds.), *Plant, soil and microbes*. Springer.
Marsalek, B., Jimenez-Cisneros, M., Karamouz, P.-A., Malmqvist, J., & Goldenfum, B. (2008). *Chocat urban water cycle processes and interactions*. Taylor & Francis Group.

Müller, A., Österlund, H., Nordqvist, K., & Marsalek, J. (2019). Viklander building surface materials as sources of micropollutants in building runoff: A pilot study. *Science of the Total Environment, 680*, 190–197.

National Water Policy 2012. (2012). Ministry of Water Resources.

Ogidi, O. I. (2023). Sustainable utilization of important medicinal plants in Africa. In *Sustainable utilization and conservation of Africa's biological resources and environment, sustainable development and biodiversity*. Springer. https://doi.org/10.1007/978-981-19-6974-4_12

Ogidi, O. I., & Akpan, U. M. (2022). Chapter 16: Aquatic biodiversity loss: Impacts of pollution and anthropogenic activities and strategies for conservation. In *Biodiversity in Africa: Potentials, threats and conservation*. Springer. https://doi.org/10.1007/978-981-19-3326-4_16

Ogidi, O. I., & Akpan, U. M. (2023). Impacts of chemical use in agricultural practices: Perspectives of soil microorganism and vegetation. In *One health implications of agrochemicals and their sustainable alternative*. Springer.

Ogidi, O. I., & Akpan, U. M. (2024). Microbial and phyto-remediation of crude oil contaminated soil. In *Eco-restoration of the environment*. CRC Press, Routledge Taylor and Francis Group.

Ogidi, O. I., & Izah, S. C. (2024). Emerging Technologies in the global south food industry: Prospects and challenges. In *Food safety and quality in global south*. Springer.

Ogidi, O. I., & Richard-Nwachukwu, N. (2024). Microbe assisted remediation of pesticide residues from soil and water. In *Eco-restoration of the environment*. CRC Press, Routledge Taylor and Francis Group.

Ogidi, O. I., Izah, S. C., & Akpan, U. M. (2023). Food quality and agrochemical use: Integrated monitoring, assessment, and management policies. In *One health implications of agrochemicals and their sustainable alternative*. Springer.

Ogidi, O. I., Emaikwu, N., & G. (2024a). Plant phenolic compound isolation and its bioinformatics approaches of molecular mechanisms in antimicrobial activities and resistance. In *Computational approaches in biomaterials, bioinformatics and biomedical engineering applications*. CRC Press, Taylor and Francis Group.

Ogidi, O. I., Angaye, T. C. N., & Carbom, H. E. (2024b). Microorganisms in the monitoring of terrestrial and aquatic ecosystems: Applications, challenges and future prospects. In *Biomonitoring of pollutants in the global south*. Springer.

Ogidi, O. I., Izah, S. C., & Tobia, P. S. (2024c). Biosurfactant production from cassava processing wastes. In *Sustainable cassava: Strategies from production through waste management*. Elsevier.

Patil, R., Deshmukh, S., & Joshi, S. (2021). Environmental regulations for pesticide use in Maharashtra and Gujarat. *Journal of Agricultural Research, 24*(4), 155–164.

Paul, Carolyn, & Wilhelmine. (2014). Lethal and sublethal effects of pesticides on aquatic organisms: The case of a freshwater shrimp exposure to roundup®. In *Pesticides – Toxic aspects*. InTech.

Preston, B. L. (2002). Indirect effects in aquatic ecotoxicology: Implications for ecological risk assessment. *Environmental Management, 29*(3), 311–323.

Pretty, J., Toulmin, C., & Williams, S. (2011). Sustainable intensification in African agriculture. *International Journal of Agricultural Sustainability, 9*(1), 5–24.

Rajesh, G., & Soman, P. (2021). Organic farming practices and environmental impacts in South India. *Journal of Environmental Monitoring, 11*(2), 85–97.

Richardson, R. J., Haller, W. T., & Petty, D. G. (2016). Aquatic plants and the ecological impacts of herbicides in aquatic environments. *Journal of Aquatic Plant Management, 54*(2), 89–104.

Robert-Sainte, P., Gromaire, M. C., De Gouvello, B., & Saad, M. (2009). Chebbo annual metallic flows in roof runoff from different materials: Test-bed scale in Paris conurbation environ. *Environmental Science & Technology, 43*(15), 5612–5618.

Rocher, V., Azimi, S., Gasperi, J., Beuvin, L., Mulle, M., & Moilleron, R. (2004). Chebbo hydrocarbons and metals in atmospheric deposition and roof runoff in Central Paris. *Water, Air, & Soil Pollution, 159*, 67–86.

Salam, M. A., AbuKhadra, M. R., & Mohamed, A. S. (2020). Effective oxidation of methyl parathion pesticide in water over recycled glass based-MCM-41 decorated by green Co3O4 nanoparticles. *Environmental Pollution (Barking, Essex: 1987), 259*(113874), 113874.

Sawyer, W. E., Jacob, D. E., Ogidi, O. I., Izah, S. C., & Popoola, A. M. (2023). Threats and conservation strategies of common edible vegetables that possess pharmacological potentials in Nigeria. In *Reference series in phytochemistry. Herbal medicine phytochemistry*. Springer.

Schäfer, T., & Schwarz, M. A. (2019). The meaningfulness of effect sizes in psychological research: Differences between sub-disciplines and the impact of potential biases. *Frontiers in psychology, 10*, 813.

Schwarzenbach, R. P., Escher, B. I., Fenner, K., Hofstetter, T. B., Johnson, C. A., von Gunten, U., & Wehrli, B. (2006). The challenge of micropollutants in aquatic systems. *Science (New York, N.Y.), 313*(5790), 1072–1077.

Sharma, A., Kumar, V., Shahzad, B., Tanveer, M., Sidhu, G. P. S., Handa, N., Kohli, S. K., Yadav, P., Bali, A. S., Parihar, R. D., Dar, O. I., Singh, K., Jasrotia, S., Bakshi, P., Ramakrishnan, M., Kumar, S., Bhardwaj, R., & Thukral, A. K. (2019). Worldwide pesticide usage and its impacts on ecosystem. *SN Applied Sciences, 1*(11).

Sharma, G. D., Thomas, A., & Paul, J. (2021). Reviving tourism industry post-COVID-19: A resilience-based framework. Tourism management perspectives, 37, 100786.

Simon, O., & Garnier-Laplace, J. (2005). Laboratory and field assessment of uranium trophic transfer efficiency in the crayfish Orconectes limosus fed the bivalve C. fluminea. *Aquatic Toxicology (Amsterdam, Netherlands), 74*(4), 372–383.

Singh, D. B., & Mandal, K. (2013). *Environmental impact of pesticides belonging to newer chemistry* (pp. 152–190). Integrated Pest Management.

Singh, N., Sharma, P., & Yadav, S. (2020). Water quality monitoring and sustainable agriculture in Punjab and Haryana. *Journal of Environmental Protection and Agriculture, 18*(3), 102–114.

Thokchom, B. (2020). Water-related problem with special reference to global climate change in India. In *Water conservation and wastewater treatment in BRICS nations* (pp. 37–60). Elsevier.

US EPA, O. W. (2014). National pollutant discharge elimination system (NPDES). https://www.epa.gov/npdes

Van Buren, M. A., Watt, W. E., & Marsalek, J. (2000). Anderson thermal enhancement of stormwater runoff by paved surfaces. *Water Research, 34*(4), 1359–1371.

van Schalkwyk, J., Pryke, J. S., Samways, M. J., & Gaigher, R. (2019). Congruence between arthropod and plant diversity in a biodiversity hotspot largely driven by underlying abiotic factors. *Ecological Applications: A Publication of the Ecological Society of America, 29*(4), e01883. https://doi.org/10.1002/eap.1883

Vymazal, J. (2011). Constructed wetlands for wastewater treatment: Five decades of experience. *Environmental Science & Technology, 45*(1), 61–69.

Zhang, C., & Kovacs, J. M. (2012). The application of small unmanned aerial systems for precision agriculture: A review. *Precision Agriculture, 13*(6), 693–712.

Part IV
Strategies for Sustainable Practice

This part establishes the interconnected dynamics of public perception, education, policy integration, economic considerations, and collaborative efforts that contribute to the shaping of sustainable agricultural practices.

Community Health and Vulnerability

Mustapha Abdulsalam, Musa Ojeba Innocent, Miracle Uwa Livinus, Abdulhakeem Idris Abdulhakeem, Fatimoh Abdulsalam Danjuma, Abdulsalam Khadijah, Tijani Abiola Tajudeen, Alege Abdulraheem Lateefat, Idowu Afeez Temitope, and Ishola Jonathan Adekunle

1 Introduction

Pesticides are used mainly in agriculture to help to fight pests and also increase food production. New information and analysis confirmed that approximately 2.5 million tons of pesticides are utilized on a global basis yearly, and vary significantly per region the agriculture type, climate, and the pest density (Tudi et al., 2021). This nevertheless, poses severe impacts to the environment and health such as deterioration of the soil, water and air pollution as well as negative health impacts to humans

M. Abdulsalam (✉) · M. O. Innocent
Department of Microbiology, Skyline University Nigeria, Kano, Nigeria

M. U. Livinus
Department of Biochemistry, Skyline University Nigeria, Kano, Nigeria

A. I. Abdulhakeem
Department of Fisheries and Aquaculture, Federal University Dutsin-Ma, Katsina, Nigeria

F. A. Danjuma
Ministry of National Guard Hospital, Riyadh, Saudi Arabia

A. Khadijah
Department of Agricultural Science, University of Ilorin, Ilorin, Nigeria

T. A. Tajudeen
Department of Soil Science, Federal University of Technology, Minna, Nigeria

A. A. Lateefat
Department of Microbiology, Kwara State University, Malete, Nigeria

I. A. Temitope
Department of Medicinal Research, Kaohsiung Medical University Hospital, Kaohsiung, Taiwan

I. J. Adekunle
Department of Public Health, Obafemi Awolowo University, Ile-Ife, Nigeria

B. R. Babaniyi, E. E. Babaniyi (eds.), *The Interplay of Pesticides and Climate Change*, https://doi.org/10.1007/978-3-031-81669-7_17

and other organisms. Charges of pesticide risks are made worse by climate change as this modifies the conditions of the environment and in the process influences the pesticides and their actions. Global warming, which occurs through the enhancement of global temperatures, change in rainfall pattern, and frequent disasters, always impacts on distribution and effectiveness if pesticides. Optimal temperatures raise the rate of evaporating of some of the pesticides, a factor that leads to the formation of aerosols and hitherto unseen hazards to the atmosphere (Boonupara et al., 2023). Times have also changed, and increased rain induces flow of pesticides into water reservoirs and thus affecting both human and animal water supplies and aquatic lives. In particular, climate change may influence pest population dynamics and activity, which may require additional pesticide treatments due to the emergence of new or increased pathogens, pests, and diseases (Skendžić et al., 2021).

Community health refers to the collective health status and quality of life of a population within a specific geographic area. It encompasses a wide range of factors, including physical, mental, and social well-being, access to healthcare, and environmental conditions. Vulnerability, in the context of community health, is the susceptibility of certain groups or individuals to adverse health outcomes due to various socioeconomic, environmental, and demographic factors (Ellena et al., 2020). Vulnerable populations, such as children, the elderly, pregnant women, agricultural workers, and low-income communities, are at a heightened risk of exposure to pesticides due to their increased contact with contaminated water, soil, or food. These groups may also face greater health risks due to preexisting conditions or limited access to healthcare. Climate change exacerbates these vulnerabilities by altering the environment in ways that can increase both direct and indirect exposure to pesticides, further compromising community health (Parsons et al., 2024).

1.1 Importance of Studying Their Interplay

Understanding the interplay between pesticide use and climate change is critical for several reasons. It allows for a comprehensive assessment of how these two factors collectively impact public health, particularly among vulnerable populations. The combination of increased pesticide use due to changing pest dynamics and heightened exposure risks resulting from altered environmental conditions presents new challenges for community health (Pathak et al., 2022). This understanding is essential for developing effective policy responses. Integrated strategies that address both pesticide regulation and climate change mitigation are necessary to protect public health and environmental integrity. Current regulatory frameworks often fail to consider the compound effects of climate change on pesticide behavior, leading to potential gaps in protection measures (Kubiak-Hardiman et al., 2023). However, studying the interplay between pesticide use and climate change highlights the need for innovative mitigation and adaptation strategies. This includes promoting climate-smart agricultural practices, strengthening community-based interventions, and enhancing public health surveillance systems to detect and respond to emerging threats promptly. A multidisciplinary approach that involves collaboration between environmental scientists, health professionals, policymakers, and local communities is key to effectively addressing these complex challenges (Rudd et al., 2018).

2 Pesticide Exposure and Human Health Risks

2.1 Routes of Exposure: Direct and Indirect

Pesticide exposure can occur through several routes, both direct and indirect.

(a) *Direct Exposure*

Direct pesticide exposure occurs when individuals come into immediate contact with pesticides during their application, handling, or use in both agricultural and non-agricultural settings. This type of exposure is particularly common among agricultural workers, pesticide applicators, and people living near treated areas. Direct exposure can happen through dermal contact, where pesticides are absorbed through the skin, inhalation of pesticide vapors or aerosols, or ingestion of contaminated food, water, or soil (Shah, 2020). These routes of exposure present immediate health risks to those in close proximity to pesticide use.

(b) *Indirect Exposure*

Indirect exposure occurs when individuals come into contact with pesticides after they have been applied. This exposure can happen through multiple pathways, such as consuming contaminated food and water, inhaling airborne pesticide residues carried by the wind, or contacting surfaces with pesticide residues. Pesticides can also persist in the environment by leaching into groundwater or running off into nearby water bodies, eventually entering the food chain and bioaccumulating in animals and plants (Musa et al., 2024a, b, c, d). Indirect exposure can affect a broader population, including those not directly involved in pesticide application, such as children, the elderly, and individuals living far from treated areas.

2.2 Acute and Chronic Health Effects of Pesticides

(a) *Acute Health Effects*

Acute health effects from pesticide exposure occur shortly after exposure and are typically reversible. These effects include headaches, dizziness, nausea, skin and eye irritation, and respiratory difficulties. In severe cases, acute poisoning can lead to seizures, coma, or even death. Acute pesticide poisoning is especially prevalent in developing countries, where protective measures and regulatory enforcement may be inadequate (Konradsen et al., 2003).

(b) *Chronic Health Effects*

Chronic health effects arise from prolonged or repeated exposure to pesticides, often at lower concentrations, over an extended period. These long-term effects include various cancers, such as non-Hodgkin lymphoma, leukemia, and prostate cancer, as well as neurodegenerative diseases like Parkinson's and Alzheimer's. Pesticides can also cause reproductive health issues, including infertility and birth defects, as well as endocrine disruption and respiratory disorders (Morris et al., 2010). Moreover, chronic exposure has been linked to developmental problems in children, such as cognitive deficits, behavioral disorders, and impaired growth (Tolins et al., 2014). Chronic exposure can occur through occupational activities, living near agricultural areas, or consuming contaminated food and water. Figure 1 illustrates the pathways of pesticide exposure (inhalation, ingestion, and dermal contact) and categorizes the resulting health effects into acute (such as neurological symptoms and skin irritation) and chronic (including cancer and endocrine disruption) impacts.

2.3 *Vulnerable Populations: Children, Elderly, and Immunocompromised Individuals*

Certain populations are more vulnerable to the adverse health effects of pesticides due to their unique physiological, developmental, or environmental conditions.

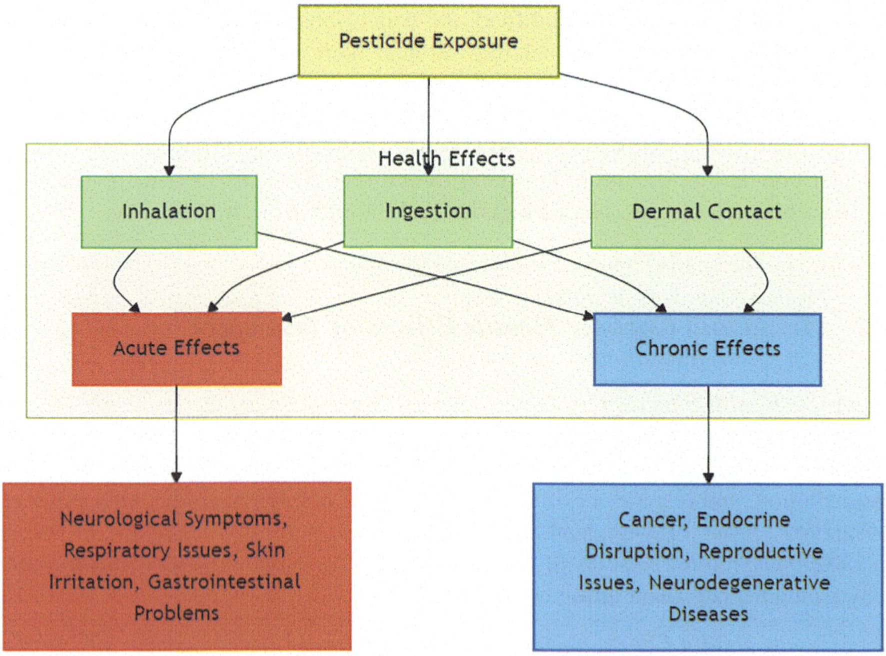

Fig. 1 Pathways of pesticide exposure and associated acute and chronic health effects

(a) Children are particularly susceptible to pesticide exposure because of their developing bodies and behaviors, such as frequent hand-to-mouth activities and playing on floors or soil where pesticide residues may linger. Due to their smaller body size and faster metabolic rate, children can absorb more pesticides relative to their body weight, leading to more severe health effects (Djekkoun et al., 2021). Prenatal and early childhood exposure to pesticides has been linked to cognitive impairments, developmental delays, and increased risks of autism spectrum disorders and attention-deficit/hyperactivity disorder (Tessari et al., 2022).
(b) The elderly are also at increased risk due to age-related changes in metabolism, immune function, and the body's ability to detoxify harmful substances. As aging often involves comorbidities like cardiovascular diseases or respiratory conditions, exposure to pesticides can exacerbate these existing health problems, leading to a higher incidence of hospitalization or mortality (Elonheimo et al., 2022). Moreover, the elderly might have prolonged exposure histories, leading to cumulative effects that heighten the risks of chronic diseases like cancer or neurodegenerative disorders (Bachmann et al., 2020).
(c) Immunocompromised individuals, including those with HIV/AIDS, cancer patients undergoing chemotherapy, and individuals with autoimmune diseases, have weakened immune systems that cannot effectively detoxify or repair damage caused by pesticide exposure. This increased susceptibility can result in more severe toxicological responses, including heightened risks of infections, organ damage, or cancers (Morad et al., 2021). Moreover, these individuals may experience longer recovery times from pesticide-related health impacts, compounding their vulnerability.

3 Impact of Climate Change on Pesticide Dynamics

3.1 Changes in Pesticide Degradation and Persistence

Climate change significantly affects the degradation and persistence of pesticides in the environment. Degradation refers to the breakdown of pesticide molecules into simpler compounds, while persistence indicates the duration that pesticides remain active in the environment. Factors like temperature, humidity, and sunlight play crucial roles in determining how quickly pesticides degrade. With rising global temperatures, certain pesticides may degrade more rapidly due to increased volatilization and microbial activity. However, this is not universally beneficial; faster degradation can sometimes lead to the formation of toxic metabolites, which may pose an even greater risk to human health and the environment (Pathak et al., 2022). Conversely, in cooler or drier climates, pesticides may persist longer due to reduced microbial degradation and limited chemical breakdown processes (Tudi et al., 2021). Higher temperatures can also increase the rate of photodegradation, which involves the breakdown of pesticides by sunlight. However, the rate and extent of photodegradation can vary significantly depending on the chemical properties of the pesticide and the intensity of sunlight (Khan & Pathak, 2020). Changes in precipitation patterns, such as increased rainfall or drought, also affect pesticide persistence. Heavy rainfall can wash pesticides off the soil surface into water bodies, reducing their effectiveness on target pests while contaminating aquatic ecosystems (Pradhan et al., 2022). On the other hand, prolonged droughts can concentrate pesticides in the soil, potentially enhancing their toxicity to plants and soil organisms (Kaka et al., 2021). Therefore, understanding the specific changes in pesticide dynamics under varying climatic conditions is essential for managing their use effectively.

3.2 Impact of Extreme Weather Events on Pesticide Spread

Extreme weather events, such as hurricanes, floods, droughts, and wildfires, are becoming more frequent and intense due to climate change, significantly impacting pesticide spread and distribution. Floods and heavy rains can transport pesticides from agricultural fields into rivers, lakes, and groundwater, causing widespread contamination of drinking water sources and aquatic habitats (Pradhan et al., 2022). Flooding events can lead to pesticide runoff into residential areas, increasing the risk of human exposure, especially in vulnerable communities lacking proper infrastructure (Ndimele et al., 2024). Droughts can also influence pesticide dynamics by increasing the concentration of pesticides in the soil, reducing their natural breakdown, and leading to higher exposure risks for plants, animals, and humans (Tudi et al., 2021). Moreover, drought conditions may cause pesticides to accumulate in specific areas, potentially reaching toxic levels that affect non-target organisms, such as pollinators, soil microorganisms, and beneficial insects (Serrão et al., 2022). Wildfires can volatilize pesticides that have been applied to vegetation, releasing them into the atmosphere and spreading them over large distances. This can result in unexpected exposure in regions far from the original application sites, posing risks to human health and the environment. Furthermore, wildfires can alter soil structure and chemistry, potentially influencing how pesticides interact with soil particles and persist in the environment (Lopez et al., 2024).

3.3 Effects on Crop Pests and Increased Pesticide Use

Climate change is also altering the dynamics of crop pests, which, in turn, affects pesticide use. Warmer temperatures and changing precipitation patterns create favorable conditions for the proliferation of various pests, including insects, weeds, and pathogens, extending their growing seasons and expanding their geographic ranges (Skendžić et al., 2021). For instance, pests such as aphids, whiteflies, and locusts thrive in warmer conditions, and climate-induced range expansions can bring these pests into new areas where they previously did not pose a threat (Srinivasa Rao et al., 2022). Moreso, climate change can influence the reproductive rates and life cycles of pests, allowing them to complete more generations within a single growing season, thereby increasing their population sizes and the intensity of infestations (Skendžić et al., 2021). As a result, farmers may need to increase pesticide applications to protect crops from these heightened pest pressures, potentially leading to a rise in pesticide resistance and a cycle of escalating pesticide use (Umina et al., 2019). The increased use of pesticides, driven by climate change, can have several adverse effects, including greater environmental contamination, negative impacts on non-target species such as pollinators and beneficial insects, and heightened health risks for agricultural workers and nearby communities (Hashimi et al., 2020). Moreover, as pests adapt to changing climates and pesticide use becomes more frequent, there is a growing need for integrated pest management (IPM) strategies that combine chemical, biological, and cultural control methods to minimize the adverse effects of pesticides on human health and the environment (Pathak et al., 2022).

4 Health Vulnerabilities in Different Communities

4.1 Rural Versus Urban Communities: Differing Exposure and Impact

Pesticide exposure and its health impacts vary significantly between rural and urban communities due to differences in lifestyle, occupation, proximity to agricultural areas, and the prevalence of environmental contaminants.

Rural communities are generally more vulnerable to pesticide exposure because of their proximity to agricultural activities. Residents, particularly agricultural workers and their families, are frequently exposed to pesticides through direct contact during application, handling, or storage, as well as through indirect routes such as drift, runoff, and contaminated water sources (Medithi, 2019). Studies have shown that people living near farms have higher concentrations of pesticide residues in their blood and urine, increasing their risk of developing acute and chronic health conditions, such as respiratory issues, neurological disorders, and cancers (Blair et al., 2015). Rural children are especially vulnerable due to behaviors such as playing on contaminated soil or helping with farm work, which can lead to significant pesticide exposure (Ferguson et al., 2017).

In contrast, urban communities face different patterns of pesticide exposure. Urban exposure often comes from pesticide use in public health pest control (such as mosquito abatement), landscaping, and household pest management (Tudi et al., 2021). Urban residents are typically less exposed to pesticides than rural counterparts; however, they may still experience health risks from pesticide residues in food and contaminated water supplies (Miyittah et al., 2024). Moreso, pesticide exposure in urban areas is often compounded by other pollutants, such as vehicular emissions and industrial discharges, creating a complex mixture of chemicals that can exacerbate health effects (Adeola, 2021). While both rural and urban communities are at risk, the health impacts are often more severe in rural areas due to higher exposure levels, limited access to healthcare, and less awareness and education about safe pesticide use (Calliera et al., 2019). In urban settings, there is a need for public awareness campaigns about safe pesticide practices, especially regarding household and garden use, to mitigate exposure risks. Figure 2 outlines the key factors that affect the health of a population by identifying various determinants, vulnerabilities, external stressors, and resilience strategies. Determinants such as socioeconomic status (SES), access to healthcare, and environmental conditions influence a community's overall well-being. Vulnerability factors like age, chronic diseases, and social marginalization further exacerbate health risks. External stressors, including climate change and pesticide exposure, amplify these vulnerabilities (Ullah & Akhtar, 2023). However, resilience strategies like community engagement and policy interventions can mitigate these risks, helping to protect and improve community health.

4.2 *Socioeconomic and Demographic Factors Influencing Vulnerability*

Socioeconomic status (SES) is a critical determinant of health vulnerability to pesticide exposure. Low-income communities often face higher exposure risks due to substandard housing, limited access to clean water, and reduced capacity to implement protective measures against pesticides (Donley et al., 2022). In agricultural settings, low-income workers may lack proper personal protective equipment (PPE) and training on safe pesticide use, increasing their exposure and associated health risks (Nguyen & Tsai, 2024). Moreover, limited healthcare access can delay the diagnosis and treatment of pesticide-related illnesses, exacerbating health outcomes in these populations (Upadhayay et al., 2020).

Other influential determinants of pesticides vulnerability include sociodemographic factors like the age, sex, and color of the people. Children and the elderly are vulnerable groups because of physiologic factors that compromise their capacity to metabolize and expel toxins. It also indicates that Furthermore, children have a higher absorption rate and a developing organ system, making them more

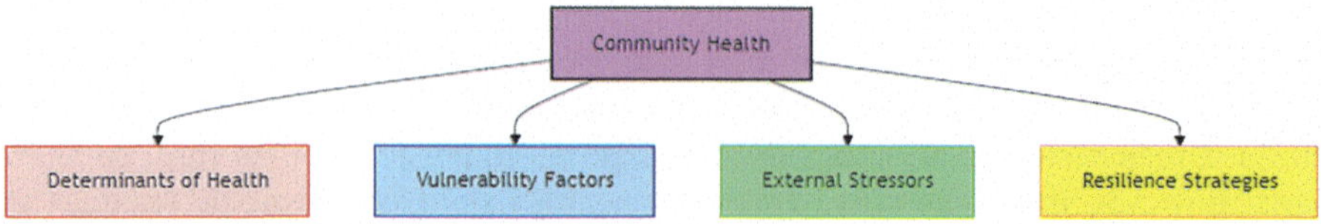

Fig. 2 Community health vulnerability framework

susceptible to neurodevelopmental alterations caused by pesticide exposure (Upadhayay et al., 2020). Persons in this age group have impaired immunity and chronic diseases which make them complicated for the chronic diseases associated with pesticide exposure (Tudi et al., 2021). Ethnicity is another determinant of vulnerability, as research has demonstrated that minority people in developing nations especially in counties having predominant agricultural practices suffer high levels of pesticide exposure owing to the risk factors associated with pesticide application in farming (Smith-Jackson et al., 2010). That is why there are language barriers, specific cultural activities, and no proper and easily accessible health education facilitating the increased risk of these groups of people and even worsening their conditions (Khatri & Assefa, 2022).

4.3 Case Studies: Regional Impacts in Different Climates

Case Study 1: Latin America (Tropical Climate)
In Latin America, where tropical climates promote year-round agricultural activity, pesticide use is prevalent. A study conducted in Nicaragua found that farmers and rural residents exposed to organophosphate pesticides experienced significant neurological symptoms, such as memory loss, depression, and fatigue (Schwartz et al., 2020). The hot and humid conditions in tropical regions can increase dermal absorption rates of pesticides, exacerbating their toxic effects (Fang et al., 2020). Moreso, limited regulation and enforcement of pesticide use, coupled with high poverty rates, contribute to increased health risks in these communities.

Case Study 2: Sub-Saharan Africa (Semi-Arid Climate)
In Sub-Saharan Africa, pesticide exposure is a growing concern due to expanding agricultural activities driven by climate change and food security needs. The semi-arid climate, characterized by intense sunlight and high temperatures, accelerates the volatilization of pesticides, increasing airborne exposure (Ofori et al., 2021). Farmers in these regions often lack access to PPE and are unaware of safe pesticide handling practices, resulting in frequent cases of acute pesticide poisoning, particularly among women and children who assist with farm tasks (Ngowi et al., 2016).

Case Study 3: North America (Temperate Climate)
In North America, the effects of pesticide exposure are particularly evident in rural communities involved in large-scale agriculture. A study in California's Central Valley, a region with a temperate climate, found that Latino agricultural workers and their families faced elevated levels of pesticide exposure, linked to adverse health outcomes such as reproductive disorders, respiratory problems, and developmental delays in children (Bradman et al., 2015). Pesticide drift from fields into residential areas is a significant concern, and ongoing climate changes, such as increased temperatures and extreme weather events, are expected to further aggravate exposure risks (Tudi et al., 2021).

5 Climate Change, Pesticides, and Vector-Borne Diseases

5.1 Role of Pesticides in Vector Control

Pesticides are essential in controlling vector-borne diseases by targeting disease-carrying organisms such as mosquitoes, ticks, and flies. Insecticides like pyrethroids, organophosphates, and carbamates are used in various forms, including sprays, treated bed nets, and larvicides (van den Berg et al., 2021). Methods like insecticide-treated bed nets (ITNs) and indoor residual spraying (IRS) are particularly effective in malaria control, significantly reducing disease incidence in endemic regions (Okumu & Moore, 2011). Pesticides also manage other vectors, such as ticks and fleas, crucial in regions where vector-borne diseases pose public health threats (Socha et al., 2022). However, the effectiveness of these measures depends on correct application, as improper use can lead to pesticide resistance, diminishing control efforts (Gill & Garg, 2014). Larvicides target mosquito larvae in breeding sites, reducing adult populations and diseases like dengue and Zika (Dambach, 2020). Despite their benefits, pesticide use raises concerns about impacts on non-target species, human health, and the environment, highlighting the need for integrated vector management strategies (Ogunah et al., 2020).

5.2 Changing Patterns of Vector-Borne Diseases Due to Climate Change

Globally, climate affects vectors for some diseases by altering their distribution, abundance and behavior, change in the transmission of the pathogens they harbor, and a particular influence on the patterns of vector-borne diseases. Sudden changes in temperature, rainfall, humidity levels, and the presence of storms affect the ecology and life cycles of vectors, potentially altering breeding areas and increasing the likelihood of cross-infections in habitats previously declared disease-free (Ogden & Lindsay, 2016). Higher temperatures can increase the growth rates of vector's populations and rates of pathogen replication within vectors that decrease extrinsic incubation period and increase transmission capacity (Kamiya et al., 2020). For example, it is revealed that elevated temperatures can enhance the frequency of bites of some mosquito species including *Aedes aegypti* and *Anopheles* which transmit dengue, Zika, and malaria, respectively (Weetman et al., 2018). It will result in increased incidences of diseases in tropical and subtropical climates and move such diseases to the temperate climate. Effects of climate change in terms of precipitation equally influence vectors and incidents of vector-borne diseases. Floods cause production of numerous places where mosquito lays eggs hence raising chances of transmission, while drought creates few places where water is found and numerous mosquito hence raising odds of the disease transmission (Ryan et al., 2020). For instance, the availably rainfall during one and rainy seasons has been found to be linked to infection of diseases such as dengue fever and chikungunya in tropical regions of the Southeast Asia and the Pacific Islands (Abdulsalam & Ila, 2023). Likewise longer warm seasons and earlier springs that are prophetic can increase the activity of vectors; for instance, tick that causes Lyme disease by taking them to areas that had limited activities by cold seasons (Ogden & Lindsay, 2016). In addition, disasters such as hurricanes and floods destroy ecosystem and human habitats to conditions suitable for vector multiplication and enhance human contact with vectors. For example, flooding can lead to increased mosquito breeding sites and outbreaks of diseases such as leptospirosis and West Nile virus (Medlock & Leach, 2015). These shifts necessitate adaptive public health strategies to address emerging threats in regions newly affected by vector-borne diseases.

5.3 Interactions Between Pesticide Use and Disease Dynamics

The relationship between pesticide use and disease dynamics is intricate. While pesticides effectively reduce vector populations and disease transmission, their misuse or overuse can lead to resistance, reducing their effectiveness in control programs (Hemingway et al., 2016). For example, pyrethroid resistance in Anopheles mosquitoes poses challenges to malaria control (Ranson & Lissenden, 2016). Pesticides can also disrupt ecosystems by killing non-target organisms, potentially increasing vector populations through "pesticide-induced pest resurgence" (Geiger et al., 2010). Furthermore, pesticide exposure poses health risks, including respiratory issues and neurological disorders, especially in areas with heavy pesticide use (Blair et al., 2015). Climate change adds another layer of complexity by affecting pesticide effectiveness and vector behavior, leading to a need for adaptive management strategies (Livinus et al., 2024).

6 Regulatory and Policy Frameworks

6.1 Current Regulations Governing Pesticide Use

Pesticides should be controlled in such a manner that the use would be checked so as to avoid any repercussions to the health of the people and the environment. On the safe use of pesticides, other countries have developed measures with regard to approvals, labeling in application, and monitoring of pesticide. In United States, under the Federal Insecticide, Fungicide and Rodenticide Act or FIFRA it will be the Environmental Protection Agency (EPA) that is charged with the responsibility of registration, amongst other tasks like setting standards for the use of pesticide. As stipulated in this act all the pesticides must be approved by the EPA before being sold or used in the country. A careful consideration of the pesticide's toxicity details for man and other organisms and the environment is conducted during the registration process (Macko et al., 2022). Pesticides are of different types depending on their lethal outcomes and availability that processes mandatory rules for quantities, precautions, and gear (Garrigou et al., 2020). Likewise, the use of pesticide in the European Union (EU) is controlled under the PPPR, the plant protection product regulation, which provides elaborate measures on approval and labeling of pesticides and monitoring. As stated by the European Food Safety Authority or EFSA, it is its responsibility to assess the risks concerning pesticide and ensure that they are safe for use before they are approved by the EFSA (EFSA, 2020). The EU regulations also require updating of the sites containing pesticide products to check on compliance with safety requirements or risks as and when they are developed (Van den Berg et al., 2021). The principle of centralization of pesticide regulation makes it possible to place the responsibility for its implementation on national regulatory agencies that should control the utilization of pesticides and prevent the increase in the degree of threat to human health and the environment. These agencies tend to work with NGOs like the Food and Agriculture Organization (FAO) and the World Health Organization (WHO) in order to develop compatibility with super regulatory measures toward rational use of pesticides (Debnath & Khan, 2017). Despite these regulatory frameworks, enforcement and compliance can vary significantly between regions, and some countries may lack comprehensive regulations or the capacity to enforce them effectively (Musa et al., 2024a, b, c, d). In developing countries, challenges such as inadequate resources, lack of training, and weak regulatory systems can undermine the effectiveness of pesticide regulations (Atreya et al., 2011).

6.2 Policies Addressing Climate Change Impacts

Policies addressing climate change are essential for mitigating its adverse effects on ecosystems, health, and agriculture. Internationally, the Paris Agreement commits to limiting global warming to below 2 °C, with efforts to keep it under 1.5 °C, and supports emission reductions and climate resilience (Lesnikowski et al., 2017). National policies often align with these goals, such as the European Green Deal, which aims for climate neutrality by 2050 through emission reductions and sustainable practices (Filipović et al., 2022). In the United States, the Climate Action Plan focuses on cutting carbon emissions, investing in clean energy, and improving climate resilience (Brunner & Klein, 1999). Sector-specific policies, like the Agricultural Act of 2014, promote climate-smart agriculture to enhance resilience. However, challenges remain, requiring policies to adapt to new climate science and address the needs of vulnerable populations (Adger et al., 2013).

6.3 Gaps and Challenges in Integrating Health, Environment, and Climate Policies

Integrating health, environment, and climate policies is challenging due to conflicting priorities, limited resources, and the need for coordinated efforts across sectors. Balancing immediate benefits of pesticide use with long-term environmental and health goals requires careful policy design (Pimentel et al., 2005). Coordination issues arise because these policies are often developed separately, leading to fragmented approaches. Effective integration demands collaboration among policymakers, health experts, and environmental scientists (Barton & Grant, 2006). Resource constraints can impede the development and enforcement of comprehensive policies, particularly in low-income regions (Sachs, 2012). Moreover, data gaps and uncertainties about the interactions between climate, health, and environmental factors complicate policy-making, highlighting the need for improved research and monitoring. Public engagement and awareness are also crucial for successful policy implementation (Abdulsalam et al., 2024b). Addressing these challenges requires a multidisciplinary approach and adaptive policy frameworks.

7 Mitigation and Adaptation Strategies

7.1 Community-Based Approaches to Reduce Exposure

Community-based approaches are essential for mitigating the health impacts of climate change and pesticide use. These strategies involve engaging local communities in efforts to reduce exposure to harmful substances and to enhance resilience to environmental changes.

(a) *Education and Awareness Campaigns*

One of the fundamental components of community-based approaches is education. Providing communities with information about the risks associated with pesticide use and climate change can empower individuals to take protective measures. For example, educating farmers about the safe use of pesticides, including proper application techniques and protective gear, can reduce the risk of pesticide exposure and its associated health effects (Garrigou et al., 2020). Awareness campaigns can also highlight the importance of climate adaptation measures, such as water conservation and sustainable agricultural practices, to build community resilience (Musa et al., 2024a, b, c, d).

(b) *Participatory Risk Assessment and Management*

Involving communities in risk assessment and management processes can enhance the effectiveness of mitigation strategies. Participatory approaches allow communities to identify local hazards, assess risks, and develop context-specific solutions. For example, community-based participatory research (CBPR) has been used to address pesticide-related health issues by involving local stakeholders in the research process, thus ensuring that interventions are relevant and culturally appropriate (Abdulsalam et al., 2024a, b, c). This approach also helps to build local capacity and foster collaboration between communities, researchers, and policymakers.

(c) *Community Monitoring and Early Warning Systems*
Community-based monitoring systems can play a crucial role in reducing exposure to environmental hazards. For instance, local volunteer networks can be trained to monitor pesticide use, track health outcomes, and report environmental changes. Such systems can provide early warnings of potential health risks and enable timely responses (Sufri et al., 2020). Early warning systems for climate-related events, such as heatwaves and floods, can also help communities prepare and adapt to changing conditions, thereby reducing vulnerability (Fisher et al., 2021).

7.2 Climate-Smart Agricultural Practices

Climate-smart agricultural practices aim to increase productivity while reducing greenhouse gas emissions and enhancing resilience to climate change. These practices are critical for adapting agriculture to changing climate conditions and minimizing the environmental impact of farming.

(a) *Conservation Tillage*
Conservation tillage involves minimal disturbance of soil, which helps to preserve soil structure, reduce erosion, and enhance soil carbon sequestration. This practice can improve soil moisture retention and reduce the need for irrigation, making agriculture more resilient to drought conditions (Lal, 2015). By reducing soil erosion and nutrient runoff, conservation tillage also helps to minimize pesticide and fertilizer losses, reducing their environmental impact (Pimentel et al., 2005).

(b) *Integrated Pest Management (IPM)*
Integrated pest management or IPM is a rational approach to pest control that depends on the biological, cultural, mechanical, and chemical techniques to maintain the pest populations at the levels below the economically tolerable limit in an integrated and environmentally friendly way. Realizing integrated pest management minimizes dependence on chemical pesticides and also has less influence on distinct species and ecosystems through monitoring of pests, natural enemies, crop rotation, and resistant crop varieties (Gurr et al., 2016). That is why IPM practices contribute to the prevention of such occurrences as pesticide resistance; diversity is achieved through various control measures, and pesticide applications are not frequently done (Hajek & Eilenberg, 2018).

(c) *Crop Diversification and Agroforestry*
Crop diversification and agroforestry are practices that enhance agricultural resilience by increasing biodiversity and reducing the risk of crop failure. Diversifying crop types and incorporating trees into agricultural systems can

improve soil fertility, enhance water retention, and provide habitat for beneficial organisms. These practices also help to buffer against climate-related stresses, such as extreme weather events and pests, by promoting a more resilient and adaptive agricultural system (Schroth et al., 2016).

(d) *Water Management Practices*

Since climate conditions are changing to extremes, water management can help put a tick on agriculture production. Measures like water harvesting, water conservation through drip irrigation, soil moisture conservation techniques make farmer's defenseless to vulnerability of drought. The application of these practices should lead to higher yields, and decreased demand for chemical inputs, including pesticides (Rockström et al., 2009).

7.3 Health Surveillance and Early Warning Systems

Health promotion and early warning indicates need for monitoring and addressing health hazards under changing climate and pesticide utilization. These systems' assist in identifying and monitoring health risks within the shortest time possible.

(a) *Integrated Health Surveillance Systems*

Integrated health surveillance systems collect and analyze data on disease incidence, environmental exposures, and other health indicators. These systems enable public health authorities to identify emerging health threats, track trends over time, and implement targeted interventions. For example, surveillance systems for vector-borne diseases can monitor changes in vector populations and disease incidence, allowing for timely response to outbreaks (Pley et al., 2021).

(b) *Early Warning Systems for Climate-Related Health Risks*

Early warning systems are designed to provide advance notice of climate-related health risks, such as heatwaves, floods, and extreme weather events. These systems use meteorological and environmental data to predict and communicate potential health impacts. For instance, heat health warning systems can alert communities to impending heatwaves and provide guidance on protective measures, such as staying hydrated and avoiding strenuous activities (Casanueva et al., 2019).

(c) *Community-Based Health Monitoring*

Community-based health surveillance requires community participation of health outcomes and setting surveillance as well as exposures. This approach may offer important information about climate change and pesticide effects relating to health in a given region. For instance, community health workers (CHWs) can ask questionnaires/ checklists to record the data on the incidence and distribution of the effects of hazardous pesticides, environmental factors in the homes, and their health insurance coverage levels. The information gathered here can assist in the establishment of hot spots to enable creators of health intercessions to identify high-risk areas (Al-Abboodi et al., 2020).

(d) *Data Sharing and Collaboration*

Effective health surveillance and early warning systems require collaboration between public health agencies, researchers, and other stakeholders. Health interventions can benefit from both sharing of data and synchronizing efforts in order to ensure better response against new health threats. Interorganizational cooperation can enhance the sharing of knowledge and resources and enhance the ability of health systems to respond to effects of climate variations and natural disasters (Nichols et al., 2019).

8 Research Gaps and Future Directions

8.1 Areas Needing Further Research on Pesticides, Climate Change, and Health

Despite substantial progress in understanding the impacts of pesticides and climate change on health, several critical research gaps remain. Addressing these gaps is essential for developing more effective mitigation and adaptation strategies.

(a) *Long-Term Health Effects*

Although there is significant research on the acute effects of pesticides on human health, human suffering, and disease, little is known about long-term and chronic consequences of exposure to pesticides. Most studies conducted were cross-sectional, so future research in this area should incorporate longitudinal designs to establish long-term health risk factors, particularly concerning cancer, neurological diseases, or reproductive disorders that might be associated with low-level pesticide exposure (Hu et al., 2015). It is important for research on the long-term impacts of pesticides to determine the effects of combined exposure to multiple pesticides and other environmental stressors (Bassil et al., 2007).

(b) *Climate Change Impacts on Pesticide Efficacy*

It is expected that climate change will have impacts on the effectiveness of the pesticides through variations in temperature, rainfall, and humidity. Further research is required on how these environmental changes have an effect on the pesticides efficacy, residual, and degradative properties. This includes understanding how changes in the weather may reshape pest resistance and the extent to which farmers may have to apply pesticides (Bras et al., 2022). An understanding of these dynamics will assist in the development of better implemented pest management strategies that will also be environmentally friendly.

(c) *Interactions between Pesticides and Climate Change*

The interactions between pesticides and climate change are complex and not yet fully understood. Research should focus on how climate-induced changes in ecosystems and agricultural systems affect pesticide behavior and impact. For example, studies are needed on how increased temperatures and altered precipitation patterns might influence pesticide runoff, soil degradation, and non-target organism exposure (Musa et al., 2024a, b, c, d). Moreso, investigating how climate change affects the behavior and resistance of pests to pesticides is crucial for adapting pest management practices (Rockström et al., 2009).

(d) *Vulnerable Populations*

More research is needed to understand how different populations, particularly vulnerable groups such as children, pregnant women, and agricultural workers, are disproportionately affected by pesticide exposure and climate change. This includes identifying the specific health risks faced by these groups and developing targeted interventions to protect them (Abdulsalam et al., 2024a, b, c). Research should also focus on socioeconomic and cultural factors that influence vulnerability and health outcomes (Fisher et al., 2021).

8.2 *Opportunities for Multidisciplinary Collaboration*

Addressing the complex issues related to pesticides, climate change, and health requires a multidisciplinary approach that brings together expertise from various fields.

(a) *Integration of Environmental Science and Public Health*

Collaboration between environmental scientists and public health experts is essential for understanding the full spectrum of health impacts related to pesticide use and climate change. Integrated research efforts can help to identify and address the pathways through which environmental changes affect human health. For instance, environmental scientists can provide insights into the mechanisms of pesticide degradation and transport, while public health experts can assess the resulting health impacts (Ogunah et al., 2020).

(b) *Collaboration with Social Sciences*

Incorporating insights from social sciences can enhance understanding of how social, economic, and cultural factors influence pesticide use and health outcomes. Social scientists can help to analyze how different communities perceive and manage pesticide risks, and how social inequalities affect exposure and vulnerability. This collaboration can inform the development of more effective and culturally appropriate interventions (Lawal et al., 2023).

(c) *Engagement with Policy Makers*

Researchers should actively engage with policymakers to ensure that scientific findings are translated into effective policies and regulations. Collaborative efforts can help to bridge the gap between research and policy, ensuring that interventions are evidence-based and address the needs of affected communities. This includes working with policymakers to develop and implement regulations that reduce pesticide use and mitigate climate change impacts (Gainforth et al., 2021).

(d) *Partnerships with Industry and Technology Developers*

Collaboration with industry and technology developers can facilitate the development and implementation of innovative solutions for monitoring and control of pesticide use and climate change impacts. Industry partnerships can help to advance research on new pesticide formulations, safer alternatives, and sustainable agricultural practices (Luiz de Oliveira et al., 2018). Engaging with technology developers can also support the integration of emerging technologies into monitoring and control systems.

8.3 Emerging Technologies and Innovations in Monitoring and Control

Emerging technologies offer new opportunities for improving the monitoring and control of pesticide use and assessing the impacts of climate change on health.

(a) *Remote Sensing and GIS*

Remote sensing and geographic information systems (GIS) are powerful tools for monitoring of environmental changes and assessing pesticide use. Remote sensing technologies can provide real-time data on land use, crop health, and environmental conditions, which can be used to track pesticide application patterns and detect potential risks (Katkani et al., 2022). GIS can help to integrate and analyze these data, enabling spatially explicit assessments of pesticide exposure and its impact on human health and ecosystems (Tiamiyu et al., 2024).

(b) *Precision Agriculture*

Precision agriculture uses advanced technologies such as GPS, sensors, and data analytics to optimize pesticide application and reduce environmental impact. Precision agriculture techniques allow for targeted pesticide application based on real-time data, reducing the need for blanket pesticide applications and minimizing the risk of runoff and non-target exposure. This approach also helps to improve pest management efficiency and reduce overall pesticide use (Magarey et al., 2019). The integration of machine learning algorithms with precision agriculture tools can further enhance the accuracy of pest detection and management strategies (Saheed et al., 2024).

(c) *Biological Control and Integrated Pest Management Innovations*

Advances in biological control methods and integrated pest management (IPM) strategies are promising for reducing reliance on chemical pesticides. Innovations in biological control include the development of new natural predators, parasites, and pathogens that target specific pests. For example, the use of genetically modified organisms (GMOs) to produce pest-resistant crops or to release sterilized pests can be effective in managing pest populations without relying solely on chemical pesticides (Van Campenhout & Eilenberg, 2021). Moreso, integrating these biological control methods with traditional IPM practices can enhance overall pest management while reducing environmental and health risks.

(d) *Wearable and Mobile Health Technologies*

Wearable technologies and mobile health applications offer new opportunities for monitoring of health outcomes and exposures in real time. Wearable sensors can track environmental exposures, such as pesticide residues, and physiological responses, such as heart rate and body temperature (Musa et al., 2024a, b, c, d). Mobile health applications can provide users with real-time information about environmental risks, health tips, and alerts. These technologies can support community-based health monitoring and early warning systems, enabling individuals and health professionals to respond more effectively to environmental health risks (Sindato et al., 2021).

(e) *Advancements in Environmental Health Risk Assessment Tools*

New tools and methodologies for assessing environmental health risks are emerging, including improved models for predicting the impacts of climate change and pesticide use on health. These tools incorporate data from multiple sources, including remote sensing, epidemiological studies, and climate models, to provide a comprehensive assessment of health risks. Advances in computational modeling and data integration can enhance the accuracy of risk assessments and support the development of targeted interventions (Saheed et al., 2023).

9 Conclusion

This chapter highlights the pressing health risks associated with pesticide exposure and the intricate interplay between pesticides, climate change, and community health. Our analysis reveals that both acute and chronic pesticide exposure can have serious health implications, particularly for vulnerable populations such as children, pregnant women, and agricultural workers. The impact of climate change further complicates these dynamics, altering pesticide efficacy and pest management practices, thereby heightening the risks faced by rural and urban communities alike. The evidence underscores the urgent need for stronger regulatory frameworks and climate-smart agricultural practices that prioritize public health. Enhanced health surveillance systems are critical for monitoring of exposure and outcomes, allowing for timely interventions. Future research should aim to fill the gaps identified in this chapter, focusing on long-term health effects, the interplay between climate change and pesticide efficacy, and the potential of innovative technologies to improve monitoring and control. By adopting a multidisciplinary approach that includes environmental science, public health, social sciences, and technology, we can more effectively address these challenges. Collaborative efforts among researchers, policymakers, and communities will be essential in developing strategies that not only mitigate health risks but also promote sustainable agricultural practices. Ultimately, fostering resilient communities that can adapt to the evolving threats posed by pesticides and climate change is imperative for safeguarding public health and ensuring a sustainable future.

References

Abdulsalam, M., & Ila, M. A. (2023). Closing the gap: Artificial intelligence integration for advancing Chikungunya Virus studies in Africa. *Biological Sciences, 3*(4), 493–502.

Abdulsalam, M., Elelu, S. A., Innocent, M. O., Ibrahim, G. O., Livinus, M. U., Lateefat, S. O., & Muhammad, A. S. (2024a). Women empowerment in environmental conservation. In *Prospects for soil regeneration and its impact on environmental protection* (pp. 351–370). Springer Nature Switzerland.

Abdulsalam, M., Hamisu, A. A., Ahmad, A. M., Wakili, F. B., Annu, F. S., & Garba, M. M. (2024b). Microbiome dynamics and strategic management strategies: Exploring synergies between integrative medicine modalities and microbial balance. *Biological Sciences, 4*(3), 725–735.

Abdulsalam, M., Salam, O. L., Muhammad, M. S., Natour, S., Garba, M. M., & Annu, S. F. (2024c). Evaluation of post-harvest microbial deterioration of ripe banana fruits in different markets. *Plant Health Archives, 2*(1), 01–06.

Adeola, F. O. (2021). Global impact of chemicals and toxic substances on human health and the environment. In *Handbook of global health* (pp. 1–30). Springer.

Adger, W. N., Barnett, J., Brown, K., Marshall, N., & O'brien, K. (2013). Cultural dimensions of climate change impacts and adaptation. *Nature Climate Change, 3*(2), 112–117.

Al-Abboodi, S. M. T., Al-Shaibani, E. J. A., & Alrubai, E. A. (2020). Preparation and characterization of nano silica prepared by different precipitation methods. *IOP Conference Series: Materials Science and Engineering, 978*(1), 012031.

Atreya, K., Sitaula, B. K., Johnsen, F. H., & Bajracharya, R. M. (2011). Continuing issues in the limitations of pesticide use in developing countries. *Journal of Agricultural and Environmental Ethics, 24*, 49–62.

Bachmann, M. C., Bellalta, S., Basoalto, R., Gómez-Valenzuela, F., Jalil, Y., Lépez, M., & von Bernhardi, R. (2020). The challenge by multiple environmental and biological factors induce inflammation in aging: Their role in the promotion of chronic disease. *Frontiers in Immunology, 11*, 570083.

Barton, H., & Grant, M. (2006). A health map for the local human habitat. *The Journal of the Royal Society for the Promotion of Health, 126*(6), 252–253.

Bassil, K. L., Vakil, C., Sanborn, M., Cole, D. C., Kaur, J. S., & Kerr, K. J. (2007). Cancer health effects of pesticides: Systematic review. *Canadian Family Physician, 53*(10), 1704–1711.

Blair, J. M., Webber, M. A., Baylay, A. J., Ogbolu, D. O., & Piddock, L. J. (2015). Molecular mechanisms of antibiotic resistance. *Nature Reviews Microbiology, 13*(1), 42–51.

Boonupara, T., Udomkun, P., Khan, E., & Kajitvichyanukul, P. (2023). Airborne pesticides from agricultural practices: A critical review of pathways, influencing factors, and human health implications. *Toxics, 11*(10), 858.

Bradman, A., Quirós-Alcalá, L., Castorina, R., Schall, R. A., Camacho, J., Holland, N. T., & Eskenazi, B. (2015). Effect of organic diet intervention on pesticide exposures in young children living in low-income urban and agricultural communities. *Environmental Health Perspectives, 123*(10), 1086–1093.

Bras, A., Roy, A., Heckel, D. G., Anderson, P., & Karlsson Green, K. (2022). Pesticide resistance in arthropods: Ecology matters too. *Ecology Letters, 25*(8), 1746–1759.

Brunner, R. D., & Klein, R. (1999). Harvesting experience: A reappraisal of the US climate change action plan. *Policy Sciences, 32*(2), 133–161.

Calliera, M., Luzzani, G., Sacchettini, G., & Capri, E. (2019). Residents perceptions of non-dietary pesticide exposure risk. Knowledge gaps and challenges for targeted awareness-raising material in Italy. *Science of the Total Environment, 685*, 775–785.

Casanueva, A., Burgstall, A., Kotlarski, S., Messeri, A., Morabito, M., Flouris, A. D., & Schwierz, C. (2019). Overview of existing heat-health warning systems in Europe. *International Journal of Environmental Research and Public Health, 16*(15), 2657.

Dambach, P. (2020). The use of aquatic predators for larval control of mosquito disease vectors: Opportunities and limitations. *Biological Control, 150*, 104357.

Debnath, M., & Khan, M. S. (2017). Health concerns of pesticides. Pesticide residue in foods: Sources, management, and control, 103–118.

Djekkoun, N., Lalau, J. D., Bach, V., Depeint, F., & Khorsi-Cauet, H. (2021). Chronic oral exposure to pesticides and their consequences on metabolic regulation: Role of the microbiota. *European Journal of Nutrition, 60*(8), 4131–4149.

Donley, N., Bullard, R. D., Economos, J., Figueroa, I., Lee, J., Liebman, A. K., & Shafiei, F. (2022). Pesticides and environmental injustice in the USA: Root causes, current regulatory reinforcement and a path forward. *BMC Public Health, 22*(1), 708.

Ellena, M., Breil, M., & Soriani, S. (2020). The heat-health nexus in the urban context: A systematic literature review exploring the socio-economic vulnerabilities and built environment characteristics. *Urban Climate, 34*, 100676.

Elonheimo, H. M., Mattila, T., Andersen, H. R., Bocca, B., Ruggieri, F., Haverinen, E., & Tolonen, H. (2022). Environmental substances associated with chronic obstructive pulmonary disease—A scoping review. *International Journal of Environmental Research and Public Health, 19*(7), 3945.

European Food Safety Authority (EFSA). (2020). The 2018 European Union report on pesticide residues in food. *EFSA Journal, 18*(4), e06057.

Fang, S. C., Rodrigues, E. G., & Christiani, D. C. (2020). Environmental health hazards in the tropics. In *Hunter's tropical medicine and emerging infectious diseases* (pp. 200–208). Elsevier.

Ferguson, A., Penney, R., & Solo-Gabriele, H. (2017). A review of the field on children's exposure to environmental contaminants: A risk assessment approach. *International Journal of Environmental Research and Public Health, 14*(3), 265.

Filipović, S., Lior, N., & Radovanović, M. (2022). The green deal–just transition and sustainable development goals Nexus. *Renewable and Sustainable Energy Reviews, 168*, 112759.

Fisher, R., Petit, F., & Porod, C. (2021). Early warning systems to strengthen the resilience of communities to extreme events. In *Handbook of disaster risk reduction for resilience: New frameworks for building resilience to disasters* (pp. 239–259). Springer International Publishing.

Gainforth, H. L., Hoekstra, F., McKay, R., McBride, C. B., Sweet, S. N., Ginis, K. A. M., & Willms, R. (2021). Integrated knowledge translation guiding principles for conducting and disseminating spinal cord injury research in partnership. *Archives of Physical Medicine and Rehabilitation, 102*(4), 656–663.

Garrigou, A., Laurent, C., Berthet, A., Colosio, C., Jas, N., Daubas-Letourneux, V., & Judon, N. (2020). Critical review of the role of PPE in the prevention of risks related to agricultural pesticide use. *Safety Science, 123*, 104527.

Geiger, F., Bengtsson, J., Berendse, F., Weisser, W. W., Emmerson, M., Morales, M. B., & Inchausti, P. (2010). Persistent negative effects of pesticides on biodiversity and biological control potential on European farmland. *Basic and Applied Ecology, 11*(2), 97–105.

Gill, H. K., & Garg, H. (2014). Pesticide: Environmental impacts and management strategies. In *Pesticides-toxic aspects* (Vol. 8, pp. 10–5772). IntechOpen.

Gurr, G. M., Lu, Z., Zheng, X., Xu, H., Zhu, P., Chen, G., & Heong, K. L. (2016). Multi-country evidence that crop diversification promotes ecological intensification of agriculture. *Nature Plants, 2*(3), 1–4.

Hajek, A. E., & Eilenberg, J. (2018). *Natural enemies: An introduction to biological control*. Cambridge University Press.

Hashimi, M. H., Hashimi, R., & Ryan, Q. (2020). Toxic effects of pesticides on humans, plants, animals, pollinators and beneficial organisms. *Asian Plant Research Journal, 5*(4), 37–47.

Hemingway, J., Ranson, H., Magill, A., Kolaczinski, J., Fornadel, C., Gimnig, J., & Hamon, N. (2016). Averting a malaria disaster: Will insecticide resistance derail malaria control? *The Lancet, 387*(10029), 1785–1788.

Hu, R., Huang, X., Huang, J., Li, Y., Zhang, C., Yin, Y., & Cui, F. (2015). Long-and short-term health effects of pesticide exposure: A cohort study from China. *PLoS One, 10*(6), e0128766.

Kaka, H., Opute, P. A., & Maboeta, M. S. (2021). Potential impacts of climate change on the toxicity of pesticides towards earthworms. *Journal of Toxicology, 2021*(1), 8527991.

Kamiya, T., Greischar, M. A., Wadhawan, K., Gilbert, B., Paaijmans, K., & Mideo, N. (2020). Temperature-dependent variation in the extrinsic incubation period elevates the risk of vector-borne disease emergence. *Epidemics, 30*, 100382.

Katkani, D., Babbar, A., Mishra, V. K., Trivedi, A., Tiwari, S., & Kumawat, R. K. (2022). A review on applications and utility of remote sensing and geographic information systems in agriculture and natural resource management. *International Journal of Environment and Climate Change, 12*(4), 1–18.

Khan, S. H., & Pathak, B. (2020). Zinc oxide based photocatalytic degradation of persistent pesticides: A comprehensive review. *Environmental Nanotechnology, Monitoring & Management, 13*, 100290.

Khatri, R. B., & Assefa, Y. (2022). Access to health services among culturally and linguistically diverse populations in the Australian universal health care system: Issues and challenges. *BMC Public Health, 22*(1), 880.

Konradsen, F., van der Hoek, W., Cole, D. C., Hutchinson, G., Daisley, H., Singh, S., & Eddleston, M. (2003). Reducing acute poisoning in developing countries—Options for restricting the availability of pesticides. *Toxicology, 192*(2–3), 249–261.

Kubiak-Hardiman, P., Haughey, S. A., Meneely, J., Miller, S., Banerjee, K., & Elliott, C. T. (2023). Identifying gaps and challenges in global pesticide legislation that impact the protection of consumer health: Rice as a case study. *Exposure and Health, 15*(3), 597–618.

Lal, R. (2015). Restoring soil quality to mitigate soil degradation. *Sustainability, 7*(5), 5875–5895.

Lawal, T. O., Mustapha Abdulsalam, A. M., & Sundararajan, S. (2023). Economic and environmental implications of sustainable agricultural practices in arid regions: A cross-disciplinary analysis of plant science, management, and economics. *International Journal, 10*(3), 3100–3114.

Lesnikowski, A., Ford, J., Biesbroek, R., Berrang-Ford, L., Maillet, M., Araos, M., & Austin, S. E. (2017). What does the Paris Agreement mean for adaptation? *Climate Policy, 17*(7), 825–831.

Livinus, M. U., Bala, S. Z., Abdulsalam, M., Innocent, M. O., Hassan, M., & Kini, P. (2024). Natural occurrences of soil dilapidation. In *Prospects for soil regeneration and its impact on environmental protection* (pp. 205–223). Springer Nature Switzerland.

Lopez, A. M., Avila, C. C. E., VanderRoest, J. P., Roth, H. K., Fendorf, S., & Borch, T. (2024). Molecular insights and impacts of wildfire-induced soil chemical changes. *Nature Reviews Earth & Environment, 5*, 1–16.

Luiz de Oliveira, J., Ramos Campos, E. V., & Fraceto, L. F. (2018). Recent developments and challenges for nanoscale formulation of botanical pesticides for use in sustainable agriculture. *Journal of Agricultural and Food Chemistry, 66*(34), 8898–8913.

Macko, M., Antoš, J., Božek, F., Konečný, J., Huzlík, J., Hegrová, J., & Kuřitka, I. (2022). Development of new health risk assessment of nanoparticles: EPA health risk assessment revised. *Nanomaterials, 13*(1), 20.

Magarey, R. D., Klammer, S. S., Chappell, T. M., Trexler, C. M., Pallipparambil, G. R., & Hain, E. F. (2019). Eco-efficiency as a strategy for optimizing the sustainability of pest management. *Pest Management Science, 75*(12), 3129–3134.

Medithi, S. (2019). A critical review to retrace the untended pathways of pesticide exposure in humans. *Indian Journal of Health Sciences and Care, 6*(3), 128–139.

Medlock, J. M., & Leach, S. A. (2015). Effect of climate change on vector-borne disease risk in the UK. *The Lancet Infectious Diseases, 15*(6), 721–730.

Miyittah, M. K., Ansah, B., Kwadzo, M., Seidu-Larry, S., & Kosivi, R. K. (2024). Assessment of pesticide exposure risks among cocoa farmers in Western region of Ghana. *International Journal of Pest Management, 70*, 1067–1085.

Morad, G., Helmink, B. A., Sharma, P., & Wargo, J. A. (2021). Hallmarks of response, resistance, and toxicity to immune checkpoint blockade. *Cell, 184*(21), 5309–5337.

Morris, L. G. T., Veeriah, S., & Chan, T. A. (2010). Genetic determinants at the interface of cancer and neurodegenerative disease. *Oncogene, 29*(24), 3453–3464.

Musa, I. O., Samuel, J. O., Adams, M., Abdulsalam, M., Abdulazeez, M., Muhammad, M. S., & Akande, S. A. (2024a). Cases of micro-and nanoplastics in food and food product. In *Emerging contaminants in food and food products* (pp. 90–105). CRC Press.

Musa, I. O., Ijah, U. J. J., Abioye, O. P., Abdulsalam, M., Pal, S. K., Abamhekhelu, I. A., & Akande, S. A. (2024b). Phytoremediation of heavy metal-contaminated soil. In *Eco-restoration of polluted environment* (pp. 35–52). CRC Press.

Musa, I. O., Isibor, P. O., Samuel, J. O., Mustapha, A., Mustapha, A., Akande, S., & Adeniji, H. (2024c). Introduction to nanotoxicology. In *Environmental nanotoxicology: Combatting the minute contaminants* (pp. 1–22). Springer Nature Switzerland.

Musa, I. O., Samuel, J. O., Adams, M., Abdulsalam, M., Nathaniel, V., Maude, A. M., & Tiamiyu, A. G. T. (2024d). Soil erosion, mineral depletion and regeneration. In *Prospects for soil regeneration and its impact on environmental protection* (pp. 159–172). Springer Nature Switzerland.

Ndimele, P. E., Ojewole, A. E., Mekuleyi, G. O., Badmos, L. A., Agosu, C. M., Olatunbosun, E. S., et al. (2024). Vulnerability, resilience and adaptation of Lagos coastal communities to flooding. *Earth Science, Systems and Society, 4*, 10087.

Ngowi, A., Mrema, E., & Kishinhi, S. (2016). Pesticide health and safety challenges facing informal sector workers: A case of small-scale agricultural workers in Tanzania. *New Solutions: A Journal of Environmental and Occupational Health Policy, 26*(2), 220–240.

Nguyen, D., & Tsai, C. S. (2024). Inadequate personal protective equipment factors and odds related to acute pesticide poisoning: A meta-analysis report. *International Journal of Environmental Research and Public Health, 21*(3), 257.

Nichols, C. R., Wright, L. D., Bainbridge, S. J., Cosby, A., Hénaff, A., Loftis, J. D., & Zarillo, G. (2019). Collaborative science to enhance coastal resilience and adaptation. *Frontiers in Marine Science, 6*, 404.

Ofori, S. A., Cobbina, S. J., & Obiri, S. (2021). Climate change, land, water, and food security: Perspectives from Sub-Saharan Africa. *Frontiers in Sustainable Food Systems, 5*, 680924.

Ogden, N. H., & Lindsay, L. R. (2016). Effects of climate and climate change on vectors and vector-borne diseases: Ticks are different. *Trends in Parasitology, 32*(8), 646–656.

Ogunah, J. A., Lalah, J. O., & Schramm, K. W. (2020). Malaria vector control strategies. What is appropriate towards sustainable global eradication? *Sustainable Chemistry and Pharmacy, 18*, 100339.

Okumu, F. O., & Moore, S. J. (2011). Combining indoor residual spraying and insecticide-treated nets for malaria control in Africa: A review of possible outcomes and an outline of suggestions for the future. *Malaria Journal, 10*, 1–13.

Parsons, E. S., Jowell, A., Veidis, E., Barry, M., & Israni, S. T. (2024). Climate change and inequality. *Pediatric Research*, 1–8. https://doi.org/10.1038/s41390-024-03153-z

Pathak, V. M., Verma, V. K., Rawat, B. S., Kaur, B., Babu, N., Sharma, A., & Cunill, J. M. (2022). Current status of pesticide effects on environment, human health and it's eco-friendly management as bioremediation: A comprehensive review. *Frontiers in Microbiology, 13*, 962619.

Pimentel, D., Zuniga, R., & Morrison, D. (2005). Update on the environmental and economic costs associated with alien-invasive species in the United States. *Ecological Economics, 52*(3), 273–288.

Pley, C., Evans, M., Lowe, R., Montgomery, H., & Yacoub, S. (2021). Digital and technological innovation in vector-borne disease surveillance to predict, detect, and control climate-driven outbreaks. *The Lancet Planetary Health, 5*(10), e739–e745.

Pradhan, S. S., Gowda, G. B., Adak, T., Guru-Pirasanna-Pandi, G., Patil, N. B., Annamalai, M., & Rath, P. C. (2022). Pesticides occurrence in water sources and decontamination techniques. In *Pesticides-updates on toxicity, efficacy and risk assessment*. IntechOpen.

Ranson, H., & Lissenden, N. (2016). Insecticide resistance in African Anopheles mosquitoes: A worsening situation that needs urgent action to maintain malaria control. *Trends in Parasitology, 32*(3), 187–196.

Rockström, J., Steffen, W., Noone, K., Persson, Å., Chapin, F. S., Lambin, E. F., & Foley, J. A. (2009). A safe operating space for humanity. *Nature, 461*(7263), 472–475.

Rudd, M. A., Moore, A. F., Rochberg, D., Bianchi-Fossati, L., Brown, M. A., D'Onofrio, D., & Worley, A. N. (2018). Climate research priorities for policy-makers, practitioners, and scientists in Georgia, USA. *Environmental Management, 62*, 190–209.

Ryan, S. J., Lippi, C. A., & Zermoglio, F. (2020). Shifting transmission risk for malaria in Africa with climate change: A framework for planning and intervention. *Malaria Journal, 19*, 1–14.

Sachs, J. D. (2012). From millennium development goals to sustainable development goals. *The Lancet, 379*(9832), 2206–2211.

Saheed, Y. K., Balogun, B. F., Odunayo, B. J., & Abdulsalam, M. (2023). Microarray gene expression data classification via Wilcoxon sign rank sum and novel Grey Wolf optimized ensemble learning models. *IEEE/ACM Transactions on Computational Biology and Bioinformatics, 20*(6), 3575–3587.

Saheed, Y. K., Salau-Ibrahim, T. T., Abdulsalam, M., Adeniji, I. A., & Balogun, B. F. (2024). Modified bi-directional long short-term memory and hyperparameter tuning of supervised machine learning models for cardiovascular heart disease prediction in mobile cloud environment. *Biomedical Signal Processing and Control, 94*, 106319.

Schroth, G., Läderach, P., Martinez-Valle, A. I., Bunn, C., & Jassogne, L. (2016). Vulnerability to climate change of cocoa in West Africa: Patterns, opportunities and limits to adaptation. *Science of the Total Environment, 556*, 231–241.

Schwartz, N. B., Aide, T. M., Graesser, J., Grau, H. R., & Uriarte, M. (2020). Reversals of reforestation across Latin America limit climate mitigation potential of tropical forests. *Frontiers in Forests and Global Change, 3*, 85.

Serrão, J. E., Plata-Rueda, A., Martínez, L. C., & Zanuncio, J. C. (2022). Side-effects of pesticides on non-target insects in agriculture: A mini-review. *The Science of Nature, 109*(2), 17.

Shah, R. (2020). Pesticides and human health. In *Emerging contaminants* (pp. 1–22). IntechOpen.

Sindato, C., Mboera, L. E., Beda, E., Mwabukusi, M., & Karimuribo, E. D. (2021). Community health workers and disease surveillance in Tanzania: Promoting the use of mobile technologies in detecting and reporting health events. *Health Security, 19*(1), 116–129.

Skendžić, S., Zovko, M., Živković, I. P., Lešić, V., & Lemić, D. (2021). The impact of climate change on agricultural insect pests. *Insects, 12*(5), 440.

Smith-Jackson, T., Wogalter, M. S., & Quintela, Y. (2010). Safety climate and pesticide risk communication disparities in crop production by ethnicity. *Human Factors and Ergonomics in Manufacturing & Service Industries, 20*(6), 511–525.

Socha, W., Kwasnik, M., Larska, M., Rola, J., & Rozek, W. (2022). Vector-borne viral diseases as a current threat for human and animal health—One health perspective. *Journal of Clinical Medicine, 11*(11), 3026.

Srinivasa Rao, M., Mani, M., Prasad, Y. G., Prabhakar, M., Sridhar, V., Vennila, S., & Singh, V. K. (2022). Climate change and pest management strategies in horticultural and agricultural ecosystems. In *Trends in horticultural entomology* (pp. 81–122). Springer.

Sufri, S., Dwirahmadi, F., Phung, D., & Rutherford, S. (2020). A systematic review of community engagement (CE) in disaster early warning systems (EWSs). *Progress in Disaster Science, 5*, 100058.

Tessari, L., Angriman, M., Díaz-Román, A., Zhang, J., Conca, A., & Cortese, S. (2022). Association between exposure to pesticides and ADHD or autism spectrum disorder: A systematic review of the literature. *Journal of Attention Disorders, 26*(1), 48–71.

Tiamiyu, B. B., Lateef, A. A., Ajadi, I., Adeyemi, S. B., Owolabi, P. O., Sagaya, A., & Mustapha, O. T. (2024). Bioclimatic modeling of current geographic distribution and future range shifts of selected edible mushrooms in Nigeria. *Sahel Journal of Life Sciences FUDMA, 2*(1), 51–61.

Tolins, M., Ruchirawat, M., & Landrigan, P. (2014). The developmental neurotoxicity of arsenic: Cognitive and behavioral consequences of early life exposure. *Annals of Global Health, 80*(4), 303–314.

Tudi, M., Daniel Ruan, H., Wang, L., Lyu, J., Sadler, R., Connell, D., & Phung, D. T. (2021). Agriculture development, pesticide application and its impact on the environment. *International Journal of Environmental Research and Public Health, 18*(3), 1112.

Ullah, I., & Akhtar, J. (2023). Exploring the effects of climate change on human health. *Cosmic Journal of Linguistics, 2*(1), 76–90.
Umina, P. A., McDonald, G., Maino, J., Edwards, O., & Hoffmann, A. A. (2019). Escalating insecticide resistance in Australian grain pests: Contributing factors, industry trends and management opportunities. *Pest Management Science, 75*(6), 1494–1506.
Upadhayay, J., Rana, M., Juyal, V., Bisht, S. S., & Joshi, R. (2020). Impact of pesticide exposure and associated health effects. In *Pesticides in crop production: Physiological and biochemical action* (pp. 69–88). John Wiley & Sons.
Van Campenhout, L., & Eilenberg, J. (2021). Microbial dynamics during industrial rearing and processing of insects. *Frontiers in Microbiology, 12*, 775603.
Van Den Berg, H., Velayudhan, R., & Yadav, R. S. (2021). Management of insecticides for use in disease vector control: Lessons from six countries in Asia and the Middle East. *PLoS Neglected Tropical Diseases, 15*(4), e0009358.
Weetman, D., Kamgang, B., Badolo, A., Moyes, C. L., Shearer, F. M., Coulibaly, M., & McCall, P. J. (2018). Aedes mosquitoes and Aedes-borne arboviruses in Africa: Current and future threats. *International Journal of Environmental Research and Public Health, 15*(2), 220.

Study of Consumers Choices on Pesticides Use and Sustainability

Mukul Machhindra Barwant, Odangowei Inetiminebi Ogidi, Channe Yogita, and Rupali Munje

1 Introduction

The use of pesticides in agricultural production is widespread and extensive. Some food activists, producers, and public health professionals argue against the widespread use of pesticides because they are unsustainable for the environment, human health, and pest management in food production. In Australia, the government has mainly given consumers the freedom to decide how to resolve this dispute. However, consumers hardly notice that pesticides are used in the production of these goods. This study traces how regulatory decisions, in which consumers play a minor role, shape consumption choices using strawberries as an example. Historically, strawberries have been an extremely pesticide-intensive crop (Parker, 2015). Concerns over the potential for chemical pesticide exposure in fresh produce, which is a crucial component of a balanced diet, have grown. Missouri customers of fresh tomatoes were surveyed online to gather information on their preferences, demographics, and attitudes toward the environment and personal health. When compared to regular tomatoes, tomatoes grown with 50% less pesticide were perceived as worth a 6% premium by the respondents. According to the findings, there might be a market for less pesticide-treated vegetables as a cost-and safety-conscious middle ground between conventional and organic goods. Academic research and corporate activity

M. M. Barwant (✉)
Department of Botany, Sanjivani Rural Education Society's, Sanjivani Arts Commerce and Science College Kopargaon, Ahmednagar, Maharashtra, India

O. I. Ogidi
Department of Biochemistry, Faculty of Basic Medical Sciences, Bayelsa Medical University, Yenagoa, Bayelsa State, Nigeria

C. Yogita · R. Munje
Department of Zoology, Sanjivani Rural Education Society's, Sanjivani Arts Commerce and Science College Kopargaon, Ahmednagar, Maharashtra, India

B. R. Babaniyi, E. E. Babaniyi (eds.), *The Interplay of Pesticides and Climate Change*, https://doi.org/10.1007/978-3-031-81669-7_18

have both highlighted shifts toward more ecologically friendly and healthier food consumption. However, consuming fresh fruits and vegetables is thought to be a vital component of a balanced diet. Exposure to potentially toxic pesticide residues is one risk. The Food and Drug Administration (FDA) of the United States reports that pesticides were detected in 55.2% of fresh vegetables and 66.3% of fresh fruits. Reduced use of pesticides in food production can help address environmental and health concerns by, for example, lowering the danger of pesticide residue and lessening the impact of pesticides on pollinators, aquatic life, and soil (Tran et al., 2023).

Information about the many facets of food product sustainability is becoming more and more necessary as a result of consumer concerns about the negative effects food production has on the environment and society. Sustainability is influenced by each component of the food chain, including cultivation, fishing, animal husbandry, processing, and transportation. Consumer concerns about the detrimental impacts of food production on the environment and society have led to an increasing need for information about the numerous aspects of food product sustainability. Every link in the food chain—cultivation, fishing, animal husbandry, processing, and transportation—has an impact on sustainability. Interest in and willingness to purchase environmental sustainability labels (Per il Clima-Legambiente, Rainforest Alliance, and EU Organic Farming). Indications of information about the origin of processed foods, specifically tomato purée. The study's findings, which were obtained through a choice experiment and conditional logistic models, show that: (i) consumers are more likely to buy products bearing environmental sustainability labels if they understand what those labels signify; (ii) consumers of all educational backgrounds place a high value on information cues regarding a product's domestic origin, in contrast to consumers with higher levels of education who tend to value environmental sustainability labels more highly; and (iii) consumers are willing to pay a premium for labels that have a greater market penetration, regardless of whether those labels are certified by private or public organizations (Aprile & Punzo, 2022).

The reduction in pesticide use is a significant objective for developed nations; however, targeted agricultural environmental initiatives have been unsuccessful in accomplishing this objective. The investigation examines the degree of enthusiasm among conventional farmers to decrease their utilization of synthetic pesticides. We achieve this by conducting a discrete choice experiment in which the presence of pests presents a potential hazard of significant output losses. We discovered that farmers' willingness to modify their practices is substantially restricted by this risk, irrespective of the effect on average profits. Additionally, the administrative burden significantly influences the decisions of producers. People are only truly motivated to take action regarding the detrimental effects of pesticides on human health and the environment when they are convinced that these effects are genuine (Chèze et al., 2020). In response to the growing global concern for environmental sustainability, there has been an increase in the attention given to consumer decisions regarding pesticide use in agriculture. Pesticides are indispensable for agricultural production and pest management; however, there are valid concerns regarding their potential impact on human health, biodiversity, and ecosystems (Popp et al., 2013).

Chemical pesticides pose a significant threat to the long-term sustainability of agriculture. In order to significantly reduce their utilization in safeguarding food supplies, it is imperative that we implement more innovative and adaptable production methods. The widespread adoption of pesticide-free production standards could potentially drive this change, as it would unite the numerous stakeholders in the food value chain. This study examined consumers' preferences for "organic" and "pesticide-free" labels by employing green tea as a case study and a choice experiment approach. The results indicate that customers' purchasing decisions are significantly influenced by the assertions made by labels and companies regarding the absence of pesticides and organic ingredients. In contrast to pesticide-free labels, individuals are prepared to pay a premium for organic labels. The claim made on the label that the product is free of organic materials and pesticides has a replacement effect. Like national brands and pesticide-free labels, national brands and organic labels are inseparable. More creative and adaptable approaches are required to support organic farming with significant pesticide reductions within the context of unsatisfactory food production. All stakeholders in the food value chain benefit from this shift, which might be based on the ideas of pesticide-free production. A non-organic, private-public standard for the production of pesticide-free wheat is being introduced by the Swiss producer organization IP-SUISSE. According to study, customers prefer organic vegetables because they don't contain pesticides. In order to supplement organic farming with extensive reductions in pesticide use within the framework of bad food production, more innovative and flexible strategies are necessary. This transition may be founded on the principles of pesticide-free production, which are advantageous to all parties involved in the food value chain. The IP-SUISSE producer organization in Switzerland is introducing a non-organic, private-public standard for the production of pesticide-free wheat. Organic vegetables are preferred by consumers due to their absence of pesticides, as indicated by research. When it comes to their ideal perceptions of organic fruit, consumers highly value and trust the promises of organic labels, such as "free of synthetic pesticides" and "free of growth hormones." Zheng et al discovered that Chinese consumers are increasingly purchasing organic agricultural products due to their concerns regarding their food.

Consequently, it is imperative to investigate consumer attitudes and behaviors regarding sustainable agricultural practices and pesticide usage. Additionally, the United Kingdom, Denmark, France, Germany, Spain, and the Netherlands have implemented voluntary initiatives to assist farmers in reducing their pesticide usage or transitioning to organic agricultural practices. The United States, Canada, Australia, the European Union, Iceland, and Norway have all implemented systems to monitor food for pesticide residues. These programs support current educational initiatives. Nevertheless, these strategies have not yielded adequate results to date. For instance, in Europe, the use of chemical inputs by farmers has increased over the past decade, with a 12% increase in France between the three-year mean of 2009–2011 and 2012–2014 (Chèze et al., 2020).

Consumers' judgments regarding pesticide utilization are contingent upon numerous variables. These encompass their comprehension of the hazards of chemical pesticides, their capacity to locate information regarding alternative pest control methods, and their perspectives on the advantages and disadvantages of sustainable farming (Detre & Gunderson, 2011). Additionally, shifts in consumer demand toward organic and eco-friendly products, driven by growing environmental consciousness, have created new challenges and opportunities for sustainable agriculture. These choices are not only driven by individual preferences but also by social, economic, and cultural factors, making consumer behavior a key driver of agricultural sustainability (Shaw et al., 2005). The usage of neonicotinoid pesticides in the ornamental horticulture sector in the United States has drawn attention because of the possible health dangers to pollinating insects. Although some merchants have declared that mandatory labeling regulations will apply to plants treated with neonicotinoids, it is unclear how specific consumers will respond when a company discloses that neonicotinoid use occurs in production and how much they will value this extra information. In this instance, we evaluate consumer preferences for eco-friendly production methods via a lab experiment, with a particular emphasis on neonicotinoid labeling. Results indicate that consumers have differing preferences for labels and information disclosure linked to neonicotinoids, despite the fact that most consumers are not familiar with them (Wei et al., 2021).

While chemical pesticides boost agricultural productivity and food production, they harm the environment and the safety and quality of agricultural products. In addition to rapidly gaining resistance to target pests, chemical pesticide overuse can have negative impacts on non-target species, such as endocrine disorders in fish, birds, and rodents (Ogidi & Tawariowei, 2023). Multiple ecosystems might be harmed, food and water sources could be compromised, and pesticide residues could spread over the globe. Some examples include increased emissions of greenhouse gases, reduced soil fertility, and high levels of nitrates in groundwater. It is believed that chemical pesticides are one of the main problems with sustainable farming. Despite the fact that reducing pesticide risks is a high priority for many nations, few have been able to achieve their goals in this area. Because there are not many viable, affordable alternatives, current policies do not always succeed in getting people to stop using pesticides in their manufacturing processes. One way to solve the issue of the quality and safety of agricultural products is to promote organic farming practices. According to Zheng et al for a product to be certified organic, it must not contain any chemically produced fertilizers, pesticides, growth regulators, or other compounds, as well as not have any residue of pesticides, growth hormones, or genetic engineering.

This explores the complex relationship between consumer choices, pesticide use, and sustainability. It delves into the factors influencing consumers' decisions regarding pesticide-laden and pesticide-free products, the role of awareness campaigns and regulatory policies, and the impact of these choices on sustainable farming practices. Furthermore, it examines the potential for consumer-driven demand to promote environmentally sustainable agricultural methods, such as organic farming and integrated pest management (IPM) highlighting both the challenges and opportunities for achieving sustainable food production systems. California shoppers favored produce that was picture perfect when given the option between it and produce that had imperfections. But these customers overwhelmingly selected the scarred oranges when they were informed that the damaged food had been grown with half as many pesticide treatments.

The produce sector thinks that consumers want their fruits and vegetables to be spotless. Environmentalists and consumer advocates assert that people are willing to give up some degree of physical beauty in exchange for less pesticide use. Why does the industry give product appearance such a high priority? The only clues in most stores are pricing and appearance. Customers frequently look for "purchase clues" before making a purchase in order to guide their choices. The only easily accessible indicators in the majority of stores are pricing and appearance. Since most retailers don't provide taste samples, customers are unable to determine how the fruits and vegetables they are contemplating will taste. Before making a purchase, consumers often seek "purchase clues" to inform their decisions. In most stores, the only readily available indicators are price and appearance. Retailers typically do not offer taste samples, leaving consumers unaware of the flavor of the fruits and vegetables they are considering. To gauge ripeness, many rely on squeezing the product to assess its firmness or softness. However, information about how the product was grown, including the farming methods used, is often unknown to most consumers.

Furthermore, consumers are generally unaware of the types or amounts of pesticide residues that may be present in the produce. While a product might look fresh and arrive at the market in good condition, its agricultural background—particularly in relation to pesticide use—remains unclear, making it difficult for consumers to make fully informed choices (Lynch, 1991). It has been demonstrated by earlier research that consumers favor organic food. Despite several efforts to cater to consumer demands for less pesticide-treated fruits and vegetables, these reductions in pesticide use were linked to IPM techniques. Nevertheless, there are no consistent or targeted decreases in pesticide use with the IPM technique. Overall, not much research has compared preferences for conventional and organic production methods with decreased pesticide use as a food characteristic (Tran et al., 2023).

2 Definition and Types of Pesticides

Pesticides are substances or combinations of substances specifically formulated to prevent, eliminate, or manage pests that pose threats to crops during cultivation and storage. These pests can include insects, fungi, rodents, and undesirable plant species, all of which have the potential to cause significant harm to agricultural production (Tano, 2011).

Definition and Scope of Pesticides

At their core, pesticides are chemicals or mixtures of chemicals designed to control, repel, or reduce the damage caused by a wide variety of pests. The spectrum of pests that pesticides target is broad, encompassing not only insects and plant pathogens but also fish, weeds, birds, nematodes (roundworms), mammals, mollusks, and various microorganisms. These organisms can threaten human food supplies, damage property, spread diseases, or otherwise be considered nuisances in both agricultural and non-agricultural settings (Ogidi & Akpan, 2024).

Common Types of Pesticides
The most commonly used pesticides are categorized based on the type of pest they target:

1. *Insecticides*: These pesticides are designed to target and eliminate insects that can damage crops, transmit diseases, or become a nuisance.
2. *Herbicides*: These help farmers keep weeds, undesirable plants, and other pests at bay, as they compete with crops for scarce resources like water, sunlight, and nutrients.
3. *Fungicides*: These combat fungal infections that can damage plants, leading to reduced crop yields and quality.
4. *Rodenticides*: These are used to manage populations of rodents that can destroy crops, contaminate food supplies, and spread diseases (Eldridge, 2008).

Less Common Types of Pesticides
Beyond these widely recognized categories, there are several less common but equally important types of pesticides:

1. *Growth Regulators*: These control the growth of plants or insects, often used to inhibit unwanted growth or to promote desirable characteristics in crops.
2. *Plant Defoliants*: These chemicals are used to cause leaves to drop from plants, which can be beneficial in harvesting certain crops or managing plant diseases.
3. *Surface Disinfectants*: These are applied to surfaces to kill or deactivate pathogens, ensuring that disease-causing microorganisms do not spread.
4. *Algaecides*: These are used in swimming pools and other water bodies to control the growth of algae and other aquatic organisms that could harm the water's quality or safety.

Role of Pesticides in Agriculture and Public Health
The agriculture sector and public health both rely heavily on pesticides. In agriculture, they are essential for protecting crops from pests and diseases, which can otherwise lead to significant losses in yield and quality. In public health, pesticides are used to control pests that can transmit diseases, such as mosquitoes, rodents, and other vectors that spread illnesses to humans.

The Green Revolution and Pesticides
The world's food supply was tripled during the Green Revolution, driven by a significant increase in agricultural output. However, this substantial increase in food production was insufficient to meet the demands of the rapidly growing human population. Among the elements that led to the Green Revolution's higher yields were the widespread usage of improved crop types. However, better management

techniques, such as the widespread use of agrochemicals like fertilizers and pesticides, had a major impact on this achievement (Ogidi et al., 2023).

Pesticides, in particular, were instrumental in controlling the wide variety of pests that threatened crops, thereby enabling higher yields and more reliable food production. The types of pesticides used are diverse and depend on the specific pests that need to be controlled. Herbicides, insecticides, fungicides, rodenticides, nematicides, and other types of pesticides have all been developed to target different pests and address various agricultural challenges. Pesticides are vital tools in modern agriculture and public health, helping to safeguard food supplies, protect property, and prevent the spread of diseases. The responsible use of these tools is crucial in order to reduce the risk of harm they may do to humans and the environment, even though they have greatly enhanced agricultural output. To this day, managing pest hazards in different contexts and maintaining food security rely heavily on the ever-evolving pesticide industry (Ogidi & Akpan, 2023).

3 Insecticides

Insecticides are a critical component in pest management strategies across various sectors, including agriculture and public health. Their effectiveness in controlling insect pests is vital for reducing crop losses and preventing the spread of diseases. Insecticides play a significant role in protecting crops and improving agricultural yields, especially in the context of global agricultural development (Dhaliwal et al., 2015; Sharma et al., 2019).

Importance of Insecticides in Pest Management

Both wealthy and developing nations suffer heavy losses due to insect pests, which pose a serious danger to agricultural productivity. Effective pest management is necessary to mitigate these losses and guarantee food security. This encompasses the management of diseases, parasites, and weeds. Insecticides have become indispensable instruments in this endeavor due to their capacity to both safeguard plants and augment crop yields (Ogidi, 2023).

Classification of Insecticides

Insecticides can be classified into several sub-classes based on their chemical composition and mode of action. Each subclass has unique characteristics, making it important to select the appropriate insecticide for specific pest control needs. Below are the main sub-classes of insecticides:

1. *Carbamates*:

   - *Mode of Action*: Carbamates are a class of insecticides that operate by interfering with the insect nervous system. Specifically, they block the enzyme acetylcholinesterase, which is responsible for breaking down the neurotransmitter acetylcholine. When this enzyme is inhibited, acetylcholine builds up

in the synapses, causing continuous nerve impulse transmission and, ultimately, paralysis and death for the insect.
- *Environmental Considerations*: Although carbamates accomplish their tasks well, they have the potential to kill non-target organisms as well, such as useful insects and other animals. Careful application and management are required to minimize environmental impact.

2. *Organochlorines*:

   - *Mode of Action*: Organochlorine insecticides disrupt the nervous system of insects by causing over-stimulation of nerve cells. This leads to convulsions and ultimately death. Organochlorines can bioaccumulate in food chains and remain in the environment for a very long time.
   - *Environmental Considerations*: Due to their persistence and potential to cause environmental harm, many organochlorines have been banned or heavily restricted in use. Their ability to remain in the environment for extended periods poses risks to wildlife and human health.

3. *Organophosphorus Compounds*:

   - *Mode of Action*: In the same way that carbamates build up in the neurological system, organophosphorus pesticides prevent acetylcholinesterase from breaking down acetylcholine. This results in paralysis and death of the insect. Organophosphorus compounds are widely used in agriculture due to their effectiveness.
   - *Environmental Considerations*: These compounds are often more toxic to non-target species, including humans, and require careful handling. Unlike organochlorines, organophosphorus insecticides have a shorter environmental persistence, which reduces their long-term impact but still necessitates cautious use (Ogidi et al., 2024a).

4. *Pyrethroids*:

   - *Mode of Action*: Pyrethroids are man-made compounds that aim to kill insects in the same way that the naturally occurring pyrethrins found in chrysanthemum flowers do. They act on the sodium channels of nerve cells, causing prolonged depolarization, which leads to paralysis and death of insects.
   - *Environmental Considerations*: Pyrethroids are favored for their rapid action and relatively low toxicity to mammals, making them popular in both household and agricultural settings. However, they can still pose risks to aquatic life and beneficial insects, so their use must be carefully managed.

5. *Neonicotinoids*:

   - *Mode of Action*: In order to eliminate insects, neonicotinoids bind to the nicotinic acetylcholine receptors of their nervous system, thereby mimicking the effects of nicotine. This binding leads to death and paralysis. Neonicotinoids are particularly effective against sucking insects, such as aphids, and are widely used in agriculture.

- *Environmental Considerations*: While neonicotinoids are effective, there are growing concerns about their impact on pollinators, particularly bees. The decline in bee populations has been linked to the use of neonicotinoids, prompting regulatory reviews and restrictions in some regions.

Insecticides are essential tools in the fight against insect pests, playing a crucial role in safeguarding crops and improving agricultural productivity. Each class of insecticides has its own unique mode of action, effectiveness, and environmental considerations, which must be carefully evaluated to ensure effective and sustainable pest management. The responsible use of insecticides is vital to minimize their environmental impact and protect non-target species, including pollinators and other beneficial organisms.

4 The Impact of Pesticides on Health and the Environment

Environmental Impact of Pesticides
The use of pesticides has raised significant concerns about their long-term impact on the environment. Despite advances in environmental risk assessment procedures, there remain substantial gaps in knowledge regarding the full extent of these impacts. Several issues contribute to these gaps:

1. *Lack of Systematic*:
    - There is a notable shortage of systematic reviews of existing data on the environmental impacts of pesticides. Without comprehensive analyses of available data, it is difficult to fully understand the risks associated with pesticide use or to develop effective strategies for mitigating these risks (Boyd, 2018; Group of Chief Scientific Advisors, 2018; Schäfer et al., 2019).
2. *Unexpected Environmental Impacts*:
    It is not uncommon for unanticipated environmental effects of pesticide use to become apparent, even with more thorough environmental risk evaluations, even years after the herbicide has been commercially available. The limits of present risk assessment techniques are brought to light by these delayed repercussions, which emphasize the necessity for continuing monitoring and review.
3. *Insufficient Monitoring*:
    Pesticide concentrations in different environmental compartments, including air, surface water, groundwater, and soil, are only systematically monitored in some regions of the world. Particularly in areas with less established monitoring systems, there are large data gaps. Furthermore, in low- and middle-income nations, monitoring of non-target organisms and ecosystems' harmful consequences is nearly nonexistent; even in many high-income countries, such monitoring is infrequent.
4. *Gaps in Post-Registration Monitoring*:

Our knowledge of the negative impacts of pesticide usage on the environment is severely lacking due to the shortcomings of prospective risk assessments and the absence of systematic monitoring after registration. Recent publications (Milner & Boyd, 2017; Rico et al., 2020) have argued for the need for more thorough monitoring of pesticide concentrations and their effects after registration.

5. *Recommendations for Improvement*:

   In order to improve environmental risk assessment, many suggestions have been put forward, such as:

- Generating additional data on pesticide toxicity and environmental behavior.
- Developing improved or new risk assessment models.
- Introducing new risk assessment paradigms.

Enhancing environmental monitoring of pesticide concentrations and their effects after registration should, however, take precedence.

Human Health Impact of Pesticides

There is a lot of need for improvement in the scientific data that supports pesticide risk assessments, even though a lot of knowledge has been gained about the health consequences of pesticides in the last few decades. Among the most important areas where information is lacking are:

1. *Complex Health Outcomes*:

   - Current toxicological evaluations often fail to fully address more complex human health outcomes. These include the following:

- *Immunotoxicity*: The impact of pesticides on the immune system.
- *Childhood Leukemias*: Possible connections between childhood leukemia and pesticide exposure.
- *Developmental Neurotoxicity*: The impact of pesticides on the growing nervous system, especially during formative years.
- *Chronic Neurological Diseases*: Associations between pesticide exposure and diseases such as Parkinson's disease, coupled with cognitive consequences and mental problems.
- *Endocrine Disorders*: Among these include reproductive senescence, metabolic syndrome, endometriosis, hormonal malignancies, and type 2 diabetes.

2. *Co-Formulants and Formulated Products*:

   - Toxicological assessments often focus on the active ingredients in pesticides, but co-formulants and fully formulated products may also pose risks. There is a need for more comprehensive evaluations of these substances.

3. *Mechanisms of Harm*:

   There has to be more research into the specific methods in which pesticides cause harm to humans, such as the adverse outcome pathways (AOPs). This knowledge is crucial for improving risk assessments.

4. *Reduction in Animal Testing*:

   The development of innovative toxicological approaches that reduce the use of animals in testing is a priority. This includes the application of in vitro toxicogenomics, which can provide insights into how chemicals affect gene expression.
5. *Combined Exposure Risks*:

   Additional research is needed to determine the potential for combined effects when exposed to various active drugs, either at the same time or in a specific order. Current risk assessment methods may not adequately account for these combined risks.
6. *Cumulative Dietary Risk Assessments*:

   Methods for cumulative dietary risk evaluations need to be further developed and standardized. These assessments should consider the combined exposure to multiple pesticides through food consumption.
7. *Protecting Sensitive Populations*:

   Protecting vulnerable populations, including those with impaired immune systems, women in agriculture and related occupations, youngsters, and those who are pregnant or nursing must be a top priority when conducting risk assessments for human health.
8. *Exposure Levels and Determinants*:

   Worker entry into treated crops and applicators using handheld or backpack sprayers should be better characterized and their pesticide exposure should be quantified. Understanding the determinants of exposure is critical for developing effective risk management strategies.
9. *Epidemiological Studies*:

   There is a need for improved characterization of pesticide exposure in epidemiological studies. Factors that contribute to exposure levels include the "real-life" actions of pesticide operators, such as the way they apply the chemical and the protective gear they wear.
10. *Nanopesticides and Biopesticides*:

    - Standardized risk assessment procedures for newer pesticide technologies, such as nanopesticides and biopesticides, are lacking. These innovations require specific risk assessment approaches due to their unique properties.

11. *Antimicrobial Resistance*:

    Risk evaluations should take into account the growing concern about the role of pesticide use in the rise of antibiotic resistance.
12. *Post-Marketing Surveillance*:

    Post-marketing surveillance of approved pesticides is becoming more acknowledged as a necessity to ensure they do not cause unforeseen health issues. This surveillance would provide valuable data to inform ongoing risk assessments and regulatory decisions. The effects of pesticides on ecosystems and people's well-being are intricate and multidimensional. There is still a lot we do not know, especially when it comes to health risk assessment and environmental monitoring. To minimize the negative consequences of pesticide usage and ensure the safety of both human and environmental populations, it is vital to address these gaps through better risk assessment models, more thorough data collecting, and improved post-registration monitoring.

5 Principles of Sustainable Agriculture: A Detailed Exploration

Sustainable agriculture is a multifaceted approach that involves not only farm management but also the broader supply chain, recognizing that the responsibility for sustainability does not rest solely with farmers. Instead, it is a shared duty involving those who purchase agricultural materials, as well as suppliers who provide products and services to farmers. Sustainable agriculture principles provide a comprehensive framework, emphasizing the interdependence of various practices. It is important to consider all interdependencies because certain practices, whether on the farm or in the supply chain, may have unintended or even conflicting consequences.

Interpretations of Sustainable Agriculture

Sustainable agriculture is interpreted in two main ways in the literature. One interpretation views sustainability as a philosophy or approach driven by an awareness of the negative impacts of traditional agricultural activities. The other sees sustainability as a set of strategies aimed at enhancing resilience and building on the existing system.

1. *Sustainability as a Philosophy*:

   Perceived as an intellectual or managerial approach, sustainability in agriculture is viewed as a philosophy. The overarching objective is to encourage the use of less harmful and more socially responsible farming methods. At the community level, this philosophical perspective can motivate people to make changes and implement sustainable behaviors.

   Nevertheless, this method has been criticized. Potential roadblocks to reaching the loftier goal of widespread sustainable food production include a dearth of generally accepted alternatives and, at times, skewed perceptions of circular or sustainable techniques. Many people have doubts about this line of thinking because they do not fully understand the principles involved or because they are biased toward circularity or sustainable practices that are supposed to replace conventional agriculture (Thompson, 2007; Hansen, 1996).

   Despite the potential for motivating change, the philosophical interpretation has struggled with practical implementation, particularly when trying to scale sustainable practices to national or global levels. The diversity in environmental conditions and agricultural practices across different regions makes it difficult to apply a one-size-fits-all approach to sustainability.
2. *Sustainability as a Set of Strategies*:

- On the other hand, some interpret sustainable agriculture as a collection of strategies aimed at making agricultural systems more resilient. This interpretation is grounded in the idea that sustainability should not prescribe a specific set of technologies or practices. Instead, it should encourage continuous innovation and adaptation in response to changing conditions.
- This approach emphasizes that sustainability is context-specific, meaning that what works in one region or system may not be appropriate in another. Thus, rather than imposing uniform models or methods, sustainable agriculture should be seen as a dynamic process of learning and improving (Pretty, 1995; Heinberg & Thompson, 1992). The strategy-focused interpretation of sustainability aligns closely with the concepts of resilience and durability. Agricultural systems can be guided by this criterion, which facilitates resilience-building and performance assessment, in the face of rapid changes and challenges. However, there are problems with using it in the real world, especially when trying to diagnose and measure sustainability (Hansen, 1996).

Alternative Agricultural Approaches
The rise of "alternative agricultures" is a result of the goal-prescribing concept of sustainability. These alternatives are often debated in terms of their sustainability credentials but include approaches such as agroecology, organic farming, biodynamic agriculture, regenerative agriculture, permaculture, environmentally sensitive farming, community-based agriculture, low-input farming, and integrated production (Pretty, 2004, 2014).

- *Characteristics of Alternative Approaches*:
  - These alternative approaches typically involve resource-conserving technologies and practices designed to improve the use of natural resources. They aim to mimic natural processes, adopt regenerative and circular perspectives, and reduce or eliminate the use of synthetic chemicals like fertilizers and pesticides (Hansen, 1996).
  - Despite their potential benefits, these approaches are often criticized for being less productive and efficient compared to conventional agriculture. Critics argue that such systems may not be able to meet the food production demands of a growing global population, particularly in the face of climate change (Pretty, 2008).
- *Sustainable Intensification*:
  - In response to concerns about the productivity of alternative approaches, some advocate for sustainable intensification. This concept suggests that it is possible to increase food production while minimizing environmental impact, thus maintaining food security for future generations (Pretty, 2004).

Challenges in Achieving Sustainable Agriculture
Sustainable agriculture is inherently complex, and several challenges must be addressed to make it a practical and effective framework:

1. *Context-Specific Approaches*:
   - The context-specific nature of sustainability makes it difficult to apply uniform practices or strategies universally. What works in one region may not work in another due to differences in climate, soil, economic conditions, and social structures (Thompson, 1992). This means that sustainable agriculture must be flexible and adaptable to different environments.
2. *Continuous Learning and Adaptation*:
   - Sustainability is not a static goal but a continuous process of learning and adaptation. As conditions change—whether due to climate change, market shifts, or technological advances—agricultural practices must evolve to remain sustainable (Pretty, 1995).
3. *Practical Application*:
   - While the philosophical and strategic interpretations of sustainability provide valuable guidance, they often fall short when it comes to practical application. Diagnosing and quantifying sustainability, particularly in diverse and dynamic agricultural systems, remain a significant challenge (Hansen, 1996).

The principles of sustainable agriculture are essential for guiding the transition to more resilient and environmentally responsible agricultural systems. However, achieving sustainability requires a nuanced understanding of the concept, recognition of its context-specific nature, and a commitment to continuous innovation and adaptation. By addressing these challenges, sustainable agriculture can contribute to a more secure and sustainable food future.

6 Integration of Sustainable Practices in Agriculture

The Issue with Pesticides in Agriculture
The widespread use of pesticides and other chemical inputs has been a significant concern in sustainable agriculture. The increased reliance on these chemicals has not only raised public concerns about their safety but also highlighted their potential to cause harm to the environment. This growing awareness has led to heightened scrutiny of pesticide products, pushing the agricultural sector to seek safer and more sustainable alternatives. One of the most promising solutions to these challenges is integrated pest management (IPM).

What Is Integrated Pest Management (IPM)?
The term "Integrated Pest Management" (IPM) refers to a method of pest control that integrates various techniques to reduce the need for synthetic pesticides without sacrificing crop protection. For sustainable pest management, integrated pest management (IPM) employs a wide range of methods, including cultural practices, biological control, habitat modification, and resistant crop cultivars.

IPM is adaptable; it can be used to reduce insect populations and increase agricultural yields in both urban and rural areas. Integrated pest management (IPM) is an approach to pest control that seeks to lessen chemical exposure and environmental degradation by combining several management strategies. The maintenance of agro-ecological systems is thus greatly aided by IPM.

The Core Principles of IPM

1. *Pest Monitoring and Identification*:
   - Effective IPM begins with the accurate identification and monitoring of pest populations. This step is crucial because not all insects or organisms are harmful, and unnecessary pesticide use can be avoided by distinguishing between beneficial and harmful species.
2. *Combination of Control Methods*:
   - *Biological Control*: In this method, parasites, diseases, or natural predators are used to keep pest populations in check. Aphid control with ladybirds or selective insect larval killing with *Bacillus thuringiensis* (Bt) are two examples.
   - *Habitat Manipulation*: Altering the environment to make it less hospitable for pests, such as by managing crop rotation, planting cover crops, or adjusting irrigation practices to reduce pest breeding grounds.
   - *Cultural Practices*: Modifying agricultural practices, such as altering planting times, using resistant crop varieties, or implementing crop rotation, to disrupt pest life cycles and reduce their impact.
   - *Mechanical and Physical Controls*: Using barriers, traps, or mechanical methods to physically remove or prevent pests from damaging crops.
   - *Chemical Control*: Using insecticides sparingly and precisely when needed to lessen their negative effects on the environment and the probability that pests will develop resistance.
3. *Environmental and Human Health Considerations*:

   The goal of integrated pest management (IPM) is to lessen the detrimental effects of toxic pesticides on humans and other species by reducing their presence in the ecosystem. Integrated pest management (IPM) provides the greatest opportunity for obtaining and maintaining long-term crop protection by combining numerous pest control methods.

Sustainable Agricultural Practices

IPM is just one aspect of a broader approach to sustainable agriculture. Other key practices include:

- *Crop Rotation*: Rotating crops helps manage weeds, diseases, insects, and other pest problems. It improves soil quality by providing alternative sources of nitrogen and helps reduce soil erosion.
- *Biological Weed Control*: Increasing the use of mechanical or biological methods to control weeds can reduce the need for chemical herbicides, thus minimizing environmental impact.

- *Soil and Water Conservation*: Efficient practices such as contour plowing, terracing, and the use of green manures help conserve soil and water resources, enhancing the sustainability of agricultural systems.
- *Efficient Use of Inputs*: Whether using natural (biological) or synthetic inputs, it is crucial to apply them in a systematic and efficient manner that avoids significant harm to humans or the environment.

Strategies for Effective Pest Management
For IPM to be effective and widely adopted, several strategies need to be implemented:

1. *Farmer and Extension Personnel Training*:
   Educating farmers and extension workers on IPM methodologies is essential for successful implementation. Training programs should focus on pest identification, monitoring techniques, and the integration of multiple control methods.
2. *Improved Availability of Critical Inputs*:
   Ensuring the availability of essential inputs such as biopesticides, beneficial organisms, and resistant crop varieties is critical for the success of IPM programs.
3. *Development of Monitoring Tools and Forewarning Systems*:
   Advancing technologies for pest monitoring and early warning systems can help farmers anticipate and manage pest outbreaks more effectively.
4. *Holistic Integration of Information*:
   Integrating all available information on pest dynamics, crop management, and environmental factors can lead to the development of bio-intensive, cost-effective IPM practices.

The integration of sustainable practices in agriculture, particularly through the adoption of integrated pest management (IPM), offers a viable solution to the challenges posed by the over-reliance on chemical pesticides. By combining multiple pest control strategies, IPM not only reduces the environmental impact of farming but also enhances the sustainability of agricultural systems. Implementing these practices requires education, access to necessary resources, and a commitment to ongoing research and development to continuously improve and adapt to changing conditions. Through these efforts, sustainable agriculture can achieve the dual goals of protecting the environment and ensuring food security for future generations.

7 Consumer Awareness and Knowledge

7.1 Levels of Consumer Awareness

Pesticide contamination in vegetables is a significant public health concern, as many vegetables available in markets are tainted with agrochemicals. Addressing this issue through consumer awareness is essential in reducing the associated health risks. This study was designed to highlight the importance of educating consumers on how to identify pesticide-contaminated leafy vegetables using appropriate methods, such as sensory detection by smell. The findings underscore the need for government and relevant institutions to conduct awareness programs that inform consumers about the health impacts of pesticide-contaminated vegetables. By empowering consumers with the knowledge to identify and avoid contaminated produce, public health can be better protected.

The Knowledge Gap Among Farmers Regarding Pesticides

Farmers, who are the primary handlers of pesticides, often lack adequate knowledge about the environmental and health risks associated with pesticide use. This knowledge gap is particularly concerning because farmers typically apply pesticides manually, without the benefit of advanced machinery that might mitigate exposure. Many paddy farmers, for instance, are unaware of the long-term environmental degradation and the threats to their own health and public health posed by improper pesticide use.

Studies in various regions, including Pulau Pinang, reveal that there is a significant deficit in farmers' awareness of safe pesticide practices. Most previous research has focused on issues like weed distribution in paddy fields and the impact of herbicides, but there has been a notable lack of focus on farmers' understanding of safe pesticide use. For example, a study by Alam et al. (2012) in the Muda irrigation area examined farmers' attitudes toward weed infestation but did not thoroughly address their awareness of pesticide safety. Similarly, research conducted by students at Universities Kebangsaan Malaysia in 1998 touched on the health effects of pesticide use among paddy farmers in Tanjung Karang, Selangor, but did not delve deeply into the farmers' knowledge, attitudes, and practices related to pesticide use.

Consumer Preferences and the Demand for Pesticide-Free Products

The risks associated with consuming vegetables that have been grown with artificial fertilizers and pesticides are becoming more widely known. This realization has resulted in a significant increase in the demand for goods that are free of pesticides. The public perceives these goods as safer, healthier, more nutrient-dense, and more environmentally friendly. Many individuals are willing to pay a premium for vegetables that are free of pesticides. This trend presents a significant opportunity for farmers who are motivated to adopt organic farming practices, as it aligns with the growing consumer demand for pesticide-free produce.

To capitalize on this trend, policy measures are recommended to promote the production and consumption of pesticide-free vegetables. Governments, organizations, and stakeholders can play a pivotal role by formulating policies that encourage sustainable farming practices and by launching effective marketing strategies and educational campaigns to raise awareness about the benefits of pesticide-free vegetables. Such policies would not only protect public health but also provide economic incentives for farmers to transition to organic farming.

The Persistent Use of Harmful Pesticides in Agriculture

Despite national and global bans on certain harmful pesticides, many of these chemicals continue to be used in regions like Udaipur, India. For example, endosulfan, a Class II organochlorine pesticide banned by the Supreme Court of India in 2011, is still being used, despite the official cessation of its production in January 2017.

Other toxic pesticides, such as profenofos, chlorpyriphos, acephate, and 2,4-D, are also commonly used by local farmers.

The continued use of these banned or restricted pesticides highlights a significant gap in the enforcement of pesticide regulations and the education of farmers about the dangers of these chemicals. Studies have shown that farmers in the Udaipur region are exposed to harmful pesticides, including those that have been banned or are otherwise restricted. For instance, imidacloprid, a pesticide used to control whiteflies in okra crops, and other chemicals like dimethoate and triazophos, are still in use, posing serious risks to both the environment and public health.

Moreover, the improper storage and disposal of pesticides further exacerbate the risks. Many farmers store pesticides in open fields and dispose of empty containers and old stocks in ways that threaten non-target animals, humans, and the environment. This practice indicates a lack of understanding among farmers about the severe consequences of pesticide misuse.

The Need for Comprehensive Education and Policy Interventions

The findings from these studies underscore the urgent need for comprehensive education programs to raise awareness among farmers about the safe handling and use of pesticides. Farmers must be educated about the recommended application rates, the proper storage and disposal of pesticides, and the long-term health and environmental risks associated with improper pesticide use. Without this knowledge, farmers are likely to continue practices that endanger both their health and the environment.

Additionally, there is a need for stricter enforcement of pesticide regulations to ensure that banned or restricted chemicals are not being used. Governments and regulatory bodies must work together to monitor pesticide use and provide farmers with safer alternatives.

In conclusion, both consumer awareness and farmer education are critical components in reducing the health risks associated with pesticide use in agriculture. By educating consumers on how to identify and avoid pesticide-contaminated produce and by providing farmers with the knowledge and resources needed to adopt safer practices, we can work toward a more sustainable and health-conscious agricultural system.

7.2 Sources of Information on Pesticides and Sustainability

Pesticides are chemical substances designed to control or eliminate pests, but their very nature means they can pose significant risks to humans, animals, and the environment. Because these chemicals are engineered to kill or adversely affect living organisms, their use must be carefully managed to mitigate potential harm.

One of the primary concerns with pesticides is that many contain nerve poisons, which function by inhibiting key enzymes in the nervous system. These enzymes are essential for normal nerve function, and their inhibition can lead to severe neurological damage in both target pests and non-target species, including humans. This characteristic underscores the need for caution in handling and using pesticides, especially in environments where humans and animals are present.

Pesticides are not limited to agricultural or industrial applications; they are also prevalent in many household products. Common household items that contain pesticides include:

1. *Insect Repellents for Personal Use*: These products are designed to keep insects like mosquitoes and ticks away from the body. While effective at preventing bites, the chemicals used can pose risks if overused or improperly applied.
2. *Cockroach Sprays and Baits*: Cockroach control products often contain potent insecticides that are effective in eliminating these pests but can be harmful if ingested or inhaled by humans or pets.
3. *Rat and Other Rodent Poisons*: Rodenticides are highly toxic substances designed to kill rats, mice, and other rodents. These poisons can be lethal to non-target animals, including pets and wildlife, and pose significant risks if accidentally ingested by children.
4. *Flea and Tick Sprays, Powders, and Pet Collars*: These products are used to control fleas and ticks on pets. While they protect pets from infestations, the chemicals involved can cause skin irritation or more severe health issues if used improperly.
5. *Kitchen, Laundry, and Bath Disinfectants and Sanitizers*: Many household disinfectants and sanitizers contain antimicrobial pesticides to kill bacteria, viruses, and fungi. While essential for maintaining hygiene, these products can be harmful if inhaled or if they come into contact with skin or eyes.
6. *Fungicides for Mold and Mildew*: Products designed to kill mold and mildew are often fungicides. These chemicals are effective in preventing the growth of harmful molds but can also pose respiratory risks if their fumes are inhaled.
7. *Lawn and Garden Products (Herbicides)*: Many products used to maintain lawns and gardens, such as weed killers, contain herbicides. These chemicals target unwanted plants but can also affect other vegetation, soil health, and nearby water sources if not used carefully (Sawyer et al., 2023; Ogidi & Emaikwu, 2024).

Given the widespread use of pesticides in everyday household products, it is crucial for consumers to be aware of the potential risks associated with their use. Proper handling, storage, and disposal of these products are essential to minimize exposure and prevent accidental poisoning. Additionally, consumers should seek out safer alternatives or natural pest control methods whenever possible to reduce the reliance on chemical pesticides and their associated risks.

7.3 Role of Education and Media

Education plays a critical role in shaping farmers' understanding of the risks associated with pesticide use, particularly regarding its effects on human health and the environment. Research has shown that farmers with at least a primary or secondary school education tend to have a significantly higher awareness of these risks compared to those with little or no formal education.

In a study, it was found that 90% of the farmers surveyed had received some level of formal education, ranging from primary school to higher educational institutions. Only a small percentage, about 6.4%, lacked any formal educational background. This correlation between education and awareness aligns with findings from Gaber and Abdel Latif (2012), who observed that farmers with formal education are generally more aware of the negative impacts of pesticide use.

Additionally, Hou and Wu contend that farmers who lack fundamental education frequently encounter difficulties in the assimilation of professional knowledge and information regarding pesticides. This discrepancy in comprehension can result in the improper application and management of pesticides, which can elevate the environmental and human health hazards. This is further supported by the research conducted by Hwang et al. (2000) and Yassin et al. (2009), which suggests that farmers with lesser educational levels are less capable of understanding the health risks associated with pesticides and the importance of adhering to safety protocols.

As a result, farmers with limited educational backgrounds are often less aware of the harmful effects of pesticide residues and the importance of adhering to standard procedures for pesticide application. This lack of awareness not only puts their health at risk but also jeopardizes the safety of the broader community and the environment. Therefore, improving educational opportunities and providing targeted training on safe pesticide use are essential steps toward enhancing farmers' ability to manage pesticides responsibly and sustainably.

8 Consumer Attitudes and Perceptions

A turning point in public opinion occurred in 1962 with the release of "Silent Spring" by Rachel Carson, which prompted an increasing social criticism of the extensive use of pesticides and other chemicals. The continuous use of synthetic pesticides is called into doubt by this critical approach, which poses important ethical problems. Striking a balance between the many costs and advantages of pesticide use is an incredibly difficult task. In this paper, we will examine the environmental, social, and economic effects of pesticides in depth in an effort to shed light on some of these complex ethical questions and add to the continuing discussion about whether or not they are morally acceptable.

8.1 Health Concerns and Safety

Training initiatives spearheaded by the Department of Agriculture and the Integrated Agricultural Development Area (IADA) must be amplified to reach farmers who apply pesticides to their crops in order to guarantee the safe control of pesticide use. Newcomers to the farming industry, especially those under the age of 30, should be the primary target of this training.

An incentive program for these younger farmers could be highly beneficial, as it would not only encourage participation but also raise awareness of the environmental and health risks associated with pesticide use. One effective method to implement this initiative is to leverage knowledge and best practices from other countries, particularly from Asian nations, where advanced and diverse approaches to pesticide management are practiced. By facilitating an exchange of ideas and experiences, farmers can gain a broader perspective and adopt more effective strategies.

Moreover, comparing work ethics and pesticide application methods between regions could provide valuable insights, helping to boost farmers' confidence and improve their practices. This approach would foster a more conscientious attitude toward the use of pesticides, ultimately leading to better environmental stewardship and health protection.

Research by Mubushar et al. (2019) lends credence to this idea by showing that landowner farmers are more likely to be cognizant of the potential dangers of pesticides to people and the environment. This type of landowner is more likely to practice responsible pesticide usage, take all required safety measures, and work to keep soil from being degraded. Farmers who lease their land, on the other hand, tend to put profit before health and the environment. This study lends credence to the idea that privately held land is more likely to have environmental protection and long-term sustainability goals.

8.2 Environmental and Ethical Considerations

Many of the benefits of pesticide use are oversimplified and center on the direct effects on crops, ignoring the significant societal, economic, and environmental costs that come with using these chemicals. Only one percent of pesticides really kill the intended pests, according to studies; the other 99.9 percent may have unintended consequences for the ecosystem. For the purpose of crafting a well-rounded pesticide policy, it is imperative that these far-reaching effects on human health and the environment must be carefully considered. A thorough examination of the complicated economic and environmental implications of pesticide dependency was undertaken by Pimentel et al. (1992) in response to this requirement. They cover a lot of ground in their analysis, including:

- *Human Health*: Evaluating the impact of pesticide exposure on human health, including potential long-term effects.
- *Domestic Animal Poisonings*: Assessing the incidence of poisoning in domestic animals resulting from pesticide use.
- *Control Expenses*: Calculating increased costs due to the destruction of natural pest control agents and the development of pesticide resistance.
- *Pollination Issues*: Investigating problems related to crop pollination and losses among honey bee populations.
- *Crop Losses*: Examining the direct losses in crops and crop products due to pesticide use.
- *Water Contamination*: Analyzing the contamination of groundwater and surface water from pesticide runoff (Iyiola et al., 2023).
- *Biodiversity Loss*: Looking at the impact on fish, wildlife, and microorganisms.
- *Government Expenditures*: Reviewing public spending on mitigating the environmental and social costs of pesticide use.

8.3 Trust in Labels and Certifications

According to the Code of Conduct, a pesticide label includes any text, graphics, or other media that is either printed on, adhered to, or affixed to the pesticide, its container, or the retail package's outer wrapping or wrapper (FAO and WHO, 2014). A label is defined by the global harmonized system (GHS) as "a collection of information elements that are attached to or printed on the immediate container or the outer packaging of a hazardous product that is pertinent to the target sector(s)".

This instruction is based on the definition that is provided by the Code of Conduct. An essential and obligatory component of any product's packaging, a label is defined here. While the label has all the necessary information, there may be further details in a Safety Data Sheet (SDS) or a separate "fold-out" booklet that comes with the container; the label will make note of these papers.

A product's label is the first and, in many cases, the only point of contact between the seller or producer and the consumer. They provide important details on the safe and effective usage of the insecticide. This contains information about the product itself, its functions, how to use it, and any limitations on its accessibility. Pesticide labels also detail potential side effects and dangers of using the product. Users can assess the potential dangers of using the product in various scenarios with the use of these data.

Labels on pesticides are a must-have resource for protecting people and the environment. Pesticide labels are mandated by law in numerous nations and are subject to certain regulations on their wording and structure. These rules make sure that the proper authorities approve all pesticide labels, including any changes or variants. To use a pesticide in a way that goes against its label is a crime, therefore it is important to follow the label's directions.

9 Factors Influencing Consumer Choices: Rupali Munje

9.1 *Economic Factors: Price and Affordability*

The use of pesticides has been critical in maintaining a steady food supply and increasing agricultural output. But because of the risks they pose to both human and environmental health, their usage is fraught with worry. Methods for determining the efficacy of pesticides in terms of both productivity gains and health risks are discussed in this article. Issues covered include resistance accumulation, hazards to human and environmental health, predator-prey dynamics, and more. The article summarizes important empirical findings and research in this area. The report also takes a look at the current pesticide policies, which vary from ones that are based on political and economic considerations like risk aversion to ones that are in line with the social optimum as proposed by economic theory. This policy is explained from the perspective of agricultural and environmental management as a whole. Furthermore, the article delves into the latest advancements in invasive species modeling and agricultural biotechnology, shedding insight into how these topics pertain to pesticide usage.

When considering the benefits and drawbacks of pesticide use and other methods of pest control, economic considerations are crucial for both public and private sector decision-makers. When is it socially beneficial to apply pesticides? That is a central question in the literature on pesticide economics. To do this, we must include the direct costs (such as the cost of the pesticide, labor, and machinery) and the indirect costs (such as the hazards to human health and environmental pollution) when calculating the value of pesticide treatments, which may include reduced crop damage (Ogidi et al., 2024b).

Many things, like externalities and uncertainties, need to be included in a thorough analysis. Economists have created useful analytical tools to assist decision-making, despite the difficulties and complexity of modeling pesticide effects. A lot of work has gone into finding ways to make pesticides safer and more effective in the last few decades, thanks to advances in science and technology. Agricultural biotechnology and integrated pest management (IPM) are two prominent instances of such developments. In order to reduce the societal costs of pesticide use and safeguard human health from unnecessary dangers, this article will analyze these instruments in conjunction with government initiatives in this area (Ogidi et al., 2024c).

Over the years, efforts to control pesticide usage have frequently been overly simplistic and naive, resulting in less-than-ideal results and putting a heavy burden on innovators. In order to affect business and market behavior, the report calls for more effective pesticide policies that are integrated with larger agriculture policies including price subsidies, output limitations, and quality requirements.

9.2 Availability and Accessibility of Sustainable Products

Sustainable pesticide products are increasingly available and accessible worldwide, driven by global initiatives and regulations aimed at mitigating the environmental and health impacts associated with traditional chemical pesticides. Key developments in this area include:

1. *Sustainable Pesticide Management Framework (SPMF)*:

   This five-year initiative, launched by Crop Life International, focuses on promoting sustainable and responsible pesticide use. It is being rolled out in several countries, including Kenya, Morocco, Thailand, and Vietnam.

   The SPMF program aims to reduce dependence on highly hazardous pesticides, support the adoption of innovative and environmentally friendly farming practices, and foster responsible pesticide use among farmers.
2. *European Union Initiatives*:

   As part of its Farm to Fork and Biodiversity strategies, the European Union has set ambitious targets to reduce both the use and risk of chemical pesticides by 50% by 2030.

   The EU's approach includes promoting biopesticides and other sustainable alternatives as part of its broader strategy to enhance food system sustainability and biodiversity.
3. *Biopesticides*:

   Biopesticides, which are derived from natural sources such as plants, bacteria, and certain minerals, are gaining popularity as a more sustainable alternative to traditional chemical pesticides.

   Despite their benefits, biopesticides face challenges including higher costs, inconsistent preparation standards, and susceptibility to environmental factors that can affect their efficacy.

Challenges

- *Cost and Accessibility*: Sustainable pesticide products, particularly biopesticides, are often more expensive and less accessible compared to conventional pesticides. This can be a significant barrier for widespread adoption, especially in regions with limited resources.
- *Regulatory Barriers*: Varying regulations across different countries can impact the availability and adoption of sustainable pesticide products. Discrepancies in regulatory standards and approval processes can hinder the global uptake of these alternatives.

Efforts are ongoing to address these challenges and improve the availability and accessibility of sustainable pesticide products. These initiatives aim to make sustainable options more feasible for farmers and more effective in reducing the environmental and health impacts of pesticide use.

9.3 Influence of Social and Cultural Factors

Social and cultural factors significantly influence both the use and impact of pesticides. Key influences include the following:

1. *Agricultural Practices*:
   - Traditional farming methods and local agricultural practices greatly affect pesticide use. In regions where monoculture—the cultivation of a single crop over large areas—is common, there is often a greater reliance on pesticides. This is because monoculture can lead to higher pest pressures, prompting increased pesticide application to manage these pests.
2. *Economic Factors*:
   - The economic conditions of farmers influence their pesticide choices. Those with limited financial resources may opt for cheaper, potentially more hazardous pesticides or might forgo protective equipment to save costs. This economic pressure can lead to higher risks for both human health and the environment.
3. *Education and Awareness*:
   - The level of education and awareness regarding the risks and proper use of pesticides varies widely. In areas with lower literacy rates or limited access to education, there may be less understanding of the dangers associated with improper pesticide use and fewer resources for safe application practices.
4. *Government Policies and Regulations*:
   - Government policies and regulations play a crucial role in managing pesticide use. Effective regulations and their enforcement can reduce harmful pesticide practices by setting standards and restrictions. Conversely, weak or poorly enforced regulations can lead to misuse and overuse of pesticides, exacerbating health and environmental issues.
5. *Cultural Beliefs and Practices*:
   - Cultural attitudes toward pest control and crop protection can influence pesticide use. In some cultures, there may be a preference for traditional pest control methods over chemical pesticides. In others, there may be a strong belief in the effectiveness of chemical solutions, which can drive higher pesticide use (Becky Mansfield & Marion Werner, 2024).
6. *Social Inequality*:
   - Social and economic inequalities can result in disproportionate pesticide exposure among marginalized communities. These groups often have limited access to safer alternatives and protective measures, leading to higher risks of pesticide-related health issues and environmental damage.

Understanding these social and cultural factors is essential for developing effective strategies to manage pesticide use and minimize its negative impacts on human health and the environment. Addressing these factors can help create more tailored and equitable solutions for pesticide management.

9.4 *Marketing and Branding Strategies*

Marketing and Branding Strategies in the Pesticide Industry

Marketing and branding are crucial for distinguishing pesticide products in a competitive market and for influencing farmer choices. Effective strategies encompass various elements:

1. *Product Differentiation and Innovation*
   - *Unique Selling Propositions (USPs)*: Highlight distinctive features of the pesticide, such as superior efficacy, reduced environmental impact, or safety enhancements. For example, a pesticide with a novel formula that combines pest control with soil enrichment can be marketed as a dual-benefit solution.
   - *Technological Advancements*: Emphasize innovations in pesticide formulations or application technologies. For instance, marketing a product with a slow-release formulation or precision application technology can appeal to farmers seeking to optimize their pest management practices and minimize environmental impact.
2. *Target Market Identification*
   - *Market Segmentation*: Segment the market based on crop type, geographic location, farm size, and other relevant factors. Tailor marketing messages to address the specific needs and challenges of each segment, such as targeting different crops or farming systems with specialized products.
   - *Customer Profiles*: Develop detailed profiles of target customers, including their preferences, pain points, and purchasing behavior. This enables the creation of more personalized and effective marketing campaigns that resonate with the specific needs of each customer group.
3. *Brand Positioning*
   - *Building Reputation and Trust*: Focus on establishing a strong brand reputation by emphasizing product quality, reliability, and safety. Positive customer reviews, endorsements from agricultural experts, and third-party certifications can enhance the brand's credibility and trustworthiness.
   - *Environmental and Health Focus*: Position the product as an environmentally friendly or safer alternative, especially in markets with heightened awareness

of sustainability and health concerns. Highlighting eco-friendly attributes or safety features can appeal to environmentally conscious consumers.

4. *Educational Marketing*
   - *Training and Support*: Offer comprehensive educational resources and training programs for farmers on best practices for pesticide use. Organize workshops, webinars, and on-site demonstrations to provide practical knowledge and build customer loyalty.
   - *Informational Materials*: Create and distribute clear, accessible information about the pesticide's benefits, usage instructions, and safety precautions through brochures, labels, and online content. This helps ensure that users understand how to use the product effectively and safely.
5. *Promotions and Incentives*
   - *Discounts and Special Offers*: Implement promotional strategies such as discounts, rebates, or bundled product offers to attract buyers and encourage bulk purchases. These incentives can drive initial adoption and boost sales.
   - *Loyalty Programs*: Develop loyalty programs or reward systems for repeat customers to encourage continued use of the brand. Offering rewards for consistent purchases can strengthen customer relationships and foster long-term brand loyalty.
6. *Distribution Channels*
   - *Strategic Partnerships*: Collaborate with agricultural cooperatives, retailers, and distributors to enhance product availability and market reach. Establishing strong relationships with key distribution partners can improve product accessibility.
   - *Online Presence*: Build a robust online presence through e-commerce platforms and digital marketing efforts. This includes maintaining an informative and user-friendly website, engaging with customers on social media, and using online advertising to reach a broader audience.
7. *Compliance and Transparency*
   - *Regulatory Adherence*: Ensure that marketing claims and product information comply with regulatory standards. Providing accurate and truthful information about product safety and efficacy is crucial for maintaining credibility and avoiding legal issues.
   - *Transparency*: Be transparent about product ingredients, environmental impact, and any potential risks. Transparency helps build trust with customers and supports a positive brand image.
8. *Cultural Sensitivity*
   - *Local Adaptation*: Adapt marketing strategies to fit local cultures and practices. This may involve translating promotional materials into local languages and tailoring content to align with regional values and preferences.

- *Community Engagement*: Engage with local farming communities to understand their needs and preferences. Tailoring marketing efforts to reflect community values and practices can enhance product acceptance and effectiveness.

9. *Sustainability Messaging*

- *Environmental Responsibility*: Highlight efforts to reduce the environmental impact of pesticide products, such as using eco-friendly packaging or supporting sustainable agriculture initiatives. Emphasizing these efforts can appeal to environmentally conscious consumers.
- *Corporate Social Responsibility (CSR)*: Promote CSR initiatives that contribute to community well-being and environmental conservation. Showcasing these initiatives can enhance the brand's reputation and appeal to socially responsible consumers.

Effective marketing and branding strategies in the pesticide industry involve a combination of innovation, customer-focused approaches, and ethical considerations. By differentiating products, understanding customer needs, and maintaining transparency, companies can build strong brands and foster lasting relationships with their customers.

10 Consumer Purchasing Behaviors: Rupali Munje

10.1 Trends in Organic and Eco-Friendly Product Purchases

Trends in Organic and Eco-Friendly Pesticide Purchases
The shift toward organic and eco-friendly pesticides reflects growing consumer and industry concerns about environmental sustainability and health impacts. Several key trends are shaping this transition:

1. *Increased Demand for Organic Products*

- *Consumer Preferences*: There is a rising consumer preference for organic and eco-friendly products, driven by concerns about health, environmental impact, and food safety. Organic pesticides, which are derived from natural sources and approved for use in organic farming, are increasingly sought after by consumers and farmers alike (Ogidi & Izah, 2024).
- *Market Growth*: The market for organic pesticides has been expanding as more farmers adopt organic practices and consumers demand products produced with minimal chemical input (Ogidi & Richard-Nwachukwu, 2024). This trend is supported by regulatory changes and incentives promoting organic farming practices.

2. *Government Regulations and Policies*

   - *Regulatory Support*: Governments are implementing stricter regulations on conventional pesticides and offering support for the adoption of eco-friendly alternatives. Policies such as subsidies for organic farming and stricter limits on synthetic pesticide use are driving the shift toward organic and eco-friendly products.
   - *Certifications and Standards*: Increasingly, certifications and standards for organic and eco-friendly products are being enforced. This includes third-party certifications like USDA Organic or EU Organic, which provide consumers with assurance of product authenticity and safety.

3. *Technological Advancements*

   - *Innovations in Formulation*: Advances in technology are leading to the development of more effective and affordable organic and eco-friendly pesticides. Innovations include the use of natural substances, biopesticides, and advanced delivery systems that enhance the efficacy and reduce the environmental impact of these products.
   - *Precision Agriculture*: The integration of precision agriculture technologies is improving the application of eco-friendly pesticides. Technologies such as drones and GPS-guided sprayers help in applying these products more accurately and efficiently, reducing waste and environmental impact (Ogidi et al., 2024d).

4. *Increased Awareness and Education*

   - *Consumer Education*: Growing awareness about the potential health and environmental risks associated with conventional pesticides is driving demand for safer alternatives. Educational campaigns and information dissemination are increasing consumer knowledge about the benefits of organic and eco-friendly pesticides.
   - *Farmer Training*: Farmers are receiving more training and resources on organic and sustainable farming practices. This includes workshops, online courses, and technical support on the use of eco-friendly pesticides, which helps in transitioning to and maintaining organic farming systems.

5. *Economic Factors*

   - *Cost Dynamics*: While organic and eco-friendly pesticides often come with higher initial costs, the long-term benefits—such as reduced environmental impact, improved soil health, and potential premium prices for organic produce—are making these products more attractive to farmers.
   - *Investment in Research*: Increased investment in research and development is driving innovation and cost reduction in the organic pesticide market. As the market matures, economies of scale and technological advancements are expected to further lower costs.

6. *Consumer Demand for Transparency*

   - *Product Labels and Claims*: Consumers are demanding greater transparency regarding the origins and contents of pesticide products. Clear labeling and honest claims about the eco-friendliness and organic certification of pesticides are becoming important factors in purchasing decisions.
   - *Corporate Responsibility*: Companies are increasingly focusing on corporate social responsibility and sustainability practices. This includes adopting eco-friendly practices in production, reducing carbon footprints, and promoting transparency in product labeling.

7. *Sustainability Trends*

   - *Environmental Impact*: There is a growing emphasis on reducing the environmental footprint of agriculture. Organic and eco-friendly pesticides are part of broader sustainability initiatives aimed at preserving biodiversity, protecting pollinators, and reducing pollution.
   - *Climate Change Considerations*: As climate change impacts agricultural practices, there is increased interest in sustainable solutions that enhance resilience and adaptability. Organic and eco-friendly pesticides are seen as part of a broader strategy to address environmental challenges.

These trends reflect a significant shift toward more sustainable and environmentally friendly pest management solutions. The increasing adoption of organic and eco-friendly pesticides is driven by a combination of consumer demand, regulatory support, technological advancements, and growing awareness of the need for sustainable agricultural practices.

10.2 Case Studies of Consumer Behavior

Case studies examining consumer behavior in pesticide use provide valuable insights into how farmers and consumers make decisions regarding pest management. These studies highlight various factors influencing pesticide choices and the impact of these choices on health, environment, and economics. Here are detailed descriptions of some notable case studies:

1. *Case Study: Transition to Organic Pesticides in the United States*

   - *Background*: In the United States, the shift toward organic farming has been driven by growing consumer demand for organic produce and concerns about the health impacts of conventional pesticides.
   - *Consumer Behavior*: Farmers transitioning to organic pesticides often do so in response to consumer preferences for organic products. This shift is influenced by the desire to meet market demand, access premium prices, and align with personal values related to environmental sustainability.

- *Findings*: Studies reveal that farmers who adopt organic pesticides experience challenges such as higher costs and limited product availability. However, the benefits include improved soil health, reduced environmental impact, and the ability to market their produce as organic, which can command higher prices.
- *Impact*: The transition has led to increased consumer awareness of pesticide-related issues and has driven market growth for organic and eco-friendly products. This case study underscores the role of consumer preferences in shaping agricultural practices.

2. *Case Study: Pesticide Use in Developing Countries*

- *Background*: In many developing countries, conventional pesticides are widely used due to their effectiveness and lower cost compared to alternatives. However, concerns about health and environmental impacts are growing.
- *Consumer Behavior*: Farmers in developing countries often rely on inexpensive, widely available pesticides due to economic constraints. Limited access to information and education about safer alternatives can lead to higher pesticide use and associated risks.
- *Findings*: Research indicates that farmers' pesticide choices are influenced by factors such as cost, availability, and perceived effectiveness. Lack of awareness and education about the risks of pesticide use and the benefits of eco-friendly alternatives contribute to continued reliance on conventional products.
- *Impact*: Efforts to promote safer pesticide use in developing countries include educational programs, subsidies for eco-friendly products, and regulatory reforms. These initiatives aim to improve health outcomes and reduce environmental damage.

3. *Case Study: Consumer Preferences for Biopesticides in Europe*

- *Background*: In Europe, there is increasing consumer interest in biopesticides as part of a broader trend toward sustainable agriculture. Biopesticides are derived from natural sources and are perceived as safer alternatives to synthetic chemicals.
- *Consumer Behavior*: European consumers are increasingly favoring biopesticides due to concerns about the potential health risks and environmental impacts of conventional pesticides. This preference is influenced by regulatory policies, environmental awareness, and product labeling.
- *Findings*: Studies show that consumers are willing to pay a premium for products treated with biopesticides and are influenced by labels that emphasize environmental and health benefits. Farmers are adopting biopesticides to meet market demand and comply with stricter regulations on pesticide use.
- *Impact*: The growing market for biopesticides in Europe has led to increased investment in research and development, driving innovation and expanding

product availability. This case study highlights the role of consumer preferences in shaping pesticide markets and regulatory policies.

4. *Case Study: Pesticide Use and Health Awareness in India*

   - *Background*: In India, pesticide use is prevalent in agriculture, but there is rising awareness of the health risks associated with excessive or improper use. Efforts are being made to educate farmers and promote safer practices.
   - *Consumer Behavior*: Indian farmers' decisions about pesticide use are influenced by factors such as cost, efficacy, and health concerns. Increasing awareness of the health risks associated with pesticides is leading to a demand for safer alternatives and better protective measures.
   - *Findings*: Research indicates that health awareness campaigns and educational programs have a significant impact on farmers' pesticide choices. Farmers who receive training on safe pesticide use and access to protective equipment are more likely to adopt safer practices.
   - *Impact*: The case study underscores the importance of education and awareness in changing consumer behavior and promoting the use of safer, eco-friendly pesticides. It highlights the need for continued efforts to improve farmer knowledge and access to safer alternatives.

These case studies illustrate the diverse factors influencing consumer behavior in pesticide use, including economic constraints, regulatory environments, health awareness, and consumer preferences. Understanding these factors is crucial for developing effective strategies to promote safer and more sustainable pesticide practices.

10.3 Barriers to Sustainable Consumption

Sustainable consumption of pesticides, which involves using products that minimize environmental and health impacts, faces several significant barriers. These obstacles can hinder the adoption of eco-friendly practices and products in pest management. Here are detailed descriptions of the key barriers:

1. Economic Constraints

   - *Higher Costs*: Sustainable and eco-friendly pesticides often come with higher price tags compared to conventional products. This can be a major barrier for farmers, especially those with limited financial resources, who may opt for cheaper, less sustainable alternatives.
   - *Investment in Transition*: Transitioning to sustainable pest management practices may require significant upfront investment in new technologies, training, and infrastructure. The financial burden of these investments can deter farmers from adopting sustainable practices.

2. *Limited Availability*
    - *Product Accessibility*: Eco-friendly and sustainable pesticides may not be readily available in all regions. In some areas, the supply chain for these products is underdeveloped, leading to limited access and higher costs.
    - *Market Penetration*: The market for sustainable pesticides is still growing, and many regions may have limited options. This can restrict farmers' choices and slow down the adoption of more sustainable solutions.
3. *Lack of Knowledge and Awareness*
    - *Education and Training*: Many farmers lack awareness or knowledge about the benefits and use of sustainable pesticides. Without proper education and training, they may be hesitant to switch from conventional methods.
    - *Information Gaps*: There can be a lack of accessible, reliable information about the effectiveness, safety, and application of sustainable pesticides. This information gap can hinder informed decision-making.
4. *Regulatory and Policy Challenges*
    - *Inconsistent Regulations*: The regulatory environment for pesticides can vary widely between regions and countries. Inconsistent or poorly enforced regulations can lead to confusion and make it difficult for farmers to navigate the requirements for using sustainable products.
    - *Policy Support*: Insufficient policy support or incentives for sustainable practices can discourage adoption. Without clear guidelines, subsidies, or incentives, farmers may find it difficult to justify the switch to eco-friendly products.
5. *Resistance to Change*
    - *Cultural and Traditional Practices*: Farmers often rely on traditional pest management practices that may be deeply ingrained in their routines. Resistance to change can stem from skepticism about the effectiveness of new methods or a preference for familiar practices.
    - *Perceived Effectiveness*: There may be skepticism about the effectiveness of sustainable pesticides compared to conventional ones. Farmers who have experienced success with traditional methods may be reluctant to adopt new approaches.
6. *Technical and Operational Challenges*
    - *Application Difficulties*: Sustainable pesticides may require different application methods or equipment, which can be challenging to implement. Farmers may face difficulties in adapting to these new requirements, especially if they lack technical support.
    - *Integration with Existing Systems*: Integrating sustainable pesticides into existing pest management systems can be complex. Farmers may need to adjust their practices and workflows, which can be a barrier if they lack the resources or expertise to do so.

7. *Market Dynamics*

    - *Demand and Supply Imbalance*: The demand for sustainable pesticides may not always match supply, leading to market imbalances. Low demand can result in fewer options and higher prices for eco-friendly products.
    - *Competitive Pressure*: In competitive agricultural markets, farmers may prioritize cost-effectiveness over sustainability to remain competitive. This can lead to continued reliance on conventional pesticides despite their environmental impacts.

8. *Infrastructural Limitations*

    - *Storage and Distribution*: Proper storage and distribution infrastructure for sustainable pesticides may be lacking in some regions. This can affect the availability and quality of products, making it difficult for farmers to access and use them effectively.
    - *Technical Support*: Adequate technical support and services for sustainable pesticide use may be insufficient. Farmers need access to expert advice and support to effectively implement and manage sustainable pest control methods.

Addressing these barriers requires a multifaceted approach, including financial incentives, improved access to products, enhanced education and training, and supportive policies. By overcoming these obstacles, the adoption of sustainable pesticide practices can be accelerated, leading to more environmentally friendly and health-conscious pest management solutions.

11 Implications for Policy and Practice: Rupali Munje

11.1 Policy Recommendations for Promoting Sustainability

To enhance sustainability in pesticide use and mitigate the environmental and health impacts of conventional methods, targeted policy interventions are essential. Here are detailed policy recommendations to promote sustainable practices:

1. *Incentivize the Adoption of Sustainable Pesticides*

    - *Subsidies and Financial Support*: Implement subsidies or financial incentives for farmers who choose sustainable pesticides and practices. This can help offset the higher costs associated with eco-friendly products and encourage their use.
    - *Grants for Research and Development*: Provide grants for research and development of new sustainable pesticide technologies. Supporting innovation can lead to the creation of more effective and affordable eco-friendly solutions.

2. *Strengthen Regulatory Frameworks*

   - *Tighten Regulations*: Develop and enforce stricter regulations on the use of conventional pesticides, including limits on application rates and restrictions on high-risk chemicals. This encourages the adoption of safer alternatives.
   - *Certification Programs*: Establish certification programs for sustainable pesticides and practices. Certifications can help consumers and farmers identify and trust products that meet sustainability standards.

3. *Promote Education and Training*

   - *Farmer Training Programs*: Create and fund training programs for farmers on sustainable pest management practices, including the safe use of eco-friendly pesticides. These programs should cover application techniques, risk management, and the benefits of sustainable practices.
   - *Public Awareness Campaigns*: Launch public awareness campaigns to educate consumers and agricultural stakeholders about the benefits of sustainable pesticides and the importance of reducing pesticide-related risks.

4. *Support Integrated Pest Management (IPM)*

   - *Encourage IPM Adoption*: Develop policies that promote integrated pest management (IPM) practices, which combine biological, cultural, and chemical methods to manage pests sustainably. Provide resources and support for farmers implementing IPM.
   - *Technical Assistance*: Offer technical assistance and guidance to farmers on implementing IPM strategies effectively. This can include expert consultations and access to IPM resources and tools.

5. *Enhance Research and Data Collection*

   - *Conduct Research*: Invest in research to evaluate the efficacy and safety of sustainable pesticides and pest management practices. Research should focus on long-term impacts on health, environment, and agricultural productivity.
   - *Data Sharing*: Facilitate the sharing of data and best practices related to sustainable pesticide use. Create platforms for collaboration and information exchange among researchers, policymakers, and farmers.

6. *Facilitate Access to Sustainable Products*

   - *Improve Distribution Networks*: Strengthen distribution networks for sustainable pesticides to ensure they are readily available to farmers. This includes supporting infrastructure for storage, transportation, and supply chain management.
   - *Market Development*: Promote the development of markets for sustainable pesticide products through partnerships with agricultural cooperatives, retailers, and distributors.

7. *Implement Risk Assessment and Management*
    - *Risk Assessment Frameworks*: Develop comprehensive risk assessment frameworks to evaluate the potential impacts of pesticide use on health and the environment. Use these assessments to guide policy decisions and regulatory actions.
    - *Risk Communication*: Improve communication about pesticide risks to farmers and the public. Provide clear, accessible information on the potential health and environmental impacts of different pesticide products.
8. *Encourage Sustainable Farming Practices*
    - *Support Sustainable Farming Initiatives*: Promote policies that support sustainable farming practices, such as organic farming, conservation tillage, and crop rotation. These practices can reduce reliance on pesticides and enhance overall sustainability.
    - *Provide Incentives for Certification*: Offer incentives for farmers to obtain certifications in sustainable farming practices, including organic or eco-friendly certifications.
9. *Monitor and Evaluate Policy Effectiveness*
    - *Establish Monitoring Systems*: Create systems to monitor and evaluate the effectiveness of policies related to sustainable pesticide use. Assess outcomes such as reductions in pesticide use, improvements in environmental quality, and impacts on health.
    - *Adjust Policies as Needed*: Use monitoring and evaluation data to make informed adjustments to policies and programs. Ensure that policies remain effective and responsive to emerging challenges and opportunities.
10. *Promote International Collaboration*
    - *Share Best Practices*: Engage in international collaboration to share best practices and lessons learned in sustainable pesticide management. Participate in global forums and partnerships to advance sustainable agriculture on a broader scale.
    - *Adopt Global Standards*: Consider adopting or aligning with international standards and agreements related to sustainable pesticide use. This can help harmonize practices and enhance global efforts toward sustainability.

By implementing these policy recommendations, governments and stakeholders can foster a transition toward more sustainable pesticide practices, reduce environmental and health risks, and promote overall agricultural sustainability.

11.2 Role of Agricultural Stakeholders

Agricultural stakeholders play a crucial role in shaping and implementing effective pesticide management strategies. Their involvement is essential for promoting sustainable practices, ensuring safety, and improving overall agricultural productivity. Here is a detailed description of the roles and contributions of various agricultural stakeholders:

1. *Farmers*
   - *Decision-Making*: Farmers are on the front lines of pesticide use and are responsible for selecting and applying pesticides. Their decisions impact not only their crops but also the surrounding environment and community health. Farmers must balance effectiveness with safety and sustainability.
   - *Adoption of Best Practices*: By adopting integrated pest management (IPM) and sustainable pesticide practices, farmers can reduce reliance on harmful chemicals and improve pest control efficiency. Their commitment to these practices can lead to better outcomes for both their operations and the environment.
   - *Education and Training*: Participating in training programs and staying informed about new technologies and best practices help farmers make more informed decisions about pesticide use and management.
2. *Agricultural Extension Services*
   - *Knowledge Dissemination*: Agricultural extension services provide farmers with critical information on pest management, including the safe and effective use of pesticides. They offer training, workshops, and one-on-one consultations to help farmers implement best practices.
   - *Technical Support*: Extension agents offer technical support and advice on pest identification, pesticide selection, and application techniques. They play a key role in promoting integrated pest management (IPM) and sustainable practices.
3. *Pesticide Manufacturers and Suppliers*
   - *Product Development*: Manufacturers are responsible for developing and producing pesticides. They can influence sustainability by investing in the development of eco-friendly products and reducing the environmental impact of their chemicals.
   - *Labeling and Safety Information*: Suppliers must ensure that pesticide labels provide clear, accurate information on safe use, application rates, and potential hazards. This helps users make informed decisions and comply with regulations.
   - *Training and Support*: Manufacturers can offer training and technical support to farmers and distributors, ensuring that their products are used correctly and effectively.

4. *Regulatory Authorities*

- *Policy and Regulation*: Regulatory authorities establish and enforce regulations related to pesticide use, including safety standards, application limits, and approval processes. They ensure that pesticides are used responsibly and that products meet safety and efficacy criteria.
- *Monitoring and Enforcement*: These bodies monitor pesticide use and compliance with regulations. They conduct inspections, manage reporting systems, and take action against violations to protect human health and the environment.

5. *Researchers and Academics*

- *Research and Innovation*: Researchers conduct studies to evaluate the efficacy, safety, and environmental impact of pesticides. They develop new technologies and methods for pest control, including sustainable and eco-friendly alternatives.
- *Knowledge Dissemination*: Academics publish findings and provide insights on best practices, emerging trends, and risk assessments. Their work helps inform policy decisions and guide the development of new practices and products.

6. *Environmental and Health Organizations*

- *Advocacy and Awareness*: These organizations advocate for safer pesticide practices and raise awareness about the environmental and health impacts of pesticide use. They often work to influence policy changes and promote sustainable alternatives.
- *Research and Education*: They conduct research on the impacts of pesticides and provide educational resources to the public and policymakers. Their work helps inform debates and drive improvements in pesticide management.

7. *Consumers*

- *Demand and Preferences*: Consumer preferences and demand for organic and eco-friendly products can drive changes in agricultural practices. By choosing products with minimal pesticide residues, consumers can influence market trends and encourage farmers to adopt more sustainable practices.
- *Advocacy and Support*: Consumers can support initiatives and policies that promote sustainable pesticide use. They can also engage in advocacy efforts to push for stricter regulations and better practices.

8. *Government and Policy Makers*

- *Policy Development*: Government bodies and policymakers play a critical role in shaping the regulatory framework for pesticide use. They create policies that promote safety, sustainability, and innovation in pest management.

- *Funding and Support*: They can provide funding for research, subsidies for sustainable practices, and support for education and training programs. Their actions can drive progress toward more sustainable agricultural practices.

9. *Agricultural Cooperatives and Associations*

- *Collective Action*: Cooperatives and associations can promote best practices and provide resources for their members. They facilitate collective action on issues such as pesticide use, training, and access to sustainable products.
- *Advocacy*: These organizations often advocate for policies that benefit their members and support sustainable agriculture. They can influence decision-making and support the implementation of new practices.

In summary, the effective management of pesticides relies on the active participation and collaboration of various agricultural stakeholders. Each group contributes to ensuring that pesticides are used safely, effectively, and sustainably, helping to protect human health, the environment, and agricultural productivity.

11.3 Strategies for Retailers and Marketers

Understanding and implementing effective strategies for retailers and marketers in the pesticide industry are crucial for promoting sustainable practices and meeting consumer demands. This study explores key strategies that can be employed to enhance the marketing and retailing of pesticides, focusing on sustainability, consumer education, and market differentiation.

1. *Market Segmentation and Targeting*

- *Identify Consumer Segments*: Segment the market based on factors such as geographic location, farm size, crop types, and consumer preferences. Tailor marketing strategies to address the specific needs and concerns of each segment, whether they are small-scale farmers, large agribusinesses, or organic growers.
- *Target Sustainable Consumers*: Focus on segments that prioritize sustainability and eco-friendly practices. Develop targeted marketing campaigns that highlight the benefits of sustainable and reduced-risk pesticides for these consumers.

2. *Product Differentiation*

- *Highlight Sustainability*: Differentiate products by emphasizing their environmental benefits, such as reduced toxicity, lower residue levels, and minimal impact on non-target species. Use eco-certifications and endorsements from environmental organizations to enhance credibility.
- *Innovative Solutions*: Promote innovative pesticide formulations and technologies, such as biopesticides or integrated pest management (IPM) solu-

tions, which offer improved performance and sustainability compared to conventional products.

3. *Consumer Education and Engagement*
   - *Educational Campaigns*: Implement educational campaigns to inform consumers about the benefits and safe use of pesticides. Use various channels, including online platforms, workshops, and informational brochures, to reach different audiences.
   - *Training Programs*: Offer training and resources to farmers on best practices for pesticide application, safety measures, and sustainability. This can help build trust and loyalty among consumers.
4. *Digital Marketing and E-Commerce*
   - *Online Presence*: Develop a strong online presence through websites, social media, and digital advertising. Provide detailed product information, customer reviews, and educational content to engage potential buyers.
   - *E-Commerce Platforms*: Utilize e-commerce platforms to make sustainable and eco-friendly pesticides more accessible. Offer online ordering, delivery services, and easy access to product information.
5. *Building Partnerships and Alliances*
   - *Collaborate with NGOs and Industry Groups*: Partner with non-governmental organizations, industry associations, and research institutions to promote sustainable pesticide practices and gain endorsements.
   - *Join Certification Programs*: Participate in certification programs that recognize sustainable products and practices. Certifications can enhance product credibility and appeal to environmentally conscious consumers.
6. *Incentives and Promotions*
   - *Offer Discounts and Rebates*: Provide financial incentives, such as discounts or rebates, to encourage the purchase of sustainable pesticides. Highlight cost savings over the long term due to reduced environmental impact and improved efficacy.
   - *Loyalty Programs*: Implement loyalty programs that reward repeat customers and encourage them to adopt sustainable practices. Offer special deals, early access to new products, or other benefits to loyal customers.
7. *Feedback and Continuous Improvement*
   - *Gather Consumer Feedback*: Collect feedback from customers to understand their needs, preferences, and concerns. Use this information to improve products, services, and marketing strategies.
   - *Adapt and Innovate*: Stay informed about market trends, technological advancements, and regulatory changes. Continuously adapt strategies to meet evolving consumer demands and industry standards.

8. *Compliance and Transparency*
   - *Ensure Regulatory Compliance*: Adhere to all relevant regulations and standards related to pesticide marketing and sales. Provide transparent information about product ingredients, safety data, and environmental impact.
   - *Build Trust*: Foster trust with consumers by being transparent about product claims and maintaining high standards of quality and safety. Address any concerns or issues promptly and openly.
9. *Promote Responsible Use*
   - *Advocate Best Practices*: Encourage responsible pesticide use by providing guidance on proper application techniques, safety precautions, and environmental considerations. Promote the use of integrated pest management (IPM) practices to reduce reliance on pesticides.
   - *Risk Communication*: Clearly communicate potential risks associated with pesticide use and how to mitigate them. Provide resources and support to help consumers make informed decisions.
10. *Evaluate and Refine Strategies*
    - *Monitor Performance*: Regularly assess the effectiveness of marketing and retail strategies. Analyze sales data, customer feedback, and market trends to evaluate performance and identify areas for improvement.
    - *Refine Approaches*: Use insights from performance evaluations to refine and optimize strategies. Implement changes based on feedback and market dynamics to better meet consumer needs and achieve business goals.

By implementing these strategies, retailers and marketers in the pesticide industry can effectively promote their products, address consumer concerns, and contribute to more sustainable and responsible pesticide use.

11.4 Community and Consumer Advocacy

Community and consumer advocacy plays a vital role in shaping pesticide policies, promoting safer practices, and protecting public health and the environment. Effective advocacy efforts can drive positive change by raising awareness, influencing policy, and encouraging responsible pesticide use. Here is a detailed look at the roles and strategies of community and consumer advocacy in pesticide use:

1. Raising Awareness
   - *Educational Campaigns*: Advocate for and participate in educational campaigns that inform the public about the risks and benefits of pesticide use. Use various platforms such as social media, community events, and informational pamphlets to reach diverse audiences.
   - *Public Workshops and Seminars*: Organize and host workshops, seminars, and community meetings to educate individuals about safe pesticide prac-

tices, alternatives, and the potential health and environmental impacts of pesticides.

2. *Influencing Policy and Regulation*

    - *Policy Advocacy*: Work to influence local, national, and international policies related to pesticide use. Advocate for stronger regulations, improved safety standards, and policies that promote sustainable and environmentally friendly practices.
    - *Lobbying Efforts*: Engage with policymakers and legislators to present evidence-based arguments for stricter pesticide regulations. Collaborate with other advocacy groups to strengthen lobbying efforts and increase impact.

3. *Supporting Research and Evidence-Based Practices*

    - *Funding and Research Support*: Support research initiatives that investigate the effects of pesticides on health and the environment. Advocate for funding and resources to advance studies on safer alternatives and sustainable practices.
    - *Promoting Evidence-Based Practices*: Use research findings to advocate for practices and policies that are grounded in scientific evidence. Encourage the adoption of practices that are proven to reduce pesticide risks and improve outcomes.

4. *Promoting Alternative Solutions*

    - *Highlighting Alternatives*: Advocate for and promote the use of alternative pest management solutions, such as integrated pest management (IPM), organic farming, and biopesticides. Educate consumers and farmers about the benefits and effectiveness of these alternatives.
    - *Support for Sustainable Products*: Encourage the development and adoption of eco-friendly pesticide products by supporting companies and initiatives that prioritize sustainability and environmental protection.

5. *Building Community Engagement*

    - *Grassroots Organizing*: Mobilize community members to participate in advocacy efforts, including attending public meetings, signing petitions, and joining local advocacy groups. Strengthening community engagement can amplify voices and drive collective action.
    - *Community Partnerships*: Collaborate with local organizations, schools, and businesses to raise awareness and support for sustainable pesticide practices. Build partnerships that enhance outreach and advocacy efforts.

6. *Consumer Education and Empowerment*

    - *Informed Decision-Making*: Educate consumers about the risks associated with pesticide use and how to make informed choices. Provide information on product labels, safe usage practices, and potential health effects.

- *Consumer Choice*: Empower consumers to demand safer and more sustainable products by providing them with the knowledge and tools to make informed purchasing decisions. Encourage consumers to support brands and products that align with their values.

7. *Monitoring and Accountability*

- *Regulation Monitoring*: Monitor the implementation and enforcement of pesticide regulations to ensure compliance and accountability. Report violations and advocate for corrective actions when necessary.
- *Transparency and Reporting*: Advocate for transparency in pesticide labeling and reporting. Ensure that consumers have access to clear and accurate information about pesticide products and their potential impacts.

8. *Addressing Social and Environmental Justice*

- *Advocating for Vulnerable Communities*: Focus on the needs of the marginalized.
- *Promoting Equity*: Address social and environmental justice issues by working toward policies that reduce pesticide exposure in low-income and minority communities. Support initiatives that aim to rectify disparities in pesticide-related health impacts.

9. *Building Awareness Through Media and Public Relations*

- *Media Campaigns*: Utilize traditional and digital media to spread awareness about pesticide issues and advocate for change. This includes writing articles, participating in interviews, and producing multimedia content to reach a broader audience.
- *Public Relations Efforts*: Develop and maintain a positive public image for advocacy efforts through strategic public relations activities. Highlight success stories, progress made, and the impact of advocacy on policy changes and community health.

10. *Collaborating with Industry Stakeholders*

- *Engaging with Industry Leaders*: Collaborate with agricultural and pesticide industry leaders to promote dialogue and find common ground on sustainable practices. Encourage industry leaders to adopt and advocate for safer and more environmentally friendly practices.
- *Partnerships with Environmental Groups*: Work with environmental organizations and other stakeholders to amplify advocacy efforts and support joint initiatives that promote sustainability and reduce pesticide risks.

11. *Developing Policy Recommendations*

- *Formulating Recommendations*: Develop comprehensive policy recommendations based on research, community feedback, and best practices. Propose actionable solutions to address identified issues and improve pesticide management.

- *Advocating for Policy Changes*: Present policy recommendations to decision-makers, including local governments, regulatory bodies, and international organizations. Advocate for the adoption and implementation of these recommendations to drive meaningful change.

12 Conclusion

Community and consumer advocacy in pesticide use are essential for promoting safer, more sustainable practices and influencing policy. By raising awareness, supporting research, advocating for alternatives, and engaging with various stakeholders, advocates can drive positive changes that benefit public health and the environment. Effective advocacy requires a collaborative approach, involving education, policy influence, and community engagement to address the challenges and opportunities associated with pesticide use.

References

Alam, M. J., Begum, I. A., Buysse, J., & Van Huylenbroeck, G. (2012). Energy consumption, carbon emissions and economic growth nexus in Bangladesh: Cointegration and dynamic causality analysis. *Energy policy, 45*, 217–225.

Aprile, M. C., & Punzo, G. (2022). How environmental sustainability labels affect food choices: Assessing consumer preferences in southern Italy. *Journal of cleaner production, 332*, 130046.

Boyd, J. R. (2018). A discourse on winning and losing (Vol. 400). Maxwell Air Force Base, AL: Air University Press.

Chèze, B., David, M., & Martinet, V. (2020). Understanding farmers' reluctance to reduce pesticide use: A choice experiment. *Ecological Economics, 167*, 106349.

Detre, J. D., & Gunderson, M. A. (2011). The triple bottom line: what is the impact on the returns to agribusiness?. *International Food and Agribusiness Management Review, 14*(4), 165–178.

Dhaliwal, G. S., Jindal, V., & Mohindru, B. (2015). Crop losses due to insect pests: global and Indian scenario. *Indian Journal of entomology, 77*(2), 165–168.

Donley, N., Bullard, R. D., Economos, J., Figueroa, I., Lee, J., Liebman, A. K., ... & Shafiei, F. (2022). Pesticides and environmental injustice in the USA: root causes, current regulatory reinforcement and a path forward. *BMC Public Health, 22*(1), 708.

Eldridge, J. J., Izzard, R. G., & Tout, C. A. (2008). The effect of massive binaries on stellar populations and supernova progenitors. *Monthly Notices of the Royal Astronomical Society, 384*(3), 1109–1118.

Gaber, S., & Abdel-Latif, S. H. (2012). Effect of education and health locus of control on safe use of pesticides: a cross sectional random study. *Journal of Occupational Medicine and Toxicology, 7*, 1–7.

Group of Chief Scientific Advisors (2018). https://era.gv.at/open-positions/seconded-national-experts-group-ofchief-scientific-advisors-rtd-03/

Hansen, B. E. (1996). Inference when a nuisance parameter is not identified under the null hypothesis. Econometrica: *Journal of the econometric society*, 413–430.

Heinberg, L. J., & Thompson, J. K. (1992). Social comparison: Gender, target importance ratings, and relation to body image disturbance. *Journal of social behavior and personality, 7*(2), 335.

Hoegh-Guldberg, O., Jacob, D., Taylor, M., Bindi, M., Brown, S., Camilloni, I., Diedhiou, A., Djalante, R., Ebi, K., Engelbrecht, F., Guiot, J., Hijioka, Y., Mehrotra, S., Payne, A., Seneviratne,

S. I., Thomas, A., Warren, R., & Zhou, G. (2018). *Global warming of 1.5°C: An IPCC special report on the impacts of global warming of 1.5°C above pre-industrial levels and related global greenhouse gas emission pathways, in the context of strengthening the global response to the threat of climate change, sustainable development, and efforts to eradicate poverty*. IPCC.

Holling, C. S. (1973). Resilience and stability of ecological systems. *Annual Review of Ecology and Systematics, 4*(1), 1–23. https://doi.org/10.1146/annurev.es.04.110173.000245

Hwang, P. A., Wang, D. W., Walsh, E. J., Krabill, W. B., & Swift, R. N. (2000). Airborne measurements of the wavenumber spectra of ocean surface waves. Part II: Directional distribution. *Journal of Physical Oceanography, 30*(11), 2768–2787.

Hu, Z. (2020). What socio-economic and political factors lead to global pesticide dependence? a critical review from a social science perspective. *International Journal of Environmental Research and Public Health, 17*(21), 8119.

Iyiola, A. O., Kolawole, A. S., Ajayi, F. O., Ogidi, O. I., & Ogwu, M. C. (2023). Sustainable water use and management for agricultural transformation in Africa. In *Water resources management for rural development: Challenges and mitigation* (Vol. 1, pp. 287–300). Published by Elsevier. https://doi.org/10.1016/B978-0-443-18778-0.00001-5

Khan, H., Sharma, C., Kathuria, V., & Bhardwaj, P. (2010). Environmental risks associated with pesticide residues in India. *Journal of Environmental Science and Health, Part B, 45*(3), 192–199. https://doi.org/10.1080/03601230903471969

Lenton, T. M., Rockström, J., Gaffney, O., Rahmstorf, S., Richardson, K., Steffen, W., & Schellnhuber, H. J. (2019). Climate tipping points – Too risky to bet against. *Nature, 575*(7784), 592–595. https://doi.org/10.1038/d41586-019-03595-0

Li, Y. (2021). Influence of temperature on pesticide transformation and degradation products: Implications for risk assessment. *Journal of Environmental Management, 297*, 113204. https://doi.org/10.1016/j.jenvman.2021.113204

Liu, Y., Chen, W., Li, Y., & Zhao, X. (2015). Environmental fate of pesticides in the aquatic system. *Journal of Environmental Management, 160*, 332–340. https://doi.org/10.1016/j.jenvman.2015.05.020

Lynch, M. (1991). Methods for the analysis of comparative data in evolutionary biology. *Evolution, 45*(5), 1065–1080.

Mansfield, B., Werner, M., Berndt, C., Shattuck, A., Galt, R., Williams, B., ... & Tittor, A. (2024). A new critical social science research agenda on pesticides. *Agriculture and Human Values, 41*(2), 395–412.

Marie, C., Céline, A., Benoit, M., & André, T. (2017). Metabolites of pesticides in surface waters and their risks to humans and aquatic systems: A case study of the Rhône River basin. *Environmental Science and Pollution Research, 24*(14), 12994–13008. https://doi.org/10.1007/s11356-017-8862-3

Martín, E., Pavlidis, K., & Karasali, H. (2020). Effects of climate change on pesticide risk and the need for integrated management in Mediterranean wetlands. *Ecological Indicators, 115*, 106382. https://doi.org/10.1016/j.ecolind.2020.106382

McCarthy, J. J., & Intergovernmental Panel on Climate Change (Eds.). (2001). *Climate change 2001: Impacts, adaptation, and vulnerability: Contribution of Working Group II to the third assessment report of the Intergovernmental Panel on Climate Change*. Cambridge University Press.

McLaughlin, P., & Dietz, T. (2008). Structure, agency and environment: Toward an integrated perspective on vulnerability. *Global Environmental Change, 18*(1), 99–111. https://doi.org/10.1016/j.gloenvcha.2007.05.003

Mieldazys, A., Tamkutė, L., Maikštėnienė, S., & Povilaitis, V. (2015). Residue levels of selected pesticides in fruits and vegetables grown in Lithuania. *Food Additives & Contaminants: Part B, 8*(4), 268–275. https://doi.org/10.1080/19393210.2015.1059730

Milner, A. M., & Boyd, I. L. (2017). *Toward pesticidovigilance. Science, 357*(6357), 1232–1234.

Mubushar, M., Aldosari, F. O., Baig, M. B., Alotaibi, B. M., & Khan, A. Q. (2019). Assessment of farmers on their knowledge regarding pesticide usage and biosafety. *Saudi journal of biological sciences, 26*(7), 1903–1910.

National Academy of Sciences. (2019). *Climate change and ecosystems*. The National Academies Press. Retrieved from https://www.ipcc.ch/report/srccl/

Nicholls, R. J., Hinkel, J., Lincke, D., & van der Pol, T. (2018). Stabilization of global temperature at 1.5°C and 2.0°C: Implications for coastal areas. *Philosophical Transactions of the Royal Society A: Mathematical, Physical and Engineering Sciences, 376*(2119), 20160448. https://doi.org/10.1098/rsta.2016.0448

Noyes, P. D., McElwee, M. K., Miller, H. D., Clark, B. W., Van Tiem, L. A., Walcott, K. C., Erwin, K. N., & Levin, E. D. (2009). The toxicology of climate change: Environmental contaminants in a warming world. *Environment International, 35*(6), 971–986. https://doi.org/10.1016/j.envint.2009.02.006

Ogden, A. E., & Innes, J. L. (2009). Application of structured decision making to an assessment of climate change vulnerabilities and adaptation options for sustainable forest management. *Ecology and Society, 14*(1), 11. https://doi.org/10.5751/ES-02715-140111

Ogidi, O. I. (2023). Sustainable utilization of important medicinal plants in Africa. In *Sustainable utilization and conservation of Africa's biological resources and environment, sustainable development and biodiversity*. Springer. https://doi.org/10.1007/978-981-19-6974-4_12

Ogidi, O. I., & Akpan, U. M. (2023). Impacts of chemical use in agricultural practices: Perspectives of soil microorganism and vegetation. In *One health implications of agrochemicals and their sustainable alternative*. Springer.

Ogidi, O. I., & Akpan, U. M. (2024). Microbial and phyto-remediation of crude oil contaminated soil. In *Eco-restoration of the environment*. CRC Press, Routledge Taylor and Francis Group.

Ogidi, O. I., & Emaikwu, N. G. (2024). Plant phenolic compound isolation and its bioinformatics approaches of molecular mechanisms in antimicrobial activities and resistance. In *Computational approaches in biomaterials, bioinformatics and biomedical engineering applications*. CRC Press, Taylor and Francis Group.

Ogidi, O. I., & Izah, S. C. (2024). Emerging technologies in the global south food industry: Prospects and challenges. In *Food safety and quality in global south*. Springer.

Ogidi, O. I., & Richard-Nwachukwu, N. (2024). Microbe assisted remediation of pesticide residues from soil and water. In *Eco-restoration of the environment*. CRC Press, Routledge Taylor and Francis Group.

Ogidi, O. I., & Tawariowei, A. M. (2023). Endocrine disrupting chemicals and their role in cancer – A review. *Molecular Sciences and Applications., 3*, 5–14. https://doi.org/10.37394/232023.2023.3.2

Ogidi, O. I., Izah, S. C., & Akpan, U. M. (2023). Food quality and agrochemical use: Integrated monitoring, assessment, and management policies. In *One health implications of agrochemicals and their sustainable alternative*. Springer.

Ogidi, O. I., Ajoko, I. T., & Tawariowei, A. M. (2024a). Sustainable application of genetic ecotoxicological techniques. In *Biomonitoring for environmental sustainability. In: Biomonitoring of pollutants in the global south*. Springer.

Ogidi, O. I., Onwuagba, C. G., & Richard-Nwachukwu, N. (2024b). Biomonitoring tools, techniques and approaches for environmental assessments. In *Biomonitoring of pollutants in the global south*. Springer.

Ogidi, O. I., Angaye, T. C. N., & Carbom, H. E. (2024c). Microorganisms in the monitoring of terrestrial and aquatic ecosystems: Applications, challenges and future prospects. In *Biomonitoring of pollutants in the global south*. Springer.

Ogidi, O. I., Izah, S. C., & Tobia, P. S. (2024d). Biosurfactant production from cassava processing wastes. In *Sustainable Cassava: Strategies from production through waste management*. Elsevier.

Oliver, T. H., Heard, M. S., Isaac, N. J., Roy, D. B., Procter, D., Eigenbrod, F., et al. (2015). Biodiversity and resilience of ecosystem functions. *Trends in Ecology & Evolution, 30*(11), 673–684. https://doi.org/10.1016/j.tree.2015.08.009

Parker, J. G., Rubin, K. H., Erath, S. A., Wojslawowicz, J. C., & Buskirk, A. A. (2015). Peer relationships, child development, and adjustment: A developmental psychopathology perspective. Developmental psychopathology: Volume one: Theory and method, 419–493.

Parmesan, C., & Yohe, G. (2003). A globally coherent fingerprint of climate change impacts across natural systems. *Nature, 421*(6918), 37–42. https://doi.org/10.1038/nature01286

Pavlidis, K., & Karasali, H. (2020). Climate variability and the importance of adaptive pest management practices for sustainable agricultural landscapes. *Agricultural Systems, 183*, 102870. https://doi.org/10.1016/j.agsy.2020.102870

Pimentel, D., Acquay, H., Biltonen, M., Rice, P., Silva, M., Nelson, J., et al. (1992). Environmental and economic costs of pesticide use. *BioScience, 42*(10), 750–760. https://doi.org/10.2307/131199

Polazzo, F., Vilas-Boas, P., & Roth, N. (2022). Resilience of aquatic ecosystems under extreme weather conditions: Impacts on species sensitivity to pesticides. *Science of the Total Environment, 827*, 154276. https://doi.org/10.1016/j.scitotenv.2022.154276

Popp, J., Pető, K., & Nagy, J. (2013). Pesticide productivity and food security. A review. *Agronomy for sustainable development, 33*, 243–255.

Pozo, K., Harner, T., Lee, S. C., Wania, F., Muir, D. C., & Jones, K. C. (2011). Assessing seasonal and spatial trends of persistent organic pollutants (POPs) in the global atmosphere using passive samplers. *Environmental Science & Technology, 45*(6), 1452–1459. https://doi.org/10.1021/es303744q

Pretty, J. N. (1995). Participatory learning for sustainable agriculture. *World development, 23*(8), 1247–1263.

Pretty, J., & Smith, D. (2004). Social capital in biodiversity conservation and management. *Conservation biology, 18*(3), 631–638.

Pretty, J. (2008). Agricultural sustainability: concepts, principles and evidence. Philosophical Transactions of the Royal Society B: *Biological Sciences, 363*(1491), 447–465.

Pretty, J., & Bharucha, Z. P. (2014). Sustainable intensification in agricultural systems. *Annals of botany, 114*(8), 1571–1596.

Qin, G., Gong, J., Fan, M., & Yu, H. (2014). Sorption and degradation of pesticides in soils and the environmental fate of pesticide residues: A review. *Journal of Agricultural and Food Chemistry, 62*(27), 6634–6648. https://doi.org/10.1021/jf5009157

Rico-Mesa, J. S., Rosas, D., Ahmadian-Tehrani, A., White, A., Anderson, A. S., & Chilton, R. (2020). The role of anticoagulation in COVID-19-induced hypercoagulability. *Current cardiology reports, 22*, 1–6.

Rodrigo, A., Vilas-Boas, P., & Silva, E. (2022). Role of constructed wetlands in reducing pesticide and metal loads from agricultural drainage: A case study in Mediterranean rice fields. *Journal of Environmental Quality, 51*(3), 721–732. https://doi.org/10.1002/jeq2.20318

Roth, N., Polazzo, F., & Vilas-Boas, P. (2022). Aquatic species sensitivity to temperature-induced pesticide exposure under climate change scenarios. *Environmental Pollution, 292*, 118333. https://doi.org/10.1016/j.envpol.2021.118333

Sarma, H. H., Borah, S. K., Dutta, N., Sultana, N., Nath, H., & Das, B. C. (2024). Innovative approaches for climate-resilient farming: Strategies against environmental shifts and climate change. *International Journal of Environment and Climate Change, 14*(9), 217–241.

Sawyer, W. E., Jacob, D. E., Ogidi, O. I., Izah, S. C., & Popoola, A. M. (2023). Threats and conservation strategies of common edible vegetables that possess pharmacological potentials in Nigeria. In *Reference series in phytochemistry. Herbal medicine phytochemistry*. Springer.

Schäfer, T., & Schwarz, M. A. (2019). The meaningfulness of effect sizes in psychological research: Differences between sub-disciplines and the impact of potential biases. *Frontiers in psychology, 10*, 813.

Scholz, N. L., Fleishman, E., Brown, L., Werner, I., Johnson, M. L., Brooks, M. L., & Schlenk, D. (2012). A perspective on modern pesticides, pelagic fish declines, and unknown ecological resilience in highly managed ecosystems. *BioScience, 62*(4), 428–434.

Schreinemachers, P., & Tipraqsa, P. (2012). Agricultural pesticides and land use intensification in high, middle, and low income countries. *Food Policy, 37*(6), 616–626. https://doi.org/10.1016/j.foodpol.2012.07.002

Schreinemachers, P., Schad, I., Tipraqsa, P., Williams, P. M., Neef, A., Riwthong, S., Sangchan, W., & Grovermann, C. (2015). Can public GAP standards reduce agricultural pesticide use?

The case of fruit and vegetable farming in northern Thailand. *Agriculture and Human Values, 32*(4), 603–619. https://doi.org/10.1007/s10460-014-9571-2
Sharma, A., Kumar, V., Shahzad, B., Tanveer, M., Sidhu, G. P. S., Handa, N., Kohli, S. K., Yadav, P., Bali, A. S., Parihar, R. D., & Thukral, A. K. (2019). Worldwide pesticide usage and its impacts on ecosystem. *SN Applied Sciences, 1*(11), 1446. https://doi.org/10.1007/s42452-019-1485-1
Shaw, J. D., Duffy, M. K., Johnson, J. L., & Lockhart, D. E. (2005). Turnover, social capital losses, and performance. *Academy of management Journal, 48*(4), 594–606.
Silva, E., Rodrigo, A., & Pavlidis, K. (2022). The impact of pesticide use and extreme weather events on Mediterranean aquatic ecosystems: Integrating climate predictions in risk assessments. *Science of the Total Environment, 828*, 154254. https://doi.org/10.1016/j.scitotenv.2022.154254
Spears, B. M., Adriaenssens, V., Burthe, S. J., Davidson, T. A., Gunn, I. D., Hawkes, J. A., et al. (2015). FORUM: Effective management of ecological resilience – Are we there yet? *Journal of Applied Ecology, 52*(5), 1311–1315. https://doi.org/10.1111/1365-2664.12497
Standish, R. J., Hobbs, R. J., Mayfield, M. M., Bestelmeyer, B. T., Suding, K. N., Battaglia, L. L., et al. (2014). Resilience in ecology: Abstraction, distraction, or where the action is? *Biological Conservation, 177*, 43–51. https://doi.org/10.1016/j.biocon.2014.06.008
Tano, S. (2014). Regional clustering of human capital: School grades and migration of university graduates. *The Annals of Regional Science, 52*, 561–581.
Thompson, T. A., Quataert, E., & Waxman, E. (2007). The starburst contribution to the extragalactic γ-ray background. *The Astrophysical Journal, 654*(1), 219.
Tran, R., Lan, J., Shuaibi, M., Wood, B. M., Goyal, S., Das, A., ... & Zitnick, C. L. (2023). The Open Catalyst 2022 (OC22) dataset and challenges for oxide electrocatalysts. *ACS Catalysis, 13*(5), 3066–3084.
Wei, G., Wang, C., Niu, W., Huan, Q., Tian, T., Zou, S., & Huang, D. (2021). Occurrence and risk assessment of currently used organophosphate pesticides in overlying water and surface sediments in Guangzhou urban waterways, China. *Environmental Science and Pollution Research, 28*(35), 48194–48206.
World Health Organization. (2014). Report: 8th FAO/WHO joint meeting on pesticide management and 10th session of the FAO panel of experts on pesticide management, 14-17 October 2014, Rome (No. WHO/CDS/NTD/WHOPES/2014.6). World Health Organization.
Wu, J. L., Hou, H., & Cheng, S. Y. (2010). The dynamic impacts of financial institutions on economic growth: Evidence from the European Union. *Journal of Macroeconomics, 32*(3), 879–891.
Yassin, H., Velthuis, A. G., Boerjan, M., & van Riel, J. (2009). Field study on broilers' first-week mortality. *Poultry science, 88*(4), 798–804.
Zheng, L. L., Zhang, J., Liu, X. Z., Tian, L., Xiong, Z. S., Xiong, X., ... & Zou, J. P. (2022). Degradation of pesticide wastewater with simultaneous resource recovery via ozonation coupled with anaerobic biochemical technology. Chemosphere, 300, 134520.

Economic Implications of Climate Change and Pesticide Use on Agriculture

Mukul Machhindra Barwant, Shreya Singh, Odangowei Inetiminebi Ogidi, Shivendra Singh, and B. Sri Sai Siddartha Naik

1 Introduction

Agriculture employs a sizable worldwide workforce and provides a large source of raw materials and nourishment. However, there are growing risks to the agriculture business due to climate change and the widespread use of pesticides. There are major consequences for agriculture as a result of climate change, which affects soil fertility, agricultural productivity, and the frequency of pests and diseases. Despite being necessary for contemporary agriculture, the use of pesticides raises a number of concerns about economic viability, environmental effect, and human health. In this chapter, the financial impacts of pesticide use and climate change on the farming industry are thoroughly examined, with a focus on the necessity of reducing the use of toxic pesticides and mitigating the adverse consequences of climate change by implementing sustainable farming practices.Understanding the interconnected nature of different challenges is essential for formulating these. A major danger to

M. M. Barwant (✉)
Department of Botany, Sanjivani Rural Education Society's, Sanjivani Arts Commerce and Science College, Kopargaon, Ahmednagar, Maharashtra, India

S. Singh
Department of Agriculture, Ramlalit Singh Mahavidyalaya, Mirzapur, Uttar Pradesh, India

O. I. Ogidi
Department of Biochemistry, Faculty of Basic Medical Sciences, Bayelsa Medical University, Yenagoa, Nigeria

S. Singh
Department of Agriculture, Udai Pratap College (An Autonomous Institution) Bhojubir, Varanasi, Uttar Pradesh, India

B. Sri Sai Siddartha Naik
Department of Agronomy, Acharya N.G Ranga Agricultural University, Guntur, Andhra Pradesh, India

B. R. Babaniyi, E. E. Babaniyi (eds.), *The Interplay of Pesticides and Climate Change*, https://doi.org/10.1007/978-3-031-81669-7_19

agriculture around the world is climate change. Changes in temperature, changes in precipitation patterns and the frequency of extreme weather events have already started to impact crop yields (Porter et al., 2014).

The efficacy of crop protection techniques significantly impacts harvest yields. The primary objective of pesticide use is to enhance food security, whereas an improved quality of life is a secondary consideration. Climate change is anticipated to impact pesticide application and agricultural yield. A detailed study is undertaken to assess the impact of altering meteorological conditions on pesticide application. Technical development, regulatory frameworks and economic conditions are evaluated, alongside their impacts on cultivated crops, indigenous pests and pesticide efficacy. This chapter aims to investigate the ways in which consumers may be exposed to pesticide residues on crops owing to climate change. Increasing temperatures and modified precipitation patterns are the primary determinants of pest and infectious disease prevalence due to climate change. Increased quantities, doses, frequency and diversity of pesticide applications are expected (Ogidi & Akpan, 2023; Ogidi et al., 2023). Climate change will lead to a reduction in pesticide concentrations due to quick degradation and increased volatilisation, both of which are greatly influenced by high moisture content, rising temperatures and direct sunlight. Increased precipitation generally facilitates pesticide distribution. Altering pesticide application methods may contribute to addressing this issue. The exposure of consumers at the lowest tier of the food chain will be influenced by modified pesticide application (Delcour et al., 2015).

Pesticides have become integral to modern agriculture to safeguard crops against pests and diseases that diminish harvest yields. Pesticides assist farmers in reducing crop losses and saving costs, although also produce unintended environmental consequences. Pimentel and Burgess (2014) assert that excessive pesticide application can lead to soil degradation, water pollution and the emergence of pest resistance to these chemicals (Ogidi & Akpan, 2024; Ogidi & Richard-Nwachukwu, 2024). Moreover, pesticide application presents health risks to consumers and agricultural labourers, resulting in increased healthcare costs and diminished production, hence exerting economic pressure on the economy. Agriculture has numerous formidable issues arising from the interaction between pesticide application and climate change. Concerns over the long-term sustainability of pesticides stem from their environmental and health ramifications, notwithstanding their claimed utility in addressing heightened insect pressures induced by climate change. Evaluating the adaptability of agricultural systems to environmental changes while reducing dependence on harmful chemicals is crucial, as is examining the economic ramifications of these factors. The current global population of 7 billion is anticipated to rise to 9.2 billion by 2050, representing an annual growth of 70 million individuals, or 30%. A 70% rise in food production demand is anticipated due to heightened population density.

The increase in dietary patterns in emerging nations towards higher-quality foods, including more intake of meat and dairy, as well as an increased use of grains for animal feed, mostly accounts for this surge (Ogidi & Izah, 2024). Supplementary farmland is available only in restricted amounts. Wild crop progenitors, organic pest adversaries and fauna depend on forests and other natural ecosystems for their survival, which will be the initial casualties of any development. Moreover, agricultural land will be repurposed for the cultivation of bio-based commodities, including fibre and biofuel, instead of food. Moreover, an increasing amount of agricultural land will be allocated to the production of bio-based products such as biofuels or fibres, rather than food cultivation (Ogidi et al., 2024). Consequently, we must cultivate food while diminishing our present consumption of water, electricity, fertilisers and pesticides, and utilising even less area. These limitations underscore the imperative for enhanced sustainable production (Popp et al., 2013).

Alterations in the statistical characteristics of the climate system over extended durations are referred to as climate change, irrespective of their source. According to scientists, alterations in precipitation, global average temperature and atmospheric carbon dioxide concentrations are all elements of climate change. Alterations in salinity and sea levels, variations in agricultural yields and soil quality, fluctuations in nitrogen deposition and shifts in plant diversity are all potential consequences of these processes (Ogidi, 2023). Temperature, light and water are the three principal determinants of an organism's development and growth. Pesticides have been extensively utilised for an extended period to enhance agricultural yields and quality by diminishing or eradicating pest populations. Agricultural crops are safeguarded against pests and illnesses and possess an edge over weeds due to their action mechanism (Ogidi & Emaikwu, 2024). Notwithstanding the potential hazards to advantageous species and human health, their utilisation is essential for maintaining present agricultural output and high quality of life standards. Climate is one of the many factors that can affect the effectiveness and application of pesticides. Given the widespread acknowledgement of climate change's primary implications, it is plausible to anticipate an influence on pesticide utilisation (Delcour et al., 2015).

Recent studies have investigated the potential financial impacts of climate change on agriculture. Climate change may positively or negatively influence agricultural productivity, contingent upon location and crop type, as indicated by studies (Lobell & Burke, 2010). Heat stress diminishes crop yields in warmer areas, whereas extended growing seasons enhance output in countries with milder winters (Schlenker & Roberts, 2009). Moore and Lobell (2015) discovered that without intervention to mitigate climate change, global agricultural output could decline by 20% by 2050. In developing nations specifically, this decline in productivity may elevate food prices, reduce farmer incomes and exacerbate food insecurity. To adapt to climate change, farmers must invest in irrigation infrastructure, drought-resistant crop varieties and advanced weather monitoring technologies. This will increase the economic costs of agriculture. Pesticides are recognised for their beneficial economic impact due to their crucial function in mitigating agricultural damage caused by pests and diseases, hence enhancing crop yields and revenues. However, frequent usage results in significant repercussions, including pesticide resistance, environmental degradation and health issues (Pimentel, 2005).

Approximately 3 million individuals are afflicted by pesticide poisoning annually, resulting in 220,000 fatalities. The bulk of these victims reside in impoverished nations, as reported by the World Health Organisation (WHO, 2020). The necessity for farmers to purchase costlier and more potent herbicides to combat resistant pests can impose a financial burden on their businesses. The pesticide treadmill results from this phenomenon that diminishes profitability and escalates production expenses (Wilson & Tisdell, 2001). The diminished ecosystem services and the costs associated with remediating polluted water sources are further economic consequences of pesticide runoff, soil degradation and biodiversity decline affecting the planet's climate (Tilman et al., 2001).

Climate change is anticipated to exacerbate insect pressures in agriculture, potentially leading to an increase in pesticide usage. Changes in precipitation patterns and elevated temperatures might affect pest life cycles, resulting in greater agricultural losses and more frequent infestations (Deutsch et al., 2018). Consequently, farmers may increase pesticide usage to protect their crops, potentially harming both the environment and the economy in the long term. Iyiola et al. (2023) projected future temperatures and looked at how they would affect things like crops, pasture and grazing, cattle, pesticide use, irrigation water demand and supply, and how assumptions about international trade would affect everything. Ultimately, we found that these effects advantaged US consumers to the detriment of producers by reducing agricultural prices. The influence of alterations in average temperature on yield variability is inconsistent.

A case study examining the interplay among weather, agriculture and climate was also included. Reilly et al. (2003) examined the potential impacts of climate change on two water sources: Chesapeake Bay and the Edward's Aquifer, which supplies water to the San Antonio, Texas region. Research indicates that the application of chemical pesticides can be markedly diminished without compromising agricultural productivity when climate adaptation strategies are coupled with pest control (IPM) techniques. Integrated pest management (IPM) aims for sustainable pest management by prioritising biological control agents, crop rotation and resistant crop varieties (Pretty, 2005). These strategies may reduce pesticide usage and enhance climate resilience, yielding beneficial outcomes for both the economy and the environment.

2 Climate Change and Agriculture

Approximately 30% of global anthropogenic emissions originate from agriculture. Therefore, the most effective approach to reducing emissions is to implement eco-efficient and eco-advanced technologies. Through eco-increment, agroecosystems have the capacity to function as a climate-friendly CO_2 sink. The objective of integrating soil and farming into any global framework for problem-solving and addressing climate-resilient agriculture is to enhance food security (Le Quéré et al., 2018). Two primary human activities that contribute to CO_2 emissions are the combustion of petroleum products, particularly concrete manufacture, and the deforestation of tropical forests. Although emissions from petroleum products are increasing, those from deforestation exhibit a consistent trend. From 2008 to 2017, over 14% of total CO_2 emissions resulted from land use alterations, whereas 16% originated from fossil fuel burning and concrete manufacturing (Archard et al., 2014). In the 1990s, moist jungles lost 8 million hectares of forest cover year, and in the 2000s, that number dropped to 7.6 million hectares, based on the gross loss of woodlands. With a range of 646–1238 Tg(C)/a, the connected carbon catastrophes in the 1990s reached 887 Tg(C)/a. They ranged from 602 to 1237 Tg(C)/a in the 2000s, with 880 Tg(C)/a being the average, with over 66% of that total occurring in humid locations. Annual evacuation assessments fail to consider soil amendments and exhibit poor comparability with emission evaluations (IPCC, 2015).

Agriculture is a significant wellspring of GHGs. However, as an industry (in created nations) and a lifestyle (in emerging nations), farming should be necessary to any worldwide plan of climate change variation and relief (Crutzen, 2002). With regard to climate change, agribusiness is a natural instead of a monetary movement. Taking into account immediate and aberrant discharges, agribusiness represents ~30% of the combined radioactive constraining (Ruddiman, 2003). The purpose of this section is to analyse how agriculture adds to global warming gases and how climate-smart farming might reduce human-caused emissions while improving soil and agroecosystem resilience to climate change (Ruddiman, 2005). About 10,000 years ago, with the advent of settled agriculture, the Anthropogenic Era began, defined by human influence on Earth comparable to geological forces. While land degradation and poor soil management were the primary causes of the Dust Bowl in the 1930s, they also accelerated the downfall of previous illustrious civilisations (Diamond, 2005).

In 2017, 9.9 Pg of carbon was emitted via the combustion of petroleum products, whereas 1.4 Pg of carbon, being approximately 14% of the total, resulted from alterations in land use. In 2011, overall emissions amounted to 1.45 Pg(Ce)/year (where Ce denotes carbon atoms), with 44% allocated to intestinal maturation, 15% to designed manures, 10% to rice growth, 7% to faeces boards and 3% to soil fertilisers (Steinbeck, 1939). In 2011, 65%, equivalent to 0.94 Pg(Ce), of the area was allocated to domesticated animals. Conversely, controlled-release fertiliser and enhanced rice contributed 0.19 and 0.15 Pg(Ce), respectively. Asia represented 44% (0.64 PgCe) of local premise emissions, the Americas 25% (0.36 PgCe) and Africa 15% (0.22 PgCe). The sequence of land use discharges was biomass, agricultural, prairie and timberland (Tubiello et al., 2014).

Agriculture has significantly contributed to greenhouse gas emissions for around 10,000 years; yet, it will also suffer adverse effects from the gradual climate change (Cline, 2008). Environmental change will adversely affect the efficiency and productivity of resource utilisation (including water and manure), and it will also increase the prevalence of irritants and bacteria. Analysis of harvest production data at the provincial level in the United States indicated that maize yields rise at temperatures up to 29 °C, soybeans at 30 °C and cotton at 32 °C (Ray et al., 2019). Yield significantly declines when temperatures exceed certain thresholds. Horticulture, as a contributor to greenhouse gas emissions and a sector inadequately prepared to address the adverse effects of climate change, must be central to any plan aimed at

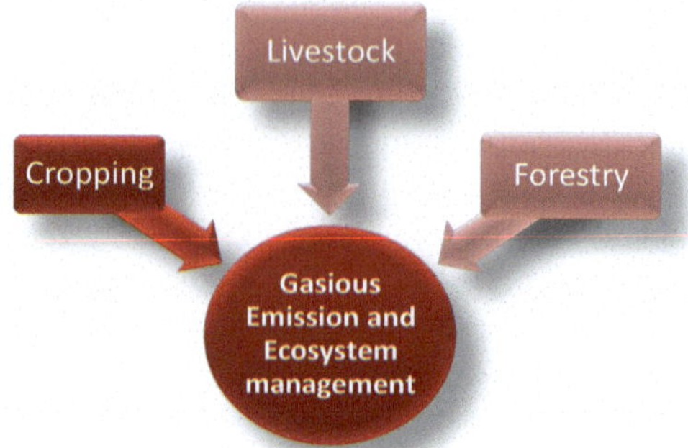

Fig. 1 Impacts of agriculture and forestry on gaseous emissions into the atmosphere

mitigating these threats to human welfare. Figure 1 represented Diagram illustrating the relationship between cropping, livestock, and forestry with gaseous emission and ecosystem management. Three arrows point from cropping, livestock, and forestry to a central circle labeled "Gaseous Emission and Ecosystem Management," indicating their contributions to environmental management (Schlenker & Roberts, 2008).

2.1 Sources of Emissions from Agriculture

A significant producer of greenhouse emissions is the extensive variety of agricultural and ranger service activities. Furrowing, wetland infiltration, intestinal maturation, composting, fertiliser management and fuel utilisation are significant horticultural practices that contribute to greenhouse gas emissions. The following land uses, in addition to agribusiness and ranger services, contributed to the median value between 2001 and 2010: 2.09 Pg(Ce)/a from net forest cover conversion to alternative landscapes; 3.27 Pg(Ce)/a from deteriorating peatlands; 4.05 Pg(Ce)/a from biomass combustion; 1.36 Pg(Ce)/a from agricultural production and cattle domestication (Lal, 2003a). In addition to livestock, rice farming is a major contributor to CH_4 emissions, which are projected to increase as a result of future climate change. Strengths in achieving environmental objectives may encompass the ongoing reduction in CO_2 emissions. Flames and biomass combustion also produce local and territorial consequences (Lal, 2003b).

Domesticated animals stay a significant wellspring of GHGs, particularly of CH_4. Overall, a cow discharges somewhere in the range of 70 and 120 kg of CH_4 each year (Crutzen & Andreae, 1999). Complete domesticated animals-based outflows might be 2.04 Pg(Ce)/a. A few evaluations survey all out emanations from domesticated animals at 8.9 Pg(Ce)/a, which is by all accounts a misjudge. Discharge of N_2O from farming is one more significant wellspring of a profoundly intense GHG (Jayasundara et al., 2014). Some estimates put the percentage of anthropogenic CO_2 emissions at 25%, anthropogenic CH_4 emissions at 65% and anthropogenic N_2O emissions at 90%, all from agricultural operations combined. A groundbreaking study on land use change has been published by the Intergovernmental Panel on Climate Change (IPCC). Despite the importance of soil carbon sequestration in mitigating anthropogenic emissions, the impact of terrestrial biological systems, such as agricultural and degraded lands, has not been adequately addressed by IDCC and other organisations. They have also neglected to assess the whole range of practices' effects on environmental transformation, as well as the consequences of existing and future land use changes. It is necessary to fix this oversight, according to Pashian et al. (2006) (Table 1).

Table 1 Factors influencing greenhouse gas emissions in agriculture and land use

Aspect	Details on impact	Emissions estimates	Factors affecting emissions	Additional notes
Land use conversion	Significant transformation of Earth's surface due to human development affecting C pool and GHG emissions	Cumulative emissions: 250 Pg(C) (last 300 years), 456 Pg(C) (8000 BCE to 1850 AD), 0.9–1.5 Pg(C)/a (current)	Variations due to land use (forests, agriculture), model use, emission factors, and types of land (tropical vs. exa-tropical)	Cumulative emissions from land use change since 1750 at 235 ± 95 Pg(C), a major source of CO_2 emissions due to deforestation, agriculture, and other anthropogenic activities
Fuel consumption in agriculture	CO_2 emissions from fuel use in farm operations such as tillage, planting, fertilisation and harvesting	For specific crops: Winter wheat (230 kg(C)/ha), sugar beet (165 kg(C)/ha), potato (201 kg(C)/ha)	Type of tillage (conventional vs. reduced/no-till), depth of ploughing, speed of machinery	Higher emissions with mechanised systems and increased ploughing depth/speed. Efficient fuel use in tillage reduces GHG emissions
N_2O emissions from fertilisers	Increased N_2O emissions from synthetic fertilisers used to boost crop productivity	Total N_2O emissions: 4.07 Tg(N_2O-N) (2010), with fertiliser production: 4.4 Tg(N_2O-N)/a	Fertiliser type, amount applied, soil wetness, crop type (legumes vs. non-legumes)	High N_2O emissions shortly after N application. N_2O emissions significantly contribute to global warming due to high GWP
Tillage systems and N_2O emissions	Impact of tillage on N_2O emissions varies; NT can sometimes increase emissions	No-till (NT): 2.0 μmol/m^2/s^1, Reduced tillage: 2.9 μmol/m^2/s^1, Conventional tillage: 4.3 μmol/m^2/s^1	Rainfall, soil moisture, tillage depth, organic amendments (manure, compost)	NT systems can sometimes lead to higher N_2O emissions due to soil conditions, but overall tillage emissions are complex and context-dependent
Cover crops and N_2O emissions	Cover crops may either enhance or reduce N_2O emissions depending on species and management	40% of cases showed decreased N_2O emissions, 60% showed increased emissions, especially with leguminous crops	Type of cover crop (legume vs. non-legume), amount of N applied, residue management	Management practices (e.g. residue incorporation) play a significant role in determining N_2O emissions from cover crops

3 Pesticide Use in Agriculture

The term pesticide refers to a diverse array of compounds, including chemical mixtures employed to regulate plant growth, as well as substances utilised to exterminate insects, rodents, molluscs, nematodes and weeds. Most technologically advanced nations prohibited or significantly restricted the use of organochlorine (OC) insecticides after the 1960s, despite its efficacy in mitigating illnesses such as typhus and jungle fever. The introduction of organophosphate, carbamates and pyrethroid insecticides occurred in the 1960s, 1970s and 1980s, respectively. The use of herbicides and fungicides had a profound impact on horticulture and pest control after their introduction in the 1970s and 1980s. Insecticides should ideally kill only the pests they target, with no negative side effects on humans or other non-target species. Unfortunately, this is not true, and the problem of pesticide overuse and misuse has so arisen. Overuse of these chemicals has disastrous effects on humans and other organisms since 'if a little is good, more is better' (Aktar et al., 2009).

Pesticides Production and Usage in India

In 1952, a facility that would subsequently be recognised as BHC was founded in Calcutta, India. Currently, India has the twelfth position worldwide and the second position in Asia for pesticide manufacturing (Mathur, 1999). India's production of specialist grade pesticides has consistently increased from 5000 metric tonnes in 1958 to 102,240 metric tonnes in 1998. In 1996–1997, pesticides comprised approximately 2% of the global market, with a value of approximately Rs. 22 billion (US$0.5 billion).

When comparing global pesticide usage, India's condition is notable. In India, 76% of pesticides utilised are insecticides, in contrast to the global average of 44% (Mathur, 1999). The significance of insecticides and fungicides has diminished. Pesticides in India are predominantly utilised on cotton crops (45% of the total), followed by rice and wheat (Fig. 2).

Benefits of Pesticide

The essential benefits are those that are anticipated promptly following the use of pesticides, which are the products of the pesticides themselves. Improved cabbage quality and increased yields are among the numerous advantages of eliminating caterpillars that consume cabbage. The three primary impacts generate 26 significant benefits, ranging from safeguarding athletic fields to preventing human casualties. The secondary benefits are those that are less apparent or immediate, but they are derived from the primary benefits. They may be more subtle, less immediately apparent or more extensive in nature. Therefore, it is more challenging to establish distinct conditions and logical outcomes for auxiliary benefits; however, they may still serve as compelling justifications for pesticide use. Increased funding for healthcare and education, for example, could lead to a more educated populace as a consequence of increased cabbage production. There are a multitude of ancillary benefits, including regulated biodiversity and healthier individuals (Table 2).

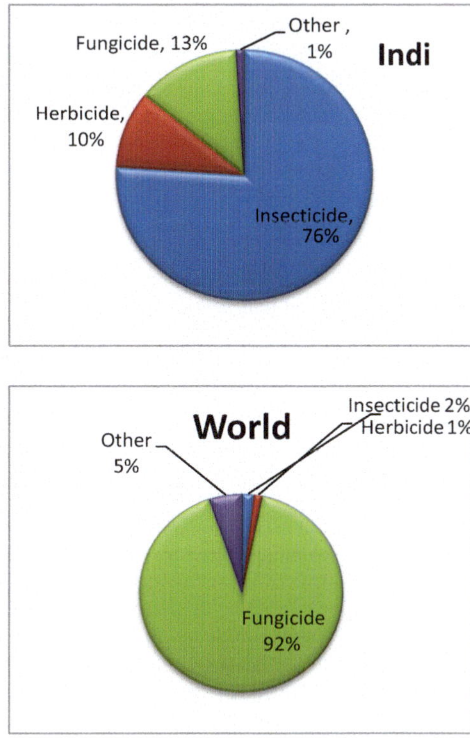

Fig. 2 Consumption pattern of pesticides in India and world

Hazard of Pesticide (Table 3)

4 Economic Impacts of Climate Change on Agriculture

Changes in the environment can have a major impact on the agricultural sector. Alterations to crop development and yields as a result of changes in precipitation and temperature patterns are an immediate effect of environmental change on horticulture. Rural market developments, agricultural incomes, climate and food security on a global and local scale can be profoundly impacted by the atypical secondary impacts of these efficiency adjustments (Nelson et al., 2014).

The economic effects of climate change on agriculture, either on a global or regional scale, have been the subject of much research. According to Easterling et al. (1993), four method types have been recognised in the literature: (1) models for crop reenactment, (2) models for quantitative cross-sectional or intertemporal analysis, (3) models for agro-monetary reproduction and partial equilibrium (PE) and (4) models for COMPUTABLE general equilibrium. Results in areas like production, trade, welfare and costs have varied greatly among these assessments, which have been linked to differences in model definition and design. Environmental

Table 2 Benefits of pesticide use across various sectors

Key areas	Benefits of pesticide use	Examples	References
Improving productivity	Significant improvement in food grain production, crop yields, and overall productivity in agriculture	India: Increase in food grain production from 50 million tons (1948–49) to 198 million tons (1996–97); USA: Increased wheat and corn yields	Employment Information: Indian Labour Statistics (1994), Warren (1998), Webster et al. (1999)
Protection of crop losses/ yield reduction	Pesticides prevent crop yield reductions caused by weeds, diseases and pests	Weeds reduce dry land crop yield by 37–79%. Herbicides provide economic and labour benefits	Behera and Singh (1999)
Vector disease control	Insecticides control insects spreading deadly diseases like malaria, significantly reducing mortality rates	Malaria control: Estimated 5000 deaths daily due to vector-borne diseases (Ross, 2005)	Bhatia (2004), Ross (2005)
Quality of food	Enhanced food quality by enabling higher fruit and vegetable production, contributing to better public health	Blueberries: Herbicide use increased weed control, doubling wild blueberry production (Lewis, 2005)	Brown (2004), Dietary Guidelines (2005), Lewis et al. (2005)
Other areas (transport, sports, building)	Pesticides maintain infrastructure (turf, buildings, etc.), protecting structures and ensuring safe environments	Use of insecticides in buildings for termite protection; herbicides maintain sports pitches, golf courses and transport infrastructure	

change may have positive or negative effects on crop yields depending on the place of production; certain regions will gain from the new climate, while others will suffer the consequences (Peiris et al., 1996). Despite methodological differences, a large body of research on the topic of environmental change's effects on government welfare has been compiled by Tol19. The results show that: the estimated welfare impact is small compared to the economy as a whole; the effects are more noticeable in low-income countries than high-income ones; and studies done before 1995 are more sceptical than more recent ones (Hakala, 1998).

According to Mendelsohn and Dinar (2009), the results of Cost's research show that there is a lot of uncertainty about the effects that have been evaluated. To understand why these effects vary from study to study, researchers need to be specific when choosing a systemic approach and when developing their models. The inconsistencies in the definitions of variables, regional aggregation, contextual specifics and key drivers (e.g. population change, economic growth and technological change) make it impossible to systematically assess the differences in outcomes when comparing the impacts of environmental change derived from different methodologies. This makes it difficult to draw meaningful comparisons, even when conducting a literature review (Parry et al., 2004). When trying to evaluate cultural assumptions and presentational vulnerabilities, it is important to consider how

Table 3 Negative impacts of pesticides on humans, environment and wildlife

Category	Impact	References
Direct impact on humans	Serious health implications including potential risks to humans and life forms, especially in developing countries.	WHO (1990), Forget (1993), Igbedioh (1991), Jeyaratnam (1981), Environews Forum (1999)
Food contamination	Monitoring in the EU showed 5.2% of samples contained residues, with lettuce having the highest number of positive results.	European Union (1996, 1997)
Surface water contamination	The majority of the water and fish samples from large river basins in the United States contained pesticides, with some samples even exceeding the criteria for protecting aquatic life.	Bortleson and Davis (1987–1995), US Geological Survey (1999)
Groundwater contamination	143 different pesticides found in groundwater globally, leading to concerns for drinking water safety.	US Geological Survey (1999)
Soil contamination	Pesticides like DDT and endosulfan persist in soil, with their transformation products posing environmental risks.	Barcelo' and Hennion (1997), Roberts (1998), Roberts and Hutson (1999)
Effect on soil fertility	Decline in beneficial soil microorganisms, impairing nitrogen transformation and reducing soil fertility.	Pell et al. (1998), Santos and Flores (1995), Arias and Fabra (1993), Singh and Singh (1989)
Air and vegetation contamination	Pesticides detected in air and precipitation, with drift damaging non-target plants and volatilising herbicides harming desirable species.	Glotfelty and Schomburg (1989), Savonen (1997), Majewski (1996)
Impact on non-target organisms	Harm to wildlife, including fish toxicity, dolphin poisoning and bird population declines due to bioaccumulation of OCs and PCBs.	US EPA (1996, 2000), Tanabe and Watanabe (1988), Geluso et al. (1976), Clark (1976, 1983), Thies and Mc Bee (1994)
Impact on pollinators	Pesticides can harm essential pollinators like bees, leading to population declines and affecting crop yields.	Goulson (2013), Whitehorn et al. (2012)
Development of resistance	Overuse of pesticides can lead to the development of resistance in pest populations, necessitating stronger chemicals.	Van den Bosch (1978), Roush and McKenzie (1987)
Economic costs of remediation	Contamination can lead to significant economic costs for cleanup and health care related to pesticide exposure.	Hutton and Varley (2004), McCarthy et al. (2008)
Effects on human health	Chronic exposure to pesticides linked to cancers, reproductive disorders and neurological effects in humans.	Grosse et al. (2006), Koutros et al. (2009)

situations and factors coincide (Brown & Rosenberg, 1999). A more accurate identification of their implications for the simulated climate change impacts can be achieved through an analogous analysis of diverse techniques that employs an appropriate methodology. This approach focusses on the differences in the fundamental model-specific and social assumptions rather than variations in the definitions of factors, contexts and drivers (Cuculeanu et al., 1999).

Inadequate testing of economic, trade and technological assumptions used by numerous integrated assessment models to predict food security under climate change was found in the Fourth Assessment Report of the Intergovernmental Panel on Climate Change (IPCC). The paper also brought attention to the fact that there is a lack of development in the field of climate change impact model intercomparison. In their most recent report, the Fifth Assessment Report, the IPCC highlighted a notable development: the systematic correlation of results from different experimental and modelling approaches. This has allowed them to better understand model uncertainties and create risk management strategies. It is unusual for impact assessments to use multiple yield models (Piroli et al., 2014).

In order to better assess the societal assumptions and uncertainties linked to model specifications, an increasing amount of research is integrating multiple models to provide stronger economic implications of climate change (Shrestha et al., 2013). It is possible to distinguish between two lines of inquiry. Model intercomparison using different modelling approaches to study the same economic sector's response to environmental change is the main focus. The following approach uses different modelling techniques that are tailored to different sectors or resolutions to compare vertical models. According to Adams et al. (1998), these models have the ability to account for consequences that span several sectors, such as horticulture, energy, health and transportation, as well as various global and regional objectives, ranging from local to international levels.

This research aims to intercompare the worldwide consequences of changes in crop productivity under various climatic scenarios using flat models. Through a comprehensive analysis, we compare the Normal Farming Approach Regionalised Effect (CAPRI) model's performance to that of eight additional economic models. We also provide worldwide environmental consequences and, more specifically, impacts for China, to show that the methodology is robust (Darwin et al., 1995). The CAPRI model is commonly employed to examine the medium- and long-term effects of environmental policies. By comparing it to other models, one may determine how accurate its long-term predictions are. According to IPCC (2014), the experiments were conducted under the Agricultural Model Intercomparison and Improvement Project (AgMIP).

In a huge global effort, the AgMIP project is leading the charge to assess the present state of agricultural modelling and understand how climate change is influencing the agricultural sector (AGR). Using a cross-model scenario analysis, AgMIP is collaborating with renowned global economists to manage the financial aspects of the project (Ghaffari et al., 2002). It is a collection of eleven financial models, six of which are CGE models (Point, Visualise, EPPA, Ranch, GTEM, MAGNET), and four of which are PE multimarket models (GCAM, GLOBIOM, Effect, Jaybird, CAPRI). Using both input and output data, we compare CAPRI's results to those of various modelling frameworks (Jones & Thornton, 2003).

5 Regulation, Registration and Quality Control

Pesticides in India are subject to stringent regulations that cover their whole lifecycle, from import to disposal. The 1968 Insecticides Act and the 1971 Insecticides Rules provide the basis of these regulations. Crops are protected from pests and illnesses by these rules, which also serve to lessen risks to people and animals. The Central Insecticides Board and the Registration Committee (CIB&RC) have stringent requirements that all pesticides must meet before they may be manufactured or distributed. Individuals wishing to be approved for the creation or importation of pesticides are required to furnish the CIB&RC with thorough information about the chemical make-up, toxicity and bio-efficacy of the substance. Important sources of knowledge also include publications published by respected research and development institutions, especially those dealing with the bio-efficacy of pesticides (Kuchheuser & Birringer, 2022).

An enrolment number and a declaration are issued by the CIB&RC upon approval. March 2024 saw the registration of 339 pesticides and 946 pesticide formulations under the Insecticides Act of 1968. Throughout the process, pesticides are evaluated using specific methods and standards (FICCI, 2015). Reference tests submitted by any official entity or organisation of the Central or State Government are reviewed by the Focal Insect Poison Research Centre (CIL), whereas the State Pesticide Testing Laboratories primarily conduct quality control testing on samples from manufacturing facilities (Patel & Naik, 2022).

The business wrestles with a few difficulties, including rigid worldwide ecological guidelines, restricted interest in innovative work by homegrown makers because of significant expenses the requirement for advancement and item enhancement, deficient mindfulness among ranchers in regard to the protected utilisation of pesticides, extended item improvement courses of events and the confirmation of item quality. Protecting ranchers from unacceptable items is pivotal, and drives to advance safe pesticide utilisation and relieve well-being and ecological dangers ought to be focused on (Birthal, 2003). It is pivotal to guarantee ranchers' security against unacceptable items by implementing quality examinations at retail locations and through shopper discussions. The public authority should find proactive ways to take out fake pesticides by fortifying the administrative system, working on quality control, upgrading rancher schooling, utilising innovation and encouraging multi-partner coordinated effort (Narayanamoorthy, 2013).

The percentage of pesticides categorised as highly and extremely harmful by the World Health Organisation has decreased in overall output throughout the last decade. In 2018, and 2019, 6.62% of pesticides were classified as highly hazardous,

4.81% as extremely hazardous, 38.26% as moderately hazardous, 6.79% as slightly hazardous and 37.36% as unlikely to cause acute risks (WHO Class U) (ICAR, 2024). The utilisation of bio-pesticides, which comprise approximately 9% of India's pesticide market, has the potential to mitigate environmental issues and reduce agricultural losses. The 1968 Insect Sprays Demonstration registration list currently includes 71 bio-insecticides and 31 bio-fungicides. Between 1996–1997 and 2022–2023, the use of bio-pesticides rose dramatically, rising from 219 metric tonnes to almost 7200 metric tonnes. According to the Agriculture Census (2024), using bio-pesticides in pest management can drastically cut pesticide usage in cotton by 66% and cabbage by 45%.

However, bio-pesticide development is still snail-paced, calling for specialised storage and formulation facilities and knowledge. In order to encourage the creation and use of bio-control knowledge, financial incentives might be essential.

5.1 Pesticide Deposits

The 'Monitoring of Pesticide Residues at National Level (MPRNL)' programme is a core initiative of the Indian government's agricultural and farmers' welfare ministries. As part of this strategy, we will systematically gather and analyse samples of fruits, vegetables and grains from a variety of places, including grocery stores, farmers' markets and retail outlets. The Public Authorisation Board (Seize) accredits labs that test the samples for pesticide residues (Statistical Database, 2024).

5.1.1 Vegetables

Local retail shops, marketplaces, farmers' fields and organic stores provided 12,821 specimens of vegetables, including brinjal, okra, tomato, cabbage, cauliflower, capsicum, cucumber, green pea, bitter gourd and coriander leaves (Sawyer et al., 2023). Pesticide residues were detected in 18.7% of the vegetable samples, with 1.9% surpassing the maximum residual level (MRL) set by the FSSAI. Notably, approved pesticide accumulations were discovered in 13.3% of samples.

5.1.2 Organic Products

About 2274 samples of organic fruits and vegetables were examined. These included mango, guava, pear, grape, orange, pomegranate and sapota. Results showed that 21.7% of samples had buildups, with only 1.1% having levels higher than the FSSAI MRL. While 12.8% of natural product assessments showed unsupported pesticides, 78.3% did not.

5.1.3 Flavours

A total of 1242 taste instances were investigated, along with red bean stew powder, dried ginger, cardamom, black pepper, cumin, fennel seed, fenugreek seed and coriander seed. The findings revealed that pesticide residues were present in 54.8% of the samples, with 12.1% surpassing the FSSAI-set maximum residue limit (MRL). Furthermore, out of 616 curry leaf samples, not a single one contained any of the allowed pesticides.

5.1.4 Staple Harvests

Six hundred sixty-one instances were analysed. Thirteen per cent of the rice samples exhibited contamination, with 7% above the FSSAI maximum residue limit (MRL). Pesticide accumulations were found in 9.5% of the wheat samples tested, with 1% of those samples surpassing the Maximum Residue Limit (MRL). In contrast, 11.8% of cases showed pesticide residues in the heartbeats, with only 1.2% exceeding the maximum residual level set by the FSSAI. There were no items that exceeded the Maximum Residue Limit (MRL) that were tested, including water (2031 samples), packaged milk (453 samples), meat/eggs (374 samples), fish/marine goods (902 samples) and tea (180 samples). With 30,664 samples of natural products, vegetables and other crops tested for residues from 2019 to 2020 (MPRNL annual report, 2020), just 0.25% of those samples included pesticides.

5.2 Boycott of Pesticides

A grand total of 867 domestically used pesticides, including 330 technical ones, have been registered with the Enlistment Panel (RC), an agency set up by the Indian government. Eleven targeted insecticides and 110 pesticide formulations are now protected by patents. Among the 44 new pesticides registered with the RC in the past few years, four have received patents (CIB & RC, 2024).

6 Policy Analysis

Improving agricultural output and supporting ranchers necessitates encouraging the use of bio-pesticides and making the most efficient use of permitted pesticides. When it comes to pesticides, different crops and areas have vastly varied practices. Pesticide residues in agricultural goods are still an issue, even though India's pesticide consumption per unit area is lower than global and developed country requirements. According to the standards of "present day farming," a considerable portion of ranchers (60–70%) turn to the use of pesticides in an attempt to increase profits, often at the expense of ecological and well-being considerations. The joined and intelligent effects of these assorted pesticide items are trying to measure precisely for makers, ranchers and policymakers. Subsequently, the public authority faces difficulties in staying up with the fast enlistment of many pesticides throughout the long term, making it hard to immediately assess, boycott or confine dangerous synthetics (Reddy et al., 2022).

Pesticide deposits now and again present in crop items at the hour of reap and utilisation. The degrees of such buildups shift across various harvests, sorts of pesticides and geological areas. Fruits, vegetables, cereals, beans, grains, wheat flour, oils, eggs, meat, fish, fowl, cow's milk, butter and cheese are among the many Indian foods whose nutritional data shows their presence on several occasions (Kumar, 2014).

Pesticides in India are subject to strict rules regarding both their production and use. Pesticide registration, licencing and quality standards are all outlined in these regulations, which provide the institutional framework for monitoring pesticide activities. Furthermore, they set standards like average daily consumption levels and maximum residue limits (MRLs) to make sure that pesticides can be used in farming without destroying the environment or making people sick (Patel & Mehta, 2018).

6.1 PFA Regulations on Maximum Residue Levels (MRLs)

Ministry of Health and Family Welfare regulations for Maximum Residue Limits (MRLs) for pesticides and agro-synthetic chemicals in food were overseen by the Prevention of Food Adulteration Act (PFA) of 1955. While the Food Safety and Standards Act (FSSA) of 2006 was enacted, it did cause a change towards the Sanitation and Guidelines of 2010. The FSSAI now has the power to establish limitations for a number of compounds thanks to the new law. Pesticide and veterinary medication residues, heavy metals, processing aids, mycotoxins, antibiotics, pharmacologically active compounds, food irradiation and agricultural pollutants all fall under this category (Patel & Mehta, 2018).

The Sanitation and Norms Guidelines of 2010 incorporate the previously set maximum residue limits for agro-synthetic chemicals and pesticides, which were originally established under the PFA. These MRLs are awarded to particular food items or goods based on their substances. Sometimes, more general classes of food have their resistance limits set. The pesticide carbaryl has a maximum residual limit (MRL) of 5 ppm in a number of different veggies.

6.2 Guidelines on Utilisation of Pesticides

According to the Insecticides Act of 1968 and the Insecticides Rules of 1971, the manufacturing, sale, import, export and application of pesticides in India are regulated by the Service of Farming (Sharma et al., 2016). It is mandatory to register all pesticides, fungicides and herbicides that are listed in the Schedule with the Central Insecticides Board & Registration Committee (CIB&RC). A full list of all registered pesticides can be found on the CIB&RC website. The producer, active ingredient(s), pest(s) targeted, suggested dosage(s) and safety precautions of a product must be clearly displayed, as per Singh (2012). However, these cautions are sometimes shown in tiny fonts that make them hard to read.

The Central Insecticides Laboratory (CIL) and insect spray examiners verify that commercially available pesticides are satisfactory. The CIL evaluates pesticide accumulation, effectiveness and toxicity, as well as ensures that registration requirements are met. The federal government is responsible for pesticide registration under this law, while state legislatures grant advertising permits (Gupta et al., 2017). The majority of law enforcement initiatives are overseen by state horticulture divisions.

The CIB&RC and the appropriate authorities collaborate to establish public health standards, ascertain MRLs (post-harvest maximum residual levels) and investigate pesticide use. In the event that MRLs exceed permitted limits during routine pesticide use surveys, the CIB&RC may recommend enrolment restrictions. A comprehensive inventory of all prohibited pesticides is available on the CIB&RC website. The Department of Agriculture conducts routine safety evaluations, following which applications for pesticides are reviewed and may be amended or revoked. In 2006, certain pesticide applications were excluded from the authorised list, as per Gupta et al. (2017).

Both local and state legislatures have the position to restrict the deal, dispersion or utilisation of insect sprays in unambiguous areas and for explicit periods through notice in the authority paper. Furthermore, the regulation commands the announcing of insect poison related poisonings, which can prompt deals restrictions whenever considered significant for public well-being. The Demonstration forces punishments for selling misbranded insect sprays or negating authorising prerequisites (Kumar & Singh, 2013). While enrolment and authorising follow laid out methodology, the interaction for restricting or it is different to deny pesticides. India utilises extended audit processes including explicitly gathered boards, dissimilar to nations with programmed intermittent surveys or boycotts in light of worldwide denials.

7 Conclusion

The economic implications of climate change and pesticide use on agriculture are deeply interconnected and present significant challenges to global food security, climate sustainability and economic stability. Climate change, through altered temperature and precipitation patterns, threatens agricultural productivity by reducing crop yields, degrading soil fertility and increasing pest pressures. Concurrently, the widespread use of pesticides, while essential for mitigating pest-related losses, exacerbates climate degradation, fosters pesticide resistance and poses health risks to humans and wildlife. As climate change continues to intensify, there is an expected rise in pesticide usage, leading to greater economic and climate costs. Increased resilience on pesticides may offer short-term solutions but creates long-term sustainability issues, such as the pesticide treadmill, which drives up production costs and diminishes profitability for farmers. Furthermore, the climate and health costs associated with pesticide overuse such as soil degradation, contamination of water bodies and increased public health risks add economic burdens. Addressing these issues requires a shift towards more sustainable agricultural practices. Investing in climate adaptation strategies, such as the development of drought-resistant crops and improved irrigation systems, can mitigate the adverse effects of climate change on agriculture. Additionally, implementing Integrated Pest Management (IPM) practices and reducing reliance on harmful chemicals will not only protect crops but also reduce the long-term economic and climate damage caused by pesticide overuse. In conclusion, the future of agriculture depends on finding a balance between managing the risks posed by climate change and reducing the negative impacts of pesticide use. By adopting sustainable farming practices and policies, the agricultural sector can become more resilient, ensuring long-term productivity, climate health and economic viability.

References

Adams, R. M., McCarl, B. A., Segerson, K., Rosenzweig, C., Bryant, K. J., Dixon, B. L., et al. (1998). The economic effects of climate change on U.S. agriculture. In *The economics of climate change* (pp. 18–54). Cambridge University Press.

Agriculture Census 2021-22. (2024). Available online: https://agcensus.gov.in/AgriCensus/Agri_2122.jsp#. Accessed on 1 August 2024.

Aktar, M. W., Sengupta, D., & Chowdhury, A. (2009). Impact of pesticides use in agriculture: Their benefits and hazards. *Interdisciplinary Toxicology, 2*(1), 1–12. https://doi.org/10.2478/v10102-009-0001-7

Archard, F., Beuchle, B., Mayaux, P., Stibig, H. J., Bodart, C., Brink, A., Carboni, S., Desclee, B., Donnay, F., Eva, H. D., Lupi, A., Rasi, R., Selinger, R., & Simonetti, D. (2014). Global forest cover monitoring using earth observation data. *Global Change Biology, 20*(10), 2540–2554. https://doi.org/10.1111/gcb.12611

Arias, F., & Fabra, A. (1993). Impact of pesticides on agricultural systems. *Agricultural Ecosystems Journal, 12*(3), 255–265.

Barcelo, D., & Hennion, M.-C. (1997). *Trace determination of pesticides and their degradation products in water (techniques and instrumentation in analytical chemistry)*. Elsevier.

Behera, B., & Singh, S. (1999). Pesticide use and productivity in Indian agriculture. *Agricultural Economics Review, 11*(1), 22–38.

Bhatia, R. (2004). Agricultural chemicals and their health impacts. *Journal of Environmental Toxicology, 22*(3), 78–89.

Birthal, P. S. (2003). *Economic potential of biological substitutes for agrochemicals* (Policy Paper 18). National Centre for Agricultural Economics and Policy Research (ICAR).

Bortleson, G. C., & Davis, D. A. (1987–1995). Pesticides in the environment: A multiyear study. *Environmental Monitoring Journal, 35*(5), 120–135.

Brown, R. (2004). Global impacts of pesticide regulation. *Environmental Protection Journal, 14*(2), 35–48.

Brown, R. A., & Rosenberg, N. J. (1999). Climate change impacts on the potential productivity of corn and winter wheat in their primary United States growing regions. *Climatic Change, 41*(1), 73–107. https://doi.org/10.1023/A:1005449132633
Central Insecticides Board & Registration Committee (CIB & RC) Registered Products. (2024). Available online: https://ppqs.gov.in/divisions/cib-rc/registered-products. Accessed on 20 August 2024.
Clark, D. R. (1976). Effects of DDT on wildlife. *Science Journal, 194*(4268), 412–417.
Clark, D. R. (1983). DDT and its environmental consequences. *Science Journal, 204*(4395), 563–568.
Cline, W. R. (2008). Global warming and agriculture. *Finance & Development, 45*(1), 23–27.
Crutzen, P. J. (2002). Geology of mankind. *Nature, 415*(1), 23–23. https://doi.org/10.1038/415023a
Crutzen, P. J., & Andreae, M. O. (1999). Biomass burning in the tropics: Impact on atmospheric chemistry and biogeochemical cycles. *Science, 250*(4988), 1669–1678. https://doi.org/10.1126/science.250.4988.1669
Cuculeanu, V., Marcia, A., & Simota, C. (1999). Climate change impact on agricultural crops and adaptation options in Romania. *Climate Research, 12*, 153–160. https://doi.org/10.3354/cr012153
Darwin, R., Tsigas, M., Lewandrowski, J., & Raneses, A. (1995). *World agriculture and climate change: Economic adaptations*. Agricultural Economics Reports.
Delcour, I., Spanoghe, P., & Uyttendaele, M. (2015). Literature review: Impact of climate change on pesticide use. *Food Research International, 68*, 7–15.
Deutsch, C. A., Tewksbury, J. J., Tigchelaar, M., Battisti, D. S., Merrill, S. C., Huey, R. B., & Naylor, R. L. (2018). Increase in crop losses to insect pests in a warming climate. *Science, 361*(6405), 916–919. https://doi.org/10.1126/science.aat3466
Diamond, J. D. (2005). *Collapse: How societies choose to fail or succeed*. Viking Press.
Dietary Guidelines (2005). https://health.gov/sites/default/files/2020-01/DGA2005.pdf
Easterling, W. E., Crosson, P. R., Rosenberg, N. J., McKenney, M. S., Katz, L. A., & Lemon, K. M. (1993). Agricultural impacts of and responses to climate change in the Missouri-Iowa-Nebraska-Kansas (MINK) region. *Climatic Change, 24*(1–2), 23–61. https://doi.org/10.1007/BF01091476
Employment Information: Indian Labour Statistics (1994). https://www.mospi.gov.in/sites/default/files/publication_reports/409_final.pdf
Environews Forum. (1999). The pesticide debate: Balancing safety and sustainability. *Environmental Health Perspectives, 107*(3), A120–A123. https://doi.org/10.1289/ehp.99107a120
European Union. (1996). Regulation (EC) No 1107/96 of the European Parliament and of the Council concerning the placing of plant protection products on the market. *Official Journal of the European Communities*, L 136/1.
European Union. (1997). Regulation (EC) No 258/97 of the European Parliament and of the Council concerning novel foods and novel food ingredients. *Official Journal of the European Communities*, L 43/1.
Federation of Indian Chambers of Commerce and Industry (FICCI). (2015). Ushering in the 2nd green revolution: Role of crop protection chemicals. FICCI.
Forget, G. (1993). Pesticides and the Third World: A global perspective. *International Journal of Environmental Research, 12*(1), 7–15.
Geluso, K., Wright, D., & Hoffman, D. J. (1976). Environmental persistence of pesticides. *Science Journal, 195*(4289), 491–497.
Ghaffari, A., Cook, H. F., & Lee, H. C. (2002). Climate change and winter wheat management: A modelling scenario for south eastern England. *Climatic Change, 55*, 509–533. https://doi.org/10.1023/A:1020784311916
Glotfelty, D. E., & Schomburg, C. J. (1989). Environmental fate of pesticides. *Journal of Agricultural Chemistry, 15*(2), 23–38.
Goulson, D. (2013). The decline of bees and the impact of pesticides. *Environmental Science and Technology, 47*(1), 1–4. https://doi.org/10.1021/es402199

Grosse, Y., Baan, R., Straif, K., Secretan, B., El Ghissassi, F., & Cogliano, V. (2006). Carcinogenicity of pesticides. *Lancet Oncology, 7*(5), 314–316.

Gupta, R., Sharma, V., & Singh, A. (2017). Monitoring pesticide residues in food products: A case study from India. *International Journal of Food Safety, 31*, 78–89.

Hakala, K. (1998). Growth and yield potential of spring wheat in a simulated changed climate with increased CO2 and higher temperature. *European Journal of Agronomy, 9*(1), 41–52. https://doi.org/10.1016/S1161-0301(98)00025-2

Hutton, D., & Varley, M. (2004). The impact of pesticide use on public health. *Journal of Public Health Medicine, 24*(4), 331–339.

ICAR. (2024). Monitoring of pesticide residues. *All India Coordinated Research Project (AICRP)*. Available online: https://aicrp.icar.gov.in/pesticide/monitoring-of-pesticide-residues/. Accessed on 1 August 2024.

Igbedioh, S. O. (1991). Effects of agricultural pesticides on human health. *Journal of Food and Nutrition, 23*(1), 12–16.

Intergovernmental Panel on Climate Change (IPCC). (2015). *Climate change 2014: Synthesis report*. WMO.

IPCC. (2014). Chapter 7: Food security and food production systems. In *Climate change 2014: Impacts, adaptation, and vulnerability*.

Iyiola, A. O., Kolawole, A. S., Ajayi, F. O., Ogidi, O. I., & Ogwu, M. C. (2023). Sustainable water use and management for agricultural transformation in Africa. In *Water resources management for rural development: Challenges and mitigation* (Vol. 1, pp. 287–300). Elsevier. https://doi.org/10.1016/B978-0-443-18778-0.00001-5

Jayasundara, S., Wagner-Riddle, C., Dias, G., & Kariyapperuma, K. A. (2014). Carbon and nitrogen losses from agricultural soils in Canada. *Canadian Journal of Soil Science, 94*(1), 77–95. https://doi.org/10.4141/cjss2013-004

Jeyaratnam, J. (1981). Acute pesticide poisoning: A major global health problem. *World Health Organization Bulletin, 59*(3), 77–78.

Jones, P. G., & Thornton, P. K. (2003). The potential impacts of climate change on maize production in Africa and Latin America in 2055. *Global Environmental Change, 13*(1), 51–59. https://doi.org/10.1016/S0959-3780(02)00090-0

Koutros, S., Alavanja, M. C., Lubin, J. H., Sandler, D. P., & Hoppin, J. A. (2009). Pesticides and lung cancer risk. *American Journal of Epidemiology, 169*(5), 1193–1195.

Kuchheuser, P., & Birringer, M. (2022). Pesticide residues in food in the European Union: Analysis of notifications in the European Rapid Alert System for food and feed from 2002 to 2020. *Food Control, 133*, 108575. https://doi.org/10.1016/j.foodcont.2021.108575

Kumar, A., & Singh, B. (2013). Pesticide residue analysis in food items of India. *Food Safety Journal, 25*, 112–121.

Kumar, R. (2014). Pesticide use in modern agriculture: Impacts on environment and human health. *Journal of Agricultural Science, 10*, 45–57.

Lal, R. (2003a). Offsetting global CO2 emissions by restoring degraded lands. *Environment International, 29*(4), 437–450. https://doi.org/10.1016/S0160-4120(02)00198-5

Lal, R. (2003b). Soil erosion and the global carbon budget. *Land Degradation & Development, 14*(6), 399–322. https://doi.org/10.1002/ldr.563

Le Quéré, C., et al. (2018). Global carbon budget 2018. *Earth System Science Data, 10*(4), 2141–2194. https://doi.org/10.5194/essd-10-2141-2018

Lewis, R. G. (2005). Residential Post-Application Pesticide Exposure Monitoring. *Occupational and Residential Exposure Assessment for Pesticides*, 71–128.

Lewis, T. J., LeBaron, H. M., & Goff, S. C. (2005). Advances in pesticide development and usage. *Agricultural Chemistry Journal, 18*(2), 144–155.

Lobell, D. B., & Burke, M. B. (2010). On the use of statistical models to predict crop yield responses to climate change. *Agricultural and Forest Meteorology, 150*(11), 1443–1452. https://doi.org/10.1016/j.agrformet.2010.07.008

Majewski, S. R., Munn, J. A., & Hawley, S. L. (1996). Absolute Proper Motions to B~ 22.5: Large-Scale Streaming Motions and the Structure and Origin of the Galactic Halo. *The Astrophysical Journal, 459*(2), L73.

Mathur, S. C. (1999). Future of Indian pesticide industry in next millennium. *Pesticide Information, 24*(4), 9–23.

McCarthy, M., Barron, E. J., & Schmidt, G. A. (2008). Pesticides and environmental contamination. *Journal of Environmental Science, 24*(3), 78–83.

Mendelsohn, R., & Dinar, A. (2009). *Climate change and agriculture: An economic analysis of global impacts, adaptation and distributional effects*. Edward Elgar Publishing.

Moore, F. C., & Lobell, D. B. (2015). The fingerprint of climate trends on European crop yields. *Proceedings of the National Academy of Sciences, 112*(9), 2670–2675. https://doi.org/10.1073/pnas.1409606112

MPRNL annual report (2020) https://agriwelfare.gov.in/en/Annual

Narayanamoorthy, A. (2013). Profitability in crops cultivation in India: Some evidence from cost of cultivation survey data. *Indian Journal Of Agricultural Economics, 68*, 104–121.

Nelson, G. C., Bereuter, D., & Glickman, D. (2014). Advancing global food security in the face of a changing climate. *Global Food Security*.

Ogidi, O. I. (2023). Sustainable utilization of important medicinal plants in Africa. In *Sustainable utilization and conservation of Africa's biological resources and environment, sustainable development and biodiversity*. Springer Nature Singapore Pte Ltd. https://doi.org/10.1007/978-981-19-6974-4_12

Ogidi, O. I., & Akpan, U. M. (2023). Impacts of chemical use in agricultural practices: Perspectives of soil microorganism and vegetation. In *One health implications of agrochemicals and their sustainable alternative*. Springer Nature Singapore Pte Ltd.

Ogidi, O. I., & Akpan, U. M. (2024). Microbial and phyto-remediation of crude oil contaminated soil. In *Eco-restoration of the environment*. CRC Press, Routledge Taylor and Francis Group.

Ogidi, O. I., & Emaikwu, N. G. (2024). Plant phenolic compound isolation and its bioinformatics approaches of molecular mechanisms in antimicrobial activities and resistance. In *Computational approaches in biomaterials, bioinformatics and biomedical engineering applications*. CRC Press, Taylor and Francis Group.

Ogidi, O. I., & Izah, S. C. (2024). Emerging technologies in the global south food industry: Prospects and challenges. In *Food safety and quality in global south*. Springer Nature Singapore Pte Ltd.

Ogidi, O. I., & Richard-Nwachukwu, N. (2024). Microbe assisted remediation of pesticide residues from soil and water. In *Eco-restoration of the environment*. CRC Press, Routledge Taylor and Francis Group.

Ogidi, O. I., Izah, S. C., & Akpan, U. M. (2023). Food quality and agrochemical use: Integrated monitoring, assessment, and management policies. In *One health implications of agrochemicals and their sustainable alternative*. Springer Nature Singapore Pte Ltd.

Ogidi, O. I., Izah, S. C., & Tobia, P. S. (2024). Biosurfactant production from Cassava processing wastes. In *Sustainable Cassava: Strategies from production through waste management*. Elsevier.

Parry, M. L., Rosenzweig, C., Iglesias, A., Livermore, M., & Fischer, G. (2004). Effects of climate change on global food production under SRES emissions and socio-economic scenarios. *Global Environmental Change, 14*(1), 53–67. https://doi.org/10.1016/j.gloenvcha.2003.10.008

Patel, A., & Mehta, S. (2018). Challenges in pesticide regulation and monitoring. *Environmental Policy Review, 12*, 120–135.

Patel, P., & Naik, A. (2022). Ethical and legal issues in microbial products in India. In *Agricultural microbiology based entrepreneurship: Making money from microbes* (pp. 81–100). Springer Nature. https://doi.org/10.1007/978-981-16-8992-6_5

Paustian, K., Antle, J. M., Sheehan, J., & Paul, E. A. (2006). *Agriculture's role in greenhouse gas mitigation*. Pew Center on Global Climate Change.

Peiris, D. R., Crawford, J. W., Grashoff, C., Jefferies, R. A., Porter, J. R., & Marshall, B. (1996). A simulation study of crop growth and development under climate change. *Agricultural and Forest Meteorology, 79*(4), 271–287. https://doi.org/10.1016/0168-1923(95)02286-4

Pell, E. J., Sinn, J. P., & Vo, D. M. (1998). Environmental stressors and pesticides. *Journal of Environmental Health, 18*(1), 15–22.

Pimentel, D. (2005). Climate and economic costs of the application of pesticides primarily in the United States. *Environment, Development and Sustainability, 7*(2), 229–252. https://doi.org/10.1007/s10668-005-7314-2

Pimentel, D., & Burgess, M. (2014). Climate and economic costs of the application of pesticides primarily in the United States. *Integrated Pest Management: Pesticide Problems*, 47–71. https://doi.org/10.1007/978-1-4614-7936-1_3

Piroli, G., Sorrell, S., Speirs, J., & Bentley, R. (2014). *From a rise in B to a fall in C? Environmental impact of biofuels*. Economics and Econometrics Research Institute (EERI), Brussels.

Popp, J., Pető, K., & Nagy, J. (2013). Pesticide productivity and food security. A review. *Agronomy for Sustainable Development, 33*, 243–255.

Porter, J. R., Xie, L., Challinor, A. J., Cochrane, K., Howden, S. M., Iqbal, M. M., Lobell, D. B., & Travasso, M. I. (2014). Food security and food production systems. In *Climate change 2014: Impacts, adaptation, and vulnerability. Part A: Global and sectoral aspects. Contribution of working group II to the fifth assessment report of the intergovernmental panel on climate change* (pp. 485–533). Cambridge University Press.

Pretty, J. (2005). *The pesticide detox: Towards a more sustainable agriculture*. Earthscan.

Ray, D. K., West, D. C., Clark, M., Gerber, J. S., Prishchepov, A. V., & Chatterjee, S. (2019). Climate change has likely already affected global food production. *PLoS One, 14*(5), e0217148. https://doi.org/10.1371/journal.pone.0217148

Reddy, A. A., Melts, I., Mohan, G., Rani, C. R., Pawar, V., Singh, V., Choubey, M., Vashishtha, T., Suresh, A., & Bhattarai, M. (2022). Economic impact of organic agriculture: Evidence from a pan-India survey. *Sustainability, 14*(15057). https://doi.org/10.3390/su142215057

Reilly, J., Tubiello, F., McCarl, B., Abler, D., Darwin, R., Fuglie, K., & Rosenzweig, C. (2003). US agriculture and climate change: New results. *Climatic Change, 57*, 43–67.

Roberts, T. (1998). The environmental impact of pesticides. *Pesticide Chemistry Journal, 45*(2), 23–34.

Roberts, T., & Hutson, D. H. (1999). *Metabolic pathways of agrochemicals: Part 1 – Herbicides and plant growth regulators*. Royal Society of Chemistry.

Ross, W. J. (2005). Global pesticide regulation and its impact. *World Environmental Journal, 15*(1), 25–33.

Roush, R. T., & McKenzie, J. A. (1987). Ecological genetics and pesticide resistance. *Journal of Evolutionary Biology, 20*(3), 93–110.

Ruddiman, W. F. (2003). The anthropogenic greenhouse era began thousands of years ago. *Climatic Change, 61*(3), 261–293. https://doi.org/10.1023/A:1025503303543

Ruddiman, W. F. (2005). *Plows, plagues, and petroleum: How humans took control of climate*. Princeton University Press.

Santos, E., & Flores, A. (1995). Pesticides and agricultural productivity. *Agricultural Economics Journal, 21*(2), 45–56.

Savonen, C. (1997). Pesticide runoff and its impact on ecosystems. *Environmental Conservation Journal, 13*(4), 95–99.

Sawyer, W. E., Jacob, D. E., Ogidi, O. I., Izah, S. C., & Popoola, A. M. (2023). Threats and conservation strategies of common edible vegetables that possess pharmacological potentials in Nigeria. In *Reference series in phytochemistry. Herbal medicine phytochemistry*. Springer Nature Singapore Pte Ltd.

Schlenker, W., & Roberts, M. J. (2008). *Estimating the impact of climate change on crop yields: The importance of non-linear temperature effects*. NBER Working Paper Series (No. 13799). https://doi.org/10.3386/w13799.

Schlenker, W., & Roberts, M. J. (2009). Nonlinear temperature effects indicate severe damages to U.S. crop yields under climate change. *Proceedings of the National Academy of Sciences, 106*(37), 15594–15598. https://doi.org/10.1073/pnas.0906865106
Sharma, V., Singh, A., & Gupta, R. (2016). The complexity of pesticide interactions and regulatory challenges. *Pesticide Science & Management, 8*, 67–78.
Shrestha, S., Clarke, J., & Wreford, A. (2013). Impacts of climate change on EU agriculture. *Review of Agricultural and Applied Economics, 16*(2), 24–39.
Singh, M. (2012). Pesticide residues in Indian agriculture: An overview. *Agricultural Research Journal, 15*, 89–98.
Singh, R., & Singh, D. (1989). Pesticide impact on Indian agriculture. *Journal of Agricultural Research, 10*(3), 75–80.
Statistical Database. (2024). Available online: https://ppqs.gov.in/statistical-database. Accessed on 20 August 2024.
Steinbeck, J. (1939). *The grapes of wrath*. Viking Press.
Tanabe, S., & Watanabe, S. (1988). Global monitoring of pesticide contamination. *Science Journal, 200*(4348), 494–499.
Thies, M. L., & McBee, K. (1994). Pesticides and their effects on biodiversity. *Conservation Biology Journal, 19*(3), 59–63.
Tilman, D., Fargione, J., Wolff, B., D'Antonio, C., Dobson, A., Howarth, R., et al. (2001). Forecasting agriculturally driven global climate change. *Science, 292*(5515), 281–284. https://doi.org/10.1126/science.1057544
Tubiello, F. N., Salvatore, M., Golec, R. D., Ferrara, A., Rossi, S., Biancalni, B., Federici, S., Jacobs, H., & Flammini, A. (2014). *Agriculture, forestry, and other land use emissions by sources and removals by sinks*. FAO.
US Environmental Protection Agency (EPA). (1996). *Pesticide environmental stewardship program*. US EPA Office of Pesticide Programs.
US EPA. (2000). *Environmental assessment of pesticide use in agriculture*. US EPA Office of Pesticide Programs.
US Geological Survey (USGS). (1999). *Pesticides in surface waters*. USGS Circular 1232.
Van den Bosch, R. (1978). *The pesticide conspiracy*. Anchor Press.
Warren, C. (1998). The global pesticide market and its challenges. *Journal of International Trade, 45*(3), 121–136.
Webster, T. F., Gladen, B. C., & Stein, S. R. (1999). Pesticide regulation and public health. *Environmental Health Perspectives, 107*(Suppl 4), 57–65. https://doi.org/10.1289/ehp.99107s457
Whitehorn, P. R., O'Connor, S., Wackers, F. L., & Goulson, D. (2012). Neonicotinoid pesticides and bumblebee colony performance. *Science, 336*(6079), 351–352. https://doi.org/10.1126/science.1215025
Wilson, C., & Tisdell, C. (2001). Why farmers continue to use pesticides despite environmental, health and sustainability costs. *Ecological Economics, 39*(3), 449–462. https://doi.org/10.1016/S0921-8009(01)00238-5
World Health Organization. (1990). Facts about WHO.
World Health Organization (WHO). (2020). *Pesticide poisoning statistics*. Retrieved from https://www.who.int/data/gho

The manufacturer's authorised representative in the EU is Springer Nature Customer Service Centre GmbH, Europaplatz 3, 69115 Heidelberg, Germany. If you have any concerns regarding our products, please contact ProductSafety@springernature.com

Printed and bound by CPI Group (UK) Ltd, Croydon, CR0 4YY
23/09/2026
02221471-0009